Springer
Tokyo
Berlin
Heidelberg
New York
Barcelona
Budapest
Hong Kong
London
Milan
Paris
Santa Clara
Singapore

T. Hori · R.W. Ridge · W. Tulecke
P. Del Tredici · J. Trémouillaux-Guiller
H. Tobe (Eds.)

Ginkgo Biloba A Global Treasure

From Biology to Medicine

With 118 figures including 13 in color

Springer

Terumitsu Hori, Dr.
Professor, Institute of Biological Sciences, University of Tsukuba
1-1-1 Tenno-dai, Tsukuba, Ibaraki, 305 Japan

Robert W. Ridge, Ph.D.
Associate Professor, Division of Natural Sciences
International Christian University
3-10-2 Osawa, Mitaka, Tokyo, 181 Japan

Walter Tulecke, Ph.D.
Professor Emeritus, Department of Science, Antioch College
795 Livermore Street, Yellow Springs, OH 45387, U.S.A.

Peter Del Tredici, Ph.D.
Director, Arnold Arboretum of Harvard University
125 Arborway, Jamaica Plain, MA 02130, U.S.A.

Jocelyne Trémouillaux-Guiller, Dr.
Laboratory of Cellular Biology and Plant Biochemistry
Faculty of Pharmaceutical Sciences, Francois Rabelais University
31 Avenue Monge, 37200 Tours, France

Hiroshi Tobe, Dr.
Professor, Faculty of Integrated Human Studies, Kyoto University
Yoshida-nihonmatsu, Sakyo-ku, Kyoto, 606-01 Japan

ISBN-13: 978-4-431-68418-3

Library of Congress Cataloging-in-Publication Data

Ginkgo biloba, a global treasure : from biology to medicine / T. Hori
. . . [et al.].
p. cm.
Includes bibliographical references and index.
ISBN-13: 978-4-431-68418-3 e-ISBN-13: 978-4-431-68416-9
DOI: 10.1007/978-4-431-68416-9
1. Ginkgo. 2. Ginkgo—Utilization. I. Hori, T. (Terumitsu),
1938– .
[DNLM: 1. Trees—anatomy & histology. 2. Plant Extracts. QK 475
G492 1997]
QK494.5.G48G55 1997
585.7—dc21
DNLM/DLC
for Library of Congress 97-27106

Printed on acid-free paper

9 8 7 6 5 4 3 2

Typesetting: Best-set Typesetter Ltd., Hong Kong

SPIN: 10681515

Foreword

It is agreed by most scientists that there is a real possibility of mankind's being plunged into crisis around the middle of the twenty-first century because of insufficient crop production to support the explosively increasing world population and because of serious pollution of the global environment. Plants should play the major role in alleviating this crisis. Consequently, plant science will become of crucial importance in the next century.

The application of plant science to crop improvement and the production of plants resistant to environmental pollution will be achieved through complete understanding of the basic functions of plants. Fundamental understanding of individual species and the evolution of plants is also important.

The Botanical Society of Japan, which was established in 1882, decided to publish a series of special issues of the *Botanical Magazine, Tokyo* (now the *Journal of Plant Research*). The aim of this series was to present overviews of recent advances in various fields of plant sciences. As a result, The Botanical Society of Japan published *Controlling Factors in Plant Development* in 1978, *Regulation of Photosynthetic Processes* in 1991, and *Cellular and Molecular Biology in Plant Cell Cultures* in 1993. The contributions to these high-quality special issues were written by leading international scientists in each field, and these publications had a great impact worldwide.

In 1997 The Botanical Society of Japan is publishing two monographs following on from the previous publications. One is *Ginkgo Biloba—A Global Treasure*, edited by Drs. T. Hori, R.W. Ridge, W. Tulecke, P. Del Tredici, J. Trémouillaux-Guiller, and H. Tobe, commemorating the 100th anniversary of the discovery of *Ginkgo* sperm by Sakugoro Hirase, a Japanese botanist, who published his findings in the *Botanical Magazine, Tokyo* (now the *Journal of Plant Research*). The other is *Evolution and Diversification of Land Plants*; edited by Drs. K. Iwatsuki and P.H. Raven. Since elucidation of the mechanisms of evolution is a major goal of plant science and of biology in general, this topic is most fitting for a publication just prior to the beginning of the twenty first century.

I trust that these two books published by The Botanical Society of Japan will provide readers with information at the very forefront of plant science.

As the president of The Botanical Society of Japan, I would like to acknowledge the editors and distinguished contributors for devoting their valuable time to these publications.

April 1997

Atsushi Komamine
President
The Botanical Society of Japan

Preface

Ask any Japanese what a *Ginkgo* is, and you will be told that it is a common tree in Japan, planted along streets, in temples and shrines, in school grounds and parks. Indeed, it would be very difficult to find a Japanese person who has never gathered its beautiful yellow leaves in autumn or who has never tasted delicious *Ginkgo* nuts. The same can be said for Chinese and Koreans. People of an older generation would tell how they used to insert two or three *Ginkgo* leaves in a book to protect the pages from being stained, and to kill worms and harmful insects, and that they used to give grilled nuts to children who wet their beds. Some may mention family crests or logos that incorporate *Ginkgo*-leaf designs. Westerners who have seen *Ginkgo* trees usually say that they love the shape of the leaves and their bright autumn color, or they may mention the disagreeable smell of the *Ginkgo* nuts when they are ripe.

Ginkgo was first officially described by Linnaeus (*Ginkgo biloba* L. 1771). There are two synonyms: *Salisburia adiantifolia* Smith (1797), and *Pterophyllus salisburiensis* Nelson (1866). *Ginkgo* has many common names in various parts of the world: icho (silver apricot), yajiao (duck foot), gong sun shu (grandfather/grandson tree), white fruit, *Ginkgo* tree, Ginkgobaum, Kew tree, maidenhair tree, Mächenhaarbaum, temple tree, arbre aux quarante ecus, noyer du Japon, and many more.

Ginkgo is dioecious, that is, the female and male structures are borne on separate trees. There are no obvious morphological attributes of *Ginkgo* to distinguish the sexes, however, apart from the emergence of the sporangia in the spring, which usually does not occur until the tree is around 30 years old. When young, the tree is usually pyramidal in shape, and when older, more spreading and broader in the crown. In some countries, such as Korea, Japan, and China, female trees are preferred because the people appreciate the taste of the nuts and have eaten them for centuries. In other countries, especially in the West where there is no tradition of eating the nuts, male trees are preferred over females because of the unpleasant smell of the fruit.

Ginkgo can grow over 35 meters high, with the main stem up to 10 meters in girth. They are long-lived. Leaves are deciduous, scattered on long shoots, or crowded at the apex of short shoots; they are stalked and are variable in size and shape. Generally, leaves are bilobed, without a midrib, and are fan-shaped. Veins are dichotomously forked and sometimes anastomose. A number of varieties are known in leaf and tree forms; these are variegata, tubiformis, triloba, pendule, macrophylla laciniata, or fastigiata. Some trees exhibit an interesting phenomenon; reproductive organs (macrosporangia or microsporangia) are directly produced on leaves without stalks or peduncles. This is called *oha-tsuki* that is "attached-to-leaves" in Japanese. The female and male trees which produce *oha-tsuki* also grow normal ovules (macrosporangia) or anthers (microsporangia) with stalks, respectively. This was first mentioned in the Japanese literature over 100 years ago. Most of these trees are designated as "living national treasures" in Japan because of their rarity and specialness.

The *Ginkgo* family, the Ginkgoaceae, have their origin in the Permian, some 200 to 225 million years ago. They are known to have reached their peak in Jurassic times, about 100 million years ago. *Ginkgo* species were very widespread, and fossil leaves have been found just about everywhere in both the southern and northern hemispheres. From the historical record it is clear that no other plant has a stronger claim to be called a living fossil, a term used by Darwin to designate survivors of the past. The tree was thriving 125 million years ago, when dinosaurs still roamed the earth, and the genus has remained virtually unchanged from that time. *G. biloba* is the sole living member of this once abundant group of plants, and it is clearly a precious link between the present and the remote past.

Ginkgo is said to survive in a wild state only in a remote mountain region in China. However, it is much cultivated throughout the world, especially in Japan, China, and Korea. Nowadays many countries have *Ginkgo* plantations where small trees are farmed for ginkgolide chemicals found in the leaves. *Ginkgo* is now recognized as a medicinal plant in Europe and the United States, as it has been since ancient times in Asia. *Ginkgo* is also a great survivor in polluted environments, growing where other trees find it difficult, and it is exceptionally resistant to fungus and insect attacks.

One hundred years ago, the Japanese botanist Sakugoro Hirase presented his discovery of the swimming sperm of *Ginkgo* in *The Botanical Magazine, Tokyo*, the organ of the Tokyo Botanical Society, now called the *Journal of Plant Research*. This great discovery was an epoch-making event in botany, because it placed *Ginkgo* as the sole living link between the lower and higher plants. To celebrate the centenary of that discovery, The Botanical Society of Japan agreed to publish this commemorative monograph.

You may find books with articles about *Ginkgo*, many of which were written in Japanese and Chinese. A great number of plant dictionaries written in various languages mention Ginkgo. There are also books on *Ginkgo* that focus on medicine, pharmacology, and chemistry. So far, however, there have been no monographs published on *Ginkgo* that cover a wide range of topics.

Articles have been contributed by specialists in *Ginkgo* throughout the world, and this is thus the first comprehensive monograph on *Ginkgo* to be published. It covers botany (Part I); paleobotany (Part II); ethnobotany (Part V); chemistry, medicine, and pharmacology (Part IV); and environmental engineering (Part III). The editors hope that the monograph will act as a stimulant for further studies and contribute to the development of new fields in *Ginkgo* research. As we think it important to summarize what research has been done on *Ginkgo* since Hirase's discovery, this monograph not only presents the state of the art in *Ginkgo* research but also reviews work published during the last 100 years. There are many references listed, some dating as far back as the mid-15th century. Reading this book, you will see that *Ginkgo* has been considered an attractive and useful tree for centuries.

We would like to acknowledge the financial support of The Botanical Society of Japan for the publication of this Volume. We thank all those who have encouraged the publication of the book, in particular, the many authors.

Spring 1997

The Editors

Contents

Phylogeny and Fossil History

Environmental Aspects

Chemistry and Clinical Application of *Ginkgo* Substances

Ginkgo and Humans

Contributors

AOKI, TAKAYUKI (p. 251)
Department of Genetic Resources I, National Institute of Agrobiological Resources, Ministry of Agriculture, Forestry and Fisheries, Japan

BRAQUET, PIERRE G. (p. 359)
Bio-Inova, France

CHUNG, GAP CHAE (p. 233)
Department of Horticulture, Chonnam National University, Korea

DEL TREDICI, PETER (p. 119)
Arnold Arboretum of Harvard University, U.S.A.

FRIEDMAN, WILLIAM E. (p. 29)
Department of EPO Biology, University of Colorado, U.S.A.

FUJIWARA, NOBUO (p. 259)
Michinoku Lakewood National Government Park Works Office, Tohoku Regional Construction Bureau, Ministry of Construction, Japan

FÜNFGELD, ERNST-WALTER (p. 323)
Weintrautstrasse 10, D-35039, Marburg, Germany

GIFFORD, ERNEST M. (p. 29)
Section of Plant Biology, University of California, U.S.A.

HAGA, MASANOBU (p. 309)
Faculty of Pharmaceutical Sciences, Health Sciences University of Hokkaido, Japan

HANDA, MARIKO (p. 259)
Environment Department, Public Works Research Institute, Ministry of Construction, Japan

HARA, NOBORU (p. 3)
Biology Laboratory, Meisei University (Ome), Japan

HASEBE, MITSUYASU (p. 173)
Botanical Gardens, Faculty of Science, University of Tokyo, Japan

He, Shan-An (p. 373)
Institute of Botany, Jiangsu Province and Academia Sinica, People's Republic of China

Hizume, Masahiro (p. 109)
Biological Institute, Faculty of Education, Ehime University, Japan

Holt, Ben (p. 223)
Department of Environmental and Plant Biology, Ohio University, U.S.A.

Honda, Hiroshi (p. 243)
Institute of Agriculture and Forestry, University of Tsukuba, Japan

Hori, Shihomi (p. 385)
2-5-23 Higashi, Tsukuba, Ibaraki, Japan

Hori, Terumitsu (p. 67, 99, 385)
Institute of Biological Sciences, University of Tsukuba, Japan

Iizuka, Yasuo (p. 259)
Planning Division, Planning and Research Administration, Public Works Research Institute, Ministry of Construction, Japan

Juretzek, Wiltrud (p. 341)
Schwabe Pharmaceuticals, Germany

Kim, Yoon Soo (p. 233)
Department of Forest Product and Technology, Chonnam National University, Korea

Kochibe, Naohisa (p. 301)
Department of Biology, Faculty of Education, Gunma University, Japan

Lee, Jae Kee (p. 233)
Department of Forest Product and Technology, Chonnam National University, Korea

Miyamura, Shin-ichi (p. 67, 99)
Institute of Biological Sciences, University of Tsukuba, Japan

Nagata, Toshiyuki (p. 413)
Department of Biological Sciences, Graduate School of Science, and Botanical Gardens, Faculty of Science, University of Tokyo, Japan

Ohba, Kihachiro (p. 159)
Institute of Agriculture and Forestry, University of Tsukuba, Japan

Pang, Zi-Jie (p. 373)
Institute of Botany, Jiangsu Province and Academia Sinica, People's Republic of China

RIDGE, ROBERT W. (p. 99)
Biology Department, Division of Natural Sciences, International Christian University, Japan

ROHR, RENÉ (p. 85)
Interactions Plantes-Champignons et Micropropagation, Bât. 405, Université Claude, France

ROTHWELL, GAR W. (p. 223)
Department of Environmental and Plant Biology, Ohio University, U.S.A.

SAHASHI, NORIO (p. 17)
School of Pharmaceutical Sciences, Toho University, Japan

SOMA, SANAE (p. 51)
Bunkyo University, Japan

TRÉMOUILLAUX-GUILLER, JOCELYNE (p. 127)
Laboratory of Cellular Biology and Plant Biochemistry, Faculty of Pharmaceutical Sciences, Francois Rabelais University, France

TSUMURA, YOSHIHIKO (p. 159)
Bio-Resources Technology Division, Forestry and Forest Products Research Institute, Japan

TULECKE, WALTER (p. 141)
Department of Science, Antioch College, U.S.A.

UEMURA, KAZUHIKO (p. 207)
Department of Geology, National Science Museum, Japan

WADA, KEIJI (p. 309)
Faculty of Pharmaceutical Sciences, Health Sciences University of Hokkaido, Japan

YIN, GU (p. 373)
Institute of Botany, Jiangsu Province and Academia Sinica, People's Republic of China

YOSHITAMA, KUNIJIRO (p. 287)
Department of Biological Science, Faculty of Science, Kumamoto University, Japan

ZHOU, ZHIYAN (p. 183)
Nanjing Institute of Geology and Palaeontology, Academia Sinica, People's Republic of China

Structure and Development

Morphology and Anatomy of Vegetative Organs in *Ginkgo biloba*

NOBORU HARA

1 Introduction

There have been many studies of *Ginkgo biloba* L. in addition to the famous study by Hirase in 1896 [1]. Before and after Hirase's study, *Ginkgo* was a plant that attracted considerable attention from other botanists in Japan. There were some studies of vegetative organs, in addition to those of reproductive organs, of *Ginkgo* in the 19th century. For example, a study in 1896 of the burls, known as "chichi" (nipples) in Japanese, by Fujii [2] is of particular interest.

We can find many primitive features of seed plants in vegetative organs of *Ginkgo*. In the present study, a few primitive vegetative features will be pointed out. For example, the shoot apex of *Ginkgo* does not show cell arrangement of the tunica and corpus, and the basic plan of the *Ginkgo* leaf is generated principally by a dichotomy, not flabellate but rather cruciate, and such development continues at least until the stage of the second bifurcation of the lamina. There are also interesting problems from the standpoint of the developmental morphology and anatomy. One example is the correlation between the developmental processes of the ordinary leaf primordium, including the lamina formation, and the malformation of cup-shaped or half-cup-shaped leaves. Another very important problem is whether a single apparent leaf of *Ginkgo* is a true leaf or a congenitally fused organ of two or more leaves. The seed-bearing leaf is very interesting. Its unique malformation has yet to be analyzed, although it was examined in 19th century, and a reasonable answer has not yet been obtained.

In this chapter, past studies of vegetative organs of *Ginkgo*, a primitive gymnosperm, will be reviewed. The author believes that studies of *Ginkgo* will continue unabated in the next century. This review will offer a degree of convenience for future studies of this plant.

Biology Laboratory, Meisei University (Ome), 2-590 Nagabuchi, Ome, Tokyo 198, Japan

There have been many studies of vegetative organs of *Ginkgo*; it is impossible to consider all earlier studies in a single brief review. Many references will be omitted or overlooked here. However, references to all relevant papers can be found in the literature cited in this chapter.

2 Shoot Apex

Foster [3], an American botanist, studied the structure of the vegetative shoot apex of *Ginkgo*. His study, reported in 1938, was outstanding, representing epoch-making progress in studies of the shoot apex. He discussed several studies of this structure; in some studies, it was concluded that the shoot apex of *Ginkgo* has three well-defined histogens (an independent protoderm, a two-layered periblem, and a central plerome), while in an other study periclinal divisions were observed in the superficial layer, and the periblem and plerome were reported to have a common group of initials.

Foster [3] considered that the two kinds of shoot, the dwarf shoot and the long shoot, presented a morphologically complicated problem, which was studied later by Gunckel and Wetmore [4, 5] and others as discussed below.

From a histogenetic standpoint, Foster [3] noted the absence of fundamental differences between the shoot apices in the two types of shoot in terms of cellular organization. He stated that differences seemed to depend basically on the relative extent and duration of activity in the rib meristem of the cortex and pith regions. In long shoots, long-term activity was observed, while internodal elongation was minimal in dwarf shoots.

Foster [3] noticed an important feature of the shoot apex of *Ginkgo*. He recognized several well-defined cellular zones in the shoot apex, as follows: zone I, the apical initial group; zone II, the central mother cells; zone III, the transition zone from zone II to zones IV and V; zone IV, the peripheral subsurface layers; and zone V, the rib meristem. However, he emphasized that the behavior of the cells of the apical initial group seemed to be entirely subordinated to the growth of the shoot apex as a whole, and was in no sense comparable to the orderly and precise segmentation that is typical of certain vascular cryptogams. He found no evidence for the existence of a single permanent apical cell. The apical initial group was, however, fundamentally similar to the single tier of initials that is characteristic of gymnosperms.

Foster [6] emphasized that features of the shoot apices of *Ginkgo* and *Cycas* clearly reveal the danger of generalizations based simply upon cellular organization in the meristems of conifers. However, Esau [7] stated that although the apical zonation in the Coniferales is less diversified than that in the cycads and is less well defined than that in *Ginkgo*, this zonation is similar fundamentally to that in the lower gymnosperms.

Foster [3, 6] laid the foundation for future studies of shoot apices of gymnosperms. His method for discriminating between cell zones in the shoot apex of

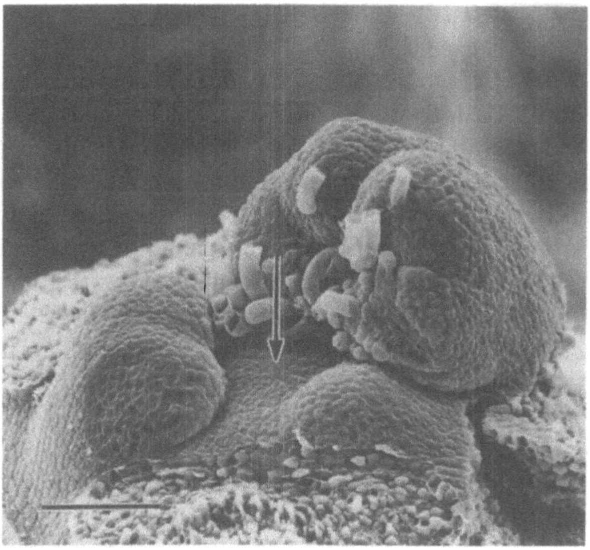

Fig. 1. The shoot apex of a long shoot. *Arrow* shows the central cell group. *Bar* = 143 μm. (From [8], with permission)

Ginkgo has influenced studies of the other vascular plants, including lower vascular plants and angiosperms, and the concept has been developed as cytohistological zonation. It is now generally accepted that three zones are recognizable in the structure of the shoot apex, namely the central zone, the peripheral meristem, and the rib meristem, although the names of the respective zones vary among authors. Of course, in gymnosperms, the central zone can be subdivided into the initial cell group and the central mother cells.

In an approximately transverse section through the surface layer, Foster [3] recognized a central group of four prominent cells which doubtless represented initials. Three or four such cells were also found by Hara in scanning electron micrographs [8] (Fig. 1). In addition, Hara recognized radial files of cells on the surface of the peripheral meristem (the surface layer of zone IV according to Foster) of the shoot apex. The leaf primordia are initiated at the periphery of the peripheral meristem.

Sakaguchi et al. [9] reported specific arrangements of cortical microtubules in the cells of the shoot apex of *Ginkgo*. They observed cortical microtubules that were arranged anticlinally in the tunica cells and randomly in the corpus cells in some angiosperms. However, the cortical microtubules were found to be arranged randomly in the shoot apex of *Ginkgo*, as well as in other gymnosperms, such as *Pinus* and *Cedrus*. It is possible that the random arrangements of microtubules in gymnosperms might occur because of failure of the tunica and corpus to differentiate.

3 Long Shoots and Dwarf Shoots

There are two types of vegetative shoot: the long shoot and the dwarf shoot (the short or spur shoot). Under natural conditions, axillary buds develop at the axil of the foliage leaves of the long shoot. A relatively large number of such buds develop into dwarf shoots, which form some foliage leaves. However, some of the axillary buds develop into long shoots. At the early stage of unfolding of the axillary bud, early in the growing season, one cannot discriminate between the long and dwarf shoots. All unfolded shoots seem to be dwarf shoots. Nonetheless some shoots soon begin to elongate, forming additional foliage leaves and developing into long shoots. Thus, the shoots of *Ginkgo* exhibit dimorphism. A series of studies on this dimorphism of *Ginkgo* was published by Gunckel and colleagues in the 1940s [4, 5, 10, 11].

The pith of the stem is formed as a result of the activity of a pith rib meristem in the shoot apex [4]. Much meristematic activity can be observed in long shoots, with little in dwarf shoots.

Gunckel and Thimann [10] examined the production of diffusible auxin by means of the *Avena* test. They recognized that long shoots produce slightly more auxin at the early stage than dwarf shoots. However, after the yield of auxin has fallen to a low level, it rises again steeply with the beginning of elongation and, thereafter, large amounts of auxin are produced by the lower, rapidly extending internodes.

Gunckel et al. [11] decapitated the terminal buds of long shoots. Decapitation of young plants caused dwarf shoot lateral buds, in particular the first and second lateral buds, to develop into long shoots. From these results, they deduced that the terminal long shoot inhibits the development of lateral buds into long shoots but it does not inhibit their development as dwarf shoots.

4 Vascular Differentiation

Gunckel and Wetmore [4, 5] demonstrated that the two separate procambial strands of a leaf develop continuously and acropetally from the procambial strands in the lower axis. Sik and Soh [12] studied the vascular differentiation in mature embryos and seedlings of *Ginkgo*. They reported that the procambium and protophloem form a continuous system that includes the epicotyl-hypocotyl root axis and cotyledons in the mature embryo, whereas the protoxylem differentiates discontinuously in the cotyledons and rarely in the upper hypocotyl. They observed that traces of the first and second leaf primordia appear at almost the same time, opposite each other at the epicotyl. There is only a single trace in the petiole of the first and second leaves, respectively. This situation differs from that reported for the foliage leaves by Gunckel and Wetmore [4, 5]. The leaves of the adult tree have a pair of traces in their petioles. The transitional stage, during which doubling of the number of traces occurs, has not been identified.

5 The Vascular Meristem, with Emphasis on the Early Ontogeny of the Vascular Cambium

Philipson et al. [13] suggested that all lateral growth of vascular plants is a "continuous process," which is uninterrupted from the apex to the mature trunk. Soh pointed out in 1972 that there had been few anatomical studies on the transition from the rather late stages of the procambium to the first appearance of the cambium. Initiation of the cambium has, indeed, been studied far less extensively than developmental changes of the differentiated cambium.

Esau [14–16] proposed that the procambium and cambium could be looked upon as two developmental stages of the same meristem, and Philipson and Ward [17] recognized that the primary growth of the vascular system merges gradually into secondary growth, and the two phases cannot be clearly distinguished.

Although a few relevant studies had been reported previously by some investigators, Soh [18] made a detailed study of the vascular meristem of *Ginkgo*, with emphasis on the problem of the transition of the procambium to the cambium. He divided the development of the vascular meristem (the procambium and cambium) into six stages, and observed each stage in detail. His results can be summarized as follows.

In tangential view, the procambium has, initially, a rather homogeneous structure. Later, it becomes organized into two distinct systems, one composed of long cells and the other of short cells (Fig. 2). The latter cells are arranged tangentially in axial files and transversely in radial files, with repeated transverse divisions. Some of the short cells elongate to intrude between neighboring cells. Therefore, long cells might be derived both from cells with a homogeneous structure at the first stage and from elongation of some of the short cells in axial files. Long cells

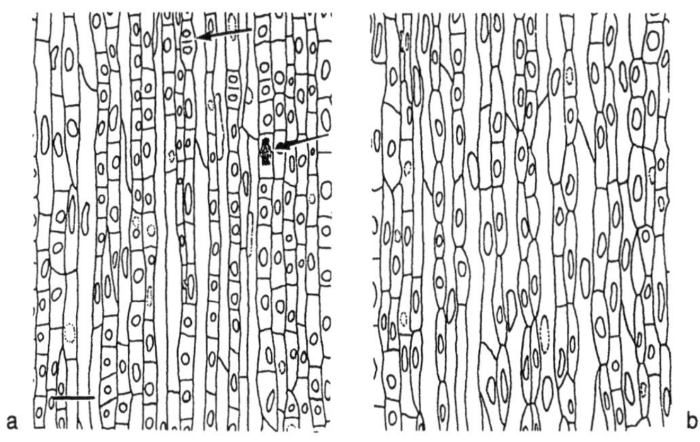

FIG. 2a,b. Tangential views of vascular meristem. **a** Two distinct systems (long and short cells) become clear in the procambium, and short cells in the axial files show transverse divisions (*arrows*). **b** Short cells in the axial files have lengthened to some extent and transverse divisions are infrequent. Bar = 40 μm. (From [18], with permission)

have predominantly tapering end walls and elongate actively. At subsequent stages, the frequency of transverse divisions of short cells in axial files decreases and these cells expand radially. The short cells in axial files are separated from one another vertically by the elongation of neighboring long cells which, in tangential view, appear to break up the axial files. Soh [18] proposed that the vascular meristem at this stage initiates the cambium. However, he noted that it is not yet clear whether the transition from the procambium to the cambium can be sharply delineated, because various other anatomical features must also be considered. This subject is quite complex, and he subsequently studied the same subject in some other species, such as *Aucuba*, *Weigela*, *Robinia*, and *Syringa* [19].

The vascular bundles within an internode of a stem are differentiated according to their phyllotactic disposition. Consequently, the early ontogeny of the fascicular cambium within a specific transverse plane of the internode is not uniform. The roots of *Ginkgo* seedlings were studied by Soh et al. [20], and they found that fusiform and ray initials of the vascular cambium were gradually derived from long and short cells of the procambium respectively. A gradual transition from the procambium to the vascular cambium of the type observed in the roots of *Ginkgo* has also been found in the stems of some seed plants, including *Ginkgo* [18]. In addition, Soh et al. [21] reported that the differentiation of the vascular cambium in *Ginkgo* proceeds bidirectionally from the cotyledonary node toward the stem and the root.

6 Ontogeny of the Leaf

Critchfield [22] recognized three ranges of leaf: early, transitional, and late leaves. He noted that most dwarf shoots produce only four to six leaves and that the distinction between dwarf and long shoots becomes evident soon after the separation of the bud scales. Additional leaves of long shoots begin to appear at the shoot tip. Three to five transitional leaves are produced on each long shoot, with most of the primordia of the transitional leaves being considered present in the winter bud. Following the transitional leaves, from 8 to 26 late leaves begin to develop. The primordia of these leaves are not present in the winter bud, and they are initiated after growth had resumed in the spring.

Critchfield recognized prominent variations in the shape of the leaf blade. The first few early leaves have broadly fan-shaped, entirely or moderately bilobed blades, with inconspicuous or nonexistent secondary notches. The blade width then shows a relative decrease and the depth of the central notch shows a relative increase in the uppermost early leaves. The late leaves of long shoots never attain a broad fan-like shape and their blades are always dissected. The central notch extends about half to two-thirds of the length of the blade, and the depth of the lateral notches ranges up to one-third of the blade length. Critchfield also discussed correlations among the form of leaves, the xylem production of veins and the production of diffusible auxin. In all cases, the leaves of *Ginkgo* have a

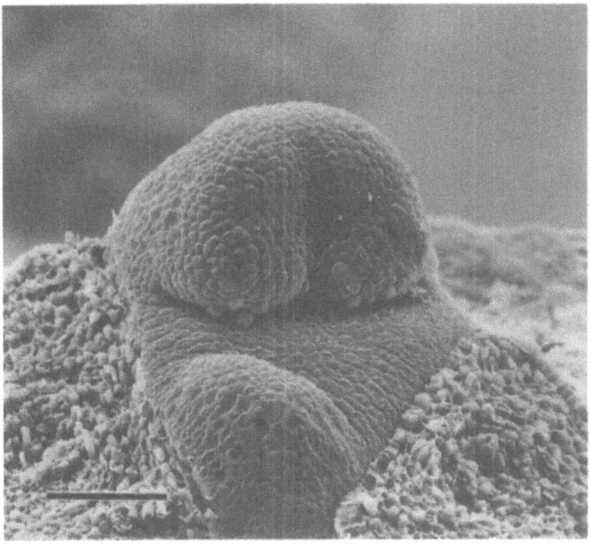

FIG. 3. Shoot tip of a long shoot. Primordia are at the buttress stage and at the shape of the first bifurcation. Bar = 155 μm. (From [8], with permission)

distinctive dichotomous venation, which is reflected by the external notched profile.

Critchfield pointed out that the late leaves develop uninterruptedly at a shoot tip, as do the leaves of the seedlings and all leaves of annual plants. The early leaves, by contrast, develop in the very different microenvironment of an enclosed bud.

Gifford and Foster [23] described the leaves of *Ginkgo*, referring to Critchfield's study and paying special attention to the formation of lobes. They recognized that on dwarf shoots and at the basal regions of long shoots, leaf blades are either entire or are divided into two lobes by a distal notch. By contrast, the blades of leaves on the upper parts of long shoots are always divided by a very deep sinus into two major lobes, each of which in turn is further dissected into segments.

Hara [8] studied the way in which the bifurcations of the leaf are initiated. The primordia of the late leaves of long shoots appear to be subject to uniformly equal processes of initiation and development (Fig. 3), but those in the winter buds are subject to various complicated processes. The former develop uninterrupted at the developing shoot tip in spring, but the latter develop within a tiny space in the winter bud, as noted also by Critchfield. Thus, differences in developmental processes might be caused by the different spatial microenvironments in which initiation and development of leaf primordia occur.

Hara provided conclusive details of the development of the leaf blade, mainly from observations of late leaf development on long shoots (Fig. 4). The first

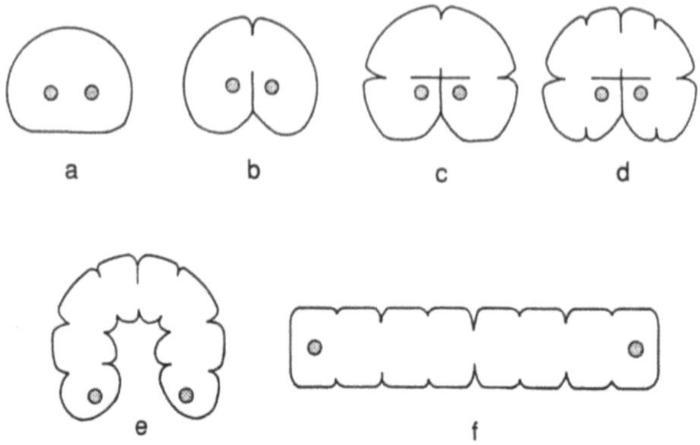

FIG. 4a–f. Diagrams of successive bifurcations of the buttress and formation of a dorsiventral blade. The lower part of each diagram shows the adaxial side of the leaf. The derivation of both edges of the fan-like form of the blade is shown by the two *dotted circles*. Both edges of the blade shown by the *dotted circles* in **f** are derived from the *dotted circles* in **a**. The first bifurcation occurs vertically (**a,b**); the second occurs approximately perpendicular to the first (**c**). The third bifurcation follows (**d**). The two lower parts separate at the planes of the first and second bifurcations (**e**); the dorsiventral profile of the blade is formed (**f**). (From [8], with permission)

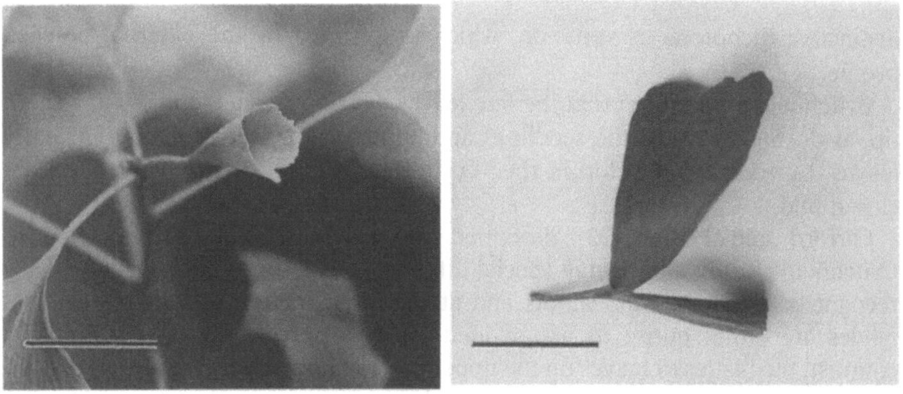

FIG. 5. Cup-shaped leaf (**a**) and cup-shaped half leaf (**b**). Bar = 3 cm. (From [25], with permission)

bifurcation occurs vertically (Fig. 4a,b) and the second occurs approximately perpendicular to the first (Fig. 4c). Then the third bifurcation occurs (Fig. 4d). Furthermore, the two lower parts, facing each other at the plane of the first bifurcation, become separated from the other part at the planes of the first and second bifurcations (Fig. 4e), and the dorsiventral profile of the blade is formed (Fig. 4f). Thus, the basic plan of the *Ginkgo* leaf is not generated principally by

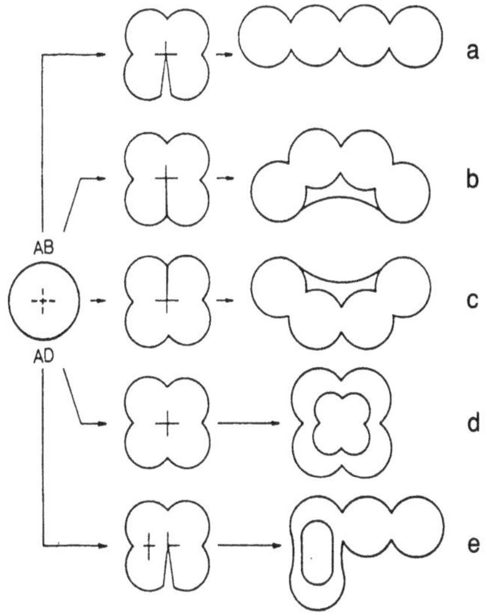

FIG. 6a–e. Deduced ontogeny of blade development. The *circle* on the left represents the primordium at a very early stage. **a** Development of the normal leaf. **b–e** development of malformed leaves (**d** cup-shaped leaf, **e** cup-shaped half-leaf). *AB*, abaxial; *AD*, adaxial. (From [25], with permission)

a flabellate dichotomy and it should, rather, be considered to be generated cruciately, at least until the stage of the second bifurcation.

Watari [24] described the cup-shaped leaves of *Ginkgo* and reviewed earlier reports that dealt with such leaves. While a normal leaf has two petiolar bundles [4, 5], Watari found a circular arrangement of bundles in the petiole of the cup-shaped leaf in cross sections. However, there have been no detailed studies of the ontogeny of cup-shaped leaves.

Hara [25] has studied the malformation of *Ginkgo* leaves (Figs. 5, 6), focusing on cup-shaped leaves and deducing that a cup-shaped leaf would be formed if the blade sector of the primordium failed to separate at the adaxial side (Fig. 6a,d). The form of the cup-shaped half leaf (Fig. 6a,e) that he found in his study suggested that the basic plan of the *Ginkgo* leaf might be formed cruciately even through the stage of the third bifurcation.

7 Venation of the Leaf

It is well known that the *Ginkgo* leaf has open dichotomous venation, which is devoid of anastomoses. However, Arnott [26] described finding vein unions in a survey of 1065 leaves, with 9.9% of leaves having one anastomosis or more. In long shoots, 13.4% of leaves had anastomoses; in dwarf shoots, 8.2% of leaves had anastomoses. In the 105 leaves with anastomoses, a total of 163 vein unions was detected. Arnott recognized that these anastomoses could be grouped into four types: A (the most common type), which comes about through the union of two veins produced by a single dichotomy; B, which is similar to type A but involves the separation of the two veins resulting from a single dichotomy; C, which occurs when the adjacent veins produced by two separate dichotomies unite; and D, which results from the union and subsequent separation of veins from two separate dichotomies.

It is generally considered that the leaf of *Ginkgo* is a primitive type among the leaves of seed plants. However, Arnott noted that, in view of the still primitive state of our knowledge of venation, it is difficult to assess how much emphasis should be placed on venation in phylogenetic considerations.

Arnott [26] found that long shoots had a significantly higher average percentage of leaves with anastomoses than dwarf shoots. In another study [27], he examined 2249 leaves and found that leaves from the median portion of long shoots had a higher percentage of anastomoses than either the basal or apical leaves. He wrote that the higher anastomosis index of these leaves was in harmony with the data on the diffusibility of auxin in long shoots reported by Gunckel and Thimann [10]. There is, however, no experimental evidence as yet to link directly the concentration of auxin and the frequency of anastomosis in *Ginkgo*.

8 Seed-Bearing Leaves

Shirai [28] found an abnormal *Ginkgo* tree, and Fujii [2] studied such trees including the tree reported by Shirai.

Fujii [2] noted that numerous examples of the progressive metamorphosis of different parts of floral organs are on record, but he found no instances of staminody and carpellody of foliage leaves in the literature on vegetable teratology. However, such unusual examples of anamorphosis, revealing various steps of progressive metamorphosis, can occasionally be found in *Ginkgo*.

Fujii studied two female trees that produced ovules on the marginal portions of the foliage leaves of dwarf shoots (Fig. 7), as well as a male tree that produced pollen sacs at the same sites. He investigated this problem in detail and compared the ovules or pollen sacs on the marginal portion of foliage leaves to those on the flower. He concluded that the ovules and the pollen sacs on the flower were both marginal formations of sporophylls. Furthermore, he concluded that the prominent edges of the stony part of the seed corresponded to the margins of the capellary leaves.

Sakisaka was also interested in the seed-bearing leaves, and he described their shapes and structures in detail in numerous reports [29–34]. He concluded [32] that the flower stalk can be considered a caulome organ, while the leaf stalk of normal, as well as capellodial leaves, and the stalklet of normal ovules are phyllome organs.

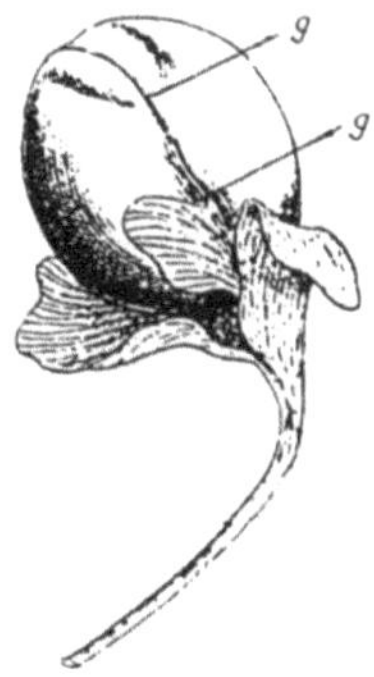

FIG. 7. Foliage leaf bearing a ripe fruit. *g*, longitudinal striation of the pulpy portion of the seed. (From [2], with permission)

9 Discussion

This chapter was written primarily to provide a review of the literature. However, an important problem that remains to be discussed is the possible correlation between the leaf trace and the early ontogeny of the leaf primordium. If we consider that the basic plan of the *Ginkgo* leaf is principally generated by a cruciate dichotomy [8], we can easily imagine that the cup-shaped leaf, which has a circular arrangement of petiolar bundles [24], might be formed by the same basic plan. However, a normal leaf is supplied by two independent traces of the stem [4, 5]. This problem raises the question of whether the apparent leaves of the adult *Ginkgo* tree are true leaves or compound organs. It seems possible that an apparent leaf might have the congenitally fused structure of two or more leaves, and an apparent petiole might show the fused condition of petioles of some true leaves. This deduction can also be made, perhaps, from malformations of cup-shaped half-leaves (Fig. 5b, 6e) since different forms are generated independently from two major lobes of a leaf. Furthermore, the form of the cup part of a cup-shaped half-leaf (Fig. 5b) is fundamentally similar to the form of the cup-shaped leaf (Fig. 5a). Thus, we can assume that each half of a leaf also has some fused structure. Therefore, it seems possible that a *Ginkgo* leaf might have a fused structure, with a few or several dichotomies.

The cup-shaped leaf, as well as the cup-shaped half-leaf, occurs very rarely. However, the malformation in this case seems to be a sporadic appearance suggesting a cruciate dichotomy, which is an ancestral feature of vascular plants; *Ginkgo* has various features typical of primitive seed plants. Therefore, we should consider malformations of *Ginkgo* leaves as a source of important information from the standpoint of phylogeny.

10 Conclusion

The vegetative organs of *Ginkgo* were studied by many Japanese botanists in the 19th century, as well as by some foreign botanists, and they have since been analyzed by botanists worldwide in the 20th century.

The cellular organization of the vegetative shoot apex of *Ginkgo* was studied by Foster [3], who laid the foundations for studies of shoot apices of gymnosperms. Long and dwarf (spur or short) shoots were studied by Gunckel and Wetmore [4, 5] and others. They recognized that large amounts of auxin are produced by the lower, rapidly extending internodes in the long shoots. Gunckel and Wetmore demonstrated that two separate procambial strands of a single leaf develop from the lower axis.

Sik and Soh [12] described the vascular differentiation in embryos and seedlings, and Soh [18] studied the differentiation of the vascular cambium.

Critchfield [22] recognized three ranges of leaf: the early, transitional, and late leaves. Hara [8] studied the way in which bifurcations of the leaf blade are initiated, and he concluded that bifurcations develop cruciately until the stage at which the second bifurcation occurs. Anastomoses of veins were studied by Arnott [26, 27], who noted that the elevated anastomosis index is in harmony with data related to the diffusibility of auxin. Fujii [2], Sakisaka [30], and others described the forms of seed-bearing leaves (leaves with ovules or with pollen sacs produced at marginal positions), and they discussed whether or not the carpellodial leaves and stalklets of normal ovules are phyllome organs.

References

1. Hirase S (1896) On the sperm of *Ginkgo biloba* (in Japanese). Bot Mag Tokyo 10:325–328
2. Fujii K (1896) On the different views hitherto proposed regarding the morphology of the flowers of *Ginkgo biloba* L. (preliminary note). Bot Mag Tokyo 10:7–8, 13–15, 104–110
3. Foster AS (1938) Structure and growth of the shoot apex in *Ginkgo*. Bull Torrey Bot Club 65:531–556
4. Gunckel JE, Wetmore RH (1946a) Studies of development in long shoots and short shoots of *Ginkgo biloba* L. I The origin and pattern of development of the cortex, pith and procambium. Amer J Bot 33:285–295
5. Gunckel JE, Wetmore RH (1946b) Studies of development in long shoots and short shoots of *Ginkgo biloba* L. II Phyllotaxis and the organization of the primary vascular system; primary phloem and primary xylem. Amer J Bot 33:532–543
6. Foster AS (1939) Problems of structure, growth and evolution in the shoot apex of seed plants. Bot Rev 5:454–470
7. Esau K (1953) Plant Anatomy. Wiley, New York
8. Hara N (1980) Morphological study on early ontogeny of the *Ginkgo* leaf. Bot Mag Tokyo 93:1–12
9. Sakaguchi S, Hogetsu T, Hara N (1990) Specific arrangements of cortical microtubules are correlated with the architecture of meristems in the shoot apices of angiosperms and gymnosperms. Bot Mag Tokyo 103:143–163
10. Gunckel JE, Thimann KV (1949) Studies of development in long shoots and short shoots of *Ginkgo biloba*. III Auxin production in shoot growth. Amer J Bot 36:145–151

11. Gunckel JE, Thimann KV, Wetmore RH (1949) Studies of development in long shoots and short shoots of *Ginkgo biloba* L. IV Growth habit, shoot expression and the mechanism of its control. Amer J Bot 36:309–316
12. Sik HS, Soh WY (1983) Vascular differentiation in the mature embryo and the seedling of *Ginkgo biloba* L (in Korean). Korean J Bot 26:207–216
13. Philipson WR, Ward JM, Butterfield BG (1971) The vascular cambium. Chapman and Hall, London
14. Esau K (1943) Origin and development of primary vascular tissues in the seed plants. Bot Rev 9:125–206
15. Esau K (1965a) Plant anatomy, 2nd edn. Wiley, New York
16. Esau K (1965b) Vascular differentiation in plants. Holt, Rinehart and Winston, New York
17. Philipson WR, Ward JM (1965) The ontogeny of the vascular cambium in the stem of seed plants. Biol Rev 40:534–579
18. Soh WY (1972) Early ontogeny of vascular cambium. I *Ginkgo biloba*. Bot Mag Tokyo 85:111–124
19. Soh WY (1990) Origin and development of cambial cells. In: Iqbal M (ed) The vascular cambium. Research Studies Press, Somerset, pp 37–62
20. Soh WY, Hong SS, Cho DY (1988) The ontogeny of the vascular cambium in *Ginkgo biloba* roots. Bot Mag Tokyo 101:39–53
21. Soh WY, Kang KD, Choi PS (1992) Early ontogeny of vascular cambium in cotyledonary node of *Ginkgo biloba* L. seedlings (in Korean). Korean J Bot 35:359–364
22. Critchfield WB (1970) Shoot growth and heterophylly in *Ginkgo biloba*. Bot Gaz 131:150–162
23. Gifford EM, Foster AS (1989) Morphology and evolution of vascular plants, 3rd edn. WH Freeman, New York
24. Watari S (1936) Morphology and structure of some ascidiate leaves (in Japanese). J Jap Bot 12:100–116
25. Hara N (1984) Early ontogeny and malformation of *Ginkgo* leaves. J Jap Bot 59:337–343
26. Arnott HJ (1959a) Anastomoses in the venation of *Ginkgo biloba*. Amer J Bot 46:405–411
27. Arnott HJ (1959b) Vein anastomoses in the leaves of long shoots of *Ginkgo biloba*. Nature 184:1336
28. Shirai M (1891) Abnormal *Ginkgo* tree (in Japanese). Bot Mag Tokyo 5:341–344
29. Sakisaka M (1926) The real nature of the epiphyllous fruits of *Ginkgo biloba* L. in Mt. Minobu, prov. Kai (in Japanese). J Jap Bot 3:168–170
30. Sakisaka M (1927) On the morphological significance of seed-bearing leaves of *Ginkgo* (in Japanese). Bot Mag Tokyo 41:273–278
31. Sakisaka M (1928) Comparison of the dwarf shoot of *Ginkgo* with the trunk of *Cycas* (in Japanese). J Jap Bot 5:101–104
32. Sakisaka M (1929a) On the seed-bearing leaves of *Ginkgo*. Jap J Bot 4:219–235
33. Sakisaka M (1929b) So-called seed-bearing leaves of *Ginkgo biloba* (in Japanese). J Jap Bot 6:30–36
34. Sakisaka M (1958) Study on *Ginkgo* (in Japanese). Kazama Shobou, Tokyo

Pollen Morphology of *Ginkgo biloba*

NORIO SAHASHI

1 Introduction

In the modern period before the 1950s when electron microscopy was not available for morphological studies of pollen, a detailed description of *Ginkgo biloba* pollen based on light microscopic observation was first given by Wodehouse [1]. He embedded the pollen in glycerin jelly without using any chemical procedure, and paid particular attention to the furrow and the rise of the furrow rim, which had been reported to be a primitive wing of Abietineae by Jeffrey [2]. Erdtman [3] established a chemical procedure, acetolysis, that was applicable to pollen analysis, and it later came into use by many researchers.

Ginkgo pollen grains have large furrows, and therefore the shape of each pollen grain changes from boatlike to almost spherical according to the degree of its dryness. Erdtman [4] illustrated this process by serial sketching. During the same period, Ikuse [5] and Wang et al. [6] published their books on pollen flora native to Japan and China, respectively. These books also included a description of *Ginkgo* pollen with figures and photographs.

Many researchers who observed *Ginkgo* pollen by light microscopy reported that it had a marked resemblance in morphology to cycad pollen. The advent of electron microscopes, however, led to the discovery of differences between these pollens. At first, Yamazaki and Takeoka [7, 8] observed the fine ornamentation on the exine surface in pollen of various gymnosperms and found a marked difference in such ornamentation between *Ginkgo* and *Cycas*. At about the sane time, Ueno [9, 10] closely observed the sporoderm stratification of the exine by transmission electron microscopy (TEM), covering general gymnosperms, and reported that *Ginkgo* and *Cycas* pollens showed entirely different structures of the exine. With the advent of scanning electron microscopy (SEM), the results of electron microscopic studies of primitive pollens have been increasingly reported. Close observations of *Ginkgo* pollen after Ueno's study, however, have been reported by only a few researchers including Rohr [11] and Audran [12, 13],

School of Pharmaceutical Sciences, Toho University, 2-2-1 Miyama, Funabashi, Chiba 274, Japan

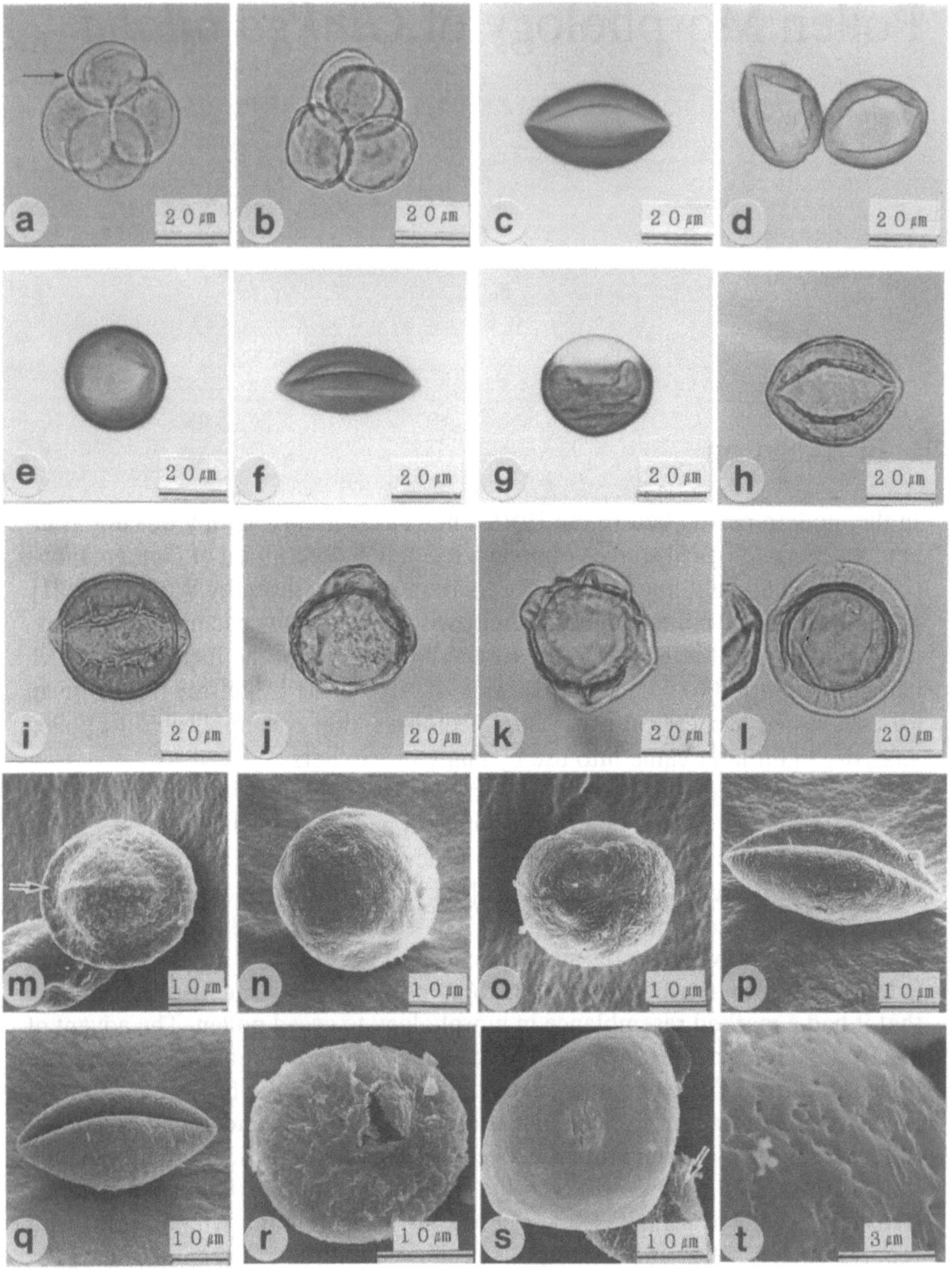

FIG. 1a–t. Pollen grains of *Ginkgo biloba* as seen with light (LM) and scanning electron microscopes (SEM). **a,b** Tetrad. **a** A late tetrad stage, cross tetrad, shows faint furrow rim (*arrow*). **b** An intermediate form between cross tetrad and a tetrahedral tetrad. **c–g** No chemical treatment: grains seen by LM. **c** Polar view: quite shrunken grain is elliptical in outline. **d** Polar view: fairly swollen grains with subhemispherical furrow. **e** Polar view: a fully expanded grain with almost rounded furrow. **f** Equatorial view: quite shrunken

who gave a detailed description of the exine, paying particular attention to the structure of the furrow and furrow rim. Sahashi and Ueno [14] morphologically compared *Ginkgo* and *Cycas* pollen and observed, for the first time, a reticulum-like ornamentation on the inside of the exine. They reported that the reticular pattern was particularly conspicuous at the inside of the furrow.

In this review of the pollen morphology of *G. biloba*, I reinvestigated both light and electron microscopic observations. In addition, the results of experiments with detachment of the exine of *Ginkgo* pollen by the bleaching procedure proposed by Mitsui [15] are shown, and differences previously found between the reticular structure on the inside of the furrow and that on the inside of the bladder of *Podocarpus macrophyllus* var. *maki* are also discussed.

2 Tetrad Formation

According to Ueno [9, 10], pollen grain tetrad formation in *Ginkgo biloba* is based on simultaneous division, and the type of tetrad is tetrahedral. Huynh [16] also described tetrahedral tetrads with some abnormal tetrads, triads, diads, and also monads. Rohr [11] and Audran [13] have studied pollen grain exine formation, especially at the tetrahedral tetrad stage. Rohr [11] found that the exine layers are deposited successively, similar to the *Taxus* nexine, and Audran [13] recognized modalities of exine formation different to that of cycads.

According to my observation, tetrad formation in *G. biloba* results in many cross tetrads (Fig. 1a) and forms that are intermediate (Fig. 1b) between cross and tetrahedral tetrads. However, the cross tetrad also forms through simultaneous division, as seen in the tetrahedral tetrad. I observed a fairly clear furrow rim at the late tetrad stage (Fig. 1a; arrow). The furrow rim can be seen clearly at

FIG. 1a–t. *Continued.* grain is fusiform in outline. **g** Equatorial view: fully expanded grain has a hemispherical aperture. **h–l** Acetolysed grains shown by LM. **h** Polar view: edge of the furrow rim, slightly undulated. **i** Polar view: furrow rim wavy, auricular projections on both sides. **j** Partly peeled furrow rim like a large auricular projection. **k** Irregularly winglike furrow rim appears around equator. **l** Entirely circular flangelike wing looks like an equatorial ridge (flange) that is seen in *Pteris* spores. **m–t** No chemical treatment or chemical treatment of grains by SEM. **m–o** Grains were treated by critical-point drying before being sputtered with gold. **m** Polar view, distal face: rounded germinal furrow (*arrow*) (anaporate). **n** Polar view, proximal face: faint linear projections. **o** Equatorial view: part of germinal furrow is bulged out beyond the furrow rim. **p–q** Non-critical-point drying treatment grains: both acetolysed (**p**) and no-chemical-treatment (**q**) grains are boat shaped with a single longitudinal furrow. **r–t** Grains after bleaching treatment. **r** Exine surface displays random cracking and partial peeling. **s** A part of exine (possibly sexine 2) detached entirely (*arrow*). **t** Enlarged surface of Fig. 2s shows faintly linear projections with minutely irregular pores (pits). (**r–t**, Photomicrographs courtesy of Prof. K. Mitsui, Nippon Dental University, Tokyo)

the mature pollen stage, but to my knowledge there has been no report of a furrow rim at the tetrad stage.

3 Germinal Aperture or Furrow

Wodehouse [1], the author of *Pollen Grains*, was the first palynologist to describe and draw in detail the pollen grains of *Ginkgo biloba*. His description of the germinal aperture (see Figs. 1c–i, m–q) is as follows: "Grains uniform in shape and size, boat-shaped with single longitudinal furrow. The furrow tends to open when moistened, the enonnous furrow reaching the entire length of the grain and involving nearly half of its surface." From his description, the germinal aperture of *G. biloba* is variable in size and shape. Erdtman [4] also gave a serial sketching for the germinal apertures at different morphological variations. Thus the terms used to describe the furrow (aperture) have varied from monocolpate [3, 17–20] to 1-sulcate [5–7, 21] or anacolpate [8, 14, 22].

In this review, I reexamined the germinal aperture of *G. biloba* pollen grains using different preparation methods. I found that soon after the grains had been mounted directly in glycerin jelly under gentle heat (less than 70°C), they were fusiform or boat shaped (Fig. 1c,f) in outline, and usually appeared quite shrunken (Sahashi and Ueno [14]). Then, after a few days or so, the fairly swollen grains each had a large and elliptical aperture (Fig. 1d). However, after the grains had been mounted directly in glycerin jelly under strong heat (almost 100°C), they expanded fully immediately, and the aperture appeared almost rounded in polar view (Fig. 1e) or hemispherical in equatorial view (Fig. 1g). However, grains after acetolysis were mostly swollen and increased in size (Fig. 1h,i). Under SEM, almost all grains that had been acetolysed or those without chemical treatment were quite shrunken and boat shaped (Fig. 1p,q) However, grains treated by critical-point drying before being sputtered with gold were quite swollen with a fully expanded germinal aperture (Fig. 1m–o).

From these observations, it is concluded that the germinal aperture of *G. biloba* has a very delicate mechanism of response to humid or dry conditions.

4 Wing or Bladder

The most interesting structures of *Ginkgo biloba* pollen grains are auricular projections (Fig. 2) or small winglike protuberances (see Figs. 1i–m,o, and 2c,d,j,k), situated on the furrow rim or near the equator (Fig. 1i). Jeffrey [2] demonstrated that "the pollen grain is winged as in *Pinus*, and the wings can be clearly discerned. It is curious that this striking feature of the pollen appears to have escaped notice." Jeffrey [23] also stated that "The microspores in the Abietineae and in the Ginkgoales also present striking points of resemblance which have only recently been completely realized." Sahashi and Ueno [14] observed some auricular projections on the grains of both *G. biloba* and *Cycas revoluta* and suggested that the projections may be remnants of an ancestral feature. Zimmerman [24] and Taylor [25, 26] supported the probable existence of bladders on pollen grains of the fossil ancestors of *C. revoluta* and the cycadalean fossil *Lasiostorobus polysacci*. In the current reinvestigation of these auricular projections, I found perine-like structures (Fig. 1j–l) in acetolysed grains. In grains that did not receive chemical treatment, I could not find such perine-like structures, which are commonly seen in the extant ferns. However, two small projections can occasionally be seen near the furrow rim (Fig. 1i) in acetolysed grains. Moreover, the small protuberances that were recognized as wings by Jeffrey [2] can also be seen at the tetrad stage (Fig. 1a; arrow). Observation using SEM reveals the expanded furrow rim in distalipolar view (Fig. 1m; arrow). This looks like an equatorial flange, and the expanded germinal aperture appears as an enormous cone. However, from the equatorial view (Fig. 1o) the furrow rim cannot be observed as a circular flangelike wing like that in Fig. 1k. On the surface of the enlarged partly distal face, the edge of the furrow rim appears to have roughly sporoderm stratification (Fig. 2d; arrow); this is the same position as that seen in axial section by TEM (Fig. 2j; arrow). Figure 2k shows a much more enlarged view, and the sporoderm stratification of the furrow rim can be clearly discerned. These are strong developments of the lamellated nexin, which is situated only at the furrow rim.

5 Ornamentation on the Proximal Face

Light microscopical descriptions of exine ornamentation on the proximal face of *Ginkgo biloba* pollens (see Figs. 1r–t and 2a,b) have varied among researchers; e.g., minutely roughened according to Wodehouse [1] and subreticulate according to Ikuse [5]. These descriptions were based on grains without chemical treatment. Ornamentation seen on chemically treated grains was described as subreticulum or a spiny wave by Yamazaki and Takeoka [7], as granulate by Shimakura [27], and as verrucate by Nakamura [19]. One of the reasons for these variable descriptions for *Ginkgo* may have been the different preparation methods employed. Frederiksen [28] compared the exine surface ornamentation on *G. biloba* pollen after two types of chemical treatment. After boiling for more than 5 min in 10% KOH, the grain length was 30–40 μm and the ornamentation was psilate, finely punctate, or weakly granulate. However, after acetolysis, the length was 38–48 μm and the ornamentation was foveolate (perforated with tiny holes). Frederiksen's acetolysis procedure was a very severe one (2 h in boiling and 8 h in cold acetolysis solution); none of the other researchers recognized foveolate ornanentation after acetolysis. However, detailed SEM or TEM observations revealed that the proximal surface was covered by irregularly arranged ridges, with minute verrucoid details among the ridges [8], spinulous [9], or reticulate with short and slightly elevated crests [11], and rugules [12].

In this study, I observed both nonacetolysed and acetolysed grains. In the former, the ornamentation on the proximal face consists of randomly arranged

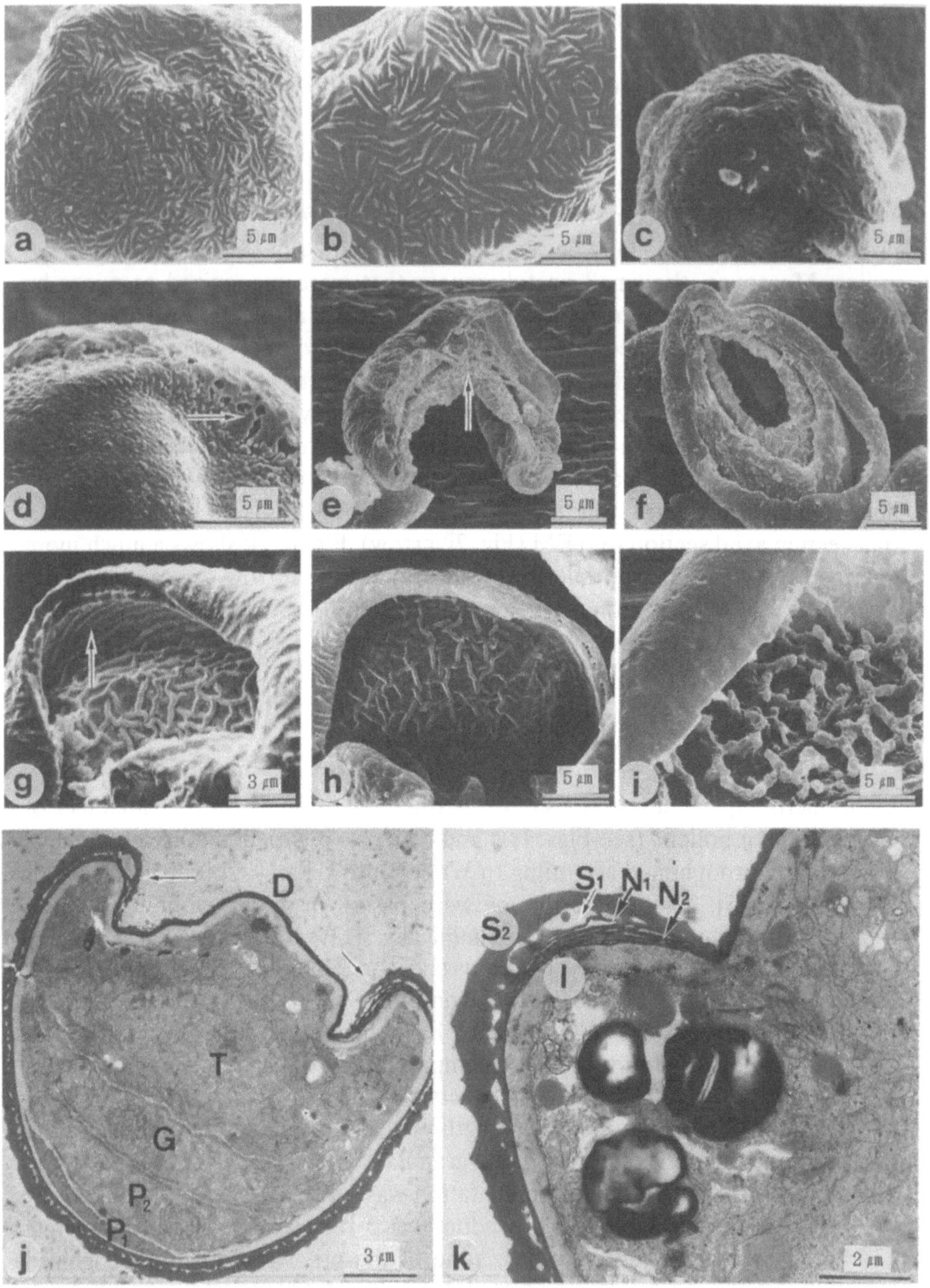

FIG. 2a–k. Pollen grains of *Ginkgo biloba* and *Podocarpus macrophyllus* var. *maki* observed under SEM and transmission electron microscopy (TEM). **a–d** Exine surface pattern on pollen grains of *G. biloba* as shown by SEM. **a** Proximal face (no chemical treatment): randomly arranged linear ridges with some granules. **b** Proximal face

linear ridges with some granules (Fig. 2a), while the latter shows randomly arranged clear linear ridges with acute apices, without irregular granules on the distal face (Fig. 2b). A further investigation was made to remove the outer exine from the grains by placing them in 20% bleaching solution (mainly Naclo) for 3–10 min. After bleaching, the exine surface displayed random cracking and partial peeling (Fig. 1r). Very often, part of the outer exine (possibly sexine 2) was detached entirely (Fig. 1s; arrow). The enlarged surface shown in Fig. 1t has faint linear projections with minutely irregular pores (pits) that are almost invisible on the exine surface without chemical treatment. Thus the results obtained by bleaching show that the exine surface has many minute pores which cause foveolate ornamentation, as seen on the surface of the exine after strong acetolysis using the method of Frederiksen [28]. This striking bleaching treatment was used for the first time on fern spores by Mitsui [15] to remove the outer spore coat (perine) from the exospore.

6 Ornamentation on the Germinal Aperture or Furrow

The ornamentation seen on the outer face of the germinal aperture of *Ginkgo* pollen grains by light microscopy is slight because of its semitransparent membrane, and usually it is almost folded up inside the grain (see Figs. 1m,o and 2d). Wodehouse [1] described this ornamentation as slight transverse corrugations occasionally seen on the floor of the furrow. On the other hand, the replica method using TEM (by Yamazaki and Takeoka [8]) shows the surface to be

FIG. 2a–k. *Continued.* (acetolysis): clear linear ridges with acute apices, without irregular granules. **c** A part of proximal face (no chemical treatment): small but clear auricular projections. **d** A part of distal face (no chemical treatment): edge of the furrow rim (*arrow*), with faint spinules or granules on the germinal aperture. **e–h** Inner face of the exine of *G. biloba* pollen grains shown by SEM. **e** Axial section: reticulum-like sculpturing on the inner face of the germinal furrow (*arrow*). **f** Tangential section: reticulum-like sculpturing on the inner face of the germinal furrow. **g** Tangential section: reticulum-like sculpturing on the inner face of the germinal aperture and somewhat clear rugose sculpturing on the inner side of the proximal part of the exine (*arrow*): **h** Tangential section: more or less tightly reticulum-like sculpture on the inner face of the germinal aperture. **i** Inner face of the bladder of *P. macrophyllus* var. *maki* pollen grain by SEM: lopho-reticulate sculpturing in the inner face of the exine, almost smooth outer surface. **j,k** Sporoderm stratification of *G. biloba* pollen grains by TEM. **j** Axial section: decreased exine in the germinal zone and two protuberances furrow rim (*arrows*), *P1,P2*, two prothallial cells; *G*, a generative cell; *T*, a tube cell; *D*, germinal aperture. **k** Axial section: sporoderms of the furrow rim (edge), showing the thicker intine (*I*), development of the lamellated nexine 2 (*N2*) and thin undulated nexine 1 (*N1*); strong development of the sexine pits (sexine 1) (*S1*) and strong development of the tectum (sexine 2) (*S2*) (**j,k**, Photomicrographs courtesy of Prof. T. Hori, University of Tsukuba)

dotted with verrucoid projections and faint linear ridges. Using SEM, Audran and Masure [12] described verrucoid components mixed with a few ridges, and Sahashi and Ueno [14], also using SEM, reported that the aperture surface of the swollen grains had more or less smooth or faint granules; namely, the outer face of the germinal aperture of nonacetolysed swollen grains showed sornewhat dense verrucoid projections (Fig. 1m) or faint spinules or granules (Figs. 1o and 2d).

7 Inner Face of the Exine Ornamentation, Especially on the Germinal Aperture

According to Sahashi and Ueno [14], the inner face of the exine of the germinal aperture on *Ginkgo* pollen grains (see Fig. 2e–i) has quite remarkable reticulum-like sculpturing. The ornamentation of the inner face of the exine on the pollen grains of some angiosperms has already been reported by Van Campo [29], who found certain kinds of endosculpture (except the reticulate form) by SEM and suggested that these were not haphazard. In the present reexamination, the inner face of the exine ornamentation of *Ginkgo* pollen grains was compared with the reticulum-like sculpturing observed on the inner face of the bladder or saccus of *Podocarpus macrophyllus* var. *maki*. An axial section of a *Ginkgo* pollen grain shows reticulum-like ornamentation on the inner face of the aperture (Fig. 2e; arrow), and also in the tangential section the grain shows clear ornamentation at the same location (Fig. 2f). An enlarged view of the inner face of the grain shows not only large reticulum-like sculpturing on the inner face of the aperture but also more or less rugose sculpturing on the inner face of the exine (Fig. 2g; arrow). Both form an enlarged reticulum-like sculpture (Fig. 2g,h), and the ridges of the reticula can be seen to cross or fuse with each other. On the other hand, Fig. 2i shows the inner face of the bladder of *Podocarpus* pollen, and a large lopho-reticulum-like pattern is clearly visible. In view of the reticulum-like sculpturing of both the germinal aperture and the bladder, I suggest that these features may be an important mechanical structure to aid bulging out for germination or for flying through the air.

8 Sporoderm Stratification by TEM

The terminology for sporoderm stratification of the pollen grain exine (see Fig. 2j,k) has been used differently even for the same membrane by different palynologists (e.g., [30, 31]). Accordingly, I have followed the original descriptions given in each of the references. I also followed the terminology used by Moore et al. [32] in 1991 (Fig. 3). The first description of sporoderm stratification in *Ginkgo* pollen was reported by Ueno [10]. According to this description, the sexine is roughly corrugate and consists of a system of densely crowded bodies,

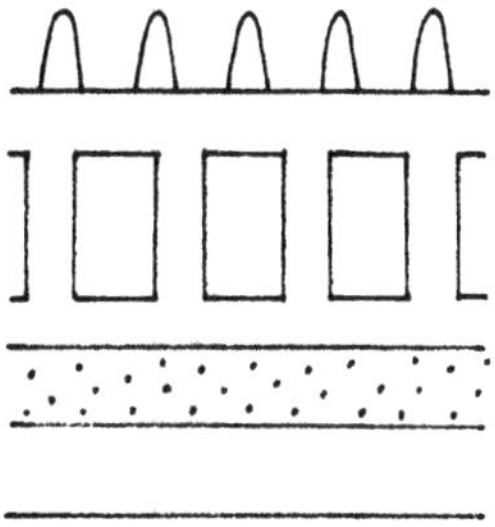

FIG. 3. Terminology of the pollen grain exine according to Moore et al. [32]. (Modified from Fig. 5.2 [32]). *1*, sculpture element, sexine 3; *2*, tectum, sexine 2; *3*, collumella, sexine 1; *4*, foot layer, nexine 1; *5*, endoexine, nexine 2; *6*, intine

gradually merging into the remarkable structure of the nexine. Rohr [11], who investigated exine formation ontogenetically, found that the exine layers of the grain are deposited successively; also, inside the longitudinal slit of the dehydrated pollen grain the sexine is lacking, the only components of the sporoderm being the two nexine layers. The most detailed investigation of sporoderm stratification in *Ginkgo* pollen was carried out by Audran and Masure [12], using SEM and TEM: the exine was composed of a tripartite sexine, the middle zone of which was full of irregular pits, and a thick and strongly lamellated nexine. Page [33] drew attention to this noticeable lamellated endoexine. Soh et al. [20] also observed sporoderm stratification and obtained results similar to those of Audran and Masure [12].

In the present TEM observation, the entire axial section (Fig. 2j) revealed the sporoderm stratification in the slightly young pollen of *G. biloba.* Part of the furrow rim (arrows in Fig. 2j) shows strong development of the exine, while the germinal aperture (D in Fig. 2j) shows a markedly reduced exine, and the rest of the exine developes normally. Inside the pollen grain, four cells can be seen clearly: two prothallial cells (P1, P2 in Fig. 2j), a generative cell (G in Fig. 2j), and a tube cell (T in Fig. 2j). An enlarged view of the furrow rim is shown in Fig. 2k. Part of this sporoderm stratification is quite distinct, namely the thicker intine (I in Fig. 2k), a darkly stained and lamellated endexine that is equivalent to nexine 2 (N2 in Fig. 2k), an irregular foot layer corresponding to nexine 1 (N1 in Fig. 2k), strong development of the sexine pits (S1 in Fig. 2k) that correspond to sexine 1 (Fig. 3), and strong development of the tectum (S2 in Fig. 2k), which corresponds to sexine 2 (Fig. 3) and the external face of the sexine 2 bristles with sharp protrusions. However, my observations suggest that the sexine 3 (Fig. 3) is quite obscure on the pollen grain of *Ginkgo*.

The most striking features are the large lamellated nexine and the strong development of the sexine pits in the furrow rim (Fig. 2k). These phenomena may be equivalent to the reduced wing or bladder in Pinaceae pollen.

9 Conclusions

The pollen grain of *Ginkgo biloba* is usually fusiform or boat shaped in outline, and it has a mono-colpate, 1-sulcate, or anacolpate germinal aperture. However, the swollen grain, especially after acetolysis, shows an almost spherical form with a large and rounded germinal aperture. Auricular projections or winglike protuberances can sometimes be seen on the furrow rim or near the equator of the grain. These projections are reminiscent of ancestral remnants, which were probably bladders on the pollen grains of a fossil ancestor. In fact, TEM observations suggest that the axial section of the edge of the furrow rim sometimes bears auricular projections, showing strong development of the lamellated nexin.

Ornamentation on the proximal face seen by light microscopy appears subreticulate to verrucate or spiny or granulate, or even foveolate after severe acetolysis. However, SEM observations show that the surface is covered with irregularly arranged ridges or is reticulate with short and slightly elevated crests. On the other hand, after bleaching treatment, the exine surface has faint linear projections with minutely irregular pores (pits) that cause foveolate ornamentation. However, the ornamentation on the germinal aperture (furrow) of the nonacetolysed swollen grain consists of somewhat dense verrucoid projections or faint spinules or granules evident by SEM.

The inner face of the exine ornamentation, especially on the germinal aperture, has quite remarkable reticulum-like sculpturing, while the inner face of the bladder (wing or saccus) of *Podocarpus macrophyllus* var. *maki* also has a lophoreticulum-like pattern. These features may be an important mechanical structure to aid bulging out for germination or for being carried through the air.

According to the sporoderm stratification revealed by TEM, the exine is composed of a thicker intine, a darkly stained and lamellated endexine (nexine 2), an irregular foot layer (nexine 1), strong development of the sexine pits (sexine 1), and strong development of the tectum (sexine 2). However, the external face of the sexine 2 bristles with sharp protrusions (sexine 3) is quite obscure on the pollen grain of *Ginkgo*.

Acknowledgments. I would like to thank Prof. T. Hori, University of Tsukuba for the TEM photomicrographs in Fig. 2j–k, and the late Prof. K. Mitsui, Nippon Dental University, for his valuable suggestions and SEM photomicrographs in Fig. 1r–t. I also wish to express my thanks to Miss S. Maeda for drawing Fig. 3.

References

1. Wodehouse RP (1935) Pollen grains. McGraw-Hill, New York
2. Jeffrey EC (1914) Spore conditions in hybrid and the mutation hypothesis of De Vries. Bot Gaz 58:322–336
3. Erdtman G (1943) An introduction to pollen analysis. In: Verdoorn F (ed) A new series of plant science books, vol 12. Chronica Botanica Company Waltham

4. Erdtman G (1957) Pollen and spore morphology/plant taxonomy. Gymnospermae, Pteridophyta, Bryophyta. In: An introduction to palynology, vol II. Almqvist & Wiksell, Stockholm
5. Ikuse M (1956) Pollen grains of Japan (in Japanese). Hirokawa, Tokyo
6. Wang F, Chien N, Zhang Y, Yang H (1995) Pollen flora of China, 2nd edn. Academia Sinica, Peking
7. Yamazaki T, Takeoka M (1959) Electron-microscope investigations on the surface structure of the pollen membrane based on the replica method. VI. Gymnospermae-l (in Japanese). Sci Rep Kyoto Prefect Univ Agric 11:86–90
8. Yamazaki T, Takeoka M (1962) Electron-microscope investigations of the fine details of the pollen grain surface in Japanese gymnosperms. Grana Palynol 3:3–12
9. Ueno J (1960) Studies on pollen grains of gymnospermae: concluding remarks to the relationships between coniferae. J Inst Poly Osaka City Univ D 11:109–136
10. Ueno J (1960) On the fine structure of the cell walls of some gymnosperm pollen. Biol J Nara Women's Univ 10:19–25
11. Rohr R (1977) Etude comparee de la formation de l'exine au cours de la microsprogenese chez une gymnospermae (*Taxus baccata*) et une prephanerogame (*Ginkgo biloba*). Cytologia (Tokyo) 42:157–167
12. Audran JC, Masure E (1978) La sculptutre et l'infrastructure du sporoderme de *Ginkgo biloba* comparees a celles des enveloppes polliniques des cycadales. Rev Palaeobot Palynol 26:363–387
13. Audran JC (1987) Comparaison des ultrastructures exiniques et des modalites de l'ontogenese pollinique chez les cycadales et ginkgoales actuelles (Prespermaphytes). Bull Soc Bot Fr Actual Bot 134:9–18
14. Sahashi N, Ueno J (1986) Pollen morphology of *Ginkgo biloba* and *Cycas revoluta*. Can J Bot 64:3075–3078
15. Mitsui K (1978) Outer spore layers of *Adiantum pedatum* L. and *Pteris vittata* L. Bull Nippon Dent Univ 7:212–218
16. Huynh KL (1976) L'arrangement du pollen dans la tetrade chez les gymnospermes. 1. Les genres *Pinus*, *Ginkgo* et *Sequoia*. Bull Soc Neuchatel Sci Nat 99:57–74
17. Tulecke WR (1954) Preservation and germination of the pollen of *Ginkgo* under sterile conditions. Bull Torrey Bot Club 81:509–512
18. Huang TC (1972) Pollen flora of Taiwan. National Taiwan University, Taipei
19. Nakamura J (1980) Diagnostic characters of pollen grains of Japan. Part I, II (in Japanese). Osaka Museum of Natural History, Osaka
20. Soh JY, Tae KH, Ko SC, Soh WY (1994) Comparative morphology of pollen grains from some Korean gymnosperms. J Plant Biol 37:203–221
21. Friedman WE (1987) Growth and development of the male gametophyte of *Ginkgo biloba* within the ovule (in vivo). Amer J Bot 74:1797–1815
22. Erdtman G, Vishnu-Mittre (1958) On terminology in pollen and spore morphology. Grana Palynol 1:6–9
23. Jeffrey EC (1917) The anatomy of woody plants. University of Chicago Press, Chicago
24. Zimmerman M (1959) Die phylogenie der pflanzen. Fischer, Stuttgart
25. Taylor TN (1970) *Lasiostrobus* gen. n., a staminate strobilus of gymospermous affinity from the Pennsylvanian of North America. Amer J Bot 57:670–690
26. Taylor TN (1981) Paleobotany: an introduction to fossil plant biology. McGraw-Hill, New York
27. Shimakura M (1973) Palynomorphs of Japanese plants (in Japanese). Osaka Museum of Natural History, Osaka

28. Frederiksen NO (1978) Preservation of cycad and *Ginkgo* pollen. Rev Palaeobot Palynol 25:163–179
29. Van Campo M (1978) La face interne de l'exine. Rev Palaeobot Palynol 26:301–311
30. Erdtman G (1952) Pollen morphology and plant taxonomy. Angiosperms. In: An introduction to palynology. Almqvist & Wiksell, Stockholm
31. Faegri K, Iversen J (1964) Textbook of pollen analysis, 2nd edn. Blackwell, Oxford
32. Moore PD, Webb JA, Collinson ME (1991) Pollen analysis, 2nd edn. Blackwell, Oxford
33. Page CN (1990) Ginkgoaceae. In: Kubitzki K (ed) The families and genera of vascular plants, vol I. Pteridophytes and gymnosperms (Kramer KU, Green PS, eds). Springer, Berlin Heidelberg New York, pp 284–289

Development of the Male Gametophyte of *Ginkgo biloba*: A Window into the Reproductive Biology of Early Seed Plants

WILLIAM E. FRIEDMAN[1] and ERNEST M. GIFFORD[2]

1 Introduction

It has been a century since the discovery of zooidogamous reproduction among seed plants by Hirase [1, 2] and Ikeno [3]. The initial observations of motile sperm in *Ginkgo* and *Cycas* represented the culmination of progress, beginning with the discovery of the pollen tube by Amici [4], in the field of plant reproductive biology during the nineteenth century. Discovery of zooidogamy in *Ginkgo* and cycads in 1896 provided a critical connection, in terms of evolutionary history, between the life cycles of non-seed plants ("cryptogams") with motile sperm and the life cycles of previously described seed plants (conifers and angiosperms) with pollen tubes and non-motile sperm [5]. Moreover, the presence of motile sperm within the male gametophytes of cycads and *Ginkgo* confirmed a prediction made almost one half century earlier by the renowned biologist Wilhelm Hofmeister [6] that flagellate sperm might be found among representatives of the seed plants [5].

General features of the male gametophyte of *Ginkgo biloba* were first established by the botanical illustrator Sakugoro Hirase in an elegant series of publications between 1894 and 1898 (Fig. 1) [1, 2, 7–12]. Hirase [11] showed that prior to fertilization, the male gametophyte of *Ginkgo* consists of a large spheroidal structure and an extensively branched system of tubes that anchor the male gametophyte within the nucellus of the ovule (Fig. 1f). Hirase also investigated the subcellular structure of the spermatogenous cell (Fig. 1e) [7, 11], and the formation of large multiflagellate sperm cells (Fig. 1g) [11]. Although Hirase is best known for his discovery of motile sperm in *Ginkgo*, his investigations of male gametophyte development and the fertilization process in *Ginkgo* were extraordinary in detail, and continue to reflect the work of a keen observer of nature.

Additional studies of the male gametophyte in *Ginkgo* have focused on microsporogenesis [13, 14], pollen development [15, 16], morphological develop-

[1]Department of EPO Biology, University of Colorado, Boulder, CO 80309, U.S.A.
[2]Section of Plant Biology, University of California, Davis, CA 95616, U.S.A.

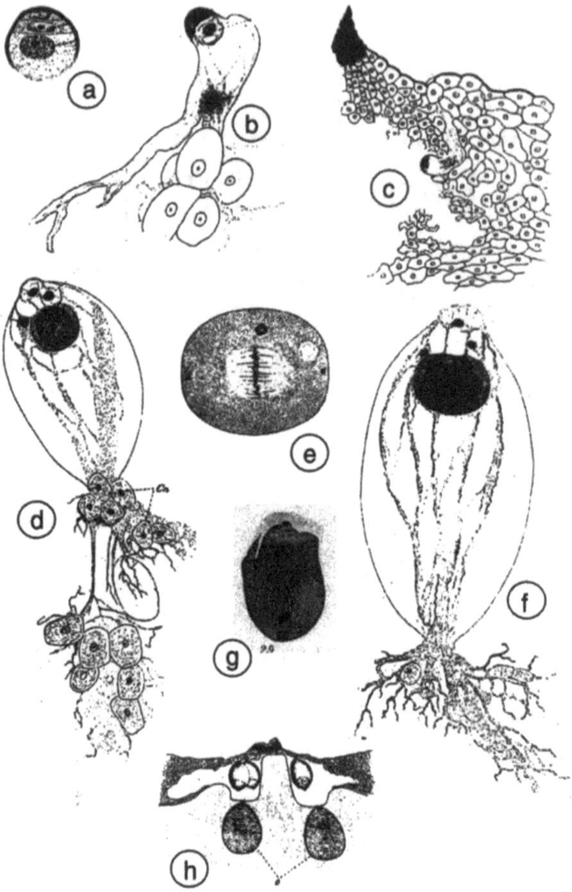

FIG. 1. Drawings of male gametophytes of *Ginkgo biloba* from Hirase's early studies. **a** Pollen grain. **b** Young male gametophyte with branched pollen tube. **c** Young male gametophyte penetrating into nucellus. **d** Proximal portion of male gametophyte approximately four months after pollination. **e** Mitosis of spermatogenous cell with blepharoplasts and osmiophilic globule. **f** Mature male gametophyte after division of spermatogenous cell into two sperm cells. **g** Multiflagellate sperm cell. **h** Drawing of three male gametophytes hanging from the nucellus above the archegonia of the female gametophyte just prior to fertilization. (**a–d,f–h** from [11]; **e** from [7])

ment [17–20], in vitro development [21–24], the unusual "ring cell" division of the generative cell [25], structure of the spermatogenous cell [26], sperm structure [18, 27–37], ultrastructure of the spermatogenous cell [38–40], and ultrastructure of the sperm cells [41]. Together with the initial studies of reproduction in *Ginkgo* by Hirase, this literature provides a broad view of the structure and development of the male gametophyte of *G. biloba*. This review will focus on post-pollination development of the male gametophyte of *G. biloba* and the fertilization process.

2 Morphological Development of the Male Gametophyte

Morphological development of the male gametophyte of *Ginkgo biloba* is divided into three phases [20, 42]. Germination of a pollen grain within the pollen chamber is characterized by an initial brief period of diffuse and roughly isodiametric growth. Initiation of tip growth and the formation of a heterotrophic pollen tube characterizes a second and extended phase of morphological development. During the period of tubular growth, the male gametophyte of *G. biloba* branches extensively within the intercellular spaces of the nucellar apex of the ovule. The final phase of morphological development involves the diffuse swelling of the unbranched proximal portion of the male gametophyte into a large saccate structure.

2.1 Early Development from Microspore to Pollen Grain

Microspores form within the microsporangia of male strobili of *G. biloba* [14] and continue to develop into pollen grains through a series of three cell divisions. The result is a four celled pollen grain with two prothallial cells, a generative cell, and a tube cell. By the time pollen is released, the first prothallial cell has degenerated [20, 39]. The mature pollen grain is monosulcate.

Pollen is released from the strobili of microsporangiate trees (*G. biloba* is dioecious) and is captured by the pollination droplet of the ovule (Fig. 2). Pollen grains adhere to the pollination droplet [11, 18–20] and subsequent male–female signalling leads to retraction of the droplet through the micropyle. The effect is to draw one or more pollen grains into the pollen chamber at the apex of the nucellus (Fig. 2).

2.2 Germination of the Pollen Grain

Approximately one week after pollen is deposited in the pollen chamber, the first signs of germination of the male gametophyte are apparent in *G. biloba* (Fig. 2d). Germination occurs through the sulcus and is manifest as a swelling of the tube cell, resulting in the formation of a slightly bulbous protuberance [20]. In vitro analysis of pollen germination confirms that the initial growth of the male gametophyte does not involve tip growth (Fig. 3), but rather a diffuse swelling from the area of the sulcus [21, 23, 24]. This pattern of early diffuse growth of the male gametophyte in *G. biloba* has not been detected in vivo in the male gametophytes of any other seed plants [20].

Shortly after germination of the pollen grain, an unbranched pollen tube develops (Figs. 2e, 4a) and penetrates into the nucellus surrounding the pollen chamber [11, 19, 20, 36]. Growth is strictly intercellular and no evidence of cell

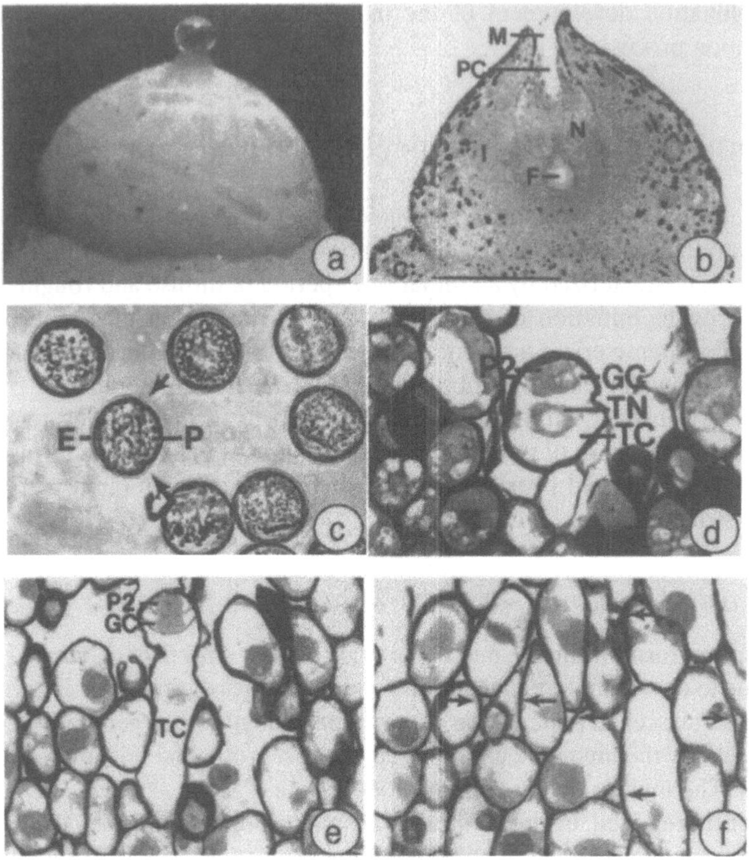

FIG. 2. Pollination and early development of the male gametophyte of *Ginkgo biloba*. **a** Ovule with pollination droplet at tip of micropyle. **b** Longitudinal section of ovule at time of pollination with micropyle (*M*), pollen chamber (*PC*), nucellus (*N*), free nuclear female gametophyte (*F*), and collar (*C*). **c** Hydrated pollen grains (*P*) with exine (*E*). *Arrows* indicate limits of sulcus. **d** Young male gametophyte just after germination in pollen chamber with second prothallial cell (*P2*), generative cell (*GC*), tube nucleus (*TN*), and tube cell (*TC*). **e** Unbranched portion of male gametophyte with second prothallial cell (*P2*), generative cell (*GC*), and tube cell (*TC*). **f** Intercellular growth of male gametophyte. *Arrows* indicate portions of a single tube cell that is ramifying between the cells of the nucellus. (**a–f** from [20], with permission)

degradation in the nucellus is apparent as the pollen tube invades these sporophytic tissues [20, 40]. At this early stage of development, the male gametophyte of *G. biloba* strongly resembles the simpler unbranched male gametophytes of many siphonogamous seed plants such as conifers, Gnetales, and angiosperms. The proximal unbranched portion of the male gametophyte may be up to 105 µm in length [20].

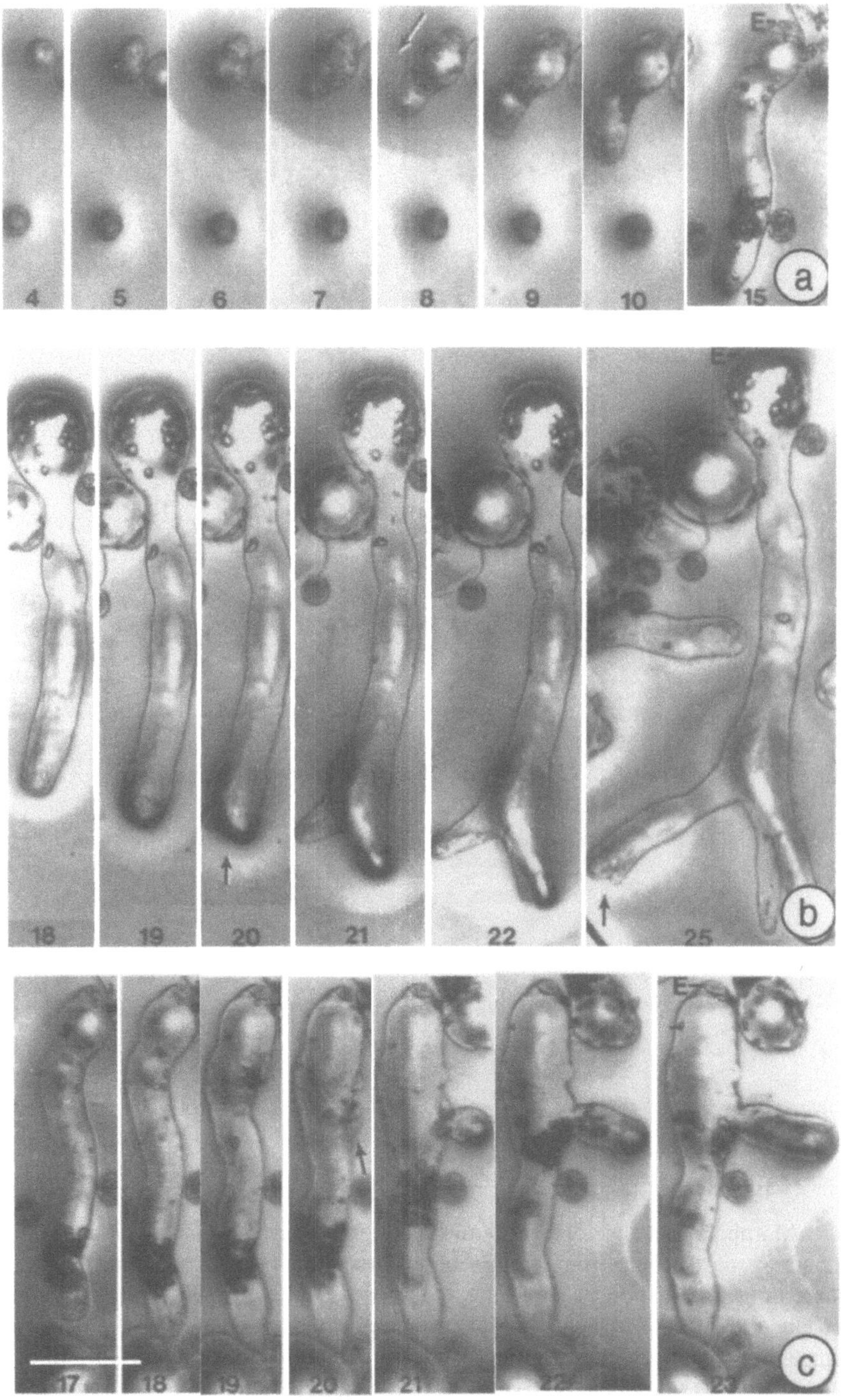

FIG. 3. In vitro development of the male gametophyte of *Ginkgo biloba*. **a** Early diffuse growth (days 4–6) followed by initiation of tip growth (days 7–15). *E*, exine. **b** Apical branching (*arrows* on days 20 and 25). **c** Subapical branching (*arrow* on day 20). *Numbers*, age of male gametophyte in days. (**a–c** from [24], with permission)

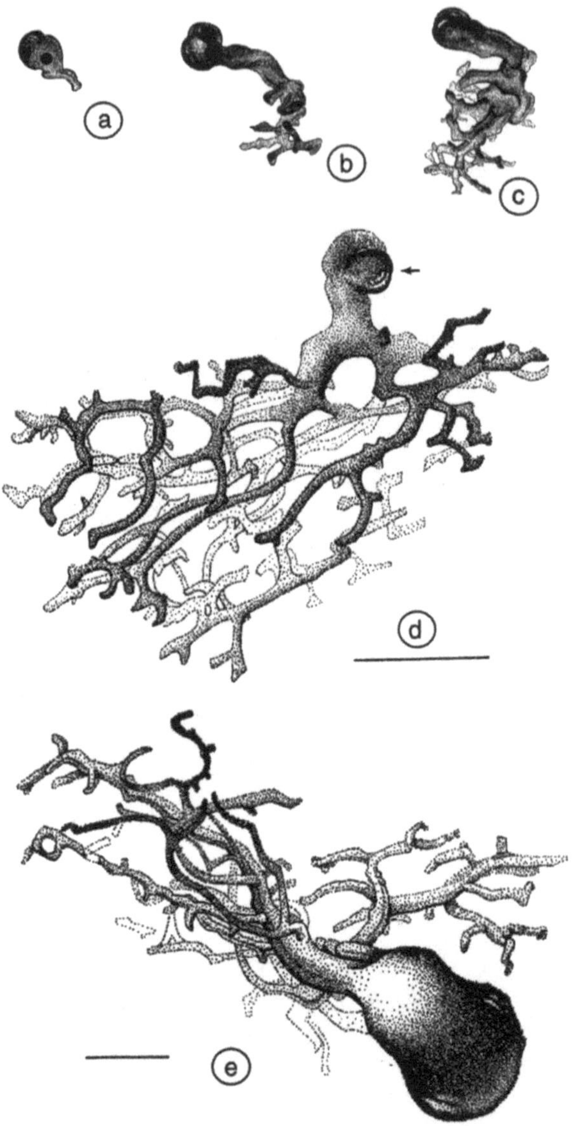

FIG. 4. Morphological development of the male gametophyte of *Ginkgo biloba* based on computer reconstructions from serial sections through the nucellus. **a** Young male gametophyte prior to initiation of branching. **b** Two-week-old male gametophyte with many intercellular haustorial branches. **c** Male gametophyte approximately 3 to 4 weeks after pollination. **d** Male gametophyte approximately 2 months after pollination. All of the branches are part of a single tube cell. *Arrow* indicates the original pollen grain. **e** Male gametophyte almost 5 months after pollination, just prior to fertilization. The unbranched portion of the pollen tube has swollen into a saccate structure that contains the spermatogenous cell. (**a–e** from [20], with permission)

2.3 Initiation of Branched Intercellular Penetration of the Nucellus

Following formation of an unbranched pollen tube, the male gametophyte of *G. biloba* begins to branch extensively [11, 17, 19, 20, 40]. Within a week or two of the penetration of the nucellus by the unbranched pollen tube, large numbers of pollen tube branches are initiated and these ramify within the intercellular spaces of the nucellus (Figs. 1a, 4). The haustoria are initially narrower in diameter than the more proximal unbranched portion of the pollen tube, and usually conform to the dimensions of the intercellular spaces in which they grow (Fig. 2f). When first formed, the haustorial pollen tube branches are typically between five and ten microns in diameter [20]. Approximately two months after pollination, the haustorial portion of the gametophyte is remarkably extensive (Fig. 4d), as can be seen in three-dimensional reconstructions of young male gametophytes derived from serial sections [20].

Nuclear divisions or cell divisions are not associated with the formation of the extensively branched male gametophyte in *G. biloba*. The entire haustorial system formed from the single tube cell (Fig. 4d) is continuous with the original unbranched portion of the pollen tube and the contents of the pollen grain [20]. By the end of its second month of growth, the male gametophyte still comprises a degenerate first prothallial cell, a viable second prothallial cell, a generative cell, and an extensive, highly branched tube cell. The tube nucleus does not appear to migrate beyond the juncture of the first branching point of the tube cell [11, 19, 20].

The strictly intercellular growth of the haustorial male gametophyte in *Ginkgo* does not lead to the breakdown of nucellar cells (as is often the case with the male gametophytes of conifers and cycads), and is thus likely to result in limited exposure to nutrients. This situation may be compensated for by the relatively high surface area to volume ratios of the male gametophyte of *Ginkgo*, compared with other extant nonflowering seed plants [20, 42]. In vitro studies [21, 24] confirm that the concentration of sugar within the growth medium is inversely correlated with the degree of branching expressed by the male gametophyte. Thus, it is possible that formation of extremely narrow diameter and highly branched haustorial pollen tubes may be a response to lower levels of nutrient availability in the nucellus, compared with the pollination droplet and pollen chamber [24].

In vitro analysis of male gametophyte development in *G. biloba* also indicates that branches may be initiated in one of two ways [24]. Time-lapse photography shows that branching in culture is typically apical and involves division of the apex into two distinct growing points (Fig. 3). Subapical (or lateral) branching (Fig. 3) occurs infrequently in vitro and only after cessation of tip growth by the male gametophyte [24]. It seems likely that branching of the male gametophyte within the ovule is usually apical and leads to the division of the growing tip into two or more separate apices [11, 20].

2.4 Late Morphological Development of the Male Gametophyte

Approximately four months after pollination (typically August), the final phase of morphological development of the male gametophyte of *G. biloba* is initiated. This period is delineated by pronounced increases in the radial dimensions of the unbranched portion of the pollen tube, which is positioned next to the pollen grain (Figs. 4, 5) [20]. Swelling of the male gametophyte appears to be opportunistic, and preferentially fills adjacent spaces in the pollen chamber. Most haustoria close to the unbranched region also increase in girth (Fig. 4e). By the time that

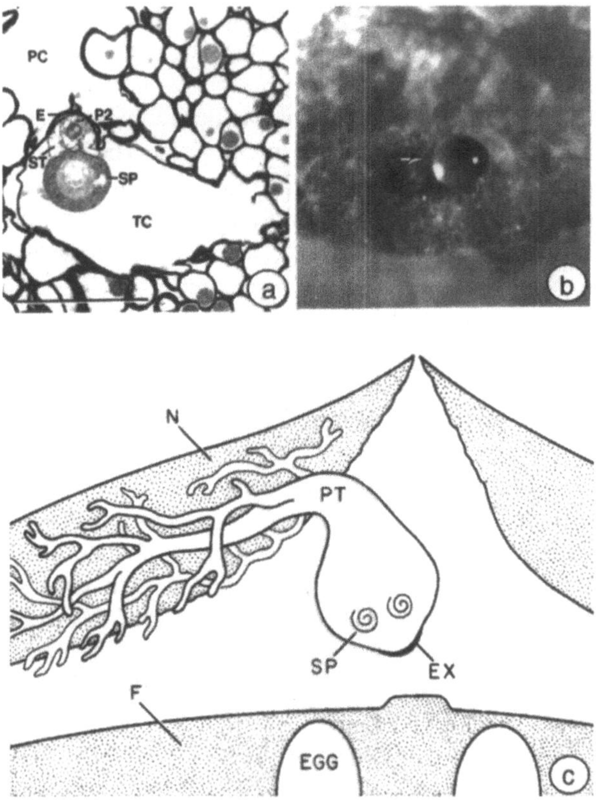

FIG. 5. Late development of the saccate portion of the male gametophyte of *Ginkgo biloba*. **a** Unbranched portion of male gametophyte next to pollen chamber (*PC*) with exine (*E*), second prothallial cell (*P2*), sterile cell (*ST*), spermatogenous cell (*SP*), and tube cell (*TC*). **b** Saccate portion of male gametophyte (*arrow*) attached to nucellus. **c** Schematic of mature male gametophyte and its branched pollen tube (*PT*), multiflagellate sperm (*SP*), and exine of pollen grain (*EX*) suspended from nucellus (*N*) above female gametophyte (*F*) at time of fertilization. The unbranched portion of the pollen tube will rupture and release the sperm to swim to the eggs. (**a,b** from [20], with permission; **c** from [42] with permission)

fertilization is imminent, the unbranched portion of the male gametophyte has developed into a large saccate structure (Figs. 1d, 1f, 4e, 5) that is suspended from the nucellus just above the archegonia of the female gametophyte (Figs. 4e, 5) [11, 18, 20, 40].

The unusual pattern of swelling of the pollen grain end of the male gametophyte in *G. biloba* is fundamentally similar to what occurs late in development in the male gametophytes of cycads [42] and appears to be unique (among extant seed plants) to these two ancient lineages of seed plants. In both groups, rupture of the swollen portion of the male gametophyte, which is situated directly over the position of the egg cells of the female gametophyte (Fig. 5c), ultimately serves to release the motile sperm to swim within the ovule to the archegonia [42].

3 Cell Divisions Within the Male Gametophyte

3.1 Formation of the Sterile Cell and Spermatogenous Cell

The mature pollen grain of *Ginkgo biloba* contains four cells: a degenerate first prothallial cell, a second prothallial cell, a generative cell, and a tube cell. In spite of the subsequent extensive morphological development of the male gametophyte, the first cell division to occur following pollination is during the fourth month of development of the male gametophyte within the ovule. This event coincides with the third and final phase of morphological development (radial growth of the pollen tube), and involves division of the generative cell to produce a sterile cell and spermatogenous cell [20].

For almost a century, it had been believed that the generative cell in the male gametophyte of *G. biloba* divided transversely to produce a sterile cell and spermatogenous cell (Fig. 6a). Webber [17] and Sterling [43] hypothesized that the adjacent positions of the second prothallial cell and spermatogenous cell at maturity (Fig. 6f) were secondarily achieved by a gradual protrusion of the second prothallial cell through the center of the sterile cell during late development of the male gametophyte (Fig. 6a). This "protrusion" theory was similarly advanced for the male gametophytes of cycads [17, 43–46], which at maturity, have a cell pattern similar to that of *Ginkgo* [25].

Recent developmental and ultrastructural data from male gametophytes of *G. biloba* growing within the ovule and in vitro [20, 23, 25] do not support the protrusion hypothesis. In vivo and in vitro studies of the generative cell division reveal that the sterile cell and spermatogenous cell arise from an unusual anticlinal ringlike division of the generative cell. During this process, a small centrally positioned cell, the spermatogenous cell, is cleaved from the parent generative cell and is consequently surrounded by a ring-shaped sterile cell (Fig. 6b,d). As a result, both the spermatogenous cell and sterile cell share common cell wall interfaces with the primary cell wall of the second prothallial cell.

Further evidence congruent with the formation of the spermatogenous cell from a ring cell division can be found in transmission electron micrographs of

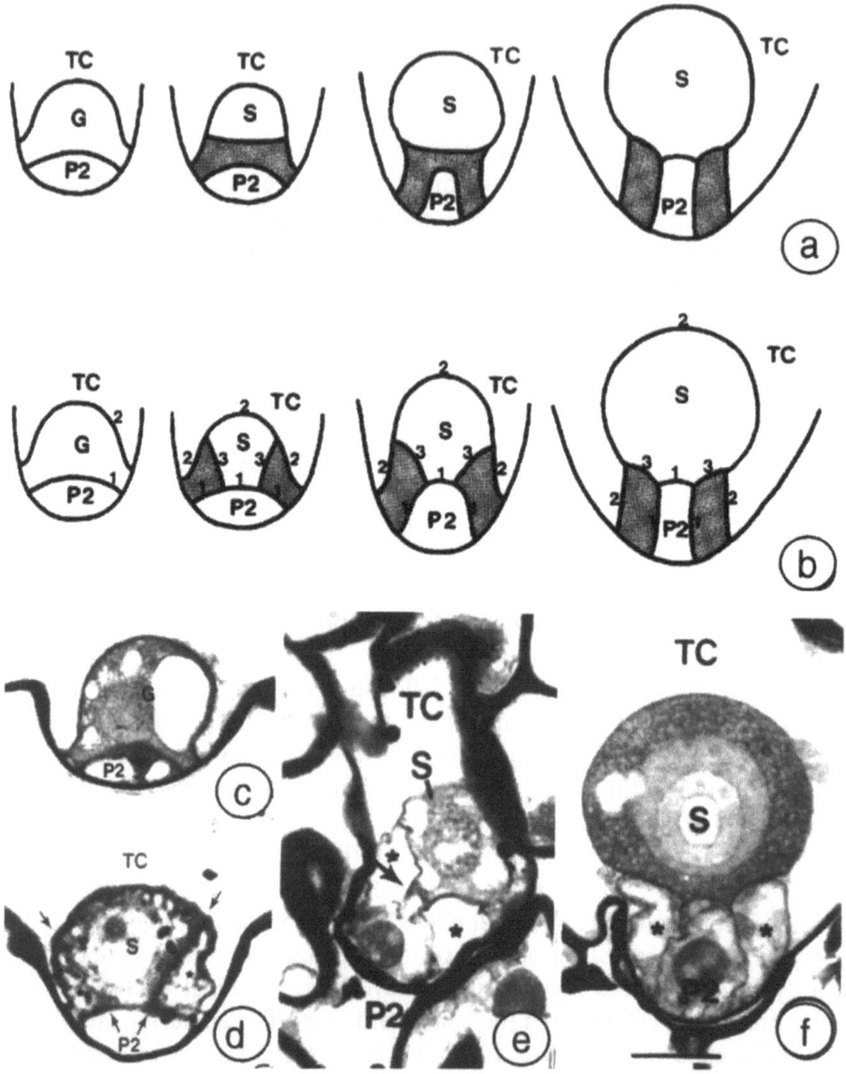

FIG. 6. Ring cell division of the generative cell to form the sterile cell and spermatogenous cell in *Ginkgo biloba*. **a** Schematic of "protrusion" theory of development of sterile cell (*stippled*) and spermatogenous cell (*S*) from generative cell (*G*) in which second prothallial cell (*P2*) protrudes through sterile cell. *TC*, tube cell. **b** Schematic of documented ring cell division of generative cell (*G*) and subsequent growth of sterile cell (*stippled*) and spermatogenous cell (*S*). The second prothallial cell (*P2*) is in contact with the spermatogenous cell from the beginning. *TC*, tube cell; *numbers* indicate different cell walls. **c** Second prothallial cell (*P2*), generative cell (*G*), and portion of tube cell from in vitro male gametophyte prior to division of generative cell. **d** Second prothallial cell (*P2*), ring-shaped sterile cell (indicated with *asterisks*), spermatogenous cell (*S*), and portion of tube cell (*TC*) from in vitro male gametophyte just after ring cell division of generative cell. **e** Median section of second prothallial cell (*P2*), ring shaped sterile cell (indicated with

male gametophytes of *G. biloba* in vivo [25] and in vitro [23]. These data clearly demonstrate that plasmodesmatal connections exist between the second prothallial cell and spermatogenous cell. Plasmodesmatal connections would not be expected if the cell wall interfaces between the second prothallial cell and spermatogenous cell had formed as a result of a protrusion of the second prothallial cell through the sterile cell [25].

During the final month of development of the male gametophyte of *G. biloba*, the spermatogenous cell expands considerably. This results in displacement of the sterile cell to a position surrounding the second prothallial cell (Fig. 6). The configuration of cells that characterizes the mature male gametophyte (Figs. 5a, 6b, 6f) thus results from changes in orientation between the walls of the second prothallial cell, sterile cell, and spermatogenous cell [25, 40].

The ringlike division of the generative cell of *Ginkgo* is unusual among land plants and is only known to occur in two additional situations: antheridial development of leptosporangiate ferns and stomatal development among certain members of the Schizaeaceae and Polypodiaceae [25]. The many similarities in development and arrangement of cells within the male gametophytes of *G. biloba* and most cycads suggest that division of the generative cell in cycads may also involve a ring cell division [25]. Although the ringlike (conically annular) divisions in the antheridia of leptosporangiate ferns and the male gametophytes of *G. biloba* are remarkably alike and result in the production of a central spermatogenous cell, it is conjectural as to whether these patterns represent a striking convergence or evolutionary homology [25].

3.2 Formation of Sperm Cells and Fertilization

Just prior to fertilization, the spermatogenous cell divides to form two multiflagellate sperm that are released to swim to the archegonia within the female gametophyte. Division of the spermatogenous cell is longitudinal [11, 30] (Fig. 1f) and a cell plate does not appear to form between the two sperm cells [37]. Each sperm cell is bounded by its own plasmalemma within the original parent cell wall of the spermatogenous cell [37, 41]. Development of an anterior flagellar band occurs in each sperm cell.

After the flagella begin to beat, the two sperm are released from the spermatogenous cell wall and swim within the saccate portion of the tube cell of the male gametophyte [30, 37]. Rupture of the saccate portion of the tube cell

FIG. 6. *Continued. asterisks*), spermatogenous cell (*S*), and portion of tube cell (*TC*) shortly after division of generative cell. *Arrow* indicates region of contact between spermatogenous cell and second prothallial cell. **f** Median section of pollen grain end of nearly mature male gametophyte with second prothallial cell (*P2*), sterile cell (indicated with *asterisks*), spermatogenous cell (*S*), and portion of tube cell (*TC*). (**a** from [20], with permission; **b–f** from [25], with permission)

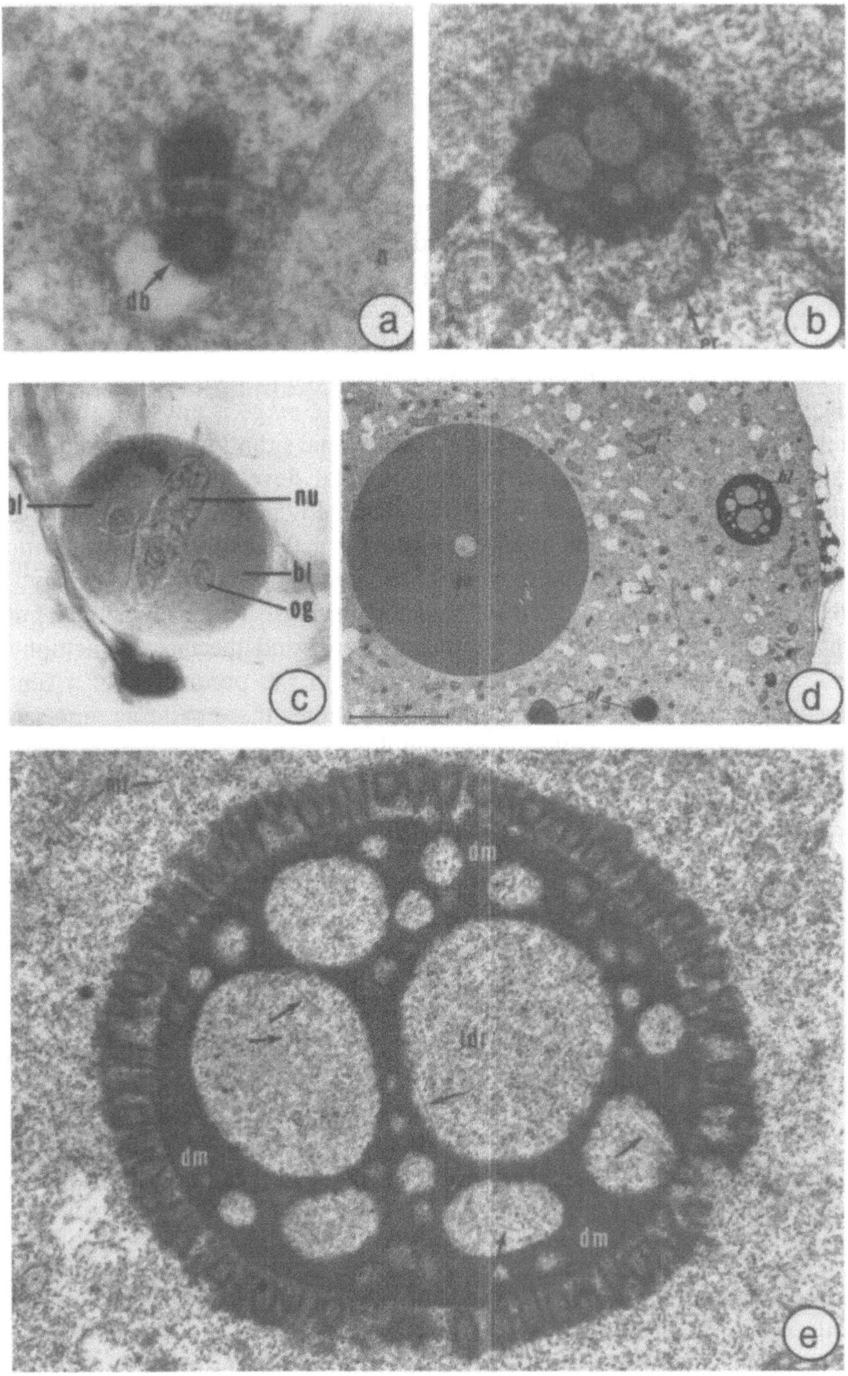
db
a
b
bl
nu
bl
og
c
d
dm
dm
dm
e

frees the sperm to swim to the archegonia within the female gametophyte (Fig. 5). Prior to release from the spermatogenous cell wall, the sperm cells are approximately 55 × 72 μm [41]. Measurements of swimming sperm after release from the spermatogenous cell range from 50 to 85 μm in width and 80 to 110 μm in length [35, 37].

4 Ultrastructure of the Male Gametophyte

4.1 The Spermatogenous Cell

Many of the initial observations of the spermatogenous cell in *Ginkgo biloba* can be traced to the pioneering works of Hirase [7, 11] and Webber [17, 36]. More recently, the structure of the spermatogenous cell, as revealed by electron microscopy, has been studied. Three publications [38–40] have explored the ultrastructural development of the spermatogenous cell and its distinctive organelles, most notably, the blepharoplasts, which are exceedingly complex organelles associated with formation of the multilayered structure, basal bodies, and flagella of the mature sperm cells.

Shortly after the spermatogenous cell is formed, a "double blepharoplast" forms de novo within the cytoplasm (Fig. 7a). This structure consists of two hemispherical bodies that are destined to separate and each develop into a separate blepharoplast within the spermatogenous cell [40]. The two halves of the double blepharoplast migrate to opposite ends of the spermatogenous cell and come to occupy a final position between the discoid nucleus and the thin outer wall of the spermatogenous cell (Fig. 7). During this migration, the first probasal bodies (centrioles) appear to develop at the periphery of the blepharoplasts [40].

At maturity, the spermatogenous cell comprises a centrally positioned discoid nucleus with one or two nucleoli, two osmiophilic globules (function unknown), a single fibrogrannular body (function unknown), and two prominent blepharoplasts (Fig. 7) [39, 40]. In addition, proplastids, mitochondria, abundant ribosomes, small vacuoles, and lipid droplets can be found within the spermatogenous cell. Endoplasmic reticulum and dictyosomes are minimally present [39].

Each mature blepharoplast is spherical and approximately 3.5 to 4.5 micrometers in diameter. There are internal differentially electron opaque regions, ap-

FIG. 7. Ultrastructure of spermatogenous cell of *Ginkgo biloba*. **a** Double blepharoplast (*db*) next to nucleus (*n*). **b** Developing blepharoplast with probasal body (*c*) and adjacent endoplasmic reticulum (*er*). **c** Light micrograph of spermatogenous cell with discoid nucleus (*nu*), two blepharoplasts (*bl*), and osmiophilic globule (*og*). **d** Electron micrograph of osmiophilic globule (*og*) and blepharoplast (*bl*) in spermatogenous cell. **e** Mature blepharoplast with numerous probasal bodies, less electron opaque region (*ldr*) and electron opaque matrix (*dm*), and nearby microtubules (*mt*). (**a,b** from [40]; **c,e** from [39]; **d** from [38]; all with permission)

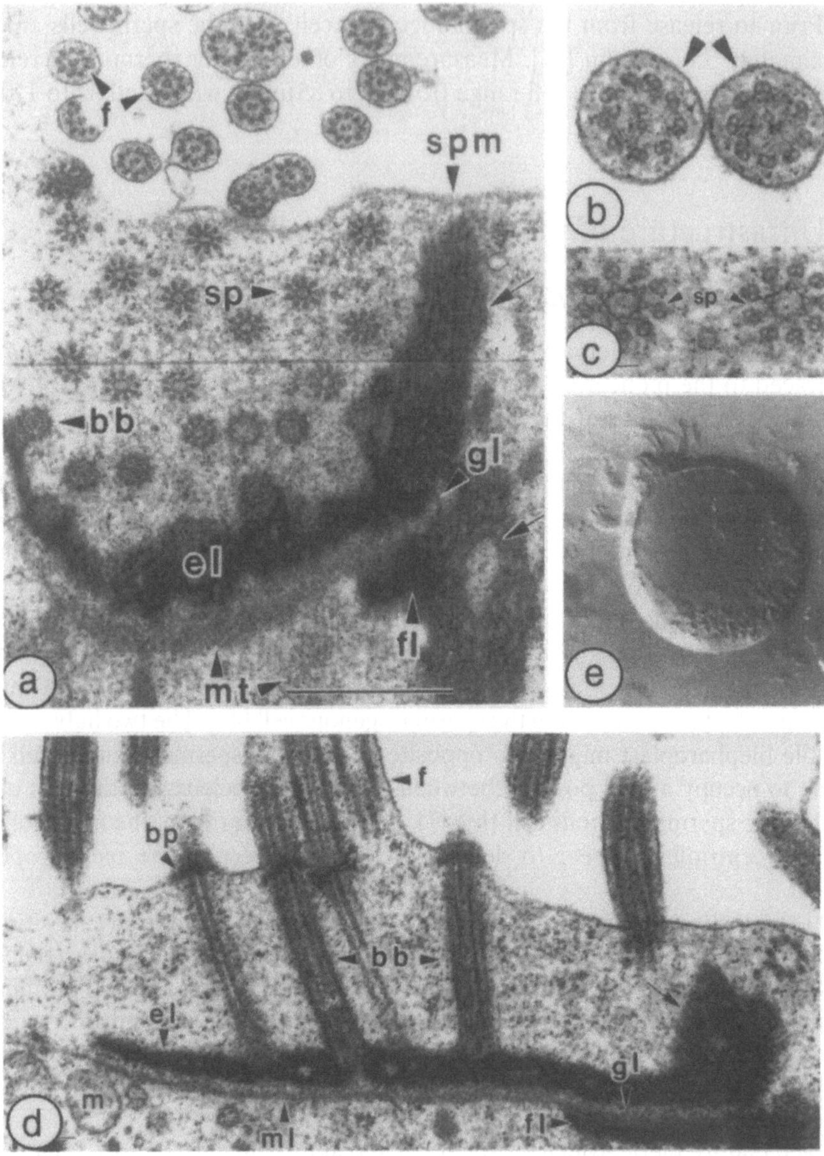

FIG. 8. Ultrastructure of multiflagellate sperm cell of *Ginkgo biloba*. **a** Multilayered structure of sperm cell with electron opaque outer layer (*el*), granular layer (*gl*), and fibrous layer (*fl*). Flagella (*f*) can be seen in cross section, as well as the stellate pattern (*sp*) of the transition region of basal bodies (*bb*). Other microtubules (*mt*) are apparent. The sperm cell plasma membrane (*spm*) is apparent. **b** Transverse section of flagella. **c** Transition regions of basal bodies with stellate pattern (*sp*). **d** Longitudinal section of the multilayered structure with rooting of basal bodies (*bb*) in electron opaque layer (*el*). Basal plate (*bp*) of flagellum (*f*) visible, as well as fibrous layer (*fl*) and granular layer (*gl*) of multilayered structure, and mitochondria (*m*). **e** Swimming sperm with anterior flagellar band. (**a–d** from [41], with permission; **e** from [47], with permission)

proximately 1000 probasal bodies (centrioles) at the periphery, and microtubules in some of the less electron opaque regions [38, 39]. Transections of the probasal bodies reveal nine separate tubules ("singlets") at the base, with a transition to nine pairs of "doublets" and ultimately nine pairs of "triplets" at the distal end [39]. A central tubule runs through the entire length of each probasal body. Each probasal body associated with a blepharoplast will develop into a basal body of a flagellum during the formation of sperm.

4.2 The Sperm Cells

The mature sperm cell of *G. biloba* (Fig. 8e) is comprised of an anterior flagellar band with approximately 1000 flagella, a large centrally placed nucleus, and additional cytoplasm with a full complement of organelles such as mitochondria, plastids and dictyosomes [41]. The flagella are anchored within the sperm cell by a spiral multilayered structure (Fig. 8d). Each membrane bound flagellum typically has nine pairs of peripheral microtubules and two individual centrally placed microtubules (a $9_2 + 2$ arrangement) (Fig. 8b) [41]. The flagellum is connected through a transition region of the basal body (with stellate pattern) which is in the cytoplasm of the sperm cell (Fig. 8c). The basal bodies have nine pairs of peripheral microtubules (a $9_2 + 0$ arrangement) and ultimately, a $9_3 + 0$ arrangement at the junction with the multilayered structure. When compared with sperm cells of cycads, the sperm of *G. biloba* appear to be ultrastructurally quite similar [41].

5 The Male Gametophyte of *Ginkgo biloba* in the Broader Context of Seed Plant Evolution

Current evidence is congruent with the hypothesis that the male gametophytes of the earliest seed plants were endosporic and fundamentally similar to the microspores of heterosporous vascular non-seed plants [42, 48–52]. The "prepollen" of lyginopterids, medullosans, early conifers, and early cordaites lacked a distal sulcus and displayed proximal sutures similar to the microspores of endosporic pteridophytes. The male gametophytes of these early seed plants probably released sperm via the rupture of the proximal suture mark of the original spore wall [48, 53–55] (Fig. 9a). These sperm (which were almost certainly multiflagellate) then swam within the ovule to the female gametophyte and its archegonia.

The evolution of exosporic haustorial growth is likely to have conferred important selective advantages to those early seed plant taxa that exhibited this pattern of male gametophyte development [42, 51, 56]. The initial and primary role of exosporic growth was to obtain nutrients (from the host sporophyte tissues) for the development of the male gametophyte during the period between pollination and fertilization [42, 51, 57, 58]. This pattern is still evident in *Ginkgo* and cycads which are the only extant zooidogamous plants with a pollen tube (Fig. 9c),

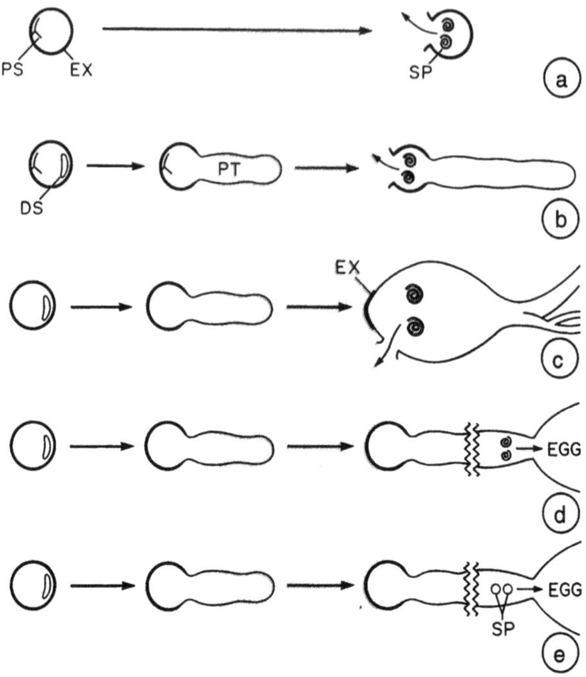

FIG. 9. Patterns of morphological and sexual development of seed plant male gametophytes. **a** Seed plants with prepollen (lyginopterids, medullosans, primitive conifers, primitive cordaites) are endosporic. Motile sperm (*SP*) are released by rupture of the proximal trilete (or monolete) suture (*PS*) in the exine (*EX*). **b** Seed plants with grains with a proximal suture and a distal sulcus (*DS*) (some conifers and cordaites). Hypothetical development may have involved distal germination of a pollen tube (*PT*), but release of motile sperm through rupture of the proximal suture. **c** Zooidogamous seed plants with pollen tubes (cycads and *Ginkgo biloba*). A pollen tube is formed from the distal sulcus. Late in development the basal end of the gametophyte swells or elongates and eventually ruptures to release multiflagellate sperm. *EX*, exine. **d** Hypothetical intermediate between zooidogamy with a pollen tube and siphonogamy with non-motile sperm. In this male gametophyte, the apical end of the pollen tube undergoes directed growth to the egg. Motile sperm swim down the length of the pollen tube and are released directly into the egg cell. No plants are known to have had this combination of male gametophyte characters. **e** Siphonogamous seed plants with non-motile sperm (*SP*) (modern conifers, Gnetales, angiosperms). In these groups of plants, the pollen tube serves as a conduit to bring non-motile sperm to the egg. (**a–e** from [42], with permission)

although other (extinct) taxa may have shared this pattern of development. With the advent of exosporic male gametophyte development among seed plants, release of swimming sperm was transferred away from the proximal face of the prepollen grain to the distally germinating pollen tube [42].

It is possible that an intermediate condition of male gametophyte development may have evolved in primitive conifers and certain cordaites whose pollen grains display not only a distal germination area (suggestive of the presence of a pollen tube) but also a proximal suture mark. In these groups, an haustorial pollen tube (whose growth was not oriented towards an egg) may have been produced from the distal face of the pollen grain, while sperm were still released through rupture of the proximal face [42, 53] (Fig. 9b).

The male gametophytes of extant conifers, Gnetales, and angiosperms are siphonogamous (Fig. 9e). In these seed plants, the pollen tube undergoes oriented growth towards the site of the egg(s) and non motile sperm are transported through the pollen tube. These are no known intermediate conditions between exosporic zooidogamous male gametophytes and siphonogamy with non motile sperm. However, the evolution of siphonogamy may have first proceeded through a stage in which the heterotrophic pollen tube acquired chemotropically oriented growth towards the site of the egg(s), while motile sperm were retained and swam through an unruptured pollen tube to the egg(s) (Fig. 9d) [42]. Such a strategy may initially have had benefits that included the competitive exclusion of the gametes of other male gametophytes [51] whose pollen tubes did not grow as quickly or were not as accurately oriented in their growth to the eggs. A final stage in the evolution of siphonogamy would later involve the loss of sperm motility. Presumably, flagella would ultimately prove unnecessary for the movement of sperm cells within the living cytoplasm of a pollen tube [20, 42].

When viewed within the context of evolutionary history, the exosporic pollen tube of seed plant male gametophytes represents a superb example of exaptation. Originally shaped by selection into a vegetative organ to acquire nutrients from the sporophyte host tissues for growth of the male gametophyte (in zooidogamous seed plants such as *G. biloba*), the pollen tube was later coopted for a sexual role to transport male gametes from the site of pollination to the site of fertilization. As such, the male gametophyte of *G. biloba* provides a critical window into an important and transitional stage in the evolutionary history of plant reproductive biology: between the earliest seed plants with endosporic male gametophytes and the siphonogamous taxa that include extant conifers, Gnetales, and angiosperms.

6 Concluding Remarks

In *Ginkgo biloba*, the mature male gametophyte consists of an extensively branched haustorial system within the micropylar apex of the nucellus and a large spherical region that forms at the pollen grain end of the pollen tube. Development of the male gametophyte of *G. biloba* is divided into three phases. Germination of a pollen grain within the pollen chamber of the nucellus is characterized by an initial brief period of diffuse and roughly isodiametric growth. A second and extended phase of morphological development is characterized by tubular growth, in which the male gametophyte of *G. biloba* branches extensively within the intercellular spaces of the nucellar apex of the ovule. The final phase of development involves the diffuse swelling of the unbranched proximal portion of the male gametophyte into a large saccate structure. The sterile cell and spermatogenous cell of the male gametophyte arise from an unusual anticlinal ringlike division of the generative cell. During this process, a small centrally positioned cell, the spermatogenous cell, is cleaved from the parent generative cell and is consequently surrounded by a ring shaped sterile cell. At maturity, the spermatogenous cell contains two prominent blepharoplasts with approximately 1000 probasal bodies (centrioles). Each probasal body associated with a blepharoplast will develop into a basal body of a flagellum during the formation of a multiflagellate sperm. The mature sperm cell of *G. biloba* is comprised of an anterior flagellar band with approximately 1000 flagella, a large centrally placed nucleus, and additional cytoplasm. Flagella are anchored within the sperm cell by a spiral multilayered structure. The male gametophyte of *G. biloba* provides considerable insight into the evolution of the seed plant male gametophyte and its developmental and reproductive biology. The male gametophytes of the earliest seed plants were endosporic and fundamentally similar to the microspores of heterosporous vascular non-seed plants. Evolution of the exosporic pollen tube of seed plant male gametophytes represents a superb example of exaptation. Originally shaped by selection to acquire nutrients from the sporophyte host tissues for growth of the male gametophyte (in zooidogamous seed plants such as *G. biloba*), the pollen tube was later coopted (as seen in conifers, Gnetales, and angiosperms) for a sexual role to transport male gametes from the site of pollination to the site of fertilization.

Acknowledgments. This research supported by a grant from the National Science Foundation (IBN-9696013) and grants-in-aid of research from Apple Computer Inc., Lasergraphics Inc., Carl Zeiss Inc., RMC Inc., Leica, Inc., and Olympus America to WEF.

References

1. Hirase S (1896a) Spermatozoid of Ginkgo biloba (in Japanese). Bot Mag Tokyo 10:171–172
2. Hirase S (1896b) On the spermatozoid of *Ginkgo biloba* (in Japanese). Bot Mag Tokyo 10:325–328
3. Ikeno MS (1896) Spermatozoid of *Cycas revoluta* (in Japanese). Bot Mag Tokyo 10:367–368
4. Amici GB (1824) Observations microscopiques sur diverses espèces de plantes. Ann Sci Nat Bot 2:41–70, 211–248
5. Ikeno S, Hirase S (1897) Spermatozoids in gymnosperms. Ann Bot 11:344–345
6. Hofmeister W (1851) Vergleichende Untersuchungender Keimung, Entfaltung, und Fruchtbildung horerer Kryptogamen und der Samenbildung der Coniferen. Fr Hofmeister, Leipzig
7. Hirase S (1894) Notes on the attraction-spheres in the pollen-cells of *Ginkgo biloba.* Bot Mag Tokyo 8:359–360
8. Hirase S (1895a) Études sur le *Ginkgo biloba*. Bot Mag Tokyo 9:238–239

9. Hirase S (1895b) Études sur la fécondation et l'embryogénie du *Ginkgo biloba*. J Coll Sci Imp Univ Tokyo 8:307–322
10. Hirase S (1897) Untersuchungen über das Verhalten des Pollens von *Ginkgo biloba*. Bot Zbl 49:33–35
11. Hirase S (1898) Études sur la fécondation et l'embryogénie du *Ginkgo biloba*. J Coll Sci Imp Univ Tokyo 12:103–149
12. Hirase S (1918) Nouvelles recherches sur la fécondation et l'embryogénie du *Ginkgo biloba*. Bot Mag Tokyo 32:139–143
13. Mann MC (1924) Microsporogenesis of *Ginkgo biloba* L. with special reference to the distribution of plastids and to cell wall formation. Univ Calif Pub Agric Sci 2:243–248
14. Wolniak SM (1976) Organelle distribution and apportionment during meiosis in the microsporocyte of *Ginkgo biloba* L. Amer J Bot 63:251–258
15. Strasburger E (1892) Über das Verhalten des Pollen und die Befructungsvorgange bei der Gymnospermen. Histol Beitrage Heft 4:1–158
16. Lee CL (1955) Fertilization in *Ginkgo biloba*. Bot Gaz 117:79–100
17. Webber HJ (1901) Spermatogenesis and fecundation of *Zamia*. Bull US Dep Agric 2:1–100
18. Favre-Duchartre M (1956) Contribution a l'etude de la reproduction chez le *Ginkgo biloba*. Rev Cytol Biol Veg 17:1–218
19. DeSloover-Colinet A (1963) Chambre pollinique et gametophyte male chez *Ginkgo biloba*. Cellule 64:129–145
20. Friedman WE (1987a) Growth and development of the male gametophyte of *Ginkgo biloba* within the ovule (in vivo). Amer J Bot 74:1797–1815
21. Tulecke WR (1957) The pollen of *Ginkgo biloba*: in vitro culture and tissue formation. Amer J Bot 44:602–608
22. Dexheimer J (1970) Recherches cytophysiologiques sur les grains de pollen. Rev Cytol Biol Veg 33:169–234
23. Rohr R (1980) Développement in vitro du pollen de *Ginkgo biloba*. Cytologia 45:481–495
24. Friedman WE (1987b) Morphogenesis and experimental aspects of growth and development of the male gametophyte of *Ginkgo biloba* in vitro. Amer J Bot 74:1816–1830
25. Friedman WE, Gifford EM (1988) Division of the generative cell and late development in the male gametophyte of *Ginkgo biloba*. Amer J Bot 75:1434–1442
26. Herzfeld S (1927) Beitrage zur Kenntnis von *Ginkgo*. Jb F Wiss Bot 66:814–862
27. Fujii K (1898) Has the spermatozoid of *Ginkgo* a tail or none? (In Japanese) Bot Mag Tokyo 12:287–290
28. Fujii K (1899a) Observation on the morphology of the pollen tube and the spermatozoid of *Ginkgo* (in Japanese). Bot Mag Tokyo 13:28–30
29. Fujii K (1899b) Remarks on Mr. Ikeno's remarks on my views regarding the morphology of the pollen-cells and the spermatozoid of *Ginkgo* (in Japanese). Bot Mag Tokyo 13:65–73
30. Fujii K (1899c) On the morphology of the spermatozoid of *Ginkgo biloba* (in Japanese). Bot Mag Tokyo 13:260–266
31. Fujii K (1900) On bicephalous spermatozoid of *Ginkgo* (in Japanese). Bot Mag Tokyo 14:16–17
32. Ikeno S (1899a) On the spermatozoid and pollen tube of *Ginkgo biloba* and *Cycas revoluta* (in Japanese). Bot Mag Tokyo 13:31–34
33. Ikeno S (1899b) Opinion of various authors on the centrosome in the pollen tube of Cycadaceae and *Ginkgo* (in Japanese). Bot Mag Tokyo 13:74–76

34. Ikeno S (1901) Contribution a l'étude de la fécondation chez le *Ginkgo biloba*. Ann Sci Nat Bot 13:305–318
35. Miyake K (1902) The spermatozoids of *Ginkgo*. J Appl Lab Meth 5:1773–1780
36. Webber HJ (1897) Notes on the fecundation of *Zamia* and the pollen tube apparatus of *Ginkgo*. Bot Gaz 24:225–235
37. Shimamura T (1937) On the spermatozoid of *Ginkgo biloba*. Cytologia (Fujii Jubilaei):416–423
38. Gifford EM, Lin J (1974) Ultrastructure des cellules spermatogenes de *Ginkgo biloba*, et particulièrement du blepharoplaste (mastigosome). C R Acad Sci (Paris) 278:2513–2515
39. Gifford EM, Lin J (1975) Light microscope and ultrastructural studies of the male gametophyte in *Ginkgo biloba*: the spermatogenous cell. Amer J Bot 62:974–981
40. Gifford EM, Larson S (1980) Developmental features of the spermatogenous cell in *Ginkgo biloba*. Amer J Bot 67:119–124
41. Li Y, Wang FH, Knox RB (1989) Ultrastructural analysis of the flagellar apparatus in sperm cells of *Ginkgo biloba*. Protoplasma 149:57–63
42. Friedman WE (1993) The evolutionary history of the seed plant male gametophyte. Trends Ecol Evol 8:15–20
43. Sterling C (1963) Structure of the male gametophyte in gymnosperms. Biol Rev 38:167–203
44. Chamberlain CJ (1909) Spermatogenesis in *Dioon edule*. Bot Gaz 47:215–236
45. Coulter JM, Chamberlain CJ (1917) Morphology of gymnosperms. Second edition. University of Chicago Press, Chicago
46. Swamy BGL (1948) Contributions to the life history of a *Cycas* from Mysore (India). Amer J Bot 35:77–88
47. Gifford EM, Foster AS (1987) Morphology and evolution of vascular plants. Third edition. WH Freeman, New York
48. Chaloner WG (1970) The evolution of miospore polarity. Geosci Man 1:47–56
49. Taylor TN, Millay MA (1979) Pollination biology and reproduction in early seed plants. Rev Palaeobot Palynol 27:329–355
50. Doyle JA (1987) Pollen evolution in seed plants: A cladistic perspective. J Palynol 23/24:7–18
51. Haig D, Westoby M (1989) Selective forces in the emergence of the seed habit. J Linn Soc 39:215–238
52. Crane PR (1990) The phylogenetic context of microsporogenesis. In: Microspores: evolution and ontogeny. Acad Pr pp 11–41
53. Millay MA, Taylor TN (1976) Evolutionary trends in fossil gymnosperm pollen. Rev Palaeobot Palynol 21:65–91
54. Kerp JHF, Poort RJ, Swinkels HAJM, Verwer R (1990) Aspects of Permian palaeobotany and palynology. IX. Conifer-dominated Rotliegend floras from the Saar-Nahe Basin (?Late Carboniferous-Early Permian; SW-Germany) with special reference to the reproductive biology of early conifers. Rev Palaeobot Palynol 62:205–248
55. Stewart WN (1951) A new *Pachytesta* from the Berryville locality of Southeastern Illinois. Amer Midland Nat 46:717–742
56. Taylor TN, Daghlian CP (1980) The morphology and ultrastructure of *Gothania* (Cordaitales) pollen. Rev Palaeobot Palynol 29:1–14

57. Steeves TA (1983) The evolution and biological significance of seeds. Can J Bot 61:3550–3560
58. Chaloner WG (1976) The evolution of adaptive features in fossil exines. In: The evolutionary significance of the exine. Academic, London, pp 1–14

Addenda

For further discussion of seed plant male gametophyte evolution see:

Rothwell GW, Serbet R (1994) Lignophyte phylogeny and the evolution of spermatophytes: a numerical cladistic analysis. Syst Bot 19:443–482.

Poort RJ, Visscher H, Dilcher DL (1996) Zoidogamy in fossil gymnosperms—the centenary of a concept, with special reference to prepollen of late Paleozoic conifers. Proc Nat Acad Sci USA 93:11713–11717

The following articles on male gametophytes of *Ginkgo* appeared after submission of this manuscript:

Wang J, Yang C, Zhai ZH (1996) The nuclear lamina in male generative cells of *Ginkgo biloba*. Sex Pl Repro 9:238–242

Moscatelli A, Cai G, Liu GQ, Tiezzi A, Cresti M (1996) Dynein-related polypeptides in pollen and pollen tubes. Sex Pl Repro 9:312–317

Development of the Female Gametophyte and the Embryogeny of *Ginkgo biloba*

Sanae Soma

1 Introduction

Trees of *Ginkgo biloba* are large, long-lived, deciduous, and dioecious. They have pale green young leaves in the spring; the leaves turn to clear yellow in the autumn. Female trees produce many edible seeds every autumn. These beautiful and valuable plants have been cultivated for more than a thousand years in China and Japan. The long-lived trees are especially well preserved in the precincts of Buddhist temples or shrines in Japan. Now they have been cultivated all over the world, but wild trees are found only in China [1]. The ovuliferous structure develops in the axil of a scale or a leaf of the short shoot of the female tree. Two to six ovuliferous structures exist in one short shoot [2]. Usually ovuliferous structures consist of a stalk and two sessile ovules on the top. More than two ovules on a stalk, or ovules on a branched stalk, were observed in certain trees [2–9] (Fig. 1). Ovuliferous structures more or less different from the usual type were found in 40% of old trees and in 25% of young trees [2]. These values mean unusual structures are not exceptional. Unusual structures were observed in the axil of a scale in 58% of old trees and in 50% of young trees, while those formed in the axil of a leaf were seen in 26% of old trees and 6% of young trees. Therefore, unusual ovuliferous structures arise much more in the axil of a scale than in that of a leaf [2].

Another pattern of unusual ovuliferous structures is the ovules borne on the leaf blade [10]. This peculiar pattern was observed mainly on old female trees in Yamanashi Prefecture in Japan. In the 100 years since Fujii observed this peculiar ovuliferous structure [10], these trees have continued to produce abundant normal and peculiar ovuliferous structures. The number of the ovules on a leaf varies from one to thirteen. These leaf ovules might be the senile form [11, 12].

Bunkyo University, 3337 Minamiogishima, Koshigaya, Saitama 343, Japan

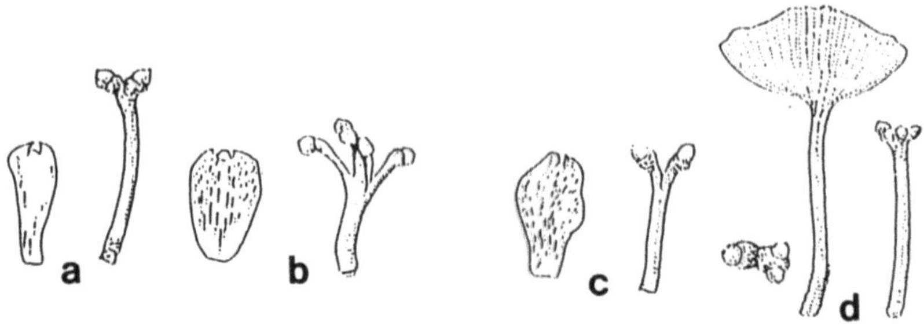

FIG. 1a–d. Ovuliferous structure and the shape of the subtending scale or leaf. **a** Usual ovuliferous structure with two sessile ovules. **b** Branched ovuliferous structure with stalked ovules. **c** Ovuliferous structure with two stalked ovules. **d** Ovuliferous structure with three sessile ovules. (From [2], with permission)

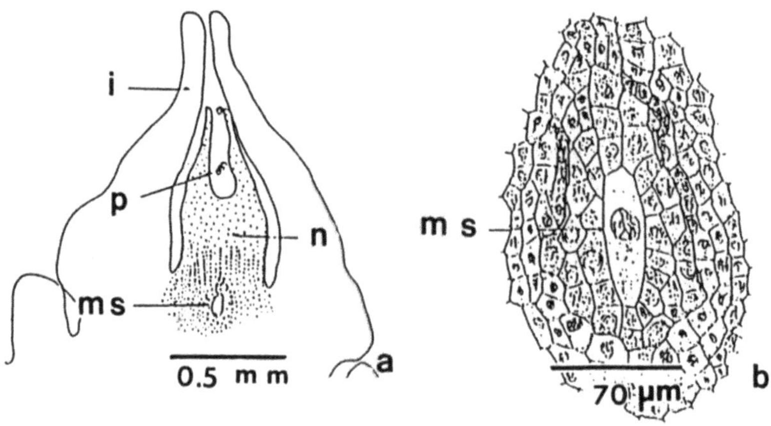

FIG. 2. **a** Vertical section of an ovule. *i*, integument; *p*, pollen; *n*, nucellus; *ms*, megaspore mother cell. (From [16], with permission). **b** Vertical section through the egg-shaped mass of cells. *ms*, megaspore mother cell. (From [15], with permission)

2 Development of Megaspore

The ovuliferous structure can be detected from October to November. According to Sprecher [13], at the early stage of the formation of an ovuliferous structure a small protuberance containing no chlorophyll, with a small concavity at the top, can be recognized. There are two swellings at the periphery of the protuberance and on these swellings the integument and the nucellus are initiated. At the end of October the nucellus is still visible from the outside. According to Takaso [14], both ovules are initiated on the abaxial side of a short stalk, then the subsequent growth of the stalk makes them face opposite to each other. The integument is initiated at two different positions, but these soon became a continuous ring. Carothers [15] investigated the young ovules, 0.25 mm in length, in the middle of April, at which time the nucellus has not yet become wholly enclosed by the integument and a nucellar beak can be observed. At the beginning of May the ovule, 0.75 mm in length, has a large pollen chamber, often containing pollen grains (Fig. 2a).

In the ovule approximately at the level of separation of the integument from the nucellus, there is an egg-shaped mass of cells which differs from the cells around it. The cells in this mass are larger than the cells around them, and have granular cytoplasm without vacuoles, and large nuclei with conspicuous nucleoli [15] (Fig. 2b). The megaspore mother cell usually appears slightly below the center of the egg-shaped mass of cells and is distinguishable from the surrounding cells by having larger size, more deeply staining wall, and no granullar cytoplasm. As a rule there is only one megaspore mother cell in the nucellus; in 50 ovules examined only one ovule had two mother cells [15]. At first the megaspore mother cell, with its large central nucleus, is not vacuolated (Fig. 3a), but later a large vacuole appears below the nucleus and sometimes a small vacuole appears above it. The mature megaspore mother cell is elongated, and below the big vacuole there exists a large dark mass of cytoplasm [15] (Fig. 3b).

The electron microscope study by Stewart and Gifford [17] showed that the immature megaspore mother cell is spherical and is surrounded by a thick complex wall. A large nucleus occupies the central region, and the organelles in the cytoplasm appear to be randomly arranged. Just before meiosis, the cell elongates along the ovular axis. The endoplasmic reticulum develops in the micropylar half of the cell, and the plastids and mitochondria migrate to the chalazal half,

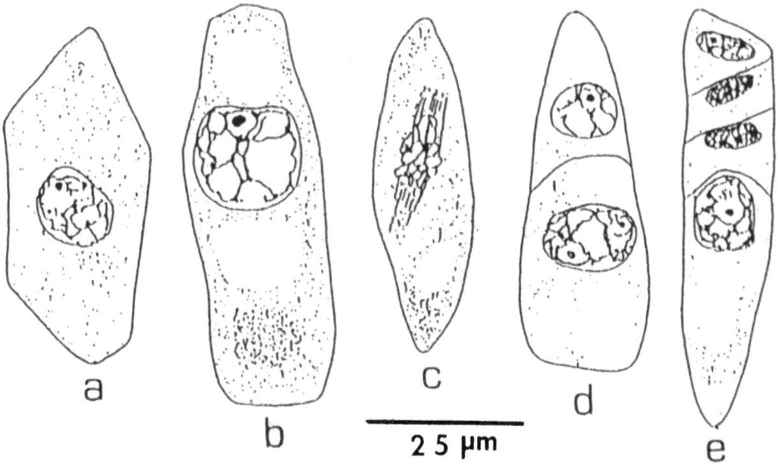

FIG. 3a–e. Megasporogenesis. **a** Young megaspore mother cell. **b** Mature megaspore mother cell, showing vacuoles and dark mass of cytoplasm. **c** The first division of meiosis. **d** Two daughter cells of the first division of meiosis. **e** Linear tetrad, the lowest cell is enlarged. (From [15], with permission)

so that there exist neither plastids nor mitochondria in the micropylar end of the cell. Therefore, after meiosis the chalazal cell of the dyad contains as many plastids and mitochondria as the megaspore mother cell.

As for the wall of the megaspore mother cell, Sprecher [6] and Carothers [15] described it as thicker than those of the cells around it. Stewart and Gifford [17] showed that the cell wall of the megaspore mother cell has a greatly thickened middle lamella. The wall itself is thinner and densely staining compared with the middle lamella. The sterile cells surrounding the megaspore mother cell are called the spongy tissue. Degeneration of the adjacent spongy tissue proceeds as the megaspore mother cell elongates, and fragments of the degenerated cells sometimes appear to become embedded in the thick jelly-like middle lamella between the megaspore mother cell and the surrounding cells [16]. The spongy tissue has a tapetal function [18, 19].

Meiosis of the megaspore mother cell was observed by Sprecher [6], Carothers [15], Coulter and Chamberlain [19], and Herzfeld [20]. In the elongated megaspore mother cell the spindle of the first division of meiosis does not appear vertically along the long axis but a little obliquely, possibly because of the existence of a big vacuole [15] (Fig. 3c). After the first division of meiosis, the cell near the chalazal end is always bigger than the one near the micropylar end (Fig. 3d). Then the second divisions of meiosis occur successively in the same directon, usually producing linear tetrad cells (Fig. 3e), but sometimes a bilateral or a mixed bilateral–linear formation of the tetrad cells can be observed [15].

It is noticed that only the lowest cell of the linear tetrad, namely the cell of the chalazal end, is functional and enlarges rapidly to become the megaspore cell [13, 15, 16]. The other three cells of the tetrad finally degenerate. According to Chamberlain [8] the development of the female gametophyte begins at the time of pollination.

A group of nucellar cells at the micropylar end enlarge and then begin to degenerate, so as to form a cavity. This is the beginning of the pollen chamber. Sprecher [13], Favre-Duchartre [16] and de Sloover-Colinet [21] noticed that megasporogenesis and the formation of the pollen chamber in the micropylar end of nucellus are synchronized with each other.

Stimulus from the pollen grain was considered to be necessary for the development of the female gametopyte [22].

3 Development of Female Gametophyte

Lee [23] describes the first division of the functional megaspore cell as occurring usually perpendicular to the axis of the ovule. This is the beginning of female gametophyte development. Then the second divisions occur parallel to the ovular axis; later divisions occur in various directions. These are free nuclear divisions. A female gametophyte 2 weeks after megasporogenesis contains 16 to 64 free nuclei. This free nuclear stage of development persists from the second week of

May to the first week of July [15]. According to Chamberlain [8] and Carothers [15], from the first to the fourth mitosis, nuclear divisions occur simultaneously. However, nuclei at metaphase and at telophase were observed at the same time during the fifth division. At the sixth division, nuclei do not divide simultaneously. Nuclei divide 13 times in succession, thus perhaps more than 8000 free nuclei are produced [24]. The ovule increases in volume. The later-produced nuclei are smaller in size because of the rapid succession of the free nuclear divisions [23]. During free nuclear divisions, the nuclei are arranged in the peripheral cytoplasm of the female gametophyte, and in the central part there exists a large vacuole [6, 23, 24, 25]. Throughout the free nuclear stage, the cytoplasm of the gametophyte adheres to the megaspore membrane. In later free nuclear divisions, divisions are not synchronized; that is, they are initiated at the chalazal end and proceed to the micropylar end [24].

During free nuclear divisions, the outer membrane of the female gametophyte, namely the megaspore wall, becomes thicker [15, 18, 19]. Favre-Duchartre [24] describes the process as follows: the membrane at first consists of a homologous cellulose–pectic endospore, later it adds a covering exine formed by the nucellus and made up of large numbers of short rods inserted perpendicularly to the plane of the intine. Pettitt [26] made detailed studies of wall thickening. He found that the foundation for the wall is constructed on the free surface of the female gametophyte during the free nuclear stage. It consists of a fibrillar matrix forming a system of radially aligned hexagonal chambers. Then the granular material begins to accumulate in the chambers, appearing first in the proximal region and then extending radially to form a series of interconnecting columns. A continuous granular basement is deposited within the undersurface of the fibrillar matrix. As development proceeds, they become electron-opaque and amorphous. Three substrata of different wall material are then formed beneath the patterned layer. The first substratum is polysaccharide-rich and the material possibly originates in the cytoplasm of the female gametophyte. The second substratum is formed when electron opaque globules are extruded across the plasma membrane. The third substratum contains polysaccharide, protein complex, or glycoprotein. It develops in contiguity with the walls which partition the megaspore cytoplasm.

Enlargement of the female gametophyte and the development of the megaspore wall are accompanied by the degeneration of nucellar tissue. An electron microscope study by Pettitt [26] shows that the innermost cells of the nucellus degenerate and that the protoplast becomes disorganized. The tapetal zone begins to be destroyed by the encroachment of the young female gametophyte [19]. Carothers [15] and Lee [23] described the process as follows: the inner cells of the spongy tissue are enlarged, then the inner walls are partly dissolved and vacuoles appear around the nuclei. The cells of the middle part of the spongy tissue contain large nuclei and slightly staining cytoplasm with minute vacuoles. The outer part is composed of smaller, slender cells with dense cytoplasm. Toward the end of the free nuclear stage, multinucleate cells are observed in the spongy tissue.

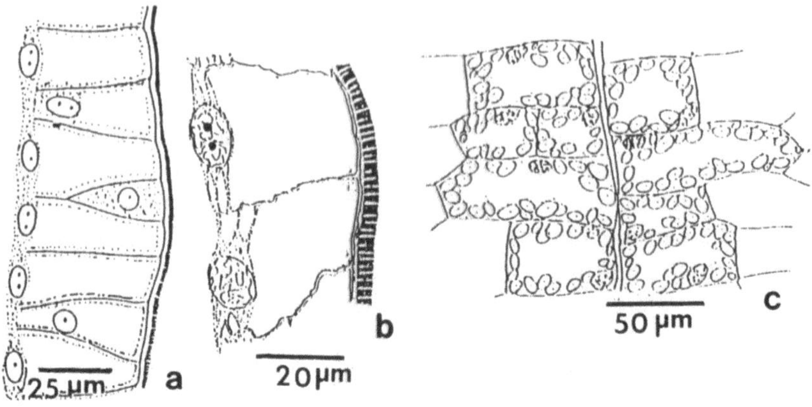

FIG. 4. **a** The beginning of cell wall formation, schematic illustration of alveoli. **b** Two alveoli. **c** Vertical section of the central part of cellular gametophyte, showing independent abutting wall. (**a,b** from [16], with permission; **c** from [15], with permission)

At the end of free nuclear division nuclei are connected by microtubules. At the fourteenth mitotic divisions anticlinal walls begin to form between the nuclei giving rise to the six-sided and tubular cells [25]. They are not completely enclosed by the end wall for some time (Fig. 4a,b). This incomplete cell was called the alveolus by Favre-Duchartre [16, 24]. The nucleus of each alveolus is located at the inner open end. Successive partitioning occurs centripetally. Wall formation is in progress by mid June [16, 23, 27]. The cellulose-pectic membrane of the newly formed peripheral cells is independent of the megaspore membrane [23, 24]. Thus the cellularization gradually proceeds from the periphery toward the central vacuole forming files of cells [24]. Sometimes a delay in cell plate formation occurs, and two or three nuclei in a single cell are observed [23].

Centripetal development of cell wall formation proceeds to the central cells of the gametophyte; the innermost cells of each file of cells do not join and the opposite files of cells form two distinct abutting walls in the middle part of the gametophyte (Fig. 4c). Therefore, a mature female gametophyte can be split easily into two portions [18, 19, 28]. The mature gametophyte contains a gibberellic-acid-like substance, the amount of which is five times as much as that contained in the integument [29]. The cellular stage of the female gametophyte shows many signs of high metabolic activity; dividing mitochondria and plastids and endoplasmic reticulum with many ribosomes are observed [30].

After cell wall formation the gametophyte becomes green and in a few weeks it is recognizable as the greenest region of the ovule [19]. Experiments by Friedman and Goliber [31] confirmed photosynthesis in the green gametophyte. The cells of the gametophyte are filled with starch grains not synthesized by the gametophyte itself, despite the fact that the gametophyte contains chlorophyll and can photosynthesize (this phenomenon is known only in this species of all the seed plants). Outside the gametophyte starch grains are abundant only in the outer region of the green integument.

All the nuclei of the female gametophyte should be haploid with regard to DNA content. Avanzi and Cionini [27], however, revealed that DNA amounts in most of the cells are not haploid but diploid or higher. Diploid mitosis is abundant in the cells of gametophyte collected from the middle to the end of June.

4 Development of the Archegonium

Certain cells in the micropylar portion of the female gametophyte divide and become smaller in size with denser protoplasm. Usually two groups of such cells are formed [23]. One superficial cell in each group becomes the archegonial initial characterized by large size and a large nucleus (Fig. 5a).

Just before the division of the archegonial initial the nucleus migrates to the micropylar end of the cell and a vacuole appears below the nucleus. Cell division occurs as soon as the nucleus reaches there (Fig. 5b). Chromatin is equally separated during the division but cytokinesis is unequal, giving rise to a large central cell and a small primary neck cell [23–25] (Fig. 5c). Usually two archegonia develop at the micropylar end of the female gametophyte, but occasionally three or four archegonia are observed [16]. Each archegonium lies in a somewhat circular groove [18].

Anticlinal division of the primary neck cell forms a pair of neck cells. The plane of this division is always parallel in the two archegonia. These neck cells then become enlarged, and project slightly above the surface of the gametophyte [23].

With the enlargement of the central cell of the archegonium, a peculiar column appears on the surface of the female gametophyte between two archegonia [18,

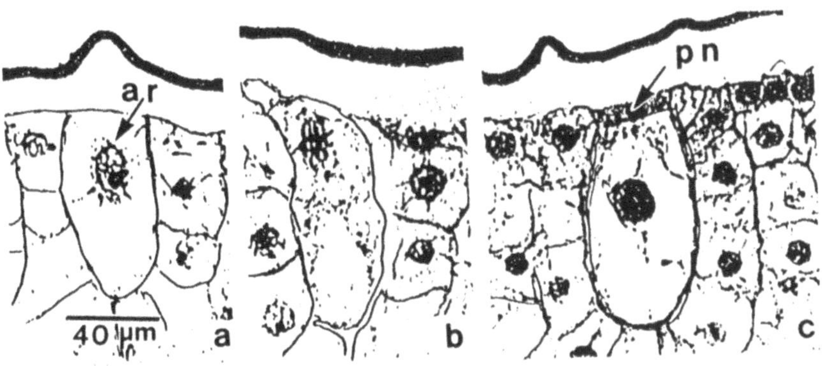

FIG. 5a–c. Vertical section of the cellular gametophyte, showing archegonial development. **a** Archegonnial initial. **b** Mitotic division in archegonial initial. **c** The primary neck cell and a large central cell. *ar*, archegonial initial; *pn*, primary neck cell. (From [23], with permission)

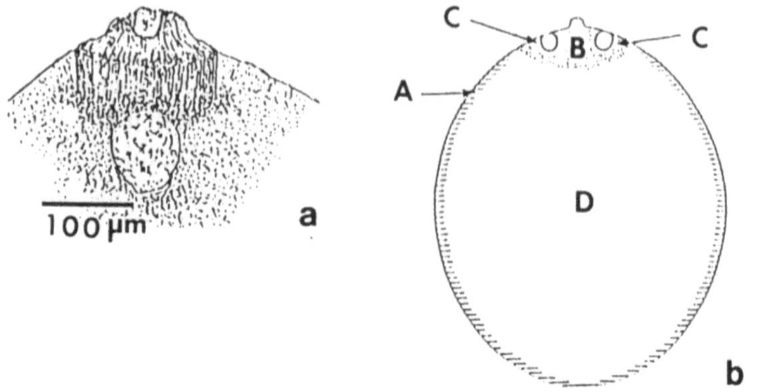

FIG. 6. **a** Dividing central cell, producing small ventral canal cell and a large egg cell. **b** Schematic illustration of vertical section of female gametophyte. *A*, peripheral zone; *B*, between the archegonia; *C*, jacket layer cells; *D*, central part. (**a** from [34], with permission; **b** from [27], with permission)

24, 25]. This column, called the "tent pole," extends toward the part of the nucellus lying below the pollen chamber at first. The term "tent" was used by Hirase [32]. In late August the nucellar tissue and the megaspore wall in this region degrade [24].

The nucleus of the central cell gradually moves toward the neck cells, and the central cell continues to enlarge. At this stage the nucleus of the central cell is colorless after Feulgen staining, but other cells of the gametophyte are Feulgen positive [33]. Before fertilization two neck cells divide obliquely to form four neck cells, and then the nucleus of the central cell divides rapidly; the spindle fibers extend to the neck cells and the other end of the fibers converge to an indistinct point (Fig. 6a). During telophase the two groups of chromatin become compacted into small nuclei, one of which becomes the egg nucleus, the other the nucleus of the ventral canal cell. Cytokinesis of this division is also unequal. The ventral canal cell degenerates soon after cytokinesis. The egg nucleus has one to two nucleoli. The nucleus moves to the center of the egg cell, and DNA appears confined within a small granular globule. A vacuole is always present in the lowermost part of the egg cell. It may occasionally be accompanied by several small vacuoles [23]. The division that forms the egg nucleus and the ventral canal nucleus has been observed by many researchers [3, 6, 16, 32, 34–36].

An electron microscope study by Camefort [37] showed that in the archegonium the numerous inclusions become isolated from the cytoplasm by the membrane of the smooth endoplasmic reticulum. Amyloplasts are abundant in the marginal region of the egg cell, and they are frequently covered by a membrane that is continuous with endoplasmic reticulum and with the plastid wall.

An autoradiographic study by Cionini and Avanzi [38] showed that the mature female gametophyte has four different groups of cells (Fig. 6b). In the peripheral portion (A in Fig. 6b) cells are uniform in size, but the size of the nucleus is heterogeneous. Tritiated thymidine incorporation is strong, but tritiated uridine uptake is weak, showing that DNA synthesis is active, while RNA synthesis is inactive in this portion. Between the archegonia (B in Fig. 6b) cells are homologous in size. Tritiated thymidine uptake is also active and tritiated uridine uptake is weak here, too. In the jacket layer (C in Fig. 6b) cells have rather large nuclei and have thick cell walls. Two different patterns of incorporation are found: intense labeling all over the nucleus, and weak labeling limited to some regions of the nucleus. Central part (D in Fig. 6b) cells and nuclei are highly uniform in size, and tritiated thymidine incorporation is limited to around the nucleolus. RNA synthesis is lowest in the gametophye. Cionini [39] showed that DNA in the nuclei of the archegonial jacket cells comprises at least two different types of DNA complex. In archegonial jacket cells of the fully differentiated gametophyte the staining of nuclei by Feulgen is weak.

The gametophyte storage tissue around the archegonia was examined at the end of August by Stopes and Fujii [40]. It contains many starch grains. They are smaller near the archegonium and a little larger in the cells of the jacket layer, while the egg cell is free from starch grains. Then the starch grains in the jacket layer cells become smaller and show a brownish staining with iodine, while small granules begin to appear in the egg cell. The starch grains then decrease in the jacket layer cells, and the number of starch grains in the egg cell increases. Finally jacket layer cells become devoid of starch grains. This emptying of starch grains spreads outward for 10 to 15 cell layers. The protein grains show similar distribution. Starch and protein are considered to be converted to soluble forms, and after entering into the egg cell they are redeposited [40].

An electron microscope study by Maugini and Fiordi [41] also shows that starch and protein grains stored in the storage tissue are transformed into soluble forms by enzymes produced mainly by tapetum cells, and then translocated to the archegonial central cell. They pass through the plasmodesma of the cell wall. Another electron microscope study by Dexheimer [42] shows that cells with lipids occupy the most superficical layer of the gametophyte. Cells with protein lipids and starch, or cells with starch, surround the middle part of gametopyte, whose cells contain little reserves.

Hirase [34] reported that a big vacuole appears at the top of the mature egg cytoplasm. At this time an acidic fluid spreads out. Lee [23] also recognized that the dense cytoplasm and the enlargement of the nuclei of the neck cells suggest a secretory function of the neck cell. Chamberlain [8], however, states that he has never been able to make observations of this fluid.

At the beginning of September, just before or after fertilization, ovules are still green in color. The integument consists of three layers. The outer fleshy layer is the only living part of the integument. It is 5–6mm thick and covered with the cutinized epidermis, which has a haplocheilic type of stomata. Epidermal cells also produce wax. Chloroplasts are present in the epidermal cells [24]. The bulk of fleshy cells, which have chloroplasts and starch grains, are formed below the epidermis. The cell sizes near the epidermis are about 40 × 15 μm, but the cells near the stony layer are as large as 200 × 130 μm. With the increase of cell size, the numbers of chloroplasts and the size of starch grains decrease. After abscission of the seed, the outer fleshy layer of the integument becomes soft and can be washed away easily. The next layer is stony and lignified. Favre-Duchartre [24] called this layer sclerotesta. This stony layer has a hard shell-like shape, is slightly flattened laterally, and is generally provided with a rib that has a suture-like function on germination. Sometimes 3 ribbed seeds can be recognized [6, 42, 43]. 3 ribbed seeds comprise 11%–31% of total seeds [44]. The inner thin layer of the integument has a withered appearance. However, its micropylar chalazal halves show different appearances. The micropylar half of this layer adheres to and lines the stony layer, but it has no connection with the nucellus. Favre-Duchartre [24] described this thin, translucent layer as parchment-like, having a golden brown color. A photomicroscope study by Favre-Duchartre [24] shows the cells laid parallel to the ovular axis. However, the chalazal half of the inner layer is always separate from the stony layer but is united with the nucellus, and is also adherent to the gametophyte. It is opaque, brittle, dark brown, and has a pearl-like consistency due to the wax formed on it. Photomicroscope studies show rows of cells with annular lignified thickenings, which for a width of about three units form ribs diverging from the hilum and going up to the equatorial plane of the ovule. The transfusion tissue forms wider and longer strands—reaching a few millimeters higher than the equatorial plane of the ovule—along the ribs of the stony layer.

5 Development of Embryo

Fertilization does not always occur simultaneously, but in the suburbs of Tokyo it occurs from the end of August to the middle of September. Hirase's famous observation [45] was carried out on the 9th of September.

After fertilization the sperm nucleus and the egg nucleus are in the egg cell. Both nuclei are distinguishable by means of staining [46]. Herzfeld [46] observed the granulated sperm nucleus on the big egg nucleus. These two kinds of nuclei were also observed by Shimamura [22], Lee [23], Ikeno [47], and Favre-Duchartre [48].

The fertilized nucleus is as large as 85–100 μm in diameter, then it decreases its size with successive divisions [49]. Nuclear division begins immediately after fertilization, and chromosomes are formed during the prophase [23]. They appear to be suspended in the center of the fertilized egg cell. During metaphase the nuclear membrane disappears completely, and the spindle fibers are well developed. The first division is always perpendicular or a little oblique to the axis of the ovule [23]. The mitotic figure of the second division is very large [50]. These free nuclear divisions occur successively seven times. Until the fifth division, spindle fibers can be observed [51]. These free nuclear divisions occur simultaneously at the beginning (Fig. 7a).

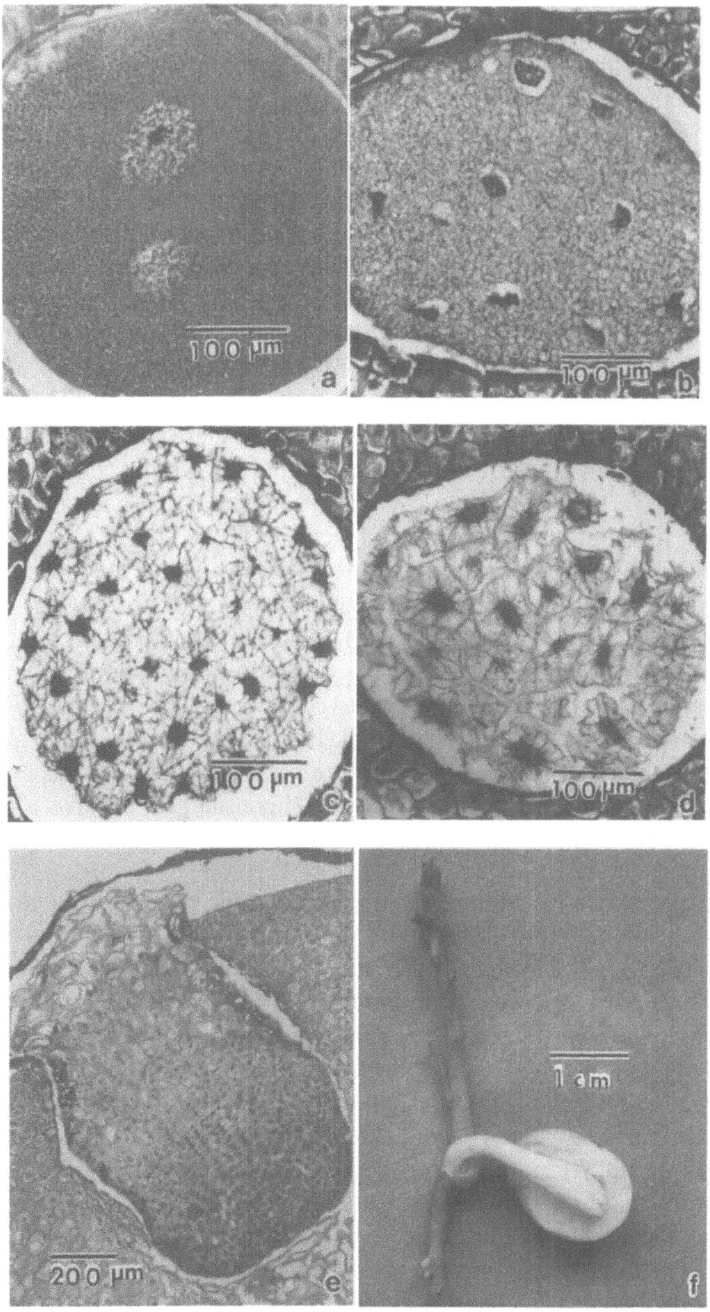

FIG. 7. **a** Vertical section of the free nuclear embryo, at four nuclear stage. **b** Free nuclear embryo at 128 nuclear stage. **c** Embryo at 256 nuclear stage, with fibrous cell wall. **d** Cell wall formation is completed. **e** Vertical section of the embryo in the gametophyte, showing vigorous cell divisions limited near the chalazal end. **f** Seedling, a part of the gametophyte is removed to show the cotyledons. *Bar*: 1cm (**a,c** from [52] with permission)

A proembryo with 128 free nuclei (formed after the seventh division) can be observed within a single day after fertilization [24]. Free nuclei are evenly distributed throughout the proembryo (Fig. 7b). However, at the 128-nuclear stage the proembryo shows somewhat different numbers of nuclei. Hirase [34] examined the number of free nuclei, finding 8 proembryos with 128 nuclei, and 1 with proembryo 123-nuclei and 1 proembryo with 133-nuclei. Soma [52] found that the nuclear number varies from 122 to 136, though the average is 128. This 128-free-nuclear stage lasts more than 12 days [24]. The different cellular structures of the proembryo at the 128-nuclear stage can be observed. Some proembryos have round nuclei in granulated cytoplasm and others have angular nuclei in bubble-like cytoplasm [23, 52]. The nucleus at this stage is 20–25 μm in diameter [49].

Cell wall formation occurs at the end of the eighth mitotic division, after which an average of 256 nuclei with thin fibrous walls can be observed [24]. A fibrous separation is formed very early between any two of these free nuclei. This fibrous separation gradually transforms into a cell plate [23]. During cell wall formation the cytoplasm of the proembryo becomes vacuolated, and each nucleus is surrounded by fine granules which are connected with the fibers radiating from the nucleus. Wall formation occurs simultaneously throughout the proembryo, and the cells produced show approximately the same size [8, 53, 54] (Fig. 7c,d).

After cell formation is completed, cells start to divide. Divisions are not simultaneous. Vigorous cell divisions take place only in the limited portion, near the chalazal end of the embryo, and the two thirds of the proembryo near the micropylar end shows fewer divisions [55] (Fig. 7e). The cells near the micropylar end become enlarged. This area is considered as a nonorganized suspensor [8] or as a reduced suspensor [55, 56]. Arnoldi [57] considers it to have a haustorial function. However, many investigators consider that the embryo of this species has no suspensor.

Behind the above-mentioned suspensor-like cells primary root meristem is initiated, which ultimately differentiates into the radicle. The chalazal end of the embryo becomes a dome-shaped meristem, and two cotyledons are initiated on its peripheral portion as two protuberances. The portion of the meristem between two cotyledons differentiates finally into the shoot apical meristem. The portion between the shoot apical meristem and the radicle is considered the transition zone by Ball [58]. Differentiation of the embryo can be observed in October. In early cotyledon development one is larger than the other and is notched at the tip. At maturity they become equal in length [8]. On the adaxial surface of the mature cotyledons stomata are recognized [59]. Usually two cotyledons are recognized in the embryo, but sometimes three cotyledons can be observed. The three cotyledons are almost equal in size [53], or one cotyledon of the two is bilobed near its basal part [60].

Cook [61, 62] examined the numbers of embryos in the seed; most seeds contain one embryo, 12% of the examined seeds have no embryo in the normal fresh tapetal gametophyte, and 2% of the seeds have two embryos.

On germination of the seed the two cotyledons are hypogeal, and the greater part remains within the seed, namely within the gametophyte (Fig. 7f). Therefore, photosynthesis in the cotyledons does not occur. Cotyledons have a role in supplying the materials necessary for the growth of the embryo [63]. Many starch grains are observed in the mature cotyledons [64].

Li [65] stored seeds in a box kept in a room through winter. The growth of the embryo could be detected throughout the winter. Eames [66] and Fabre-Duchartre [24] also recognize that the embryo of this species has no dormancy, if favorable conditions prevail.

The mature female gametophyte with embryo is edible. Del Tredici [67] examined the seed: carbohydrate occupies 38% of the total weight, most of it in the form of starch, but a small amount of sucrose, glucose, and fructose are present. He ascribes to these sugars the sweet taste of the nuts.

References

1. Seward AC, Gowan J (1900) The maidenhair tree (*Ginkgo biloba* L). Ann Bot 14:109–154
2. Karstens WKH (1945) Variability of the female reproductive organs in *Ginkgo biloba* L. Blumea 5:532–553
3. Strasburger E (1879) Die Angiospermen und die Gymnospermen. Fischer, Jena
4. Wettstein R von (1899) Die weibliche Blüte von *Ginkgo*. Öster Bot Zeit 49:417–425
5. Čelakovský LJ (1900) Die Vermehrung der Sporangien von *Ginkgo biloba* L. Öster Bot Zeit 50:229–236, 276–283, 336–341
6. Sprecher A (1907) Le *Ginkgo biloba* L. These Univ de Genéva
7. De Haan HRM (1920) Contribution to the knowledge of the morphological value and the phylogeny of the ovule and its integuments. Recueil des trav bot néerl 17:219–324
8. Chamberlain CJ (1935) Gymnosperms, Structure and Evolution. University of Chicago Press, Chicago
9. Mehra PN (1938) Some abnormalities in the female strobilus of *Ginkgo biloba* L. Proc Indian Acad Sci B8:211–217
10. Fujii K (1896) On the different views hitherto proposed regarding the morphology of the flowers of *Ginkgo biloba* L. Bot Mag Tokyo 10:7–8, 13–15, 104–110
11. Sakisaka M (1928) On the seed-bearing leaves of *Ginkgo*. J Jpn Bot 4:219–235
12. Sakisaka M (1927) On the morphological significance of seed-bearing leaves of *Ginkgo*. Bot Mag Tokyo 41:273–278
13. Sprecher A (1906) L'origine du sac embryonnaire de *Ginkgo biloba*. Arch des sci phys et nat Genéve 4me période t 21:74–77
14. Takaso T (1980) A developmental study of the integument in Gymnosperms (1) *Ginkgo biloba* L. J Jpn Bot 55:14–27
15. Carothers IE (1907) Development of ovule and female gametophyte in *Ginkgo biloba*. Bot Gaz 43:116–130
16. Favre-Duchartre M (1956) Contribution à l'étude de la reproduction chez le *Ginkgo biloba*. Rev Cytol et Biol Vegétales Paris 17:1–214
17. Stewart KD, Gifford EM, Jr. (1967) Ultrastructure of the developing megaspore mother cell of *Ginkgo biloba*. Amer J Bot 54:375–383
18. Bold HC, Alexopoulos CJ, Delevoryas T (1987) Morphology of plants and fungi, fifth edn. Harper and Row, New York

19. Coulter JM, Chamberlain CJ (1917) Morphology of Gymnosperms. University of Chicago Press, Chicage
20. Herzfeld S (1927) Beiträge zur Kenntnis von *Ginkgo*. Jahrb Wiss Bot 66:814–862
21. De Sloover-Colinet A (1963) Chambre pollinique et gamétophyte mâle chez *Ginkgo biloba*. Cellule 129–145
22. Shimamura T (1935) Über die Bestäubung und Befruchtung bei *Ginkgo biloba* L. Jpn J Genet 11:180–184
23. Lee CL (1955) Fertilization in *Ginkgo biloba*. Bot Gaz 117:79–100
24. Favre-Duchartre M (1958) Ginkgo, an oviparous plant. Phytomorphology 8:377–390
25. Gifford EM Jr, Foster AS (1989) Morphology and evolution of vascular plants, 3rd edn. WH Freeman, New York
26. Pettitt JM (1977) The megaspore wall in gymnosperms: ultrastructure in some zooidogamous forms. Proc R Soc Lond B195:497–515
27. Avanzi S, Cionini PG (1971) A DNA cytophotometric investigation on the development of the female gametophyte of *Ginkgo biloba*. Caryologia 24:105–116
28. Foster AS, Gifford EM Jr. (1959) Comparative morphology of vascular plants. WH Freeman, San Francisco
29. Banergee SN, Radforth NW (1967) Gibberellin-like substances in the fruits of *Ginkgo biloba* L. Plant and Cell Physiol 8:207–209
30. Dexheimer J (1973) Quelques aspects ultrastructuraux du prothalle femelle alvéolaire du *Ginkgo biloba*. CR Acad Sci Paris 276:2789–2792
31. Friedman WE, Goliber TE (1986) Photosynthesis in the female gametophyte of *Ginkgo biloba*. Amer J Bot 73:1261–1266
32. Hirase S (1898) Études sur la Fécondation et l' Embryogénie du *Ginkgo biloba*. J Coll Sci Imp Univ Tokyo 12:103–149
33. Shimamura T (1935) Zur Cytologie des Befruchtungsvorganges bei Cycas und Ginkgo unter Benutzung der Feulgenschen Nuclealreaktion. Cytologia 6:465–473
34. Hirase S (1895) Études sur la fécondation et l' embryogénie du *Ginkgo biloba*. J Coll Sci Imp Univ Tokyo 8 part 2:307–322
35. Hirase S (1918) Nouvelles recherches sur la fécondation et l' embryogénie du *Ginkgo biloba*. Bot Mag Tokyo 32:139–143
36. Hirase S (1918) Further studies on the fertilization and embryogeny in *Ginkgo biloba* (in Japanese). Bot Mag Tokyo 32:83–108
37. Camefort H (1965) L' organisation du protoplasme dans le gamète femelle, ou oosphère, du *Ginkgo Biloba* L. J Microscopie 4:531–546
38. Cionini PG, Avanzi S (1971) An autoradiographic study on the development of the female gametophyte in *Ginkgo biloba*. Giorn Bot Ital 105:281–294
39. Cionini PG (1971) A DNA cytophotometric study on cell nuclei of the archegonial jacket in the female gametophyte of *Ginkgo biloba*. Caryologia 24:493–500
40. Stopes MC, Fujii K (1906) The nutritive relations of the surrounding tissues to the archegonia in Gymnosperms. Beih Bot Centr 20:1–24
41. Maugini E, Fiordi AC (1970) Passagio di materiale dalle cellule del tappeto alla cellula centrale dell' archegonio e al proembryone di *Ginkgo biloba* L. Caryologia 23:415–440
42. Dexheimer J (1973) Étude ultrastructurale du gametophyte femelle de *Ginkgo biloba*. 1. Les cellules a reserves. Caryologia 25 (Suppl): 85–96
43. Favre-Duchartre M (1943) Sur le comportement des ovules de *Ginkgo biloba*. Bull Soc Bot Fr 90:111–116
44. Kirchheimer F (1945) Ein Beitrag zur Morphologie der Samen von *Ginkgo biloba* L. Bull Soc Bot Swisse 55:304–312

45. Hirase S (1896) On the spermatozoid of *Ginkgo biloba* (in Japanese). Bot Mag Tokyo 10:325–328
46. Herzfeld S (1926) Neue Beiträge zur Kenntnis des Befruchtungs-vorganges von *Ginkgo biloba.* (Vorläufige Mitteilung) Öster Bot Zeit S: 158–161
47. Ikeno S (1901) Contribution à l' étude de la fécondation chez le *Ginkgo biloba.* Ann Sci nat Bot 8 13:305–318
48. Favre-Duchartre M (1950) Contribution à l' étude de la fecondation et du development syncytial du proembryon chez *Ginkgo biloba.* CR Acad Sci Paris 230:569–570
49. Shimamura T (1931) A note on the mitotic division in the proembryo of *Ginkgo*, with special reference to chromatin elimination (in Japanese). Bot Mag Tokyo 45:525–530
50. Shimamura T (1928) On the formation of proembryo of *Ginkgo biloba* L. Bot Mag Tokyo 42:71–76
51. Herzfeld S (1928) Über die Kernteilungen im Proembryo von *Ginkgo biloba.* Jahrb Wiss Bot 69:264–294
52. Soma S (1991) Numbers of nuclei and the nuclear location in the proembryo of *Ginkgo biloba* (in Japanese with English summary). Ann Rep Fac Educ Bunkyo Univ 25:126–137
53. Lyon HL (1904) The embryogeny of *Ginkgo.* Minn Bot Stud 3:275–290
54. Bierhorst DW (1971) Morphology of vascular plants. Macmillan, New York
55. Sugihara (1992) Embryogeny of Gymnosperms (in Japanese). University of Tokyo Press, Tokyo
56. Wang F, Chen Z (1983) A contribution to the embryology of Ginkgo with a discussion on the affinity of the Ginkgoales (in Chinese with English abstract). Acta Bot Sin 25:199–207
57. Arnoldi W (1903) Beiträge zur Morphologie der Gymnospermen. 6. Ueber den Bau der Zellkerne im Embryo von Ginkgo biloba. 7. Die Embryobildung bei *Ginkgo biloba* (in Russian with German abstract). Ann Inst Agron et Forest à Nowo-Alexandria 16:117–139
58. Ball E (1954) Experiments on the embryo of *Ginkgo biloba.* 8th Cong Intern Bot Paris Rap et Comm Sect 8:265–267
59. Wigglesworth G (1903) The cotyledons of *Ginkgo biloba* and *Cycas revoluta.* Ann Bot 17:789–791
60. Soma S (1995) Hypocotyl and Epicotyl in *Ginkgo biloba*, with special reference to the bilobed cotyledon (in Japanese). Ann Rep Educ Bunkyo Univ 29:52–56
61. Cook MT (1902) Polyembryony in *Ginkgo.* Bot Gaz 34:64–65
62. Cook MT (1903) Polyembryony in *Ginkgo.* Bot Gaz 36:142
63. Bulard C (1952) Culture aseptique d'embryons de *Ginkgo biloba*: Róle des cotylédons dans l'absorption du et la croissance de la tige. CR Acad Sci Paris 255:739–741
64. Hills TG, Fraine E de (1909) On the seedling structure of Gymnosperms 3. Ginkgoaceae. Ann Bot 23:433–458
65. Li TT (1934) The development of embryo of *Ginkgo biloba.* Sci Rep Natl Tsing Hua Univ Ser B. 2:29–35
66. Eames AJ (1955) The seed and *Ginkgo.* J Arn Arb 36:165–170
67. Del Tredici P (1991) *Ginkgo* and people—a thousand years of interaction. Arnoldia 51:2–15

Contribution to the Knowledge of Fertilization of Gymnosperms with Flagellated Sperm Cells: *Ginkgo biloba* and *Cycas revoluta*

TERUMITSU HORI and SHIN-ICHI MIYAMURA

1 Introduction

It was Strasburger [1] who first drew attention to fertilization in *Ginkgo biloba*. He followed the development of the embryonic sac and embryo as well as the formation of the female gametophyte. Much later, as a result of his study on fertilization in *Juniperus virginiana*, he argued in favor of the existence of canal cells in *Ginkgo* and contributed to the knowledge of the development of the embryo [2]; however, he did not make any observations on fertilization himself. His last work on *Ginkgo* was on the development of the pollen [3].

In 1893 Hirase started his studies on fertilization in *Ginkgo* [4]. His first paper corrected the idea held by European botanists that the embryo of *Ginkgo* develops after the ripened seed falls from the tree [4]. In 1895 Hirase [5] published precise information on the development of the archegonium of the female gametophyte, the formation of the canal cells, and the embryo, but no detailed information on fertilization. Afterwards he devoted his studies to the development of the pollen tube and the sperm [6–12].

It was in 1896 that Hirase [8–11] for the first time observed living sperms of *Ginkgo* coming out of the pollen tube. Since then the morphology and structure of *Ginkgo* sperms and pollen tubes have been studied in detail by Webber [13], Hirase [12], Fujii [14–18], Miyake [19, 20], Ikeno [21–23], Bessey [24], Herzfeld [25], Shimamura [26], Lee [27], Favre-Duchartre [28], Tulecke [29], Gifford and Lin [30], Gifford and Larson [31], Rohr [32], Friedman [33, 34], Friedman and Gifford [35], Li et al. [36], Hori et al. [37], and Hori and Miyamura [38].

More extensive works on the life history of *G. biloba* were done by Hirase [5, 12, 39–41] and others, but detailed studies on fertilization were done by Ikeno [42], Hirase [41], Shimamura [43–45], Herzfeld [25], Lee [27], Favre-Duchartre [28, 46], De Sloover-Colinet [47], Camefort [48], and Soma et al. [49]. However, there remain many unresolved problems and contradictory observations. This chapter reviews additional, new findings of late developmental stages of gameto-

Institute of Biological Sciences, University of Tsukuba, 1-1-1 Tennodai, Tsukuba, Ibaraki 305, Japan

phytes and the fertilization process investigated by transmission electron microscopy (TEM), high-speed video, and various combinations of light microscopy. The materials for TEM were prepared and stained according to standard procedures.

2 Male Gametophyte and Sperm

In many textbooks that describe the sexual reproduction of *Ginkgo biloba* the male gametophyte is shown as the basal end of the male gametophyte freely suspended down into the pollen or archegonial chamber, with the haustorial ends located in the nucellus. Hence, one has the impression that the sperms swim down toward the archegonium which is usually shown lying below the pollen tubes. The arrangement in the ovule is actually the reverse, because the micropyle of the *Ginkgo* ovule faces down toward the ground in autumn. Hence, the sperm cells have to swim up from the pollen tube to the archegonium in the ovule.

2.1 Pollination and Sperm Formation

The early ontogeny of the male gametophyte of *Ginkgo* resembles those of the cycads except for producing two prothallial cells. After the second prothallial cell has been formed, the antheridial initial [50] divides into a generative cell and a tube cell. At this four-celled stage, the young male gametophyte (the pollen grain) is released into the air by the dehiscence of the microsporangium wall [3, 20, 25, 28, 51] and is carried to the megasporangiate (female) tree by the wind. The pollen adheres to the mucilaginous pollination droplet that is exuded from the micropyle of the ovule. Since the pollen of other plants (for example, *Pinus*) is often observed in the ovules of *Ginkgo*, both pollination droplet and ovule seem to have no specific selectivity for pollen. However, these foreign pollen grains have not been observed germinating in *Ginkgo* ovules. Retraction of the pollination droplet brings the *Ginkgo* and foreign pollens into the pollen chamber where a much branched, rhizoid-like haustorial pollen tube of *Ginkgo* pollen is developed at one pole of the pollen grain in the nucellar tissue [33, 34].

The generative cell divides into the stalk cell (or sterile cell) and spermatogenous cell [50]. Previously, it was thought that the stalk cell surrounds the second prothallial cell [25, 28], so that the spermatogenous cell lies distal to the complex of the stalk cell and the secondary prothallial cell. However, it has been shown by Friedman and Gifford ([35] and this volume) that an anticlinal ring-like division of the generative cell produces the stalk cell.

As the final stage leading to formation of sperm approaches, the spermatogenous cell becomes elliptical in shape and the nucleus becomes compressed to fusiform. The major axis of the nucleus is at right angles to the major axis of the cell. At the end of the development, the spermatogenous cell, which is encircled by a very thin cell wall that is hardly discernible [28], reaches a diameter between 50 and 80 μm.

2.2 Structure of the Spermatogenous Cell

About 2 weeks before fertilization two distinct cell organelles are organized in the spermatogenous cell [6]: the blepharoplasts [13] (corresponding to variously termed structures [52], e.g., attraction spheres [6] or centrosome-like bodies [53], etc.); and the osmiophilic globules [30] (corresponding to oil-like globules [26], nucleolus-like bodies [13, 25] or vacuole-like structures [27]) (Fig. 1a).

The blepharoplast is a distinct organ of the protoplasm of the spermatogenous cell, having as its primary function the formation of the flagella of the motile sperms [13]. Two blepharoplasts occur uniformly in each spermatogenous cell, located in the cytoplasm on opposite sides of the nucleus approximately midway between the nucleus and the cell wall. They are spherical and range in diameter 2–2.5 μm [13], with 3 μm as the maximum size [25], to 3.5–4.5 μm [30], or 4–4.5 μm at maturity [27] in *Ginkgo*. The ultrastructure of the blepharoplast is described in work on *Zamia* [54] and *Ginkgo* [30, 31, 37, 38]. The blepharoplast consists of a peripheral layer of numerous probasal bodies and an inner spherical dense body with many less dense globules that contain microtubules [30, 37, 38]. Our examination by electron microscopy shows that each probasal body is 320 nm in length and 160 nm in diameter, and the central body is 3.8 μm in diameter (Fig. 1b). Based on these dimensions, it is calculated that approximately 5400 probasal bodies make up the peripheral layer of a blepharoplast. This number is different from that of Gifford and Lin [30]. They counted approximately 1000 probasal bodies. This contradiction may be due to different stages of maturation, since the dimensions of the blepharoplast central body shown by them is approximately 2.8 × 3.2 μm, while ours is 3.8 μm in diameter.

The blepharoplasts had been thought to be excreted from the nucleus [52, 55, 56], but recently it was revealed that they are formed from one protoblepharoplast that is differentiated de novo near the nuclear envelope from a microtubules-organizing center [30, 31].

Between the blepharoplast and the nuclear envelope on each side of the nucleus are regular spherical globules, of nearly uniform size, 17 to 20 μm in diameter. The nature of the two globules was discussed by Hirase [5, 6], Webber [13], and Herzfeld [57], who interpreted them as additional nuclei. However, electronmicroscopy shows that they are osmiophilic globules [30] that are not surrounded by a limiting membrane (Fig. 1a).

2.3 Cell Division Leading to Formation of Two Sperms

Concomitant with mitosis of the spermatogenous cell to form the two sperms, the blepharoplast disorganizes (Fig. 1c) and the probasal bodies become aligned on a band (corresponding to the multilayered structure [MLS] in modern terminology) arranged in the form of a helicoid spiral on one side of each sperm cell as basal bodies of the flagella. The whole content of the spermatogenous cell is used to form two equal daughter cells that become motile sperms [7–11, 24–28].

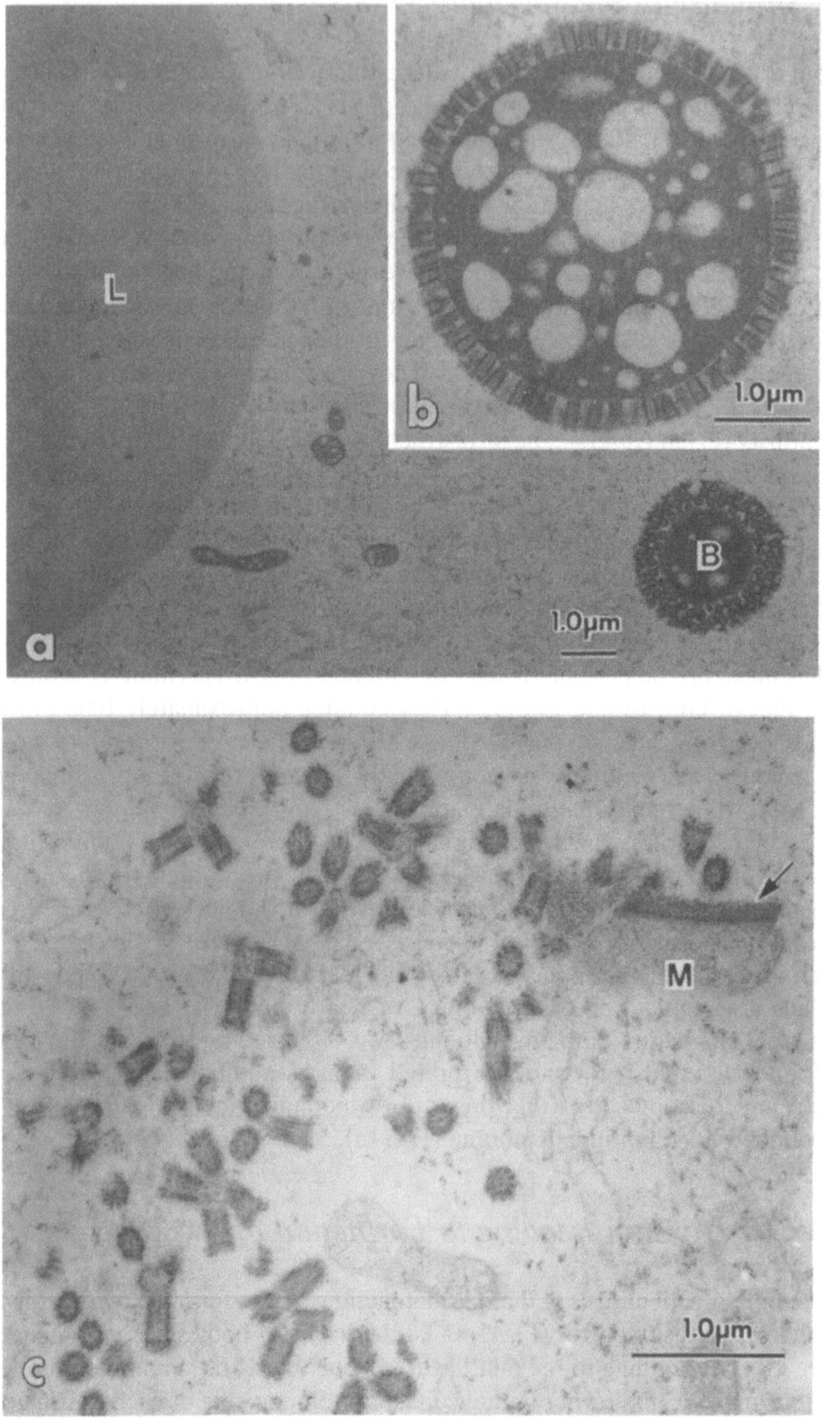
L
b
1.0µm
B
1.0µm
a
M
1.0µm
c

In *Dioon edula* [55] and *Ceratozamia mexicana* [58], Chamberlain observed that two sperms were formed within the two sperm mother cells by the division of the generative cell. In *Ginkgo* there is no wall between the two sperm cells since they develop within the common cell wall of the spermatogenous cell [17, 19, 27]. The nearly formed sperms, which are approximately hemispherical and appressed against each other along their flat faces, are surrounded by this cell wall. Wolniak [59] reported that centripetal wall formation occurs during meiosis in the microsporocyte of *Ginkgo*. This suggests that partitioning of the cytoplasm in *Ginkgo* occurs by furrowing of the cell membrane. We also observed the centripetally proceeding division furrow of the cell membrane during division of the spermatogenous cell leading to the formation of two sperm cells. It may be noteworthy that a primitive type of cytokinetic mechanism is preserved in a primitive seed plant.

2.4 Structure of the Sperms and Their Behavior

The nucleus occupies the greater part of the sperm body. The spiral band of flagella makes an indentation that becomes a deep helicoid furrow on the outer surface of the sperm body.

Figure 2 shows two sperms that are motile within the pollen tube. An anterior protrusion of the sperm apex, termed the nose in this chapter, is clearly seen in Fig. 2a. When the sperms swim inside the spermatogenous cell wall, they actively vibrate their noses as they seem to break the cell wall. A spiral band of flagella extends from the nose posteriorly. There is a question as to whether the flagellar bands are arranged in two and a half [17, 20] or three spiral whorls [12, 24]. Present observation seems to support the latter conclusion.

Shimamura described the direction of the turn of the flagellar spiral band as left to right (from the cell apex to the open end) (Text-fig. 4 in [27]), that is, in a clockwise direction when viewed from the apex of the cell body. But it is right to left in another figure (Text-fig. 2 in [27]). The same situation is found in papers by Fujii [17] (right to left in Figs. 4 and 5b and left to right in Fig. 9) and Miyake [20] (right to left in Fig. 10 and left to right in Fig. 8). Figure 27 of Hirase [12] is drawn as right to left, and Fig. 55d of Favre-Duchartre [28] is left to right. Our analysis of many sperms indicates the possible presence of two types of spiral turns in *Ginkgo*. As far as we have confirmed with a pair of sperms produced in a single spermatogenous cell, however, they show the same turn, right to left (Fig. 2b). In addition we confirmed that in *Cycas revoluta* it is constantly right to left in

FIG. 1a–c. Electron micrographs of *Ginkgo biloba*. **a** A part of a large osmiophilic globule (*L*) and blepharoplast (*B*) in the spermatogenous cell. **b** A blepharoplast consisting of a central dense globule with many transparent less dense globules and a peripheral layer of probasal bodies in the spermatogenous cell. **c** Scattered probasal bodies resulting from disorganization of blepharoplast and an incipient multilayered structure (*arrow*) with attached mitochondrion (*M*) before maturation of sperm of *G. biloba*

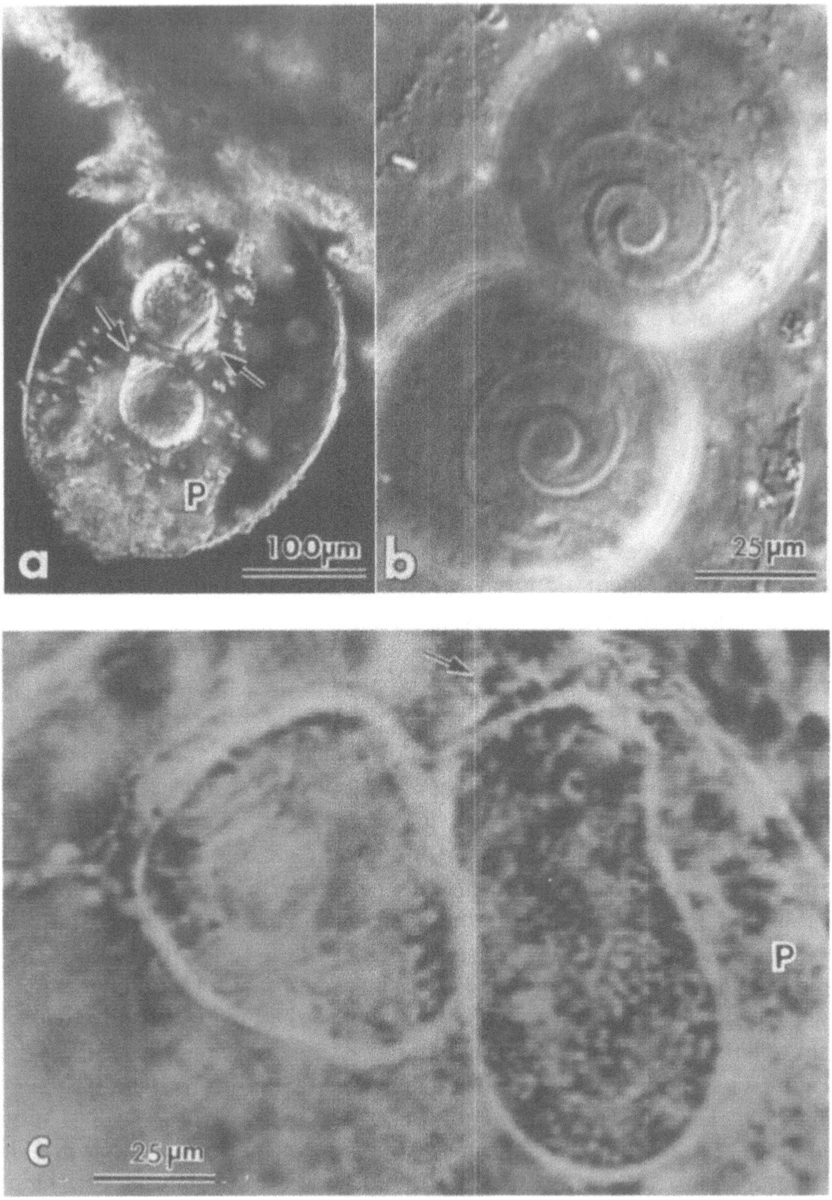

FIG. 2a–c. Light micrographs of *Ginkgo biloba*, showing two sperms swimming in the pollen tube (*P*). **a** anterior nose-like projections (*arrows*) are prominent. **b** The direction of the turn of the flagellar spiral band, which extends from the cell center outwards, is clearly found to be identical on a pair of sperms produced in a spermatogenous cell. **c** A rather round-shaped and an elongated sperm in the pollen tube (*P*). *Arrow* indicates an anterior nose-like projection

five pairs of sperms gained from five pollen tubes. This may at least suggest that sperms produced in a spermatogenous cell have the same pattern of spiral turn.

When *Ginkgo* sperms begin to swim within the mother cell wall, they gently vibrate their flagella and part of their body, especially the cell anterior. Soon they revolve independently of each other in the mother cell. When the mother cell wall ruptures, two sperms are released into the cytoplasm of the pollen tube, in which they variously change their shapes in response to the space available for swimming (Fig. 2c). The sperms of *C. revoluta* flow out either through a tiny gap formed at the junction of the cell wall of the pollen tube and the stalk cell, or through the secondary prothallial cell [60]. But in the case of *Ginkgo* we could not confirm its precise pathway, because the release of sperms from the pollen tube in sucrose solution is very fast. The flagellated sperm has an extraordinary elasticity to protect the organization of the cell body, since it is strongly stretched during its passage through narrow gaps when it is forcefully liberated from the pollen tube and when it enters into the archegonium through a small gap between the neck cells. In 10% sucrose solution sperms first move forward spirally and the flagella may act like a propeller to pull the sperm body forward.

3 Female Gametophyte and Egg Cell

3.1 Development of the Tent Pole

As the time for fertilization approaches, the tent pole [12], which is a central column formed by a localized projection of prothallial tissue between the archegonia, grows toward the pollen chamber and finally ruptures the megasporangial wall overlaying the nucellus [28]. This establishes a continuity between the pollen and archegonial chambers and allows the sperms to approach the archegonia.

Herzfeld [25] reported that no "column" was found in her material except for a small projection. This might be due to observations on younger female gametophytes, or on materials from an eccentric tree that does not differentiate a normal tent pole, or one that produces a small type of tent pole.

3.2 Structure of the Archegonium

3.2.1 Formation of the Archegonial Neck Opening

Since the two-celled condition of the archegonial neck was observed very early [1], it has been considered a feature in common with members of the Cycadales [27]. The two neck cells, however, divide obliquely, forming a group of four cells [27, 61]. Later they all become swollen and together form a hemispherical projection.

3.2.2 Formation of the Receptive Cavity

The nucleus of the central cell of an archegonium undergoes a rapid division just prior to fertilization [5, 27, 41, 42]. After this, an unequal cytokinesis cuts off the small ventral-canal cell and a large egg cell. The ventral-canal cell usually degenerates soon, resulting in formation of a cavity at the neckcell end of the archegonium [41, 43]. The cavity was termed "la chambre receptive" [41] or receptive cavity [26] in *Ginkgo*, or the receptive vacuole in *Ginkgo* [43] and *Pinus* [62]. The receptive cavity has been considered to be formed by coalescence of several vacuoles in the egg cytoplasm and extrusion outside the egg cell [27, 41]. But our observations suggest that the degeneration of the ventral-canal cell may contribute most to the formation of the receptive cavity. The cavity may be useful for storing liquid in the archegonium in which sperms swim for a while before fertilization. This swimming behavior may be necessary for recovery of sperm body organization which is enormously stretched during passage through an intercellular gap between the neck cells.

3.2.3 Centering of Egg Nucleus

After the central cell divides, the juvenile egg nucleus begins to move toward the cell center in the egg cell. During this movement the nucleus expands and develops a special structure that functions as the receptor apparatus of the male nucleus (RAMN) for sexual nuclear fusion as described below (Section 4.2). In spite of the large size of the nucleus, the chromatin contained within is slight and diffuse. With the Feulgen stain [27] or DAPI (DNA specific fluorochrome 4′-6-diamidino-2-phenylindole) staining (in the present study) the nucleus appears nearly colorless [45].

3.2.4 Number of the Archegonia Produced

Usually two mature archegonia are produced in an ovule, but sometimes one or three, or more, may be present [28]. The number of archegonia produced is not correlated with the number of the pollen tubes accommodated in the pollen chamber [27]. Sometimes abnormal archegonia are produced: an archegonium without the neck cells, or the production of the neck cells without an archegonium.

4 Fertilization

4.1 Entrance of the Sperm into the Archegonium

4.1.1 Opening of the Archegonial Neck

Opening of the archegonial neck is essential for entrance of the sperm cell into the archegonium. Different mechanisms have been proposed for formation of the

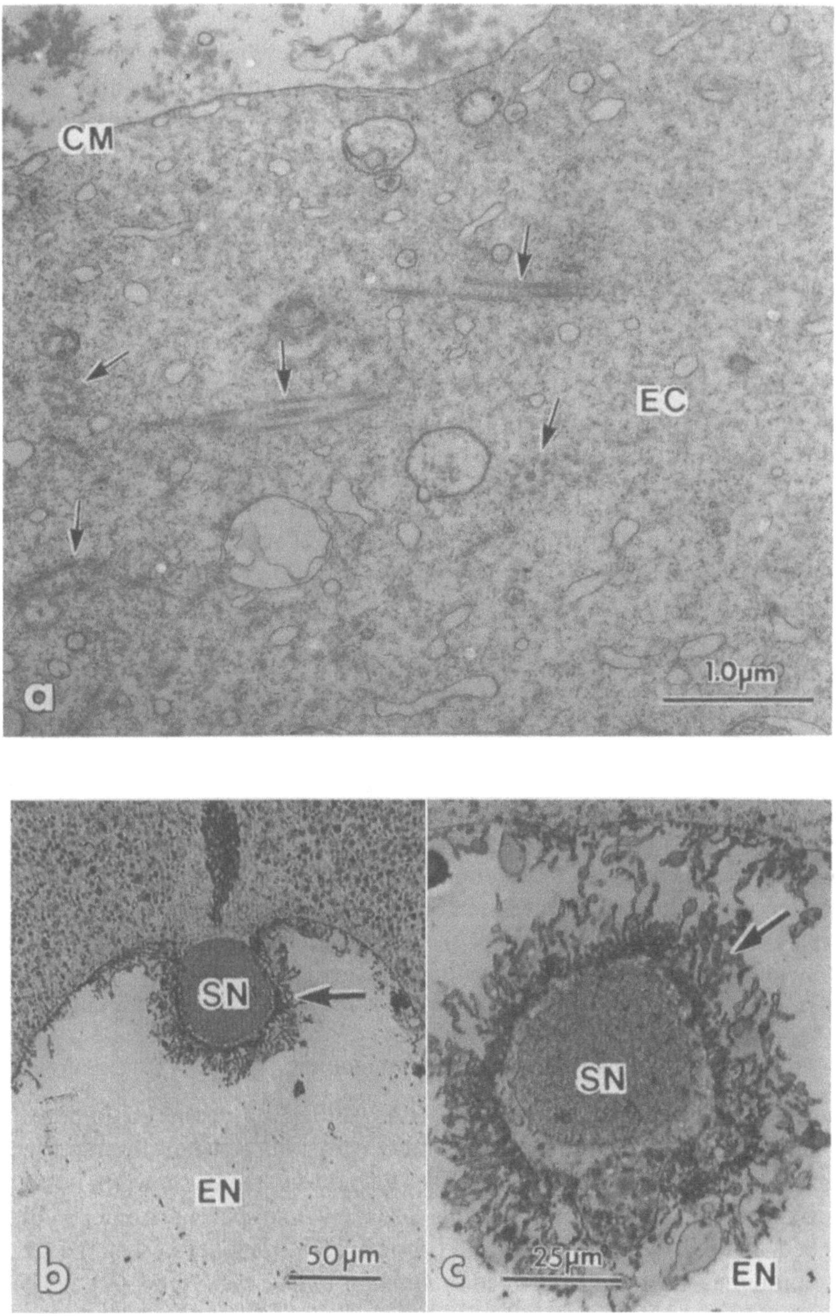

FIG. 3a–c. *Cycas revoluta*. **a** Flagellar axonemes (*arrows*) incorporated in the periphery of the egg cytoplasm (*EC*) after the union of the sperm with egg cell. *CM*, cell membrane of the egg cell. **b–c** Accomodation of the sperm nucleus (*SN*) in the the receptor apparatus of the male nucleus (*arrows*) of the egg nucleus (EN)

neck opening. Herzfeld [25] suggested that four neck cells are pushed apart by the pressure of vacuolar fluid secreted from the egg cytoplasm just before fertilization. Lee [27] observed that a part of the egg cytoplasm forms a beak which separates neck cells. As soon as the sperm becomes attached to the beak, the elongated egg nucleus extended into the beak withdraws toward the center of the egg cell and the beak retreats, thus forming a small path at the top of the archegonium. However, our observation does not substantiate this; neither the pressure of the fluid nor the protrusion of a cytoplasmic beak causes opening formation; it is mainly through the swelling and reflexing of the neck cells themselves [63].

4.1.2 Entry of the Sperm into the Archegonium and the Union of the Gametes

The neckcell opening is narrower than the diameter of the sperm body, varying from one fifth to one tenth of the swimming sperm. The archegonial neck consists of four cells. The sperm cell does not enter directly into the cytoplasm of the egg, but into the receptive cavity [26, 40, 43, 62], where the sperm may swim for a while before fertilization.

Ikeno [42] suggested that the *Ginkgo* sperm gets rid of the cytoplasm inside of the egg cell. This may mean that the entire sperm cell including the cell membrane completely enters into the cytoplasm of the egg cell. Schnarf [64] also stated that the sperm nucleus slips out of its protoplasm and penetrates the egg. Shimamura [43] mentioned that the male nucleus slips from the flagella-bearing cytoplasm, which may be left in the receptive cavity, and enters into the egg cytoplasm. However these descriptions seem to be incomplete from the viewpoint of modern biology. The cell membrane of the sperm is fused with that of the egg cell and the whole cytoplasm of the sperm, including flagellar axonemes, is directly incorporated into the egg cytoplasm (Fig. 3a).

4.1.3 Number of Sperms Entering the Archegonium

Ikeno [42] suggested that probably only one of the two sperms could penetrate into the egg. But, superfluous sperms have occasionally been observed in the fertilized archegonium [25, 41, 43, 65]. The present study shows that lagging sperms do occasionally penetrate; in this case the whole sperm cell enters, not just the head as described previously [27]; however, it degenerates in the cavity. Polyspermy has not been confirmed in *Ginkgo*, unlike cycads [66–69].

4.1.4 Approach to the Egg Nucleus

As was pointed out previously [28, 42], the journey of the male nucleus toward the egg nucleus may proceed so rapidly that no one has observed the intermediate stages of the approach, except Hirase (Fig. 11 in [41]). This suggests two possibilities: one is that the male nucleus travels very fast toward the female nucleus; the other is that the distance between the point where the sperm fuses to the egg and the egg nucleus is rather short, as mentioned below.

4.2 Differentiation of the Receptor Apparatus of the Male Nucleus (RAMN) and Fertilization

The process of fusion of both sperm and egg nuclei in *Ginkgo* has been observed in detail [28, 41, 42, 45]. After the sperm cell fuses with the egg cell the male nucleus moves in the egg cytoplasm toward the egg nucleus to make contact with it, forming a diploid zygote that gives rise to the future embryo [70]. As soon as they come into contact, the sperm nucleus begins to press on the envelope of the female nucleus; under the influence of this pressure, part of the envelope depresses more and more on contact with the male nucleus, finally dissolving [42] so that the substances of the two sexual nuclei are completely fused.

By applying modern methods of sectioning and staining we studied fertilization of *G. biloba* and *Cycas revoluta*. A distinct RAMN is differentiated in the female nucleus of *C. revoluta* [38, 71]. The RAMN is a specially differentiated invagination in the egg nucleus on the side toward the archegonial neck. The inner surface of nuclear envelope at the depression is adorned with material that is probably the fine foldings of the inner nuclear membrane (Fig. 3b,c). After the sperm nucleus plunges into the invagination, the two nuclei begin to fuse. Thus, the invagination may function as a specialized receptor region of the sperm nucleus and as a site of fusion in *C. revouta* [71]. This structure may correspond to "Empfängnisshöhle (fecundation depression)" which Ikeno [72] designated at the moment of the fertilization in *C. revoluta*. He described that the female nucleus produces a distinct depression or crater at the site near the neck cells, before the sperm nucleus comes in contact with the egg nucleus.

Early workers thought that the depression of the female nucleus is caused by the repression of the male nucleus [28, 41, 42, 63, 73]. Ikeno concluded that in the female nucleus of *G. biloba* there is no specially preferred region for fusion with the sperm nucleus; the male nucleus sometimes was found above the female nucleus, sometimes on the side [42]. Lee [27] also believed that the sperm head moves around the egg nucleus and may penetrate it in any part; the point of entry need not be confined to any particular region. However, these observations may not negate the differentiation of the RAMN in *Ginkgo*. Our preliminary observations on the fusion of the male and female nuclei suggest the presence of the RAMN similar to that of *C. revoluta*, though it seems to be a less complicated structure.

4.3 Comparison of Reproductive Structures in Gymnosperms

Ikeno [72] noted that the manner of fertilization in *Cycas revoluta* represented a new type unknown until that time. He later made a more extensive investigation of the fertilization of *Ginkgo biloba* [42], and pointed out that the mode of fusion of the sexual nuclei in *Ginkgo* and *Cycas* is different to that of other angiosperms. At the time of nuclear union the male and female nuclei are conspicuously unequal in size and volume in *Ginkgo* [28, 42] and *Cycas* (Fig. 3b), with the female nucleus sometimes being about 10 times larger than the male. Such a difference had never been observed in other gymnosperms [42].

Table 1 is a summary of the sizes of egg cell, male and female nuclei and the neck-egg nucleus (NEN) distance, which is the shortest distance between the archegonial neck and the female nucleus, or distance from the surface of egg cell provided with the receptive cavity to the female nucleus. In cycads, *Ginkgo*, and some representatives of the Pinaceae, this may represent the shortest pathway for a male nucleus approaching the female nucleus in the egg cytoplasm.

Since Ikeno pointed out the peculiarity of nuclear fusion in *C. revoluta* [72], the formation of a similar depression on the female nucleus at fertilization was

TABLE 1. Comparison of some sexual features of cycads, *Ginkgo*, and conifers

	Dimension (μm)				
	Egg cell	Female nucleus	Male nucleus		
Plants	Length width	Length width	Length width	NEN distance[a]	References
Cycadophyta					
Ceratozamia mexicana	3407 × 981	ND	ND	ND	[68, 73, 74]
Zamia (4 spp.)[b]	3350 × 1123	689 × 449	140 × 140	1289	[69, 80–83]
Macrozamia (2 spp.)[b]	2425 × 1032	500 × 428	112 × 138	1070	[72, 84]
Cycas (3 spp.)[b]	1710 × 720	264 × 232	127 × 129	763	[70, 85]
Bowenia serrulata	1805 × 613	159 × 157	73 × 73	947	[71]
Stangeria paradoxa	874 × 874	ND	ND	529	[59]
Pinaceae					
Larix (3 spp.)[b]	1270 × 387	107 × 133	31 × 43	335	[86–89]
Pseudotsuga (3 spp.)[b]	789 × 240	185 × 103	ND	223	[90–92]
Keteleeria (2 spp.)[b]	707 × 187	ND	ND	ND	[93, 94]
Pinus (8 spp.)[b]	596 × 341	98 × 107	35 × 53	168	[62, 75, 77, 95–101]
Abies (3 spp.)[b]	567 × 150	136 × 73	34 × 26	291	[76, 102–104]
Pseudolarix kampferi	557 × 150	138 × 124	ND	114	[105]
Picea (3 spp.)[b]	503 × 199	116 × 87	ND	190	[106–108]
Tsuga (3 spp.)[b]	462 × 147	94 × 60	ND	190	[109–111]
Cedrus deodara	ND × 210	84 × 70	18 × 28	106	[112]
Ginkgophyta					
Ginkgo biloba	405 × 464	117 × 94	27 × 20	119	[27, 28, 41–44, 46]

NEN, neck-egg nucleus; ND, no data.
[a] Distance between the neck cells and anterior edge of the egg nucleus.
[b] Parentheses indicate the number of species used for measurement of reproductive features of the genus.

observed in *Ginkgo* [41] and other non-flagellated gymnosperms [62, 75–79]. This suggests a rather wide occurrence of localized nuclear envelope fusion of sexual nuclei in gymnosperms, as Hirase discussed [79]. All published figures showing the union of male and female nuclei in cycads [13, 66–69, 73, 74] are similar to that of *C. revoluta* [72].

Table 1 shows that in every aspect the sexual features of cycads are exceptionally large compared with other gymnosperms. According to our observations the sperm nucleus of *C. revoluta* seems to proceed straight to the egg nucleus located in the cell center (Fig. 3b), which is far away from the site where the sperm fused to the egg cell. The question arises how the male nucleus reaches the female nucleus without fail in the large egg cells of cycads, and probably of *Ginkgo*. The substance that attracts the male nucleus toward the egg nucleus in *G. biloba* and *C. revoluta* has not been identified. In *C. revoluta* there are wide spaces on both sides of the egg nucleus through which the male nucleus may move around the female nucleus into the cell posterior. The RAMN differentiated in the female nucleus may function to attract the male nucleus in *C. revoluta*.

The egg cell of *Ginkgo* is nearly round in shape, in contrast to the ellipsoidal shape in the cycads and the Pinaceae. But as the dimensions of the male nucleus are smaller, and the space on both sides of the egg nucleus, when viewed in section, is larger, it seems that the male nucleus could pass through into the cell, posterior to the egg nucleus, resulting in no fertilization. Thus, the presence of the RAMN in the egg nucleus of *Ginkgo* may also reasonably be expected. The NEN distance of *Ginkgo* is shorter than that of cycads and of most of the Pinaceae. This may explain why it is difficult to observe a sperm nucleus that is midway on its journey toward the egg nucleus after the union of sperm and egg cell, as discussed in Section 4.1.4.

The reproductive features of the egg cell in the Pinaceae are intermediate between those of the cycads and *Ginkgo* (Table 1), but in most aspects their dimensions are larger than those of most other gymnosperms. In some representatives of the Pinaceae a rather large space is found on both sides of the egg nucleus as in cycads. In addition the size of the male nucleus in the Pinaceae is smaller than the spaces on the sides of the egg nucleus. Therefore, there is a possibility that a structure similar to the RAMN differentiates in some members of the Pinaceae.

Taking account of differentiation of the RAMN in *C. revoluta* and possibly in *G. biloba* [38], it can be assumed that the RAMN differentiates in all other cycads, and possibly in some non-flagellated gymnosperms, as Hirase suggested [79].

5 Concluding Remarks

In spite of a number of fertilization studies performed on *Ginkgo* and cycads, there are many problems that remain to be analyzed. The occurrence of multiflagellated sperm is actually unique in seed plants and even in gymnosperms. The question of why the sperms of *Ginkgo* and the cycads are provided with multiflagella remains to be answered. If they were biflagellated, they might not pass through the tiny gap of the archegonial neck, because the mode of biflagellate beating would be inadequate to propel the large sperms of *Ginkgo* and *cycads* during their passage through the small gap. The same situation is suspected in the case of the multiflagellated fern sperms. The beating motion of spirally arranged flagella may allow entrance of the sperm into the archegonium, since such an arrangment of multiflagella could be expected to act like a drill that pulls the sperm cell body forward.

The unique area of nuclear envelope differentiation (RAMN) on the egg nucleus, the presence of which was first noticed by Ikeno [68] in *C. revoluta*, was revealed to be a rather complex structure and not a simple invagination. The RAMN is currently known, possibly, only in the flagellated members of gymnosperms: *Ginkgo* and the cycads. However, the ontogeny and the significance of these structures during sexual fusion in *Ginkgo* remain to be solved.

References

1. Strasburger E (1872) Die Coniferen und die Gnetaceen. Jena
2. Strasburger E (1878) Über Befruchtung und Zellteilung. Jena
3. Strasburger E (1892) Über das Verhalten des Pollens und die Befruchtungsvorgänge bei den Gymnospermen. Histol Beitr 4, Jena
4. Hirase S (1894) On the fecundation of *Ginkgo biloba* (in Japanese). Bot Mag Tokyo 8:7–9
5. Hirase S (1895) Études sur le fécondation et l'embryogénie du *Ginkgo biloba* (1). J Coll Sc Imp Univ Tokio 8:307–322
6. Hirase S (1894) Notes on the attraction-spheres in the pollen-cells of *Ginkgo biloba*. Bot Mag Tokyo 8:359–360
7. Hirase S (1896a) Spermatozoid of *Ginkgo biloba* (in Japanese). Bot Mag Tokyo 10:171–172
8. Hirase S (1896b) Observation on the behavior of the spermatozoid of *Ginkgo biloba* (in Japanese). Bot Mag Tokyo 10:366
9. Hirase S (1896c) On the spermatozoid of *Ginkgo biloba* (in Japanese). Bot Mag Tokyo 10:325–328
10. Hirase S (1897) Untersuchungen über das Verhalten des Pollens von *Ginkgo biloba*. Bot Centbl 69:33–35
11. Ikeno S, Hirase S (1897) Spermatozoids in gymnosperms. Ann Bot 11:344–345
12. Hirase S (1898) Études sur la fécondation et l'embryogenie du *Ginkgo biloba* (second mémoire). J Coll Sc Univ Imp Tokyo 12:103–149
13. Webber HJ (1897b) Notes on the fecundation of *Zamia* and the pollen tube apparatus of *Ginkgo*. Bot Gaz 24:225–235
14. Fujii K (1898) Has the spermatozoid of *Ginkgo* a tail or none? (in Japanese). Bot Mag Tokyo 12:287–290
15. Fujii K (1899a) Observations on the morphology of the pollen tube and the spermatozoid of *Ginkgo biloba* (in Japanese). Bot Mag Tokyo 13:28–30
16. Fujii K (1899b) Remarks on Mr. Ikeno's remarks on my views regarding the morphology of the pollen-cells and the spermatozoid of *Ginkgo* (in Japanese). Bot Mag Tokyo 13:65–73

17. Fujii K (1899c) On the morphology of the spermatozoid of *Ginkgo biloba* (Preliminary note) (in Japanese). Bot Mag Tokyo 13:260–266
18. Fujii K (1900) On the bicephalous spermatozoid of *Ginkgo biloba* (in Japanese). Bot Mag Tokyo 14:16–17
19. Miyake K (1898) On the spermatozoid of *Ginkgo biloba* (in Japanese). Bot Mag Tokyo 12:333–339
20. Miyake K (1902) The spermatozoid of *Ginkgo*. Notes on the morphology and methods. J Appl Micro Lab Metho 5:1773–1780
21. Ikeno S (1899a) On the spermatozoid and pollen tube of *Ginkgo biloba* and *Cycas revoluta* (in Japanese). Bot Mag Tokyo 13:31–34
22. Ikeno S (1899b) Opinion of various authors on the centrosome in the pollen tube of Cycadaceae and *Ginkgo biloba* (in Japanese). Bot Mag Tokyo 13:74–76
23. Ikeno S (1928) Spermatozoids of *Ginkgo biloba* and *Cycas revoluta* (in Japanese). Shizenkagaku 3:14–28
24. Bessey EA (1901) Notes on the spermatozoids of *Ginkgo*. Science 13:255
25. Herzfeld S (1927) Beiträge zur Kenntnis von *Ginkgo*. Jahrb wiss Bot 66:814–862
26. Shimamura T (1937) On the spermatozoid of *Ginkgo biloba*. Cytologia (Fujii Jubi Vol): 416–423
27. Lee CL (1955) Fertilization in *Ginkgo biloba*. Bot Gaz 117:79–100
28. Favre-Duchartre M (1956) Contribution à l'étude de la reproduction chez le *Ginkgo biloba*. Rev Cytol Biol Veg 17:1–218
29. Tulecke WR (1957) The pollen of *Ginkgo biloba*: in vitro culture and tissue formation. Amer J Bot 44:602–608
30. Gifford EM, Lin J (1975) Light microscope and ultrastructural studies on the male gametophyte in *Ginkgo biloba*: the spermatogenous cell. Amer J Bot 62:974–981
31. Gifford EM, Larson S (1980) Developmental features of the spermatogenous cell in *Ginkgo biloba*. Amer J Bot 67:119–124
32. Rohr R (1980) Dévelopment in vitro du pollen de *Ginkgo biloba* L. Cytologia 45:481–495
33. Friedman WF (1987a) Growth and developpement of the male gametophyte of *Ginkgo biloba* within the ovule (in vivo). Amer J Bot 74:1797–1815
34. Friedman WF (1987b) Morphogenesis and experimental aspects of growth and development of the male gametophyte of *Ginkgo biloba* in vitro. Amer J Bot 74:1816–1830
35. Friedman WF, Gifford EM (1987) Division of the generative cell and late development in the male gametophyte of *Ginkgo biloba*. Amer J Bot 75:1434–1442
36. Li Y, Wang FH, Knox RB (1989) Ultrastructural analysis of the flagellar apparatus in sperm cells of *Ginkgo biloba*. Protoplasma 149:57–63
37. Hori T, Ridge R, Michikawa M, Miyamura S (1988) Ultrastructure of the sperm of *Ginkgo biloba* (in Japanese). Proc of 53rd Ann Meet Bot Soc Jpn, 13–15 October, Okayama, p 213
38. Hori T, Miyamura S (1996) Further studies on the gametes and fertilization of *Ginkgo biloba* and *Cycas revoluta*. Abs of Cent Comme Sym Discovering *Ginkgo Sperm*, "Biology of *Ginkgo biloba*." 8 Sep 1996, University of Tokyo, Tokyo, p 13
39. Hirase S (1895) Études sur le *Ginkgo biloba*. (Note préliminaire). Bot Mag Tokyo 9:240–239
40. Hirase S (1918) Nouvelles recherches sur la fécondation et l'embryogénie du *Ginkgo biloba*. Bot Mag Tokyo 32:139–144

41. Hirase S (1918) Further studies on the fertilization and embryogeny in *Ginkgo biloba* (in Japanese). Bot Mag Tokyo 32:83–108
42. Ikeno S (1901) Contribution a l'étude de la fécondation chez le *Ginkgo biloba*. Ann Sc Nat Bot ser. 8 13:305–316
43. Shimamura T (1928) On the formation of proembryo of *Ginkgo biloba* L (in Japanese). Bot Mag Tokyo 42:71–76, 77–82
44. Shimamura T (1931) A note on the mitotic division in the proembryo of *Ginkgo*, with special reference to a chromatin-elimination (in Japanese with English resume). Bot Mag Tokyo 45:525–530
45. Shimamura T (1935) Zur Cytologie des Befruchtungsvorganges bei *Cycas* und *Ginkgo* unter Benutzung der Feulgenschen Nuklearreaktion. Cytologia 6:465–473
46. Favre-Duchartre M (1958) *Ginkgo*, an oviparous plant. Plantmorphology 8:377–390
47. De Sloover-Colinet A (1962) Chambre pollinique et gamétophyte mâle chez *Ginkgo biloba*. Cellule 64:129–145
48. Camefort H (1965) L'organisation du protoplasme dans le gamète femelle, ou oosphère, du *Ginkgo biloba* L. J Microscopie 4:531–546
49. Soma S, Kaneko F, Kasai H, Yokomizo C (1986) Fertilization in *Ginkgo biloba* (in Japanese). Ann Rep Fac Edu Bunkyo Univ No. 20:67–74
50. Sterling C (1963) Structure of the male gametophyte in gymnosperms. Biol Rev 38:167–203
51. Jeffrey EC (1914) Spore conditions in hybrids and the mutation hypothesis of de Vries. Bot Gaz 58:322–336
52. Chamberlain CJ (1898) The homology of the blepharoplast. Bot Gaz 26:431–435
53. Webber H (1897) Peculiar structures occurring in the pollen tube of *Zamia*. Bot Gaz 23:453–459
54. Mizukami I, Gall J (1966) Centriole replication II. Sperm formation in the fern, *Marsilea*, and the cycad, *Zamia*. J Cell Biol 29:97–111
55. Chamberlain CJ (1909) Spermatogenesis in *Dioon edula*. Bot Gaz 47:215–236
56. Kuwada Y, Maeda T (1929) On the structure of the cytoplasm around the blepharoplast in *Cycas revoluta*. Mem Coll Sci Kyoto Imp Univ 4:165–174
57. Herzfeld S (1928) Über die Kernteilungen in Proembryo von *Ginkgo biloba*. Jahrb Wiss Bot 66:264–294
58. Chamberlain CJ (1912) Morphology of *Ceratozamia*. Bot Gaz 53:1–19
59. Wolniak SM (1976) Organelle distribution and apportionment during meiosis in the microsporocyte of *Ginkgo biloba* L. Amer J Bot 63:251–258
60. Hori T, Miyamura (1991) Analysis of the releasing process of the sperms in *Cycas revoluta* by high-speed video (in Japanese). Proc of 56th Ann Meet Bot Soc Jpn, 13–15 September, Tokyo, p 125
61. Foster AS, Gifford EM (1974) Comparative morphology of vascular plants, 2nd edn. WH Freeman, San Francisco, 751pp
62. Ferguson MC (1901) The development of the egg and fertilization in *Pinus Strobus*. Ann Bot 15:435–479
63. Hori T, Miyamura S (1990) Analysis of the fertilization process of *Cycas revoluta* by high-speed video (in Japanese). Proc of 55th Ann Meet Bot Soc Jpn, 2–4 October, Shizuoka, p 137
64. Schnarf K (1933) Embyologie der Gymnospermen. In: Hand Pflanzenana 10/2, Berlin
65. Hori T (1989) Ultrastructural reinvestigation on fertilization of the gymnosperm, *Cycas revoluta* (in Japanese). Proc of 54th Ann Meet Bot Soc Jpn, 27–29 September, Sendai, p 174

66. Chamberlain CJ (1916) *Stangeria paradoxa*. Bot Gaz 61:353–372
67. Lawson A (1926) A contribution to the life history of *Bowenia*. Trans Roy Soc Edin 54:357–393
68. Swamy BGL (1948) Contributions to the life history of a cycas from Mysore (India). Amer J Bot 35:77–88
69. de Silva BLT, Tambiah MS (1952) A contribution to the life history of *Cycas rumphii* Miq. Ceylon J Sci (A) 12:223–244
70. Gifford EM, Foster AS (1989) Morphology and evolution of vascular plants, 3rd edn. WH Freeman, New York, 626pp
71. Hori T, Miyamura S (1993) Cytological and video analysis of fertilization in *Cycas revoluta*. Abs of XV Int Bot Con, 28 August-3 September, Yokohama, p 96
72. Ikeno S (1898) Untersuchungen über die Entwickelung der Geschlechtsorgane und den Vorgang der Befruchtung bei *Cycas revoluta*. Jahrb Wiss Bot 32:557–602
73. Webber HJ (1901) Spermatogenesis and fecundation of *Zamia*. US Dept Agricul Ber Plant Ind Bull Washington No. 2:1–92
74. Brough P, Taylor MH (1940) An investigation of the life cycle of *Macrozamia spiralis* Miq. Proc Linn Soc New South Wales 65:494–524
75. Blackman VH (1898) On the cytological features of fertilization and related phenomena in *Pinus silvestris* L. Phil Trans Roy Soc London Ser B 190:395–426
76. Jäger L (1899) Beiträge zur Kenntnis der Endospermbildung und zur Embryologie von *Taxus baccata* L. Flora 86:241–288
77. Arnoldi W (1900) Beiträge zur morphologie der Gymnospermen: III. Embryogenie von *Cephalotaxus Fortunei*. Flora 87:46–63
78. Miyake K (1903) Contribution to the fertilization and embryogeny of *Abies balsamea*. Beih Bot Centr 14:134–144
79. Hirase S (1918) On the fertilization of *Pinus Thumbergii* Parl. Bot Mag Tokyo 32:343–356
80. Norstog K, Overstreet R (1965) Some observations on the gametophytes of *Zamia integrifolia*. Phytomorphology 15:46–49
81. Norstog K (1975) The motility of cycad spermatozoids in relation to structure and function. In: Duckett JG, Racey PA (eds) The biology of the male gamete. Academic, London, pp 135–142
82. Bryan GS, Evans RI (1956) Chromatin behavior in the development and maturation of the egg nucleus of *Zamia umbrosa*. Amer J Bot 43:640–646
83. Bryan GS, Evans RI (1957) Types of development from the central nucleus of *Zamia umbrosa*. Amer J Bot 44:404–415
84. Baird AM (1939) A contribution to the life history of *Macrozamia reidlei*. J Roy Soc W Aust 25:153–175
85. Lang WH (1900) Studies in the development and morphology of cycadean sporangia: II. The ovule of *Stangeria paradoxa*. Ann Bot 14:281–306
86. Barner H, Christiansen H (1960) The formation of pollen, the pollination mechanism, and the determination of the most favourable time for controlled pollinations in *Larix*. Silv Gen 9:1–11
87. Camefort H (1967) Observations sur les mitochondries et les plastes de la cellule centrale et de l'oosphère du *Larix decidua* Mill (*Larix europea* D.C.). C R Acad Sc Paris 265:1293–1294
88. Camefort H (1967) Fécondation et formation d'un néocytoplasme chez le *Larix decidua* Mill. (*Larix europea* D.C.). C R Acad Sc Paris 265:1784–1787

89. Owens JN, Molder M (1979) Sexual reproduction of *Larix occidentalis*. Can J Bot 57:2673–2690
90. Allen GS (1943) The embryogeny of *Pseudotsuga taxifolia* (Lamb.) Britt. Amer J Bot 30:655–661
91. Allen GS (1946) Embryogeny and development of the apical meristem of *Pseudotsuga*. I. Fertilization and early embryogeny. Amer J Bot 33:666–677
92. Owens JN, Morris SJ (1990) Cytological basis for cytoplasmic inheritance in *Pseudotsuga menziesii*. I. Pollen tube and archegonial development. Amer J Bot 77:433–445
93. Sugiura Y (1943) Embryological observations on *Keteleeria davidiana* Beissner var. *formosana* Hayata. Sci Rep Tohoku Imp Univ 17:215–222
94. Wang FH (1948) Life history of *Keteleeria*. I. Strobili, development of the gametophytes and fertilization in *Keteleeria evelyniana*. Amer J Bot 35:21–27
95. Haupt AW (1941) Oögenesis and fertilization in *Pinus lambertiana* and *P. monophylla*. Bot Gaz 102:482–498
96. Shimamura T (1956) Cytochemical studies on the fertilization and proembryo of *Pinus Thunbergii* (in Japanese). Bot Mag Tokyo 69:524–529
97. Konar RN, Ramchandani S (1958) The morphology and embryology of *Pinus wallichiana* Jack. Phytomorphology 8:328–346
98. McWilliam JR, Mergen F (1958) Cytology of fertilization in *Pinus*. Bot Gaz 119:246–249
99. Konar RN (1962) Some observations on the life history of *Pinus geradiana* Wall. Phytomorphology 12:196–201
100. Camefort H (1965) Étude en microscopie électronique de la fécondation et de la première mitose du zygote chez le *Pinus laricio* Poir. var. *austriaca* (*P. nigra* Arn.). C R Acad Sc Paris 261:4479–4482
101. Lill BS (1976) Ovule and seed development *in Pinus radiata*: postmeiotic development, fertilization, and embryogeny. Can J Bot 54:2141–2154
102. Hutchinson AH (1915) Fertilization in *Abies balsamea*. Bot Gaz 60:457–472
103. Sugiura Y (1947) The embryogeny of *Abies firma* Siebold et Zuccarini. Bot Mag Tokyo 60:58–62
104. Singh H, Owens JM (1981) Sexual reproduction in subalpine fir (*Abies lasiocarpa*). Can J Bot 59:2650–2666
105. Miyake K, Yasui K (1911) On the gametophyte and embryo of *Pseudolarix*. Ann Bot 25:639–647
106. Miyaka K (1903) On the development of the seuxl organs and fertilization in *Picea excelsa*. Ann Bot 17:351–372
107. Owens JN, Molder M (1979) Sexual reproduction of white spruce (*Picea glauca*). Can J Bot 57:152–169
108. Owens JN, Molder M (1980) Sexual reproduction of Sitka spruce (*Picea sitchensis*). Can J Bot 58:886–901
109. Murrill WA (1900) The development of the archegonium and fertilization in the hemlock spruce (*Tsuga canadensis*, Carr.). Ann Bot 14:583–607
110. Stanlake EA, Owens JN (1974) Female gametophyte and embryo development in western hemlock (*Tsuga heterophylla*). Can J Bot 52:885–893
111. Owens JN, Molder M (1975) Sexual reproduction of mountain hemlock (*Tsuga mertensiana*). Can J Bot 53:1811–1826
112. Chowdhury CR (1961) The morphology and embryology of *Cedrus deodara* (Roxb.) Loud. Phytomorphology 11:283–304

Ultrastructure of *Ginkgo biloba*

René Rohr

1 Introduction

This chapter describes aspects of *Ginkgo* biology that were investigated at the ultrastructural level using transmission electron microscopy (TEM) and scanning electron microscopy (SEM). TEM techniques were mainly used in the study of male and female organ ontogeny and of gametophyte development in situ and in vitro, as well as in studies of secretory structures and photosynthetic apparatus. Studies of mycorrhizae were done using both TEM and SEM; leaf surface, cuticle structure, and resistance to air pollution were investigated with SEM only. Generally, the material was prepared and stained according to standard procedures, except in the case of mycorrhizae for which specifically adapted techniques were used to improve SEM observation of the endophyte in situ as well as for the cytochemical localization of phenols with TEM. Most SEM was carried out on air-dried material coated with gold. TEM investigations were usually performed on material fixed in glutaraldehyde-osmium.

2 Sexual Reproduction

In this chapter, we are mainly concerned with the successive stages of ontogeny that have been studied at the ultrastructural level. See chapters by Friedman and Gifford, Soma, Hori and Miyamura for a complete review of sexual reproduction.

2.1 Pollen Ontogeny

2.1.1 Microsporogenesis

The pollen of *Ginkgo* is shed at the four-celled stage. It contains two prothallial cells, a vegetative cell, and a generative cell. The most interesting ultrastructural

Interactions Plantes-Champignons et Micropropagation, Bât. 405, Université Claude Bernard Lyon 1, 69622 Villeurbanne Cedex, France

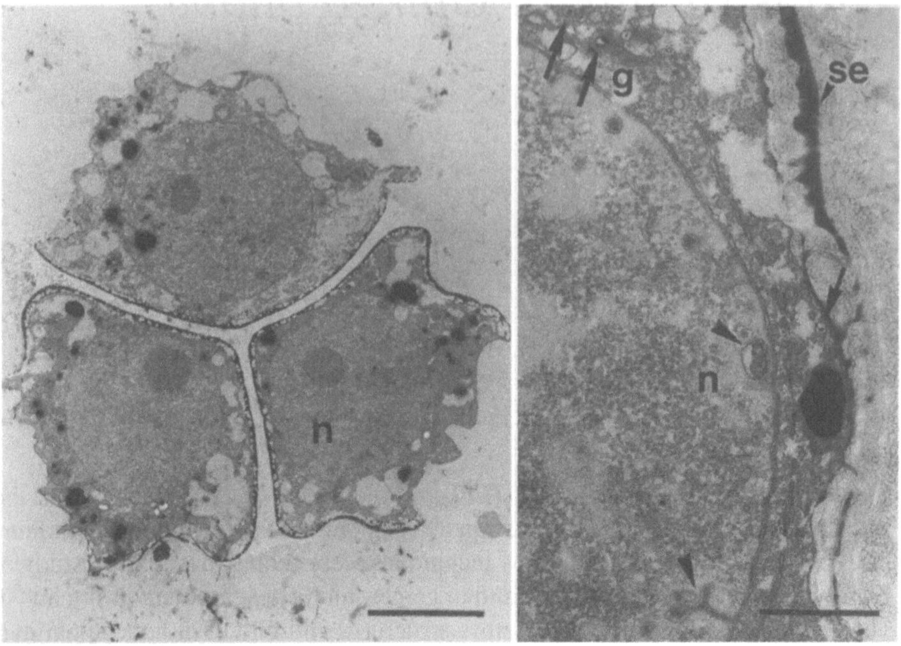

a b

FIG. 1a,b. Early stage of exine formation around microspores. **a** Sexine deposition is limited to the internal cell walls of the tetrad. Scale bar, 5 μm. **b** Detail shows nuclear membrane invaginations (*arrowheads*) and dictyosome vesicles containing osmiophilic products (*arrows*). *g*, Golgi body; *n*, nucleus; *se*, sexine. Scale bar, 1 μm. (From [1], with permission)

feature of microsporogenesis is exine deposition [1]. At the early tetrad stage, sexine deposition is limited to the inner cell walls of the tetrad (Fig. 1a). The nuclear envelope and Golgi apparatus show a particularly elevated activity, which is probably related to sexine deposition (Fig. 1b). The subsequent nexine deposition involves the entire surface of the microspore.

The pollen grains of cycads and *Ginkgo* are boat shaped, monosulcate, and bilaterally symmetrical (Fig. 2a; [2]), but notable differences exist regarding pollen surface ornamentation and exine stratification. Moreover, the modalities and chronology of exine deposition are different in each of the groups [3]. Inside the longitudinal slit of the dehydrated *Ginkgo* pollen grain, the sexine is lacking, the only components of the sporoderm being the two nexine layers (Fig. 2b; arrowheads). During the germinating process the pollen tube bursts out of this thinner area [1]. Reticulate ornamentations on the inner surface of the germinal zone have been reported from SEM observations but not explained [4]. In our study, the ornamentations do not appear as it would be expected in this area of the dry pollen grain (Fig. 2b; arrowheads).

a

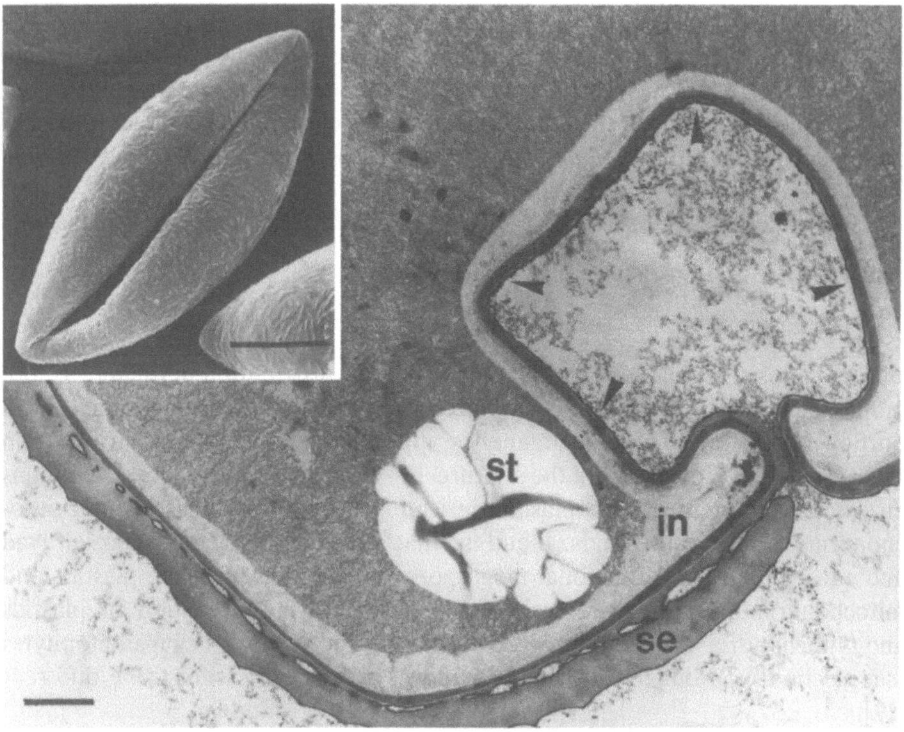

b

FIG. 2a,b. Dehydrated pollen of *Ginkgo*. **a** General view. Exine ornamentation revealed by scanning electron microscopy (SEM) consists of short reticulate tufts. **b** Transverse section shows longitudinal slit where sexine (*se*) is lacking (*arrowheads*). *st*, starch; *in*, intine layer. Fixation, $KMnO_4$. Scale bars, 1 µm. (From [1], with permission)

2.1.2 Spermatogenesis

The spermatogenous cell is produced by division of the generative cell. The presence of plasmodesmata between the spermatogenous cell and the second prothallial cell lends support to a new interpretation of the type of division found in the generative cell, only known to occur in cycad spermatogenesis and in antheridial development of leptosporangiate ferns [5].

The existence of a flagellar apparatus in sperm cells is one of the most interesting features of *Ginkgo* biology. Just before mitosis, the sperm mother cell contains a large central nucleus as well as two blepharoplasts and two types of cytoplasmic bodies, the chemical composition of which has not been determined. One of these is represented by two large osmiophilic globules [6, 7] and the other by a fibrogranular body [8]. The blepharoplasts originate from a double hemispherical structure. Their origins are de novo in the spermatogenous cell, as in the case for the osmiophilic and fibrogranular bodies [8].

The ultrastructure of the sperm cell and its flagellar apparatus was studied in sperm cells recovered from ovules 5 months after pollination [9]. The flagellar apparatus is a microtubular system made of numerous flagella connected through basal bodies to a multilayered structure (MLS). The flagella are extracellular, but a long transition region of the basal bodies is intracellular [9]. The organization of the flagellar apparatus of *Ginkgo* is similar to that existing in the cycad *Zamia* [10].

2.2 Ovule Ontogeny

2.2.1 Megasporogenesis

Before meiosis, organelles exhibit polarity in their distribution in the megaspore mother cell (MMC) [11]. In the immature MMC, the nucleus occupies the central portion of the cell. The cell organelles appear to be randomly distributed in the cytoplasm. However, in the mature MMC, the micropylar half contains an extensive system of endoplasmic reticulum. In contrast, plastids and mitochondria migrate to the chalazal end, so that after meiosis I, the chalazial dyad and consequently the functional megaspore receive almost all the plastids and mitochondria. It is quite remarkable to note that a segregation of plastids and mitochondria has been observed in the cultured *Ginkgo* megagametophytes (see Section 3.2), but the distribution and its significance seem to be different [12].

The change in ultrastructure of the megaspore membrane during development of *Ginkgo* ovules was studied using TEM and was compared with megaspore membranes of the Middle Jurassic *Ginkgo yamensis*. The megaspore membrane of mature ovules of *G. biloba* developed differently than that of the fossil [13].

2.2.2 Megagametogenesis

The localized distribution of endoplasmic reticulum (ER) around plastids was reported in different cell types of developing ovules of *Ginkgo*. It was observed in the tapetum and in the central cell of the archegonium [14], in the megagametophyte [14, 15], and in the egg [16]. Such patterns appear to be associated with plastid division. A close association between plastids and ER was also observed in haploid tissue derived from female gametophyte culture [17], but in this case the phenomenon did not appear to be connected with plastid partitioning.

The cytoplasm of the egg is abundantly filled with microbody-like inclusions, the significance of which remains to be established [16]. At maturity, the megagametophyte tissues show an extreme degree of cell differentiation. A thin outer layer of small isodiametric cells, filled with lipid droplets, surrounds a cortical zone rich in proteolipids and a central region filled with starch [18, 19]. The micropylar region contains choroplasts with an unique granum.

3 Gametophyte Culture

The in vitro culture of *Ginkgo* gametophytic tissue has been carried out for two different purposes: (1) to study spermatogenesis continuously, which is much too difficult in vivo because of the inaccessibility of male gametophytes in the ovular tissue as well as the problem associated with the time required to produce gametes, 3.5–5 months [2, 7, 9, 18]; and (2) to analyze the gametophyte potential, such as embryogenic capacity, which has been recently demonstrated with microspores [20] and female prothalli collected 8–11 weeks after pollination [21].

3.1 Microgametophyte Culture

In vitro culture is an efficient alternative to in vivo study. Early stages of pollen germination, including haustorium ultrastructure as well as spermatogenous cell ultrastructure and flagellar apparatus initiation, have been characterized with TEM [1, 22]. The improvement of culture conditions should increase the number

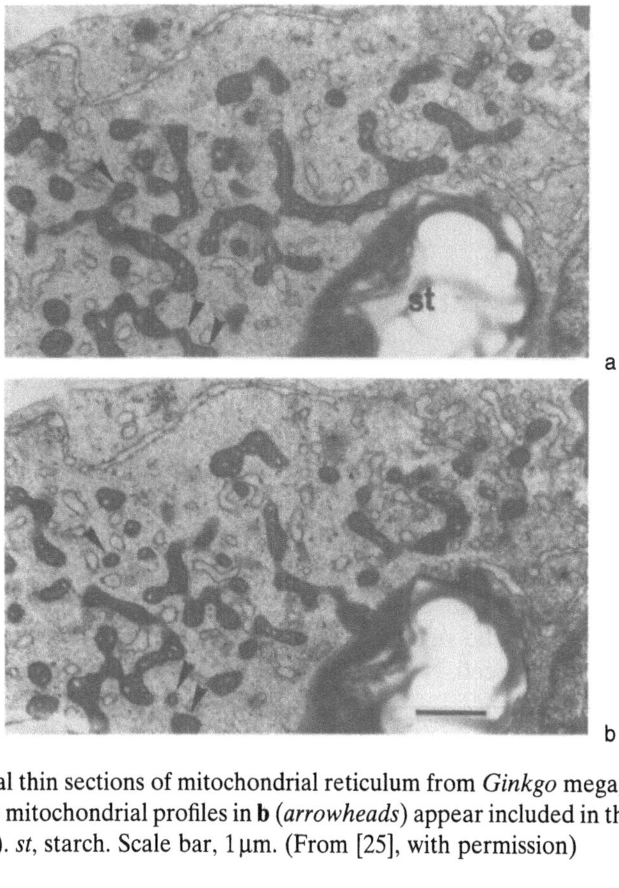

FIG. 3a,b. Serial thin sections of mitochondrial reticulum from *Ginkgo* megagametophyte culture. Isolate mitochondrial profiles in **b** (*arrowheads*) appear included in the network in **a** (*arrowheads*). *st*, starch. Scale bar, 1 μm. (From [25], with permission)

of mature sperm cells in good condition that can be completely analyzed at the level of their kinetic apparatus.

3.2 Megagametophyte Culture

Even though most cell types in the mature endosperm of *Ginkgo* are specialized for storage of various products, they are nevertheless able to dedifferentiate and to proliferate in culture [23]. A subculturable haploid tissue was produced by the female gametophyte [24], and changes in its ultrastructure were followed during the first 10 months of development [17]. Chloroplasts contained a unique granum that occasionally included a small starch grain. This type of organization is not found in the sporophytic tissue and seems therefore to be linked to the haploid stage [17]. A novel type of mitochondrial reticulum, never before seen in vascular plants, was seen in a 10-month-old tissue line. Using serial thin sections (Fig. 3) and thick sections [25], a composite image was created. The presence of similarly branched mitochondria was also observed in the stem secretory cavities of *Ginkgo*, which would suggest that mitochondria in the sporophytic tissue of *Ginkgo* can fuse and build complexes as well [26]. The ability of the haploid tissue of *Ginkgo* to differentiate tracheary elements reveals developmental possibilities that have been reported in fern prothallia and in cultured gametophytes of gymnosperms but which have never been expressed in natural conditions (see [17]). The results recently obtained in gametophyte embryogenesis [20, 21] are promising, suggesting that the lack of continued organized growth reported earlier [27] is most likely attributable to inappropriate culture conditions rather than to deleterious recessive genes expressed at the haploid stage.

4 Secretory Apparatus

Ginkgo is of particular importance in pharmaceutical research because of the number and diversity of secondary metabolites produced by its tissues (see chapters by Yoshitama, Fünfgeld, Juretzek, and Braquet, this volume).

In a preliminary study of a green tissue culture from the female gametophyte of *Ginkgo*, the presence of several specialized cell types, including resin cells, was reported [28]. From TEM observations of parenchyma cells made at different stages in resin formation, it was concluded that resin production started with the deposition of electron-dense material in the peripheral cytoplasm, close to the cell wall. Eventually the opaque deposit occupied large areas of the cell and finally filled the entire cell. Neither the chemical nature nor the site of biosynthesis was known, however.

Because the localization of these phytochemicals at the tissue level and their relation with the secretory apparatus of the plant were not clear [26], multidisciplinary studies combining chemical analysis and ultrastructure attempted to identify the nature of these products and characterize the ultrastructural organization of the secretory cells. Secretory cavities occur in various organs

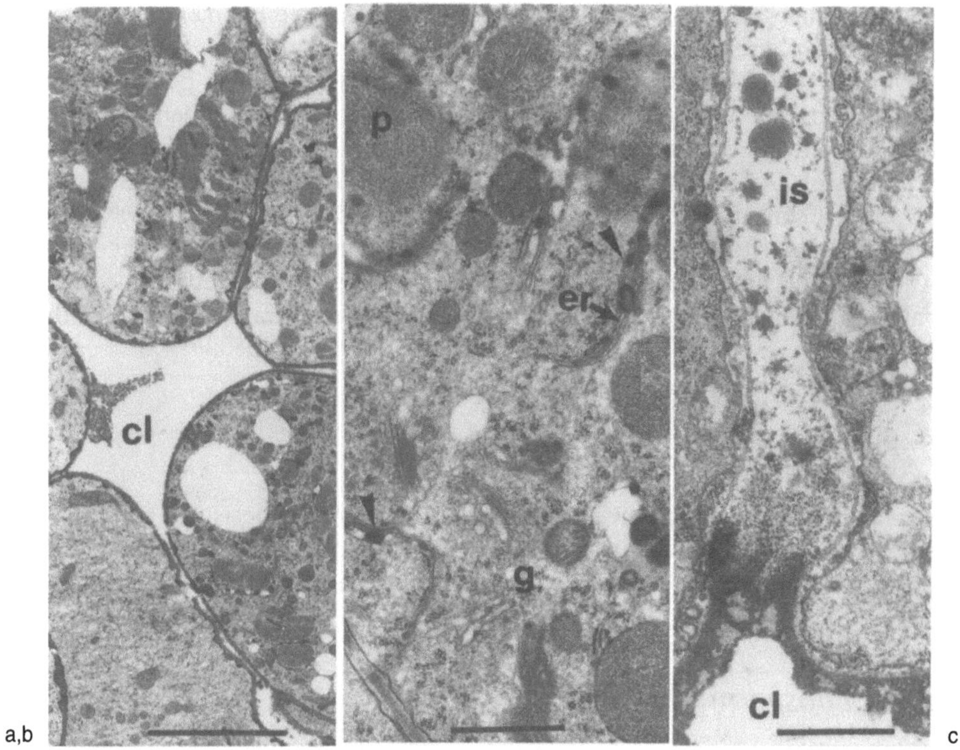

FIG. 4. **a** Survey picture of the secretory cavity in transverse section. *cl*, cavity lumen. Scale bar, 5 μm. **b** Organization of the cytoplasm of a secretory cell. Osmiophilic products are visible in the endoplasmic reticulum (ER) luminal spaces (*arrowheads*). *p*, plastid; *er*, endoplasmic reticulum (*arrow*); *g*, Golgi body. **c** Degradation of the middle lamellae and release of products in the intercellular space (*is*) and cavity lumen (*cl*). Scale bars in **b** and **c**, 1 μm. (From [26], with permission)

of the plant, i.e., leaves, stem, ovule fleshy integument, and microsporophylls. Their organization and differentiation were studied in the stem with light microscopy on resin-embedded material, and the ultrastructural characteristics of the active epithelial cells were described [26]. It was demonstrated that the process of secretory pocket formation begins schizogenously (Fig. 4a) and ends lysogenously (Fig. 4c), although in other organs, such as microsporophylls, the formation of the cavities was merely lysogenous [29]. As regards the ultrastructure of stem secretory cells, two features were reported. The first, leucoplasts and paired endoplasmic reticulum cisternae, built complexes that played a key role in the synthesis of the secretory product. Osmiophilic droplets, which were assumed to be this lipophilic secretory product, were conspicuous in the space of swollen ER cisternae (Fig. 4b, arrowheads). The second feature, a lytic process starting in the middle lamellae, released polysaccharides in the intercellular spaces or in the cavity lumen (Fig. 4c). A similar observation was made in resin duct cells of maritime pine [30], which suggested that the double-paired ER membranes were the system of discharge from the cell for the secretory product.

More recently, Cartayrade et al. [31] and Cartayrade (1996, personal communication) demonstrated the existence of two different biosynthetic pathways of isopentenylpyrophosphate (IPP), the precursor of terpenes, involved in the synthesis of *Ginkgo* sitosterol 1 and ginkgolides. This would assume a strict compartmentalization of the cell processes involved in the two pathways. The biogenesis of sitosterol 1 is expected to take place in the cytosolic phase while the IPP units involved in ginkgolide synthesis would likely reside in organelles, perhaps plastids. There is a need for ultrastructural verification of this hypothesis.

5 Photosynthetic Apparatus

The ability to carry out light-independent chlorophyll synthesis and chloroplast development is not universal among gymnosperms. *Ginkgo* is completely dependent on light for chlorophyll synthesis and chloroplast development [32, 33].

5.1 Endosperm

The megagametophyte of *Ginkgo* is the only ovulate plant gametophyte to contain chlorophyll. Chloroplasts can be seen in the micropylar region [34] and in the cells adjacent to the surface [19]. The photosynthetic apparatus of the plastids found in situ in the peripheral region of the megagametophyte consists of a unique granum with as many as 40 thylakoid membranes. Plastids located in deeper tissues appeared less differentiated, had fewer thylakoid membranes, and contained prolamellar bodies [34].

According to Friedman and Goliber [35], the ability of the gametophyte of *Ginkgo* to produce chlorophyll and engage in photosynthesis results from a combination of two factors, the absorption of sufficient light (20–70 µmol photons $m^{-2} s^{-1}$) and the ability of the gametophyte to respond.

5.2 Sporophyte

Chloroplast development was studied at the molecular level and with TEM during greening of *Ginkgo* dark-grown seedlings transferred to constant white light [36]. This study showed that the upper shoot portion of greening *Ginkgo* seedlings followed a developmental program different from that of the lower shoot portion. Etioplasts were visible in the upper part of dark-grown seedlings. Between day 7 and day 10 of greening, there was an increase in the grana stacking and in starch accumulation of chloroplasts. Conversely, a substantial loss of cell organelles, including chloroplasts, was observed in the lower portion by 10 days of greening. The lower portion failed to green and continued a program of development into woody tissue.

6 Mycorrhizae

The morphology of *Ginkgo* vesicular-arbuscular mycorrhizae (VAM) was investigated in natural and controlled conditions using light microscopy [37]. The most interesting morphological feature was the abundance of intracellular hyphae; intercellular hyphae were rarely observed. Likewise, in the other gymnosperms only intracellular hyphae were found. It was suggested that in this group of plants the fungus penetrated from cell to cell (Fig. 5a, arrowheads), while spreading of infection via intercellular hyphae was a common feature in angiosperms.

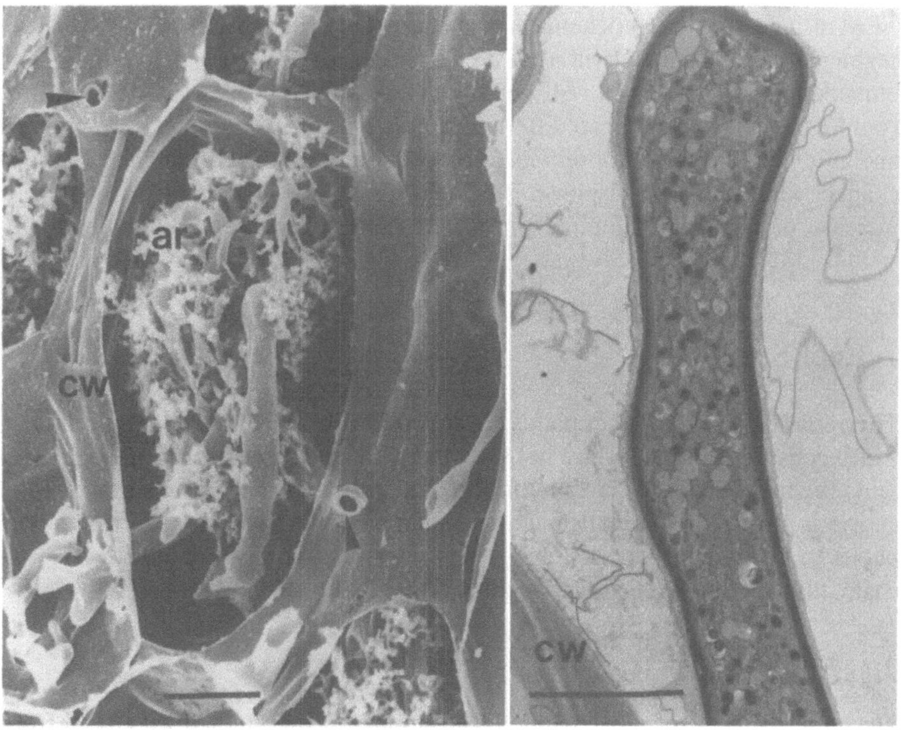

FIG. 5. **a** SEM view of arbuscules of *Glomus caledonium* in a *Ginkgo* root. Note the cell-to-cell passage of the mother hypha (*arrowheads*). *ar*, arbuscule; *cw*, cell wall. Scale bar, 10 µm. **b** A coiled hypha colonizes a cortical cell in the outer part of the root. The host cell is characterized by thick walls (*cw*). The hypha is surrounded by the host plasma membrane. The interface space is filled with an electron-dense material. The fungus is strongly electron opaque. Scale bar, represents 5 µm. (Courtesy of Bonfante-Fasolo, unpublished)

SEM and TEM were used as complementary methods for studying the ultrastructural characteristics of the infection of *Ginkgo* plantlet roots infected by four different species of *Glomus* [38]. A special procedure was adopted to obtain internal SEM views of infected root cells. Fixed roots were cut in a freezing microtome and treated with a sodium hypochlorite solution to remove the host cell protoplasm (Fig. 5a). That paper [38] describes in detail the pattern of *Ginkgo* root colonization. After penetration through the rhizoderm, the endophyte developed enlarged coiled hyphae in the outer cortex (Fig. 5b). Two types of unidentified electron-dense granules and bacteria-like inclusions were observed in the fungal cytoplasm. The nature of these remains unknown. In the inner cortex, coiled hyphae also penetrated inside (Fig. 5a), giving rise to lateral branches that differentiated into intercalary arbuscules. This pattern of arbuscule formation differs significantly from that in angiosperms, in which arbuscules originate from dichotomous branches of intercellular hyphae (see references in [38]).

Cell wall-bound phenols, which are involved in resistance to fungal enzymes as well as being a barrier against fungal penetration, were localized in *Ginkgo* with TEM, using a specific cytochemical procedure. Glutaraldehyde- (GA-) fixed root segments were incubated in 10% $FeSO_4$ for 2 days before dehydration and embedding [39]. These cell wall-bound phenols only appeared to be constitutive compounds in *Ginkgo*, as no difference in phenol distribution between infected and uninfected roots was observed. In coil-bearing cells the host did not show any change, while the host response is evident in the arbuscule-infected cells; e.g., increase of host plasmalemma surface was interpreted as a compatibility response [38].

7 Leaf Surface

The aerial surfaces of higher plants are covered with an extracellular lipid cuticle, which constitutes the interface between the plant and its environment and represents the first barrier to the penetration of xenobiotics. The outer surface of the cuticle is covered by epicuticular waxes that can take many forms, from amorphous to crystalline deposits, and which consist of complex mixtures of long-chain aliphatic and cyclic components [40].

7.1 Epicuticular Wax

The cuticle surface of the expanded leaf of *Ginkgo* was studied with SEM from air-dried and gold-coated material and was correlated with the chemical composition of epicuticular wax [41, 42]. A continuous wax layer could be observed on both sides of the leaf surface. A dense arrangement of rodlike crystalloids covered adaxial and adaxial leaf surface. They were identified as small tubules, 0.5–1.1 μm long and 0.1–0.2 μm wide [40, 41]. The opening was clearly visible at a magnification of ×20000. The presence of similar wax tubules was also reported on the ovule surface of *Ginkgo* by Barthlott and Wollenweber (1981; cited by [41]). The epicuticular wax of *Ginkgo* has an intermediate composition between that of a conifer and that of a deciduous broadleaf tree. It contains many of the usual wax lipids of angiosperm trees and, in addition to these, the secondary alcohol nonacosan-10-ol, which is a characteristic component of conifer needle waxes. A clear correlation between nonacosan-10-ol and the occurrence of crystalline wax tubules was observed, as also reported previously for conyferyl waxes [41].

7.2 Resistance to Air Pollution

To see whether ambient air pollutants induce changes in the leaf wax structure, the chemical and surface characteristics of various conifer needles and *Ginkgo* leaves exposed to urban and industrial air pollution were compared [43]. Needles of two pine species and of Norway spruce and *Ginkgo* leaves were harvested in October and observed with SEM. A degradation of stomatal wax crystalloids was found in the conifer samples, while *Ginkgo* leaves collected in areas of heavy road traffic did not show any significant damage of wax structure, in spite of higher amounts, i.e., five- to tenfold, of sulfur in the foliage. This suggests to the authors that maidenhair trees are resistant to air pollution and therefore could be planted in urban areas along avenues and at roadsides even where other trees die [27].

8 Conclusion

The unique position of *Ginkgo* in plant evolution is underscored by a number of ultrastructural features summarized in this chapter. The occurrence of multiflagellate sperm in an ovulate plant is extremely rare, shared only with cycads. The pollen morphology and ultrastructure is in many ways typically gymnosperm, but the ontogeny of the sperm in the ovular tissue is of an intermediate unique siphonogamous type. Fertilization by motile sperm is reminiscent of that of lower archegoniates. Another significantly different aspect is the capacity of the female gametophyte of *Ginkgo* to photosynthesize.

In spite of the number of ultrastructural studies carried out on *Ginkgo*, many questions remain unanswered. The tridimensional structure of the flagellar apparatus remains to be analyzed and interpreted. Two approaches that would be useful include the immunolocalization of flagellar proteins and reconstruction of the apparatus using serial sections, perhaps using pollen cultured in vitro. Other applications of cytoimmunological methods could include studies on cell compartmentalization of secondary metabolite biosynthesis, as well as the characterization of a number of unidentified cell inclusions observed during male and female organ differentiation and in mycorrhizae.

Acknowledgments. We are grateful to the Beaufour laboratories, France, for their support.

References

1. Rohr R (1977) Etude comparée de la formation de l'exine au cours de la microsporogénèse chez une gymnosperme (*Taxus baccata*) et une préphanérogame (*Ginkgo biloba*). Cytologia (Tokyo) 42:157–167
2. Wang FH (1990) Microsporogenesis in cycads and *Ginkgo*. In: Microspores: evolution and ontogeny. Academic, New York, pp 147–155
3. Audran JC (1987) Comparaison des infrastructures exiniques et des modalités de l'ontogenèse pollinique chez les Cycadales et les Ginkgoales actuelles (préspermaphytes). Bull Soc Bot Fr Actual Bot 134:9–18
4. Sahashi N, Ueno J (1986) Pollen morphology of *Ginkgo biloba* and *Cycas revoluta*. Can J Bot 64:3075–3078
5. Friedman WE, Gifford EM (1988) Division of the generative cell and late development in the male gametophyte of *Ginkgo biloba*. Amer J Bot 75:1434–1442
6. Gifford EM, Lin J (1974) Etude des cellules spermatogènes de *Ginkgo biloba*, et particulièrement du blépharoplaste (mastigosome). C R Acad Sci Paris 278:2513–2515
7. Gifford EM, Lin J (1975) Light microscope and ultrastructural studies of the male gametophyte in *Ginkgo biloba*: the spermatogenous cell. Amer J Bot 62:974–981
8. Gifford EM, Larson S (1980) Developmental features of the spermatogenous cell in *Ginkgo biloba*. Amer J Bot 67:119–124
9. Li Y, Wang FH, Knox RB (1989) Ultrastructural analysis of the flagella apparatus in sperm cells of *Ginkgo biloba*. Protoplasma 149:57–63
10. Norstog K (1974) Fine structure of the spermatozoid of *Zamia*: the vierergruppe. Amer J Bot 61:449–456
11. Stewart KD, Gifford EM (1967) Ultrastructure of the developing megaspore mother cell of *Ginkgo biloba*. Amer J Bot 54:375–383
12. Rohr R (1978) Existence d'un réticulum mitochondrial dans les cellules d'une culture de tissu haploïde d'une plante vasculaire. Biol Cell 33:88–92
13. Zhou Z (1993) Comparative ultrastructure of fossil and living ginkgoacean megaspore membranes. Rev Palaeobot Palynol 78:167–168
14. Fiordi AC, Maugini A (1972) Observation of certain plastids of the ovule of *Ginkgo biloba* L. Caryologia 25:189–197
15. Camefort H, Schaeffer J (1965) Différenciation et organisation des chloroplastes dans le parenchyme foliaire et dans le prothalle femelle, ou endosperme, du *Ginkgo biloba* L. C R Acad Sci Paris 260:6177–6180
16. Camefort H (1965) L'organisation du protoplasme dans le gamète femelle, ou oosphère, du *Ginkgo biloba* L. J Microsc (Oxf) 4:531–546
17. Rohr R (1978) Ultrastructure du tissu haploïde de *Ginkgo biloba* L. Z Pflanzenphysiol 87:365–378
18. Favre-Duchartre M (1956) Contribution à l'étude de la reproduction chez le *Ginkgo biloba*. Rev Cytol Biol Veg 17:1–218
19. Dexheimer J (1973) Etude ultrastructurale du gamétophyte femelle de *Ginkgo biloba*. I. Les cellules à réserves. Caryologia (Suppl) 25:85–89
20. Laurain D, Tremouillaux-Guiller J, Chenieux JC (1993) Embryogenesis from microspores of *Ginkgo biloba* L., a medicinal woody species. Plant Cell Rep 12:501–505

21. Fontanel A, Serraf I, Pétiard V (1994) Essais de régénération chez le *Ginkgo biloba* L.: aptitude à l'embryogenèse du mégagamétophyte selon le stade de développement. C R Acad Sci Paris 317:149–155
22. Rohr R (1980) Development in vitro du pollen de *Ginkgo biloba* L. Cytologia (Tokyo) 45:481–495
23. Rohr R (1980) Dédifférenciation en culture in vitro des cellules à réserves dans l'endosperme de *Ginkgo biloba*. Z Pflanzenphysiol 96:423–434
24. Rohr R (1977) Evolution en culture in vitro des prothalles femelles âgés chez le *Ginkgo biloba* L. Z Pflanzenphysiol 85:61–69
25. Rohr R (1979) Reconstruction tridimensionnelle du chondriome de *Ginkgo* en culture tissulaire; étude au moyen de coupes épaisses et de coupes fines sériées. Can J Bot 57:332–340
26. Cartayrade A, Bourgeois G, Balz JP, Carde JP (1990) The secretory apparatus of *Ginkgo biloba*: structure, differentiation and analysis of the secretory product. Trees 4:171–178
27. Rohr R (1989) Maidenhair tree (*Ginkgo biloba* L.). In: Bajaj YPS (ed) Biotechnology in agriculture and forestry, vol 5. Trees. II. Springer, Berlin Heidelberg New York
28. Tulecke W (1967) Studies on tissue cultures derived from *Ginkgo biloba* L. Phytomorphology 17:381–386
29. Ameele RJ (1980) Develomental anatomy of secretory cavities in the microsporophylls of *Ginkgo biloba*. Amer J Bot 67:912–917
30. Charon J, Launay J, Carde JP (1996) Spatial organization and volume density of leucoplasts in pine secretory cells. Protoplasma 138:45–53
31. Cartayrade A, Schwarz M, Jaun B, Arigoni D (1994) Detection of two independent mechanistic pathways for the early steps of isoprenoid biosynthesis in *Ginkgo biloba*. In: Proceedings of the 2nd symposium of the European network on plant terpenoids, 23–27 January 1994, Strasbourg
32. Mariani P, Rascio N (1982) Plastid ultrastructure in etiolated seedlings of *Ginkgo biloba* L. Caryologia 35:390–391
33. Rascio NP, Mariani P, Orsenigo M (1984) Photosynthetic apparatus differentiation in *Ginkgo biloba*. In: Sybesma C (ed) Advances in photosynthesis research, vol IV. Nijhoff/Junk, The Hague, pp 661–664
34. Dexheimer J (1973) Etude ultrastructurale du gamétophyte femelle de *Ginkgo biloba*. II. Les cellules du sommet prothallien. Rev Cytol Biol Veg 36:269–290
35. Friedman WE, Goliber TE (1986) Photosynthesis in the female gametophyte of *Ginkgo biloba*. Amer J Bot 73:1261–1266
36. Chinn E, Silversthorne J, Hohtola A (1995) Light-regulated and organ-specific expression of types 1, 2, and 3 light-harvesting complex *b* mRNAs in *Ginkgo biloba*. Plant Physiol (Rockv) 107:593–602
37. Fontana A (1985) Vesicular-arbuscular mycorrhizas of *Ginkgo biloba* L. in natural and controlled conditions. New Phytol 99:441–447
38. Bonfante-Fasolo P, Fontana A (1985) VAM fungi in *Ginkgo biloba* roots: their interactions at cellular level. Symbiosis 1:53–67
39. Codignola A, Verotta L, Spanu P, Maffei M, Scannerini S, Bonfante-Fasolo P (1989) Cell wall bound-phenols in roots of vesicular-arbuscular mycorrhizal plants. New Phytol 112:221–228
40. Backer EA (1982) Chemistry and morphology of epicuticular waxes. In: Cuttler DF, Alain EA, Price CE (eds) The plant cuticle. Academic, London, pp 139–165

41. Gülz PG, Müller E, Schmitz K, Marner FJ, Güth S (1992) Chemical composition and surface structures of epicuticular leaf waxes of *Ginkgo biloba*, *Magnolia grandiflora* and *Liriodendron tulipifera*. Z Naturforsch Sect C Biosci 47:516–526
42. Gülz PG (1994) Epicuticular leaf waxes in the evolution of the plant kingdom. J Plant Physiol 143:453–464
43. Kim YS, Lee JK (1990) Chemical and structural characteristics of conifer needles exposed to ambient air pollution. Eur J For Pathol 20:193–200

Analysis of Flagellar Movement in *Ginkgo biloba* Sperm by High Speed Video Microscopy

ROBERT W. RIDGE[1], TERUMITSU HORI[2], and SHIN-ICHI MIYAMURA[2]

1 Introduction

The discovery of a freely swimming sperm in *Ginkgo biloba* at the end of the nineteenth century by Sakugoro Hirase [1–3] was one of the most significant moments in botany, because it led to the establishment of *Ginkgo* as the true and sole link between the non-flowering primitive plants and the advanced seed plants. Hirase, as both amateur botanist and artist, was able to beautifully depict the sperm [2] and fertilization process in *Ginkgo* [3], and although he was able to describe the ornamentation of the spiral, from which emanate thousands of flagella, the instruments of the day prevented him from more fully understanding the complexity of the apparatus.

The term flagella is used through this article. However, the distinction between eucaryotic flagella and cilia is somewhat arbitrary, since flagella can behave as stiff oars in some organisms, solely or for part of the time (as in *Chlamydomonas* when it is moving forward, but not when moving in reverse), and cilia can produce undulating bends. However, both cilia and flagella work on the same molecular principle and the common method of classification is loosely based on size and number.

There is very little published about *Ginkgo* sperm. The overwhelming reason for this is the great difficulty in isolating them, as mentioned in the Material and Methods section below. However, work done on the ultrastructure of *Ginkgo* sperm has shown that it has many similarities to cycad sperm, and that the basis of the spiral is a multilayered structure that binds the flagellar basal units in a band [4, 5].

We have used a high-speed video recording system [6] to investigate flagellar movement in *Ginkgo*, and show that waves of flagellar beat are initiated at the center of the spiral while at the same time the whole apparatus oscillates on an

[1]Biology Department, Division of Natural Sciences, International Christian University, 3-10-2 Osawa, Mitaka, Tokyo 181, Japan
[2]Institute of Biological Sciences, University of Tsukuba, 1-1-1 Tennodai, Tsukuba, Ibaraki 305, Japan

axis defined by a line central through the apparatus and through the sperm. The oscillation seems to be the result of the power of the flagellar beat. Hirase was unfortunate indeed to have missed the beautiful movements we have seen on our video screen.

2 Material and Methods

Ginkgo sperm appear during a short period from late August to early September in the central part of Japan. For a given ovule, the sperm is present before fertilization for only a few hours. Mature ovules were collected from mid-August and many were examined daily to find the onset of the fertilization period, which is known to vary from year to year, place to place, and quite possibly tree to tree. Ovules were sliced and the gametophyte plus nucellus were pulled out of the integument. The nucellus was cut around the archegonial chamber and removed, revealing the archegonial neck cells. The removed nucellus was mounted in 10% sucrose for observation of the pollen–archegonial chamber complex and of male gametophytes. They were thus viewed as they might appear by looking upward from the archegonial chamber. Divided spermatogenous cells were examined for extended periods to observe their development and subsequent release into the pendulous end of the pollen tube, and then into the pollen chamber. The sperm would normally enter the archegonial chamber liquid at this stage.

Video images were obtained using the NAC Microscopic High Speed Video MHS-200 system (Yokohama, Japan), which uses stroboscopic light to achieve 200 frames per second, attached to a Nikon Optiphot microscope (Tokyo, Japan) equipped with Differential Interference Contrast optics. Video tapes were played back at slow speed for flagellar movement analysis. Figures were made by selecting representative sequences and photographing still frames directly from the monitor screen, using a Nikon 55 mm Macro lens and Kodak Technical Pan film (Rochester, NY).

3 Observations and Discussion

The *Ginkgo* sperm is a large single cell (about 80 microns in diameter, Fig. 1a) with a prominent spiral flagella apparatus that covers approximately 15% to 20% of the surface of the cell (Fig. 1). The flagella beat in waves and allow the sperm to move and maneuver by a cilia-like power stroke and by altering the angle of the spiral relative to the body of the sperm (Fig. 2). The sperm moves forward spiral first.

3.1 Spiral Formation

During spiral formation, there are waves of cytoplasmic movement within the cell, and the spiral appears to be made from a series of additions to the outer end

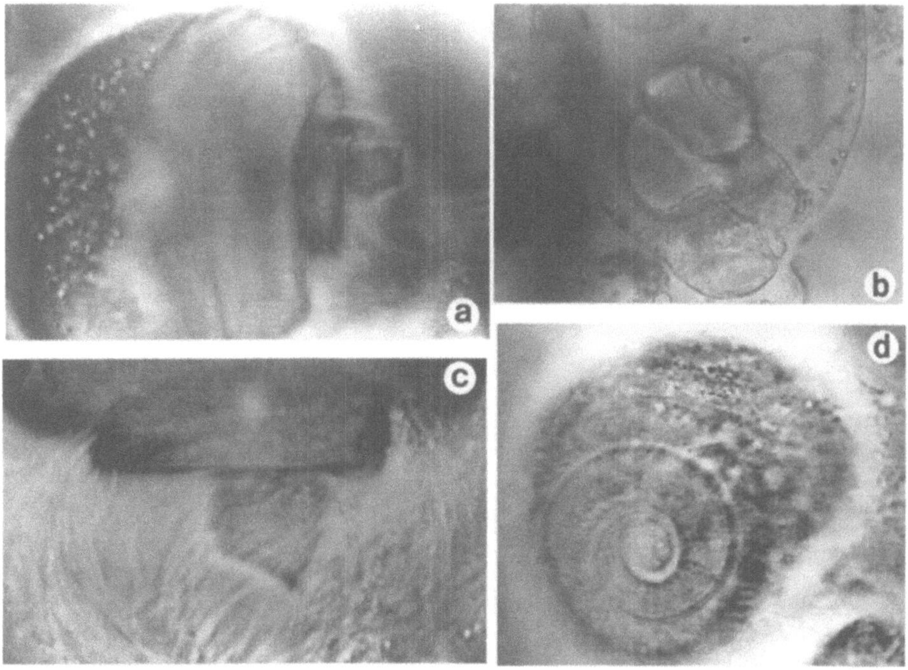

FIG. 1a–d. Micrographs of *Ginkgo* sperm. **a** Side view showing body of sperm and spiral apparatus. The sperm is about 80 microns in diameter. Note the prominence of the spiral center and the flagella. **b** Two sperm still under development in the pollen tube sac. **c** Close view of spiral center. Motion of the flagella is reduced to a blur, even in this 5 μs video picture. **d** Face view of spiral apparatus; flagella are just visible around the spiral

of the spiral; it is thus formed from the center outwards. In Fig. 3, a series of micrographs at 100-ms intervals show the gradual appearance of a swelling (arrowed in the micrograph) at the outer end of the developing spiral, and the incorporation of spiral band material at that place. These swellings appear at regular intervals (10–15 s) lasting approximately 2 s each time. Larger swellings also occur within the already-completed spiral band. The completed part of the spiral is very flexible at this stage and flexes during building. Once the spiral is complete (about 2.5 turns) it assumes a more well defined appearance—though still flexible—and all flagella commence beating in a vigorous manner. During construction, the flagella of the inner turn of the spiral appear and vibrate (but not beat) even before completion of the spiral. Thus the whole apparatus is not built in separate stages, but as a flowing concatenation. This spiral formation occurs while the sperm are still within the spermatogenous cell and attached to their partner sperm (Fig. 1b). After completion, the spiral center forms a distinct prominence (Fig. 1a,c).

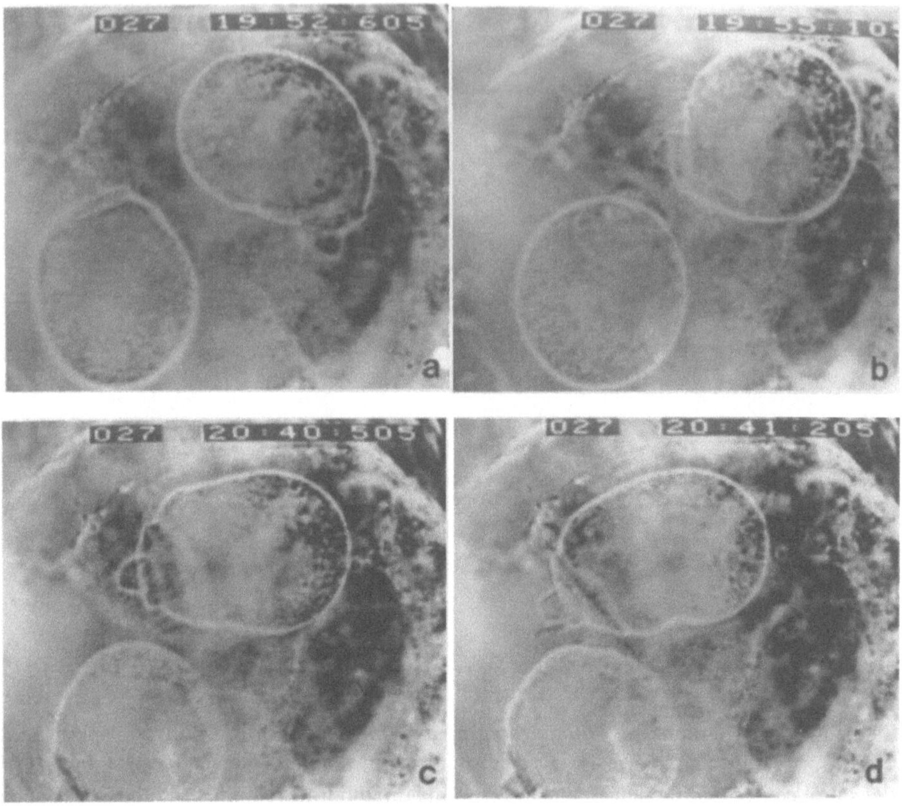

Fig. 2a–d. Series of micrographs showing how the sperm rotates and changes direction. The complete spiral apparatus changes angle relative to the body of the sperm, which rotates on an internal axis. *Upper sperm*: **a** Beginning of turn. **b** Finished turn 2.5 s later. **c** Straightened up sperm, position maintained for 45 s. **d** Another change in direction. Notice that the *lower sperm* has also turned

3.2 Flagellar Power Stroke

Flagella were observed to beat with an effective stroke and an ineffective recovery stroke, as is normally described for cilia, and the sperm moves forward, spiral first. Figure 4 is a series of frames taken at 5-ms intervals over 255 ms, showing a side view of the sperm at low magnification. Flagella at fully extended (completed) effective stroke can be seen in frame numbers 29:17:040 and 29:17:200. The flagella of this sample were thus beating every 160 ms, or about six times per second, at the time of recording. However, the frequency of the beat was variable, and beat speeds of up to 28 times per second (every 35 ms) were recorded.

On the recovery stroke, the fully extended flagella bunch back toward the spiral center, and can thus be seen at the end of recovery stroke in frame

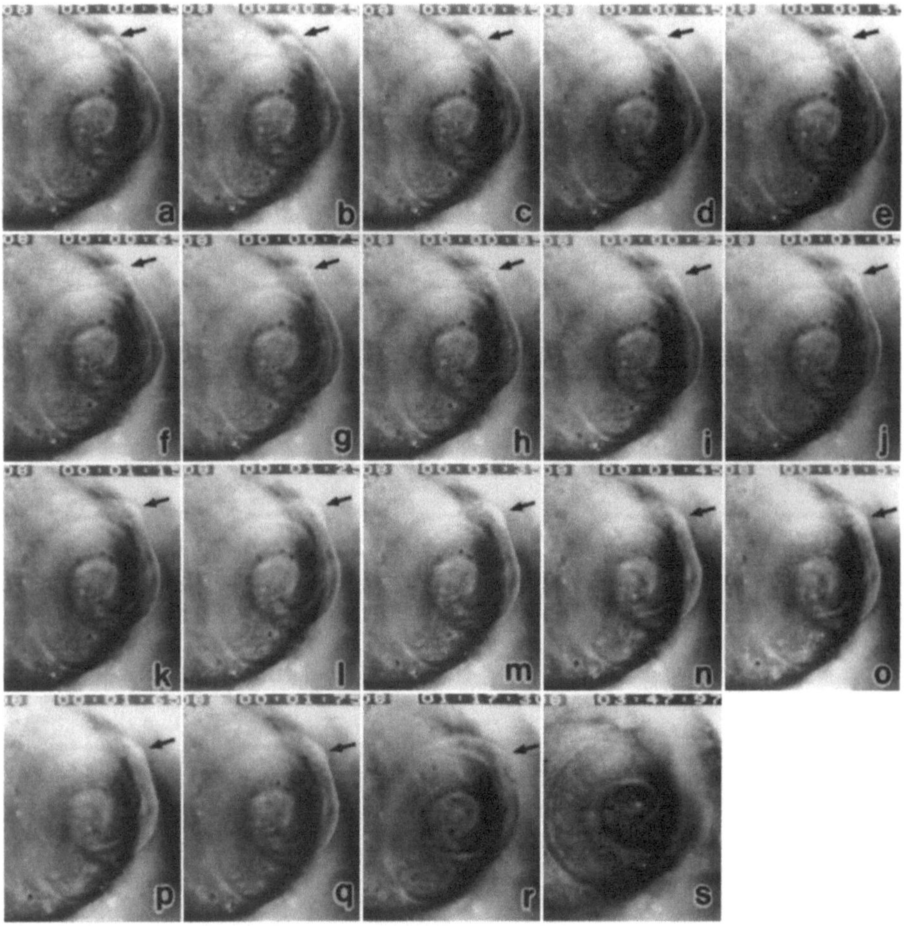

FIG. 3a–s. Spiral formation in *Ginkgo*. All micrographs are of the same sperm. *Ginkgo* sperm is about 80 to 85 μm in diameter. **a–q** A series of 17 micrographs, taken at 100-ms intervals, close to the completion of spiral formation. A series of swellings or processes derived from the cytoplasm raise the surface of the cell, presumably incorporating band material (*arrow*, i.e., the multilayer structure plus basal bodies) seen on the *right hand side* of the cell. The series here depicts the incorporation of one of these swellings over 1.6 s, although on average the appearance and incorporation lasted about 2 s. The spiral apparatus is very flexible at this stage. **r** Micrograph taken 75.55 s after the last frame of the series of 17 (**q**), showing further development of the spiral apparatus. The inner flagella of the spiral are already vibrating at this stage. **s** Micrograph showing the spiral apparatus building completed and "hardening" of the spiral (it loses its flexibility)

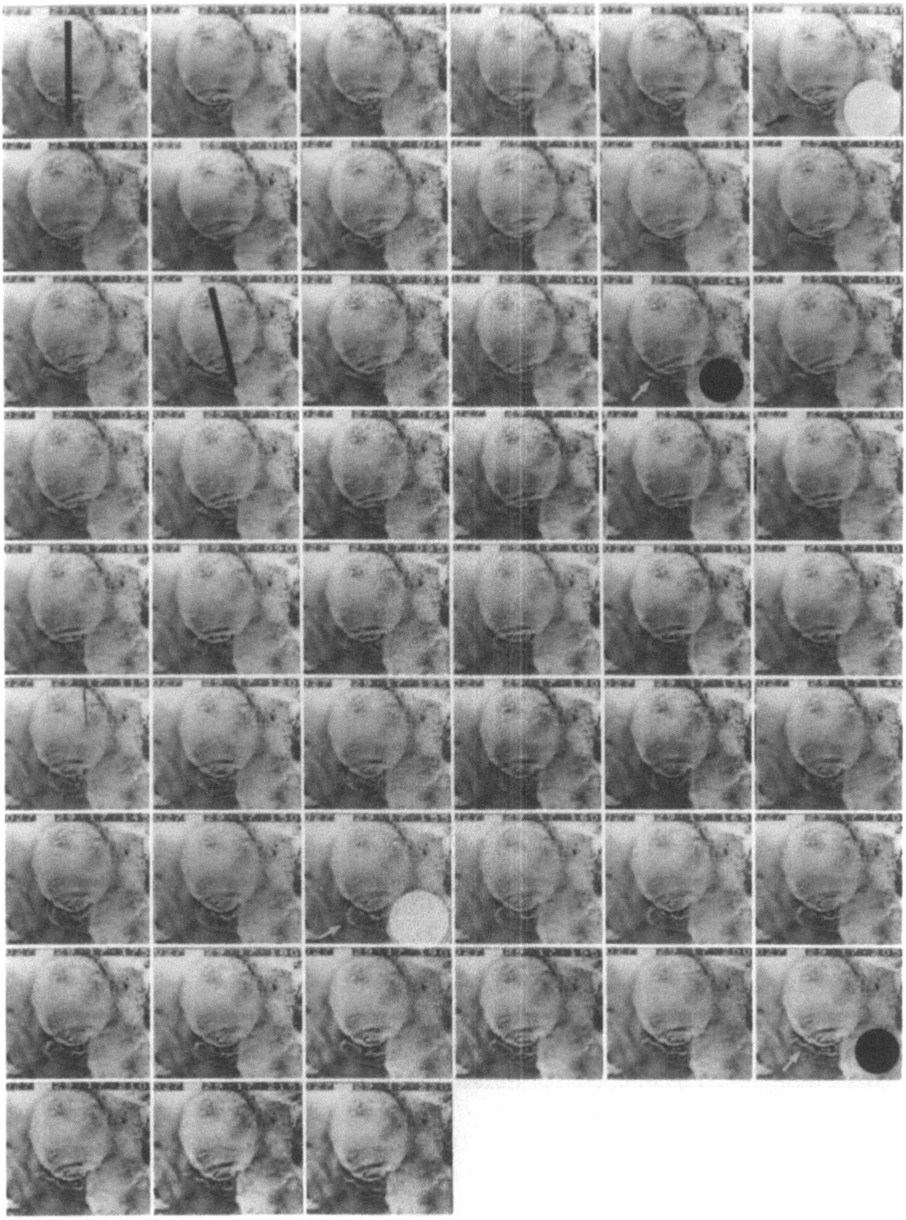

FIG. 4. Flagellar power stroke. A series of micrographs of two effective strokes taken at 5-ms intervals of the side view of an actively moving *Ginkgo* sperm, to demonstrate the action of the flagella in sperm movement. The sperm moves with spiral apparatus leading. The movement is described in the text. Notice that the sperm has also moved with reciprocation, and some of the micrographs have *lines* drawn through the center to indicate the reciprocation. Micrographs with *white dots* indicate the start of the effective stroke (flagella shown by *arrows*). Micrographs with *black dots* indicate the end of the effective stroke

29:17:105. The flagella seem to be tangled at this time but they quickly come together to form an integrated unit as the effective stroke is initiated. Compare frames 29:17:085–110 with frames 29 : 17 : 115–150 in which the flagella have commenced the effective stroke. The power of the effective stroke in this example causes the whole body of the sperm to appear to oscillate on its axis by 5 to 6 degrees each way (about 11 degrees total angular movement) when viewed from the side. This body movement is more complex, however, as described below. The oscillation varies according to the speed of the beat, the oscillation angle being larger when the beat is more frequent.

3.3 *Flagellar Waves*

The power stroke of the flagella does not occur simultaneously throughout the spiral, but was observed as a travelling wave along it. Flagellar beat is initiated in the center of the spiral, and after travelling for one turn, the next wave is initiated. Thus at any one moment, two waves can be seen travelling along the

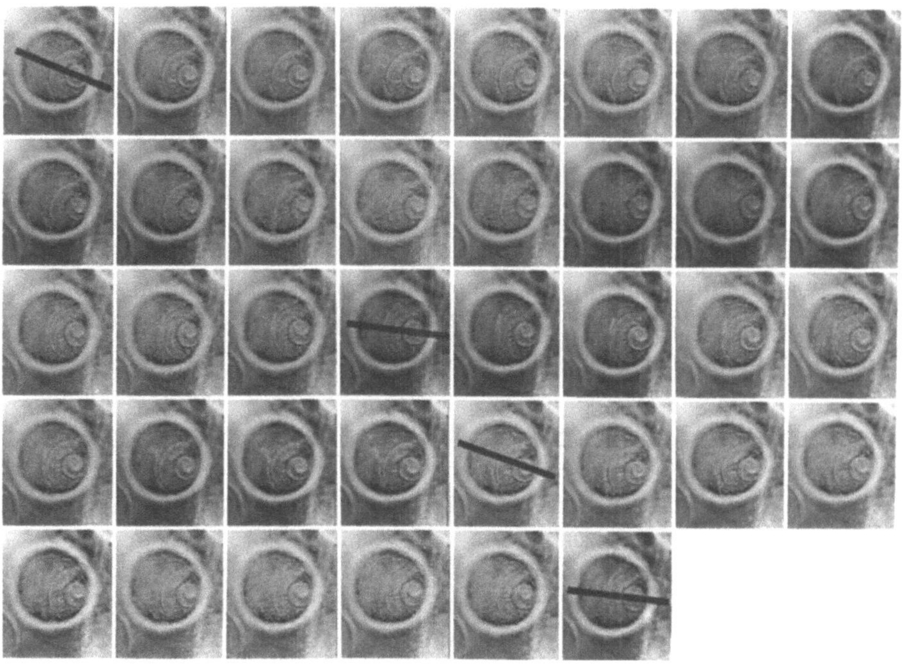

FIG. 5. Circular body oscillation. A series of micrographs taken at 5-ms intervals (view from top left to right, and then to next row, as in written text) to show that the whole body of the sperm oscillates as the flagella waves are generated. Some of the sperms have a *line* drawn through them to facilitate observation of oscillation, and the oscillation of this sample is shown in Fig. 6. Sideways view of oscillation is relatively easy to observe, but observation of circular oscillation requires close examination of the micrographs. This oscillation reflects the power of the flagella effective stroke

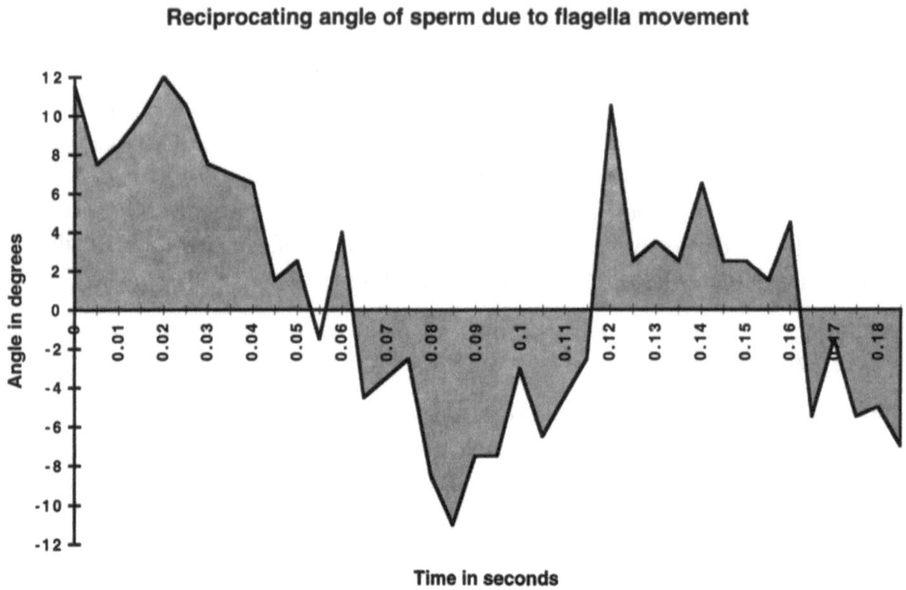

FIG. 6. Graph of an example of body oscillation due to flagella movement; data taken from micrographs in Fig. 5. The angle varies by up to 12 degrees either side of a pole drawn through the center of the sperm

spiral from the center to the outside. The effect of the flagellar wave is to cause a circular oscillation in the body of the sperm, so that if a stick were inserted into the center of the spiral, the end of the stick would scribe a circle in space (Figs. 5, 6). When viewed from the side, the sperm appears to oscillate from side to side. A maximum oscillation of 22.5 degrees total angular movement has been recorded, occurring in less than one tenth of a second (Figs. 4, 5, 6).

4 Concluding Comments

These are the first observations of *Ginkgo* sperm taken with a high-speed video system. Even though the cycad sperm is very similar, there are no reports on its swimming rhythm. *Ginkgo* sperm moves forward by a ciliary beat that emanates as a wave from the center of a 2.5-turn spiral, and the force of this makes the body of the sperm oscillate in reaction. The spiral is flexible and can protrude prominently at the center. Yet despite possessing this complex apparatus, the sperm has only to swim a few tens of microns to reach the archegonia of the ovule.

References

1. Hirase S (1896) On the spermatozoids of *Ginkgo biloba* (in Japanese). Bot Mag Tokyo 10:325–328

2. Hirase S (1898) Études sur la Fécondation et l'Embryogénie du *Ginkgo biloba* (Second mémoire). J Coll Sci Imp Univ Tokyo 12:103–149
3. Hirase S (1918) Further studies on fertilisation and embryogeny in *Ginkgo biloba* (in Japanese). Bot Mag Tokyo 32:83–108
4. Hori T, Ridge RW, Michikawa M, Miyamura S (1988) Ultrastructure of *Ginkgo biloba* sperm. In: Proc 53rd Ann Meet Bot Soc Jpn, 13–15 Dec, Okayama
5. Li Y, Wang FH, Knox RB (1989) Ultrastructural analysis of the flagellar apparatus in sperm cells of *Ginkgo biloba*. Protoplasma 149:57–63
6. Ridge RW, Hori T, Miyamura S (1989) High speed video analysis of living *Ginkgo* sperm. In: Proc 2nd Ann Meet Jpnese Soc Plant Morph, Tohoku Univ, Sendai

Chromosomes of *Ginkgo biloba*

MASAHIRO HIZUME

1 Introduction

This review is an attempt to summarize all cytogenetic studies in *Ginkgo biloba* including chromosome number, karyotype, chromosome banding, physical mapping, and genome size. Because several karyotypes and idiograms in which chromosomes were arranged in various orders have been reported, I wish to propose a standard karyotype for future study. I show some results of fluorescent banding and physical mapping of ribosomal RNA genes. These results are discussed with the sex chromosomes. The techniques of chromosome observation are appended in the last section for future chromosome studies.

2 Chromosome Number

The somatic diploid chromosome number is $2n = 24$ (Fig. 1a), and the haploid chromosome number is $n = 12$ (Fig. 1b,c) [1–23]. In early studies, $n = 16$ or $2n = 16$ was reported [24–26a]. Although a natural polyploid plant has been unknown until now, an artificial tetraploid plant was obtained from seed treated with colchicine [27]. The haploid chromosome number was observed in tissues cultured from a female gametophyte [27, 28] and the 24, 48, and other heteroploid numbers were also observed as well as the haploid number in the tissues cultured from pollen [29, 30]. Twelve bivalents (Fig. 1c) and haploid chromosomes appear in pollen mother cells (PMCs) at the metaphase of the first meiotic division and the anaphase of the first and second division [1, 11, 31–34].

Biological Institute, Faculty of Education, Ehime University, 3 Bunkyo, Matsuyama, Ehime 790, Japan

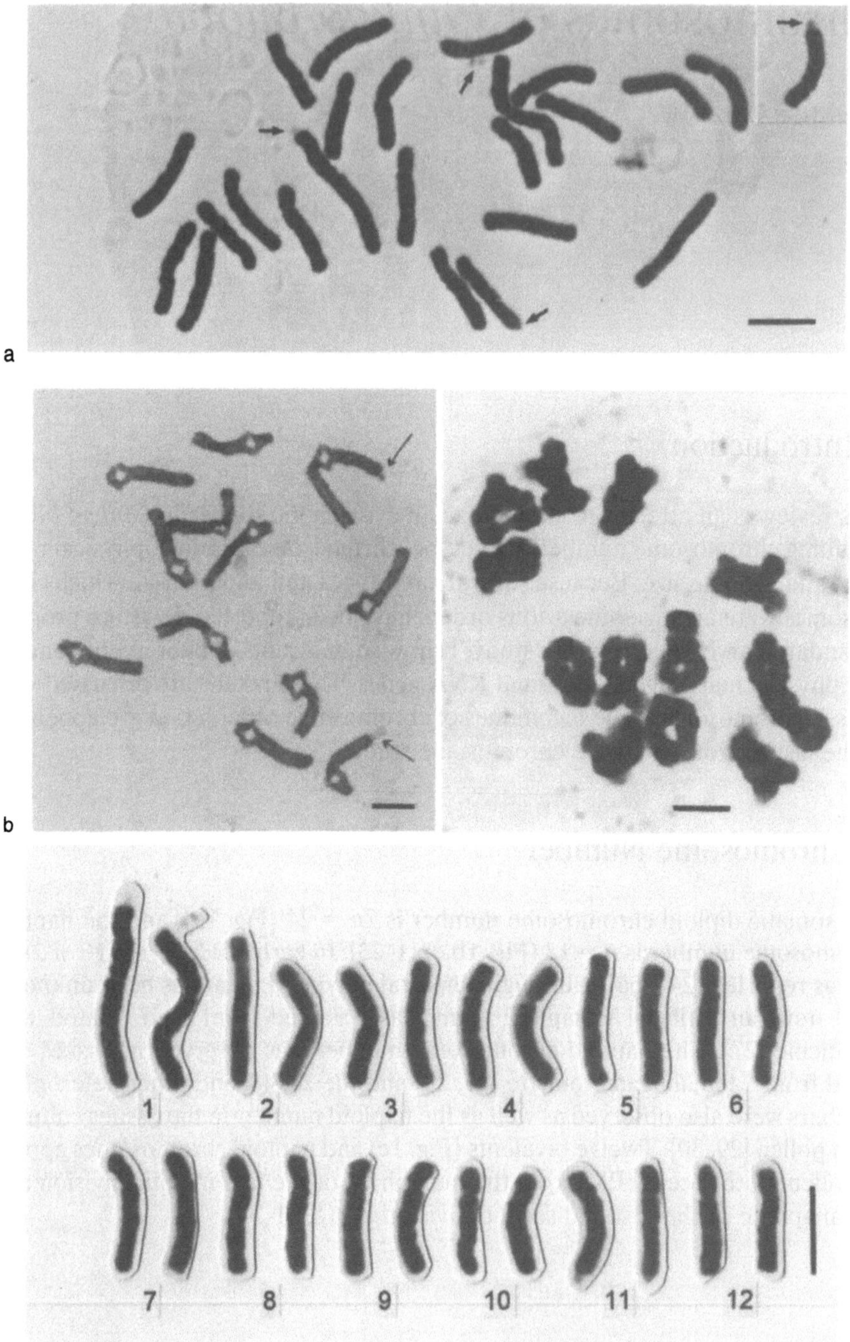

FIG. 1a–d. Chromosomes of *Ginkgo biloba*. **a** Somatic diploid chromosomes ($2n = 24$) of a young leaf in a male tree. **b** Haploid chromosomes ($n = 12$) in a ovule (after [11]). **c** Meiotic chromosomes in a PMC at metaphase of the first division. **d** Standard karyotype. Arrows show secondary constrictions or satellites. Bar = 5 μm

3 Karyotype

The chromosome complement of *Ginkgo biloba* is a bimodal and asymmetrical karyotype [8, 10–15, 17, 22, 23]. The karyotypes or idiograms presented previously are generally similar to each other but the order and name of individual chromosomes in the karyotype are different among the reports. Standardization of karyotype is necessary for further research of chromosome and genome in *G. biloba*. The standard karyotype, composed of the following chromosome pairs, is presented in Fig. 1d.

No. 1:	Long, submetacentric chromosome with a small satellite at the end of short arm
No. 2:	Short subtelocentric chromosome, the most asymmetrical
No. 3:	Short and metacentric chromosome
Nos. 4–10:	Short and subtelocentric chromosomes, indistinguishable from each other
No. 11:	Short and subtelocentric chromosome with a small satellite at the end of long arm
No. 12:	The shortest and subtelocentric chromosome

4 Banding

Banding techniques (Fig. 2) such as C-banding [17, 19, 23], silver staining [23], G-banding [21], and fluorescence banding have been applied to the chromosomes of *Ginkgo biloba*. The procedures for banding are present in the last section, Section 9 of this chapter. The C-bands (Fig. 2a) and chromomycin A_3 (CMA) bands (Fig. 2b) appear at the secondary constriction and satellite of four chromosomes (no. 1 and 11). Three or four silver deposits indicating active nucleolar organizing regions (Ag-NORs) occur at the secondary constrictions (Fig. 2c,d). The three bandings indicate that chromosomes of *G. biloba* poorly differentiate along their long axes. The G-banding reported by Chen [21] is expected to be useful for identification of subtelocentric chromosome pairs indistinguishable from each other.

5 Karyotype Variation: Are Sex Chromosomes Present?

Newcomer [12] reported differently shaped long chromosomes in pollen mitoses. Tanaka et al. [10] and Newcomer [12] reported the satellite on the short arm of the subtelocentric chromosome (no. 11). Lee [11] and Pollock [13] observed no satellite in one of the short nucleolar chromosomes in male plants and suggested from this heteromorphy that the XY-type sex determination system was present. Ho [15] observed carefully the satellites in male and female plants and concluded that the satellite was not associated with sex. On the basis of observation of the C- and Ag-positive large satellites that appear on one of the long chromosomes

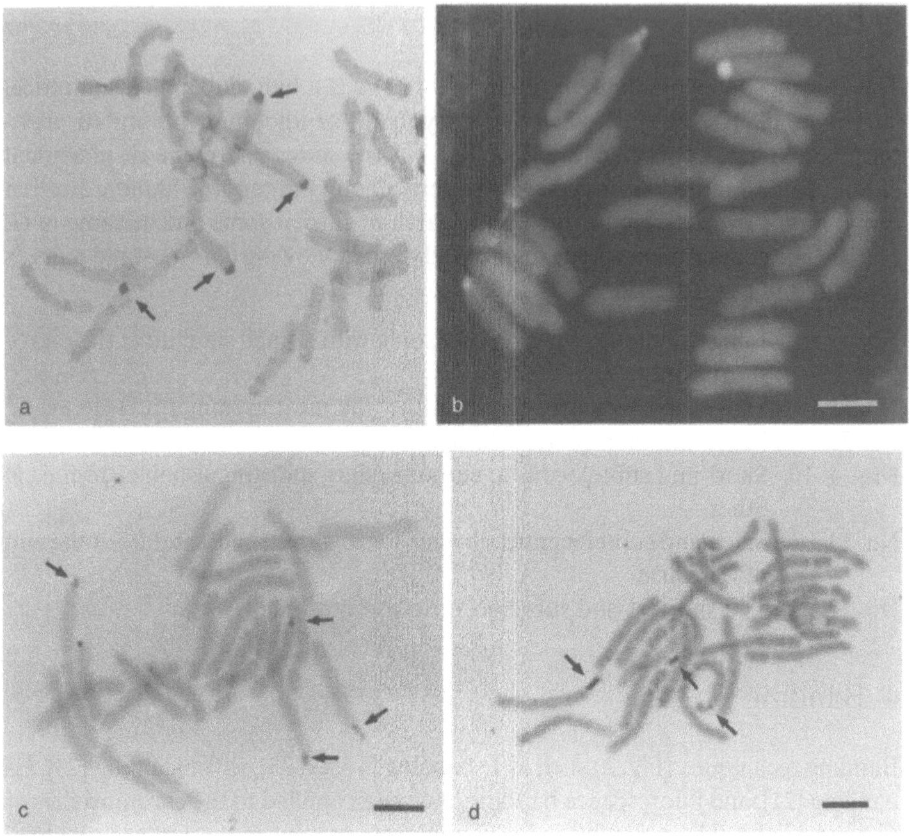

FIG. 2a–d. Chromosome bandings. **a** C-banded somatic chromosomes (after [22]). **b** Fluorescence banding with CMA. **c,d** Silver-stained chromosomes possessed four (**c**) and three (**d**) silver deposits, respectively. *Arrow* shows a C-band or Ag-NOR. Bar = 5 μm

(no. 1) in female plants, Chen et al. [22, 23] proposed the ZW-type sex determination system.

Variation of satellites occurs in *Ginkgo* plants. The number of satellites and of nucleoli per nucleus were reexamined in 10 male and 15 female plants at the Johoku campus of Ehime University. All plants excepting two had three or four satellites in a chromosome complement and formed at most three or four nucleoli in a nucleus, corresponding to the number of satellites. These variations might be caused by deletion or inactivation of the NOR. Because both male and female plants show the same satellite variation, the karyotypic variation involving the satellite is concluded not to be related to sex.

6 Physical Mapping of rDNA by In Situ Hybridization

A tandem array of 18*S*-5.8*S*-26*S* rRNA genes is located at the NOR. In situ hybridization (ISH) using a rDNA probe is applied to examine the involvement

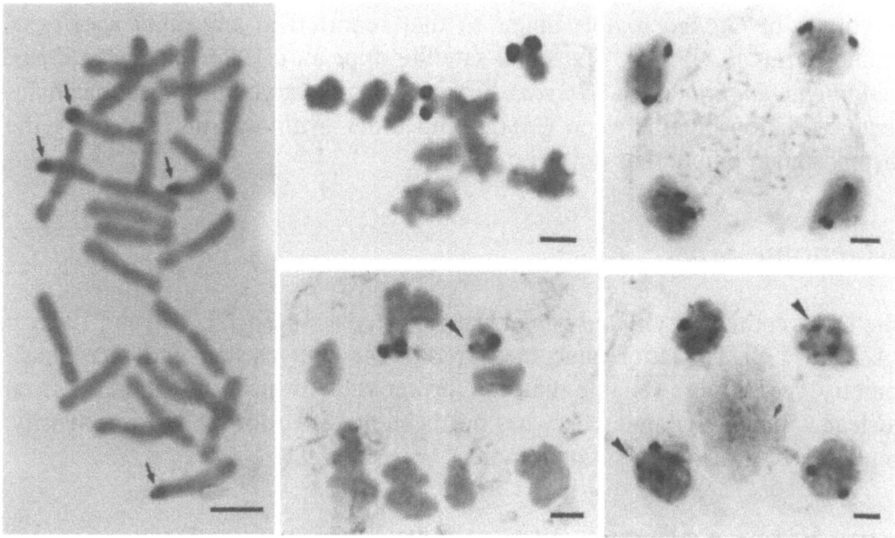

FIG. 3a–e. In situ hybridization (ISH) of the chromosomes with rDNA probe. **a** ISH of somatic metaphase chromosomes. **b–e** ISH of meiotic chromosomes (**b,c**) and tetrad (**d,e**) of PMC from trees with four (**b,d**) and three (**c,e**) satellites. *Arrows* show ISH signals in **a**; *arrowheads* show the small signal in **c** and **e**. Bar, 5 µm

of rDNA with the NOR variation observed in this species. After ISH with the 18*S* rDNA probe in the mitotic chromosomes, four hybridization signals appeared at the end of the short arm of the long chromosomes (no. 1) and of the long arm of the short chromosomes (no. 11) (arrows in Fig. 3a). The loci are at the same position as the secondary constrictions and the small satellite or C- and CMA-bands. In PMCs collected from the same plant, the large bivalent and one of the small bivalents showed four similar-sized signals on each homologous chromosome (Fig. 3b). In the tetrads, each nucleus contained two similar-sized signals (Fig. 3d). In the plant forming maximally three nucleoli in a nucleus and with three satellites, four signals appeared but one signal of the small bivalent was smaller than the others (Fig. 3c). Two nuclei of the tetrads had the small and large signals (Fig. 3e). Three secondary constrictions were stained with silver staining in the plant (Fig. 2d), suggesting that the small locus of rDNA is transcriptionally inactive. These ISH results indicate that the amount of rDNA varied among plants.

7 Comparison of *Ginkgo* Karyotype with Other Gymnosperms

The diploid chromosome number $2n = 24$ (as in *Ginkgo biloba*) is reported in most species of Pinaceae except for *Pseudotsuga menziesii* and *Pseudolarix amabilis*, *Taxus*, and Cephalotaxaceae [35]. The bimodal and asymmetrical karyotype of *Ginkgo* is not similar to that reported in any other species of gymnosperms [8, 36, 37]. The small satellite appears in *Sequoia sempervirens*, *Cunninghamia*, *Ephedra*, and *Cycas*. Similarity in the karyotype indicating phylogenetic relationships between *Ginkgo* and other gymnosperms, excepting the chromosome number, has not been studied.

8 Genome Size

The DNA content of *Ginkgo biloba* was reported as 54.2 pg/2C [16] and 19.86 ± 1.42 pg/2C [38]. The latter value, measured by the Feulgen stain and microdensitometry, seems more reliable than the former value, which was estimated from nuclear volume. Genomic DNA has not been analyzed for its buoyant density, satellite DNA, or reassociation kinetics.

9 Techniques for Chromosome Observation

9.1 *Somatic Chromosomes*

Young leaves and root tips are pretreated in 0.05% colchicine or 2 m*M* 8-hydroxyquinoline at 20°C for 4–6 h, fixed in cold ethanol–acetic acid (3:1) for 1 h more, and then stored at −20°C. After maceration with 1 *N* HCl–45% acetic acid (2:1) at 60°C for 20 s, the materials are stained and squashed in 2% acetoorcein. For C-banding and fluorescence banding, the fixed materials are macerated in 45% acetic acid at 60°C for 5 min and squashed in 45% acetic acid. The preparations are removed from the coverslip by a razor blade after freezing on a block of dry ice and air-drying. For ISH and silver staining, the fixed materials are treated with 2% cellulase Onozuka RS (Yakult, Tokyo, Japan) and 0.5% pectolyase Y-23 (Seishin Farma, Tokyo, Japan) in 0.1 *M* sodium acetate buffer (pH 4.5) at 37°C for 1 h and rinsed with distilled water. The chromosomes are prepared by the flame-drying technique or squashing after removal from the coverslip by the dry ice method.

Haploid chromosomes are obtained from young pollen in the anthers and in ovules at the time of initiation of the free nuclear stage of the female gametophyte. The anthers and ovules are fixed in ethanol–acetic acid, and the mitotic cells are stained and squashed with acetoorcein.

9.2 *Meiotic Chromosomes*

The PMCs collected in early April are suspended by squeezing the male strobili with a forceps in 0.075 *M* KCl for 10 min and are then fixed in the ethanol–acetic acid. The fixed PMCs can be stored in a deep freezer for several months more. Before making chromosome preparations, the fixative is renewed. One or two drops of the fixative containing the PMCs are put on a glass slide. After the fixative is allowed to evaporate, the PMCs are squashed in acetoorcein or 45% acetic acid. Some preparations are removed from the coverslip by the dry ice method and air-dried for ISH or fluorescence banding [39].

9.3 Silver Staining

The chromosomes prepared by enzymic maceration are stained by the addition of 50 µl of 50% silver nitrate and covered with a piece of nylon mesh. The preparations are incubated at 65°C in a moist chamber until the chromosome arms are yellow. After incubation the chromosomes are thoroughly washed with distilled water [40].

9.4 C-Banding

The preparations are treated with 5% $Ba(OH)_2$ at 55°C for 8 min, rinsed twice in distilled water for 10 min, then treated with 2× SSC (sodium saline citrate) at 60°C for 1 h and rinsed in distilled water. The chromosomes are stained with 2% Giemsa in phosphate buffer, pH 7.0, for 10 min and rinsed with distilled water [17, 19, 22].

9.5 Fluorescence Banding

The preparations are soaked in McIlvaine citric acid–Na_2HPO_4 buffer, pH 7.0, for 15 min, stained with 0.1 mg/ml distamycin A for 10 min, and rinsed with buffer containing 5 m*M* $MgCl_2$. The chromosomes are stained with 0.1 mg/ml CMA in the Mg buffer for 10 min, rinsed with buffer, and treated with 0.1 µg/ml 4′,6′-diamindino-2-phenylindole (DAPI) for 5 min. After rinsing with buffer, the chromosomes are mounted in glycerol and stored overnight in a refrigerator before observation under a fluorescence microscope equipped with a blue filter [41].

9.6 In Situ Hybridization Using a Biotin-Labeled rDNA Probe

Genomic DNA is extracted from leaves by the cetyltrimethylammoniumbromide (CTAB) method [42, 43]. The fragments of rDNA are amplified by the polymerase chain reaction using primer sets specific for the 18*S* rRNA gene [44] and labeled with biotin by nick translation.

The probe solution contains the labeled DNA (0.4 µg/ml), 50% formamide, and 2× SSC. After treatment with 0.1 mg/ml RNase and postfixation in 4% paraformaldehyde, the preparations are air-dried through an ethanol series. Then, 10 µl of probe solution is added to the preparation, which is covered with a coverslip and denatured for 10 min on a hot plate set at 76°C. After overnight incubation at 37°C, the preparation is washed in 2× SSC and the hybridized probe is detected by the avidin–peroxidase–diaminobenzidine system. The chromosomes are counterstained with 2% Giemsa or 0.1% neutral red.

10 Conclusions

The chromosomes of *Ginkgo biloba* possess the following features:

1. The chromosome numbers are $n = 12$ and $2n = 24$.
2. The karyotype is composed of one pair of long submetacentric chromosomes, one pair of short metacentric chromosomes, and ten pairs of short subtelocentric chromosomes.
3. Small satellites appear at the end of short arm of the no. 1 pair and the long arm of the no. 11 pair. The satellite varied in occurrence and size among trees.
4. C-bands, CMA-bands, and Ag-NORs appeared at the secondary constriction and satellite.
5. 18*S* rDNA is mapped at this banded region.
6. The sex chromosome reported previously was not confirmed.

References

1. Ishikawa M (1910) Über die Zahl der chromosome von *Ginkgo biloba* L. Bot Mag Tokyo 24:225–226
2. Eichhorn A (1928) Sur la mitose somatique et le satellite du *Ginkgo biloba*. C R Soc Biol Paris 99:1787–1788
3. Herzfeld S (1927) Beiträge zur Kenntnis von *Ginkgo*. Jahrb Wiss Bot 66:814–862
4. Herzfeld S (1928) Über die Kernteilung im proembryo von *Ginkgo biloba*. Jahrb Wiss Bot 69:264–294
5. Simamura T (1928) On the formation of the proembryo of *Ginkgo biloba* L. Bot Mag Tokyo 42:71–76
6. Simamura T (1931) A note on mitotic division in the proembryo of *Ginkgo*, with special reference to a chromatin elimination. Bot Mag Tokyo 45:525–530
7. Sinoto Y (1929) Chromosome studies in some dioecious plants, with special reference to the allosomes. Cytologia (Tokyo) 1:109–191
8. Sax KC, Sax HJ (1933) Chromosome number and morphology in the conifers. J Arnold Arbor Harv Univ 14:356–375
9. von Resend F, Rijo L (1948) Structure of chromosomes as observed in root tips. V. Olistherochromatin, chromatic aggulutination and mutations. Port Acta Biol 2:117–147
10. Tanaka N, Takemasa N, Sinoto Y (1952) Karyotype analysis in gymnospermae. I. Karyotype and chromosome bridge in the young leaf meristem of *Ginkgo biloba* L. Cytologia (Tokyo) 17:112–123
11. Lee CL (1954) Sex chromosomes in *Ginkgo biloba*. Amer J Bot 41:545–549
12. Newcomer EH (1954) The karyotype and possible sex chromosomes of *Ginkgo biloba*. Amer J Bot 41:542–545
13. Pollock EG (1957) The sex chromosomes of the maidenhair tree. J Hered 48:290–294
14. Czeika G, Schiman H (1960) Die Chromosomen von *Ginkgo biloba*. Österr Bot Zeit 107:1–5
15. Ho T (1963) The nucleolar chromosomes of the maiden-hair tree. J Hered 54:67–74
16. Price HJ, Sparrow AH, Nauman AF (1974) Evolutionary and development considerations of the variability of nuclear parameters in higher plants. I. Genome volume,

interphase chromosome volume, and estimated DNA content of 236 gymnosperms. Brookhaven Symp Biol 25:390–421
17. Tanaka R, Hizume M (1980) C-banding techniqe for some species of gymnosperms. Bot Mag Tokyo 93:169–170
18. Terasaka O (1982) Nuclear differentiation of male gametephytes in gymnosperms. Cytologia (Tokyo) 47:27–46
19. Chen KY (1983) A giemsa C-banding technique for chromosomes of *Ginkgo biloba* (in Chinese). Bot Res 1:267–270
20. Vachova M (1986) Karyological studies of the Slovak Czechoslovakia flora. XXI. Acta Fac Rerum Nat Univ Comenianae Bot 34:27–32
21. Chen R (1988) G-band staining of chromosomes in some higher plants with special reference to *Lilium davidii* (in Japanese). La Kromosomo II-50:1635–1651
22. Chen R, Song W, Li X, Chen Y (1989) Study on the sex chromosomes of *Ginkgo biloba*. In: Hong D-Y (ed) Plant chromosome research 1987. Organizing committee of the Sino-Japanese symposium on plant chromosomes, Beijing and Hiroshima, pp 381–386
23. Chen R, Song W, Li X, An Z (1993) A study on the sex chromosomes of *Ginkgo biloba*. Cathaya 5:41–48
24. Carothers JE (1907) Development of ovule and female gametophyte of *Ginkgo biloba*. Bot Gaz 43:116–130
25. Chamberlain CJ (1935) Gymnosperms: structure and evolution. University of Chicago Press, Chicago
26. Illick JT (1941) Chromosomes of *Ginkgo biloba* during early developmental stages of the male gametophyte. Anat Rec 81:78–79
26a. Kim CS (1982) Studies on artificial polyploid forest trees. XVI. Breeding of colchitetraploid tree named "Gyeong-Sin Eun Hang" (*Ginkgo biloba* Linne). J Gyeong Song Bac Univ Biol Agric 21:21–24
27. Tuleche W (1964) A haploid tissue culture from the female gametephyte of *Ginkgo biloba* L. Nature 203:94–95
28. Tulecke W (1967) Studies on tissue cultures derived from *Ginkgo biloba* L. Phytomorphology 17:381–386
29. Tulecke W (1953) A tissue derived from the pollen of *Ginkgo biloba*. Science 117: 599–600
30. Tulecke W (1957) The pollen of *Ginkgo biloba* in vitro culture and tissue formation. Amer J Bot 44:602–608
31. Cardiff JD (1906) Studies of synapsis and reduction. Bull Torrey Bot Club 3: 271–306
32. Mann MC (1924) Microsporogenesis of *Ginkgo biloba* L. with special reference to the distribution of the plastids and to cell wall formation. Univ Calif Publ Agric Sci 2:243–248
33. Alejandro NG (1954) Study of meiosis in *Ginkgo biloba* L. with special reference to the cytoplasmic granules. Nat Appl Sci Bull 14:265–289
34. Favre-Duchartre M (1956) Contribution à l'étude de la reproduction chez le *Ginkgo biloba*. Rev Cytol Biol Veg 17:1–219
35. Khoshoo TN (1961) Chromosome numbers in gymnosperms. Silvae Genet 10:1–9
36. Mehra PN, Khoshoo TN (1956) Cytology of conifers. I. J Genet 54:165–180
37. Mehra PN, Khoshoo TN (1956) Cytology of conifers. II. J Genet 54:181–185
38. Ohri D, Khoshoo TN (1986) Genome size in gymnosperms. Plant Syst Evol 153: 119–132

39. Hizume M, Kondo T, Miyake T (1989) Meiosis in pollen mother cells of 15 *Cryptomeria japonica* clones with special emphasis on the behavior of the 6th chromosome pair heteromorphic to chromomycin A_3 band. Jpn J Genet 64:287–294
40. Hizume M, Sato S, Tanaka A (1980) A highly reproducible method of nucleolar organizing regions staining in plant. Stain Technol 55:87–90
41. Hizume M, Ohgiku A, Tanaka A (1983) Chromosome banding in the genus *Pinus*. I. Identification of chromosomes in *P. nigra* by fluorescent banding method. Bot Mag Tokyo 96:273–276
42. Hizume M, Ishida F, Kondo K (1992) Differential staining and *in situ* hybridization of nucleolar organizers and centromeres in *Cycas revoluta* chromosomes. Jpn J Genet 67:381–387
43. Murray M, Thompson WF (1980) Rapid isolation of high-molecular-weight plant DNA. Nucleic Acid Res 8:4321–4325
44. Hizume M (1994) Chromosomal localization of 18*S* and 5*S* rRNA genes in garlic, *Allium sativum*. Chromosome Inform Serv 56:6–8

Lignotuber Development in *Ginkgo biloba*

Peter Del Tredici

1 Introduction

Ginkgo biloba is famous for its powers of survival. On the evolutionary time scale, the genus can be traced back at least to the middle Jurassic, while species that are similar to *G. biloba* date back to the Cretaceous. On the human time scale, the longevity of individual trees is equally impressive, particularly in Asia where ancient specimens, often approaching or exceeding a thousand years in age, can be seen growing in the vicinity of Buddhist or Taoist temples. In modern times, the ginkgo tree has been widely planted in many cities, where it flourishes in the midst of extreme air pollution and extensive pavement.

It has long been considered a mystery why *G. biloba* should have survived as a wild plant in China when all other members of the division Ginkgophyta became extinct long ago. One reason that has been proposed is that *Ginkgo* possesses the ability, unusual among gymnosperms, to sprout out from an underground burl, or lignotuber, following traumatic injury to the primary stem [1]. Anatomical studies on lignotuber development in a number of woody plants, including *Eucalyptus* spp. [2], *Arbutus unedo* [3], *Quercus suber* [4], *G. biloba* [1], and *Sequoia sempervirens* [5], have shown them to be genetically determined structures that develop from buds located in the axils of both cotyledons and a few of the leaves immediately above them. Over time, lignotubers can become quite large and can contribute to the survival of plants in three ways: (1) they are a site for the production and storage of suppressed shoot buds that can sprout out following injury to the primary stem; (2) they are a site for the storage of carbohydrates and mineral nutrients, which may allow for the rapid growth of these suppressed buds following stress or damage; and (3) for plants growing on steep slopes, they can function as a kind of clasping organ that anchors the plant to the rocky substrate [3, 5, 6]. Lignotuber-producing species are most commonly found in Mediterranean-type ecosystems that are characterized by hot, dry summers and periodic fires [7–9]. The purpose of this chapter is to review

Arnold Arboretum of Harvard University, 125 Arborway, Jamaica Plain, MA 02130, U.S.A.

the literature on lignotuber development in *G. biloba* and to present experimental data on the effect of seedling orientation on their initiation and development.

2 Literature Review

All *Ginkgo* seedlings produce lignotubers (originally called "basal chichi") as part of their normal ontogeny from buds located in the axils of the cotyledons. Within 6 weeks of germination, these bud clusters become embedded in the cortex of the stem, and their subsequent growth and development occur below the surface of the bark. When stimulated by some traumatic event, one of these bud clusters typically grows down from the trunk to form a woody, rhizome-like basal lignotuber, which is capable of generating both aerial shoots and adventitious roots [1].

The ecological importance of lignotubers in *Ginkgo* was first described from field work on a "semi-wild" population of trees growing on West Tian Mu Mountain in Zhejiang province, China [6]. The study reported that vegetative reproduction from basal lignotubers played an important role in the long-term persistence of *Ginkgo* on the steep, rocky slopes where they were growing. Of the 167 trees located on the mountain, 40% (67 individuals) possessed at least two trunks greater than 10cm in diameter. Of these, 73% (49 individuals) had an intact primary trunk, indicating that logging was not the cause of the multitrunked condition.

Lignotubers were also observed on seedings in cultivation for leaf production in Sumter, SC (U.S.A.), where plants were periodically cut down to ground level to facilitate mechanical harvesting [1]. Most of the seedlings responded to this severe treatment by producing branched lignotubers that gave rise to new shoots as well as new roots. On plants that had been cut back to ground level two or more times, at 4- to 5-year intervals, numerous secondary shoots originated from the proximal end of the massive lignotuber. It was also observed that among unpruned, untransplanted seedlings that had been raised under crowded nursery conditions, 11% of 1000 randomly selected 3-year-old plants spontaneously produced a downward growing lignotuber longer than 1 cm. In all these seedlings, only one of the cotyledonary buds formed a lignotuber, and none of the lignotubers had produced any shoots [1].

Mature specimens of *Ginkgo* can sometimes produce induced or aerial lignotubers along the trunk and branches. Known as "chichi" in Japan and "zhong ru" in China, these unusual woody growths were first described in 1895 by Fujii [10], who considered them a "pathological formation" that developed in association with an embedded shoot bud. Fujii showed that aerial lignotubers, despite their rootlike appearance, were essentially leafless, downward-growing shoots that were covered with developmentally suppressed vegetative buds. Fujii also noted that aerial lignotubers grew to as much as 2 m in length and possessed the ability to form both roots and leafy shoots on contacting soil. A more recent study has elucidated the anatomy of aerial lignotubers as well as their anomalous cell structure but failed to shed light on what, if any, environmental factors induced their formation [11]. A recent review of the horticultural literature indicated that aerial lignotuber production in *Ginkgo* is a developmental (i.e., epigenetic) phenomenon that is characteristic of the senescent phase of the tree's life cycle and can be perpetuated through vegetative propagation [12].

3 Materials and Methods

To test the effect of seedling orientation on lignotuber development, experiments were conducted in the Dana Greenhouses of the Arnold Arboretum of Harvard University, located in Boston, MA. Locally produced *Ginkgo* seeds were sown in a warm greenhouse (heated to a minimum temperature of 13°C) in late winter. After one season of growth, 34 dormant seedlings were planted vertically in plastic containers 9 × 9 × 12 cm with a soil depth of 10 cm. During transplanting, great care was taken to make sure that as few roots as possible were broken. A similar group of 34 dormant seedlings were laid horizontally in large wooden flats 10 × 50 cm and covered with 2 cm of soil. As in the case of the vertical seedlings, the total soil depth was 10 cm. Despite the different-shaped containers, the volume of soil per plant was approximately equal for both the vertical and the horizontal treatments (approximately 800 cm^3). The seedlings were then allowed to grow for one season (April through November 1994) in the greenhouse where they were watered as needed and fertilized every 2 weeks.

In January 1995, the dormant seedlings were removed from their containers and washed free of soil. They were measured for initial height, final height, number of leaves produced, and stem diameter at the cotyledonary node. In addition, the plants were scored for the presence or absence of a visible lignotuber at both cotyledonary nodes and along the stem. A stem swelling was defined as a visible lignotuber if it protruded from the stem by 2 mm or more. Fresh weight and dry weight of the following plant parts were measured: leaves; root system (including all adventitious roots that formed on the stems of the horizontal plants); cotyledonary node (a 1-cm-long segment that included developing lignotubers); the segment of stem produced during the first year of growth, exclusive of the cotyledonary node (year-1 stem); the segment of stem produced during the second year of growth (year-2 stem); and last, all visible lignotubers produced along the stem.

Field experiments were conducted on the plantation of Garnay, Inc., in Sumter, SC, which produces *Ginkgo* leaves used for the production of a medical extract. Two-year-old seedlings, which had been raised without transplanting in outdoor seedbeds, were lined out 40 cm apart in rows that were 1 m apart; 100 seedlings were planted in normal, vertical orientation and 100 were laid horizontally in a shallow trench and covered with 5 cm of soil. Both groups were then left to grow for 2 years, during which time they were periodically watered with overhead irrigation, fertilized, and weeded. At the end of two full growing seasons, 56 surviving horizontally oriented plants and 52 vertically oriented plants were harvested and the following stem increments measured: initial stem length, consisting of stem growth at the time of planting; experimental stem length, consisting of stem and branch growth made during years 3 and 4; and the length and number of all visible lignotubers.

All data were subjected to analysis of variance (ANOVA) using the general linear means procedure of the SAS statistical software program [13]. Within treatment groups, Pearson correlation coefficients were calculated using the same program.

4 Results

The results of both the greenhouse and the field experiments (Table 1) showed that orienting seedlings horizontally in the soil increased lignotuber formation. Among the horizontally oriented greenhouse plants, 50% developed one or more

TABLE 1. Effect of horizontal orientation on the growth of *Ginkgo biloba* seedlings

Experiment	Horizontal seedlings	Vertical seedlings	*P* value
Green house experiment (*n*)	34	34	
Cotyledonary node lignotuber number (per plant)	0.35 ± 0.07	0.15 ± 0.22	*
Cotyledonary node lignotuber length (cm)	0.11 ± 0.03	0.06 ± 0.07	—
Cotyledonary node fresh weight (g)	0.88 ± 0.07	0.83 ± 0.07	—
Stem lignotuber number (per plant)	1.12 ± 0.17	0	***
Stem lignotuber length (cm)	0.30 ± 0.05	0	***
Stem lignotuber fresh weight (g)	0.05 ± 0.01	0	***
Total lignotuber number (per plant)	1.47 ± 0.22	0.15 ± 0.22	***
Total lignotuber length (cm)	0.41 ± 0.07	0.06 ± 0.07	***
Stem lignotuber + cotyledonary node fresh weight (g)	0.97 ± 0.07	0.83 ± 0.07	—
Year 1 stem length (cm)	12.38 ± 0.35	11.44 ± 0.35	—
Year 1 stem fresh weight (g)	4.04 ± 0.25	2.48 ± 0.25	**
Year 1 stem + cotyledonary node fresh weight (g)	4.92 ± 0.31	3.31 ± 0.31	**
Year 2 stem length (cm)	25.79 ± 1.63	22.24 ± 1.63	—
Year 2 stem fresh weight (g)	3.58 ± 0.33	2.80 ± 0.33	—
Leaf fresh weight (g)	7.19 ± 0.63	8.06 ± 0.61	—
Stem + lignotuber fresh weight (g)	8.55 ± 0.61	6.11 ± 0.61	**
Root fresh weight (g)	11.59 ± 1.32	17.56 ± 1.32	**
Total fresh weight (g)	27.34 ± 2.46	31.73 ± 2.46	—
Field experiment (*n*)	56	52	
Total lignotuber number (per plant)	1.46 ± 0.15	0.92 ± 0.16	*
Total lignotuber length (cm)	2.07 ± 0.28	1.78 ± 0.29	—
Initial stem length (cm)	17.59 ± 0.56	19.06 ± 0.58	—
Experimental stem length (cm)	59.25 ± 8.12	173.31 ± 8.43	**

Data are mean value ± standard error per plant.
*, $P < .05$; **, $P < .01$; ***, $P < .001$.

lignotubers along some portion of their buried stems above the cotyledonary node, while none of the vertically oriented plants did. It is also worth noting that the orientation of these stem lignotubers, as well as those at the cotyledonary node, was always perpendicular to the orientation of the stem, i.e., downward (Fig. 1). In terms of lignotuber production at the cotyledonary node, the difference between treatments was also significant: 11 of the horizontally oriented greenhouse plants produced one lignotuber and none produced two; among vertically oriented plants, 4 produced one lignotuber at the cotyledonary node and 1 produced two.

Horizontal orientation also affected the fresh weight values of the year-1 stem segments of greenhouse seedlings. At the start of the experiment, there was no statistically significant difference in stem length between the two treatment groups, but after 1 year of growth, the mean fresh weight of the same stem segments was significantly greater in horizontally oriented plants than in vertically oriented plants ($P < .001$). In contrast, both length and weight of the year-2 stem segments of both groups were not significantly different.

Correlation analysis highlighted some interesting trends in lignotuber formation within treatment groups (Table 2). Among horizontally oriented greenhouse plants, both the length and the number of lignotubers that developed along the stem were significantly and positively correlated with experimental stem length. Among vertically oriented plants, lignotuber number and length were significantly and positively correlated with both experimental stem growth and total stem growth. For both horizontally and vertically oriented greenhouse plants, stem lignotuber fresh weight was highly positively correlated ($P < .001$) with stem, leaf, root, and total fresh weight. Dry weight data were omitted from the table because dry weight trends were statistically equivalent to those observed in fresh weight data for all parameters analyzed.

TABLE 2. Correlation analysis of mean number and length of *Ginkgo* lignotubers with stem lengths in two different growth orientation treatments

Treatment	Horizontal seedlings		Vertical seedlings	
	Mean total lignotuber number	Mean total lignotuber length	Mean total lignotuber number	Mean total lignotuber length
Greenhouse experiment				
Initial stem length	0.28	0.18	0.30	0.24
Experimental stem length	0.51**	0.59***	0.35*	0.38*
Total stem length	0.53**	0.58***	0.37*	0.38*
Field experiment				
Initial stem length	0.01	0.11	0.23	0.15
Experimental stem length	0.21	0.49***	0.20	0.32*
Total stem length	0.20	0.47***	0.21	0.32*

Values equal Pearson correlation coefficients.
*, $P < .05$; **, $P < .01$; ***, $P < .001$.

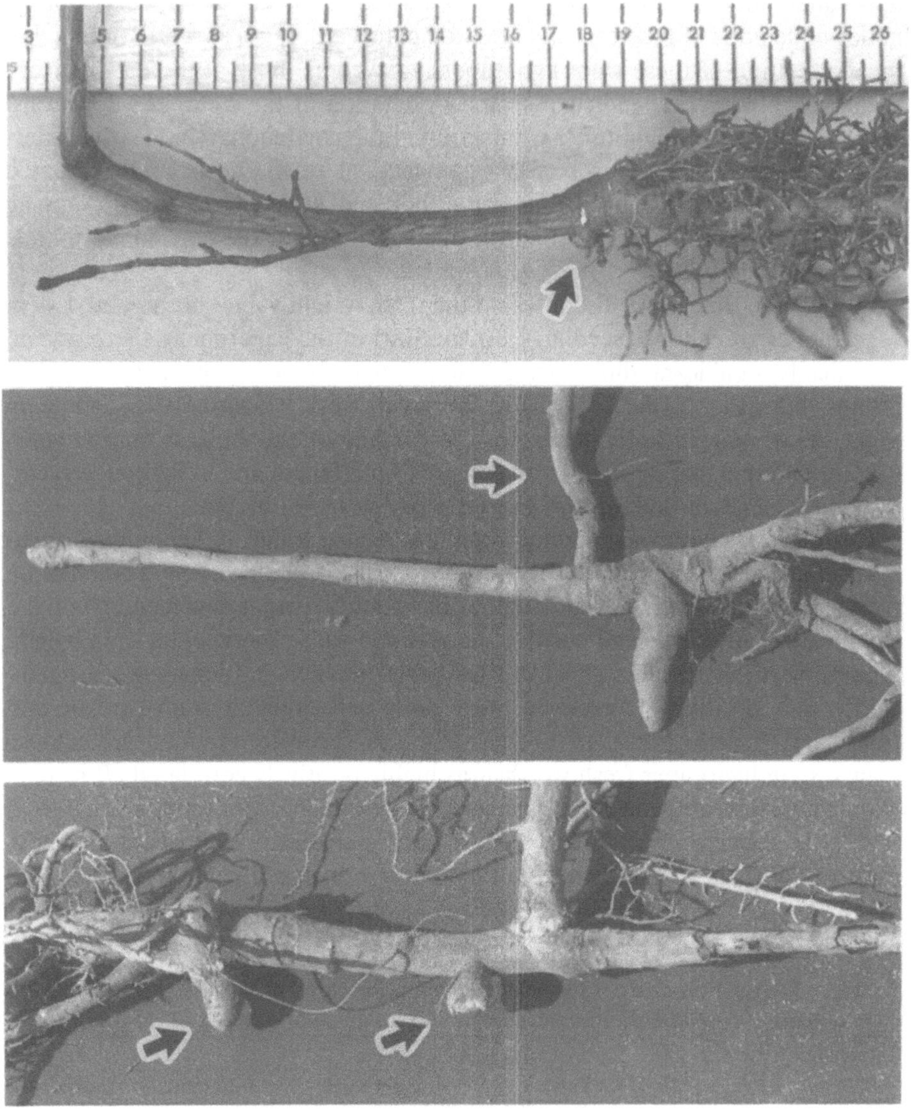

a

b

c

FIG. 1. **a** One of the 2-year-old horizontal *Ginkgo* seedlings grown in the greenhouse. Note the adventitious roots that formed along the horizontal stem; the lignotuber which developed at the cotyledonary node (*arrow*); and the year-2 growth that originated from the terminal bud. Scale is in centimeters. **b** One of the 4-year-old, horizontal *Ginkgo* seedlings grown outdoors at Sumter, SC (U.S.A.). *Arrow* shows the vertical, 2-year-old replacement shoot that originated from an axillary bud. Note the positive geotropic orientation of the 3-cm-long lignotuber that originated at the cotyledonary node. **c** One of the 4-year-old, horizontal *Ginkgo* seedlings grown outdoors at Sumter, SC. The vertical replacement shoot formed just below the junction of the year-1 and year-2 growth. Note the downward-growing lignotubers at the cotyledonary node and on the stem (*arrows*)

The field-grown plants showed growth responses that were substantially different from those of the greenhouse plants, probably reflecting the more stressful growing conditions that prevailed out-of-doors. For one thing, the vertically oriented, field-grown seedlings made two and a half times more stem growth than their horizontally oriented counterparts ($P < .001$). For another, only 60% of the horizontally oriented field-grown plants produced a vertical shoot from the original terminal bud, while 97% of the greenhouse plants did (Fig. 1a). Despite these differences, the data on lignotuber development in the field experiment paralleled the results of the greenhouse experiment in that the total number of lignotubers formed on the horizontally oriented plants was significantly greater than the number formed on the vertically oriented plants ($P = .014$) (Table 1). In addition, 30% of the horizontally oriented plants formed downward-growing lignotubers along their buried stems above the cotyledonary node (Figs. 1b,c). Within treatment groups, total lignotuber length was significantly and positively correlated with experimental and total stem length in both the horizontally oriented ($P < .001$) and vertically oriented ($P = .0211$) plants.

5 Conclusions

1. The horizontal treatment resulted in an increase in total lignotuber number and length in the greenhouse experiment and an increase in lignotuber number in the field experiment.
2. Correlation analyses of the data within treatment groups showed that the extent of lignotuber development in any given plant was correlated with its vigor as measured by total stem growth and fresh weight. In plants cultivated in the greenhouse, both total lignotuber number and length were positively correlated with total stem growth, while in field-grown plants only total lignotuber number was positively correlated with stem growth.
3. Regardless of the orientation of the seedlings, both stem and cotyledonary node lignotubers always grew with a strong downward orientation relative to the stem, indicating a positive geotropic growth response.
4. The statistically significant increase in weight of the buried year-1 stem segments of the horizontally oriented plants relative to those of the vertically oriented plants indicated that seedling orientation may affect carbohydrate allocation patterns. Further evidence for altered allocation was provided by the fact that fresh weights of both the stem and the roots were highly significantly different between treatment groups, while both leaf and total plant fresh weight were not. The production of stem lignotubers by the horizontally oriented seedlings was probably another manifestation of carbohydrate reallocation induced by horizontal orientation.
5. The data presented in this chapter suggest that lignotuber development in *Ginkgo* is the result of a complex interaction between environmental stress and individual seedling vigor.

Acknowledgments. The author thanks Kim Tripp of Smith College for her advice and assistance with the statistical analysis; J.P. Balz and C. Johnson of Garnay, Inc., for their cooperation in the field studies in South Carolina; and the greenhouse staff of the Arnold Arboretum of Harvard University for technical assistance.

References

1. Del Tredici P (1992) Natural regeneration of *Ginkgo biloba* from downward growing cotyledonary buds (basal chichi). Amer J Bot 79:522–530
2. Carr DJ, Jahnke R, Carr SGM (1984) Initiation, development, and anatomy of lignotubers in some species of *Eucalyptus*. Aust J Bot 32:415–437
3. Sealy JR (1949) The swollen stem-base in *Arbutus unedo*. Kew Bull 4:241–251
4. Molinas ML, Verdaguer D (1993) Lignotuber ontogeny in the cork-oak (*Quercus suber*: Fagaceae). II. Germination and young seedling. Amer J Bot 80:182–191
5. Del Tredici P (1997) Lignotuber formation in *Sequoia sempervirens*: developmental morphology and ecological significance. In: Edelin C (ed) 3rd international congress, "The Tree," 11–15 September 1995, Montpellier, France. Naturalia Monspeliensia, (in press)
6. Del Tredici P, Ling H, Yang G (1992) The *Ginkgos* of Tian Mu Shan. Conserv Biol 6:202–209
7. James S (1984) Lignotubers and burls—their structure, function and ecological significance in Mediterranean ecosystems. Bot Rev 50:225–266
8. Mesleard F, Lepart J (1989) Continuous basal sprouting from a lignotuber: *Arbutus unedo* L. and *Erica arborea* L., as woody Mediterranean examples. Oecologia (Berl) 80:127–131
9. Canadell J, Zedler PH (1994) Underground structures of woody plants in Mediterranean ecosystems of Australia, California, and Chile. In: Kalin Arroyo MT, Zedler PH, Fox MD (eds) Ecology and biogeography of Mediterranean ecosystems in Chile, California, and Australia. Springer, Berlin Heidelberg New York, pp 177–210
10. Fujii K (1895) On the nature and origin of so-called "chichi" (nipple) of *Ginkgo biloba* L. Bot Mag Tokyo 9:444–440
11. Li Z, Lin J (1991) Wood anatomy of the stalactite-like branches of *Ginkgo*. Int Assoc Wood Anat Bull 12:251–255
12. Del Tredici P (1993) *Ginkgo* chichi in nature, legend, and cultivation. Int Bonsai 15:20–25
13. SAS (1988) SAS procedures guide, 6.03. SAS Institute, Cary, NC

Cyclic Embryogenesis from Male and Female Protoplasts

JOCELYNE TRÉMOUILLAUX-GUILLER

1 Introduction

Ginkgo biloba (the maidenhair tree) of the family Ginkgoaceae is the single extant species of Order Ginkgoales, which appeared in the Permian era. This tree is distinguished by several remarkable features, such as its zoidogamous fertilization process, discovered by Hirase (1896) [1]. To assure growth of the zygotic embryo, the ovules of *G. biloba* store an abundance of reserve substances [2]. This species also produces molecules, ginkgolides and bilobalide, that are unique in the plant kingdom and show pharmaceutical properties. Ginkgolide B acts as an antagonist to platelet-activating factor (PAF), and has been reported as useful for the treatment of inflammatory reactions and immune diseases [3]. Unfortunately, the leaves of *Ginkgo* also contain small amounts of terpenoids, and the accumulation of metabolites is modified, depending on the conditions of culture and climate [4, 5]. In vitro cultures can provide an alternative strategy for increasing the ginkgolide accumulation. Indeed, numerous procedures used for in vitro culture are known to enhance biochemical variability and are potentially able to provide new molecules never detected in the mother plant [5, 6]. Nevertheless, the capacity to accumulate secondary metabolites seems to be connected with morphological differentiation [7] but is often low for a given cell line. To solve this problem, the establishment of differentiated tissue cultures can be investigated.

2 Brief Review of In Vitro Cultures

Few methodologies for the establishment of *Ginkgo biloba* cell cultures have been reported, with the exception of classical procedures leading to haploid tissue cultures [8–10] and some studies concerning the in vitro zygotic embryo development [11, 12]. To elucidate the ginkgolide biosynthesis pathways, some in

Laboratory of Cellular Biology and Plant Biochemistry, Faculty of Pharmaceutical Sciences, Francois Rabelais University, 31 Avenue Monge, 37200 Tours, France

vitro cultures were initiated starting in 1971 [13], but ginkolides A and B were not detected. More recently, calli and cell suspensions derived from zygotic embryos [14] or leaves were found to produce ginkgolides [15]. Subsequently, new investigations of gametophytic embryogenesis from isolated microspores [16, 17], immature prothallus-derived protoplasts [18, 19], and mature prothallus [20], as well as of somatic embryogenesis from mature [21] and immature zygotic embryos [22], were published. Various aspects of in vitro culture studies of *G. biloba* were also reviewed [23, 24].

3 Gametophytic Embryogenesis

3.1 Objectives

It would be of interest to select cell, organ, and plantlet cultures producing high terpenoid content or new molecules. Several strategies, such as cell cloning, have been devised to improve metabolite production [25–27]. Considered as a substantial source of genetic diversity and somaclonal variation [28], protoplast biotechnology offers a unique opportunity to regenerate plants, particularly when the protoplasts are prepared from embryogenic cell cultures [29]. Such a strategy has successfully been used in conifers [30–32]. Moreover, biochemical variants showing therapeutic properties have been selected from isogenic diploids [33]. This study reports the establishment of fast-growing gametophytic suspension cultures, the isolation, at high frequency, of totipotent protoplasts from previous competent cell suspensions, and protoplast-cloning procedures for developing embryos with a view to their regeneration into plantlets.

3.2 Establishment of Gametophytic Suspension Culture

3.2.1 Cultures from an Embryogenic Microspore-Derived Cell Line

Mesoblasts bearing male inflorescences were collected from a single tree in the Botanical Garden of Tours (from March 23 to April 3, 1992) to obtain microsporophylls containing uninucleate microspores. Isolated microspores directly developed into embryogenic colonies and single embryos when they were cultured in BN (Bourgin and Nitsch 1967 [34]) liquid medium with or without growth regulators [16].

The embryogenic colonies were successively transferred onto various solid media: BN medium supplemented with 11.61 μM kinetin and 0.0057 M sucrose (Gamborg et al. 1976 [35]), B5 solid medium with or without growth regulators, and B5 solid medium supplemented with 1.07 μM naphthalene acetic acid (NAA) and 8.87 μM benzylaminopurine (BA). The colonies then showed retarded growth. In May 1993, a white cell suspension (called MS1) was initiated from 0.92 g of friable embryogenic clusters introduced into 25 ml MT (Murashige and Tucker 1969 [36]) liquid medium containing 10.74 μM NAA and 0.93 μM kinetin (Table 1). The best growth was achieved after 3 weeks in agitated liquid medium,

TABLE 1. Modified liquid and solid media of Bourgin and Nitsch [34] and Nitsch and Nitsch [37] used for the *Ginkgo biloba* protoplast culture isolated from two gametophytic cell suspensions

	Liquid media												Solid media	
Medium	BN (0)	BN (1)	BN (2)	BN (3)	BN (4)	BN (5)	BN (6)	BN (7)	BN (8)	BN (9)	N0(L)	BN0(L)	N0(S)	BN0(S)
Mineral medium	BN	BN	BN	BN	BN	BN	BN	BN	BN	BN	N	BN	N	BN
Glucose (*M*)	0.25	0.25	0.25	0.25	0.25	0.25	0.25	0.25	0.25	0.25	0.40	0.35	0.40	0.35
Sorbitol (*M*)	0.35	0.35	0.35	0.35	0.35	0.35	0.35	0.35	0.35	0.35	0.10	0.20	0.10	0.20
NAA (μ*M*)	—	—	5.37	10.74	5.37	—	—	—	—	—	—	—	—	—
2,4-D (μ*M*)	—	—	—	—	—	—	—	—	—	9	—	—	—	—
Kinetin (μ*M*)	—	—	0.93	2.32	—	4.8	4.8	—	4.8	4.18	—	—	—	—
BA (μ*M*)	—	—	—	—	—	—	—	4.4	4.4	—	—	—	—	—
Coconut milk (g/l)	—	10	—	—	10	10	—	—	—	—	—	—		

BN, mineral medium of Bourgin and Nitsch [34] supplemented with vitamins (MS) of Murashige and Tucker [36] medium, including BN(0), BN(1), to BN(9); BN(L), BN liquid medium; BN(s), BN solid medium of Bourgin and Nitsch [34] with mineral salts and vitamins (BN); N, mineral salts and vitamins of Nitsch and Nitsch [37] liquid [BN0(L)] and solid medium [BN0(S)]. NAA, naphthalene acetic acid; BA, benzylaminopurine. For all media, the pH was adjusted to 6 before autoclaving.

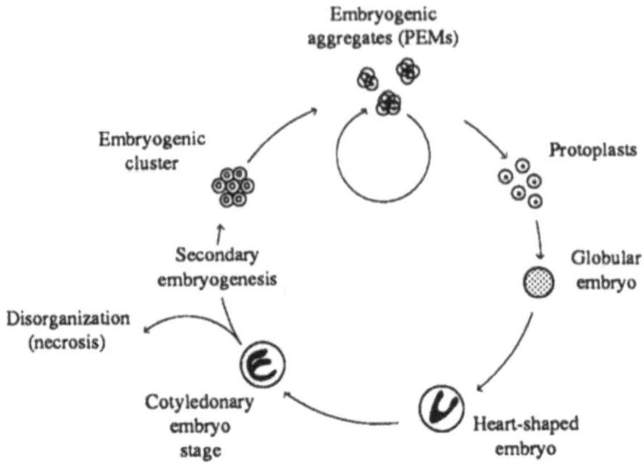

FIG. 1. Diagram of cyclic pattern of male and female gametophytic embryogenesis in protoplast cultures of *Ginkgo biloba*

and biomass increased by a factor of 5. By a single cycle similar to that established from protoplasts (summarized in Fig. 1), one fast-growing cream-colored cell line composed of clusters of small cells referred to as proembryogenic masses (PEMs) was obtained after several months of subculture.

3.2.2 Cultures from Prothallus-Derived Cell Lines

Ovules were harvested during a short period of time (July 9–18, 1994) from a female tree in the Botanical Garden of Tours. Fresh ovules were sterilized as described previously [18], and the prothalli (i.e., megagametophytes) were carefully removed. After cutting, mature prothallus halves were placed on various solid MT media, MT/2 and MT (half- and full-strength mineral MT medium, respectively). Both media were supplemented with 30 g/l sucrose, 8 g/l agar, and various combinations of growth regulators: NAA (26.85 μ*M*), NAA (10.74 μ*M*) + kinetin (kin) (0.93 μ*M*), NAA (10.74 μ*M*) + BA (0.88 μ*M*), NAA (1.07 μ*M*) + BA (8.87 μ*M*), NAA (1.07 μ*M*) + *Zea* (9.12 μ*M*), BA (8.87 μ*M*), or *Zea* (9.12 μ*M*). After 2 months of culture, white or green globular embryos developed directly from the surface cells at the periphery of some prothalli and developed up to the heart stage, as previously described [20]. When some prothalli turned brown, a few calli were initiated with a prothallus half-section. A cell suspension was prepared (November 15, 1993) by introducing 0.65 g of one yellowish-green callus into 25 ml MT liquid medium (see Table 1) supplemented with NAA (10.74 μ*M*) and kin (0.93 μ*M*), placed in an Erlenmeyer flask on a shaker set at 75 rpm and maintained at 24°C in the dark. As subcultures proceeded, compact cellular aggregates became small and dispersed; a fast proliferating cell line showing PEMs was established (Fig. 2a).

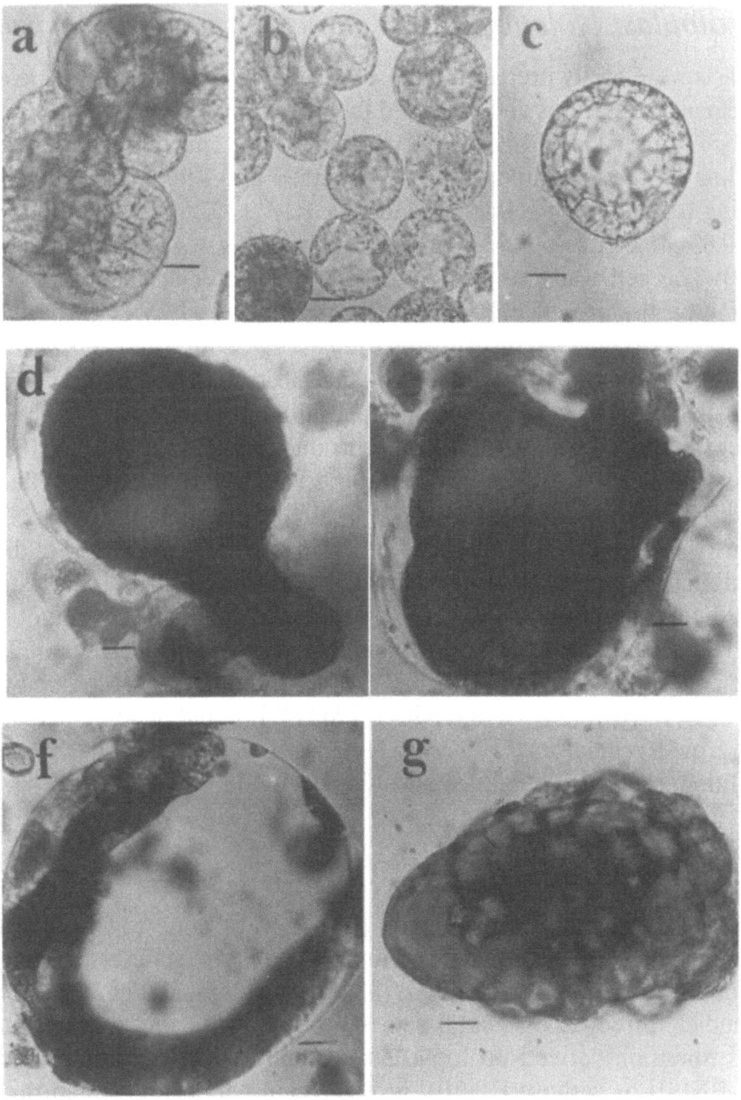

FIG. 2a–g. Embryo formation from female cell suspension-derived protoplasts. **a** An aggregate belonging to the female cell suspension. **b** Isolated protoplasts. **c** Microclone formation after 1 week. **d** A globular embryo with a small suspensor and two secondary embryos. **e** A protoplastderived heart-shaped embryo. **f** Female embryo at the curled cotyledonary stage. **g** A voluminous structure surrounded by a thick fibrous coating was visible after 1 week. Bar, 25 μm

3.3 Protoplast Isolation from Gametophytic Cell Suspensions

Protoplasts were easily prepared from 4- to 6-day-old male and female embryogenic suspension cultures and collected by filtration through a nylon sieve. From both subcultures, 620 ± 0.14, 650 ± 0.09, or 1100 ± 0.2 mg of fresh cell aggregates (depending on the experiments) were dispersed into 4 ml of enzymatic mixture containing 0.5 g% (w/v) cellulase Onozuka R10 + 0.25 g% (w/v) macerozyme R10 + 0.034 *M* $CaCl_2$ + 0.5 *M* sorbitol and/or 0.05 g% (w/v) pectolyase Sigma (pH 5.5). The cell wall was completely digested after an incubation of 2–8 h, depending on the presence or absence of pectolyase, at 25°C in the dark. The protoplasts were collected by centrifugation at 80 × *g* for 3 min and washed twice in a solution containing 0.7 *M* mannitol and 0.034 *M* $CaCl_2$. Large quantities of protoplasts, $1.6 \times 10^6 \pm 4.4$ protoclones per milliliter (P/ml) per g fresh weight of cells, were produced with a very good reproductibility in the ratio of 100% (Fig. 2). After purification, a viability test by Evans blue staining or fluorescein diacetate (FDA) revealed that 90%–95% of the protoplasts remained viable. Freshly isolated spherical protoplasts varied in size (58.8 ± 13.8 µ*M* in diameter) and exhibited a clear cytoplasm with small uniformly distributed plastids (Fig. 2b). Purified protoplasts were gently introduced in liquid or immobilized in gelrite (0.20 g%) or agarose (0.50 g%) solidified media at densities of 2.7 ± 2.8 10^4 P/ml and plated in small Petri dishes (55-mm-diameter). Modified BN [34] (BN(0) or BN), N(0) (Nitsch and Nitsch 1969 [37]) liquid or solid media, with or without growth regulators (GR), were tested (see Table 1). For each experimental condition, replicates of 5, 8, 9, and 12 were used.

3.4 Protoplast Culture

Additional experiments on protoplast culture were done to improve embryo maturation. Plant growth regulators, nutrient salts, or culture medium consistency can affect the potential embryo [38] or developmental embryo pattern. In these studies, three experimental sets are reported: in the first set, male and female suspension-derived protoplasts were cultured in BN liquid media with [BN(1)–BN(9)] or without [BN(0)] growth regulators (Table 1); in the second and third set, female protoplast cultures were placed in BN(0) or N(0) liquid and solidified media without any plant growth regulators (Table 1).

After 24–48 h, the viability of protoplasts ranged from 72.5% ± 20% (first set) to 75.5% ± 16.4% (second set) and 76.10% ± 11.20% (third set). Cell wall regeneration, indicated by changes in the shape of the cells, took place within 24, 36, or 48 h in all trial media. Simultaneously, first divisions started as early as 36–48 h for 14% ± 2.5% to 58% ± 9% of the viable protoplasts, showing that *G. biloba* protoplasts were able to divide within a very short period of time. Such a capacity has rarely been observed in woody plant protoplasts, because they usually require a longer period of time to enter into division and need a hormonal supply during the early stages of culture, as has been described for gymnosperms [29–32, 39]. Likewise, protoplasts prepared from crown gall cells showed an absolute requirement of exogenous hormones to start cell division [40], with the exception of some rare protoplasts of plant tumors [41, 42, 43] and explants [44, 18]. The *G. biloba* protoplasts divided by internal division or partition. From this latter process, the initial division could be oblique or transverse and symmetrical or asymmetrical, i.e., similar to that observed in the zygote, giving rise to a small apical cell (a future embryo) and a basal cell (a future suspensor) [45, 46]. An asymmetrical division of the egg cell is not a prerequisite for embryo development in all plants [47]. In *Ginkgo* protoplasts, the early stages of embryo development [47, 48] could be observed (Fig. 2c–f).

3.5 Embryo Formation

In the 3-day-old cultures the protoplast-derived cells increased notably in size and, by internal divisions, developed into microclones (i.e., globules of small round cells) before proembryo formation (Fig. 2c). After a week in BN liquid media (first set), the female microclone numbers were estimated respectively, at $3.3 \pm 0.7 \times 10^3$/ml and 0.96 ± 0.5–$3.8 \pm 1.1 \times 10^3$/ml without or with growth regulators (Table 2). The microclones evolved into proembryos and then into globular embryos 2 weeks later (Fig. 2d). From male protoplasts (first set) plated in the same BN media, the microclone and proembryo numbers varied from $2.15 \pm 0.20 \times 10^3$/ml (without GR) to 1.65 ± 0.35–$1.80 \pm 0.30 \times 10^3$/ml (with various GR) (Table 3). Gametophytic embryogenesis occurred directly from protoplasts without the intervention of a callus phase. A few proembryos began the first differentiation step by producing a protoderm (future epidermis) (Fig. 2). The

TABLE 2. Numbers and efficiencies (%) of microclones and embryos developed from *G. biloba* protoplast cultures derived from female cell suspension after 1 week or 2, 3, or 4 months of culture

Liquid media	Microclones (10^3/ml), 1 week	Embryos/ml			Embryo efficiency (%)		
		2 months	3 months	4 months	2 months	3 months	4 months
BN(0)	3.3 ± 0.7	227	297	197	2	2.7	1.80
BN(1)	1.6 ± 0.6	117	57	64	1	0.52	0.60
BN(2)	3.8 ± 1.1	204	70	96	1.85	0.64	0.87
BN(3)	2.6 ± 1.3	17	61	53	0.15	0.55	0.50
BN(4)	2.1 ± 1.3	ND	54	ND	ND	0.50	ND
BN(5)	2.4 ± 0.2	79	86	103	0.71	0.80	0.94
BN(6)	2.7 ± 0.8	73	96	109	0.70	0.87	1
BN(7)	0.96 ± 0.5	ND	ND	ND	ND	ND	ND
BN(8)	2.7 ± 0.6	65	68	ND	0.60	0.62	ND
BN(9)	1.7 ± 0.4	41	38	ND	0.37	0.34	ND

BN, Bourgin and Nitsch [34] modified liquid medium. Embryo efficiency, the number of embryos divided by the number of protoplasts originally cultured ×100. For each BN modified liquid medium, BN(0) and BN(1)–BN(9), averages of 12 and 5–8 replicates, respectively, were realized. ND, not determined.

TABLE 3. Numbers and efficiencies (%) of embryos developed from *G. biloba* protoplast cultures derived from male cell suspension after 4 and 8 weeks

Liquid media	Microclones (10^3/ml) and proembryos, 3 weeks	Embryos/ml		Embryo efficiency (%)	
		4 months	8 months	4 months	8 months
BN(0)	3500 ± 200	245	143	0.98	0.57
BN(1)	2450 ± 30	162	101	0.65	0.40
BN(2)	3500 ± 200	223	156	0.90	0.62
BN(4)	2800 ± 350	143	126	0.60	0.50
BN(5)	ND	99	76	0.40	0.30
BN(8)	2950 ± 500	286	194	1.15	0.78

BN, Bourgin and Nitsch [34] modified liquid medium. Embryo efficiency, the number of embryos divided by the number of the protoplasts originally cultured ×100. For each BN modified liquid medium, BN(0) and BN(1)–BN(8), averages of 8 and 5–6 replicates, respectively, were male. ND, not determined.

others evolved into embryos forming a dark central core of tightly packed cells that subsequently became brown. Proembryos or embryos with polymorphous suspensor-like structures were observed.

3.6 Embryo Development

After 1 month of culture, embryos were seen in all tested media, and their numbers and efficiencies were estimated after 2–3 or 4–8 months. The efficiency (%) was defined as the number of embryos divided by the number of protoplasts ×100.

3.6.1 First Set

In BN liquid media (see Table 1), the numbers of female protoplast-derived embryos were, respectively, 227, 297, and 197 embryos per milliliter (E/ml) (without GR) and 41–204, 38–96, and 53–109 E/ml (with GR); their respective efficiencies were 2%, 2.7%, and 1.8% (without GR) and 0.76 ± 0.50, 0.60 ± 0.20, and 0.80 ± 0.20 (with GR) (Table 2) in 2, 3, or 4 months of the culture. The numbers and efficiences, respectively, of the male embryos were estimated at 245 and 143 E/ml; 0.98% and 0.60% (without GR) and 99–286 and 76–194 E/ml; and 0.74 ± 0.30 and 0.52% ± 0.18% (with GR) after 4–8 months (Table 3). The present results are in agreement with those previously obtained [17, 19]. The highest embryo numbers and efficiencies were found in BN(0) hormone-free media. Nevertheless, BN(2) medium supplemented with NAA (5.37 μ*M*) + kin (0.93 μ*M*) permitted some embryo development. After 2 months of culture, the embryo growth slowed down in all media. Subsequently, embryos that were 3–4 months older were observed at different developmental stages and their frequencies counted for male or female cultures with [BN(1)–BN(8)] or without [BN(0)] hormones (Table 4). The more complete embryo development and highest maturation frequencies seemed be correlated with the absence of exogenous growth

TABLE 4. Various developmental embryo stages from male and female suspension-derived protoplasts after 3 and 4 months

Developmental stages	Male embryos/ml after 3 months					Female embryos/ml after 4 months				
	G	O	$H_{(s)}$	T	Co	G	O	$H_{(s)}$	T	Co
BN(0)	92	38	9.5	3.2	3	70	70	32	18.5	6.5
BN(1)	60	32	9.5	—	—	14	4	3	6	—
BN(2)	60	38	24	2	2	27	27	29	11	2
BN(3)	ND	ND	ND	ND	ND	13	16	19	5	—
BN(4)	76	48	2	—	—	ND	ND	ND	ND	ND
BN(5)	54	22	—	—	—	6	68	13	13	3
BN(6)	ND	ND	ND	ND	ND	19	48	19	19	4
BN(8)	108	67	13	3	3	ND	ND	ND	ND	ND

BN, Bourgin and Nitsch [34] modified liquid medium; G, globular; O, oblong; $H_{(s)}$, heart shape; T, torpedo shape; Co, cotyledonary stage. ND, not determined. (−) (not detected).

regulators (Table 4). The presence of these can be interesting, however, to increase the formation of biochemical variants. Careful microscopic examination showed that female embryos developed as linear and curled cotyledon stages (Fig. 2f) similar to the zygotic embryo [47].

3.6.2 Second Set

To compare the effect of solid and liquid media on embryogenesis, female protoplasts were cultured in BN(0)L liquid and BN(0)S solid hormone-free medium. Between 2 and 3 months of culture, the embryo numbers and their efficiencies [368 ± 56 E/ml, 3.8% ± 0.70% in BN(0)L, and 585 ± 209 E/ml, 6% ± 2.20% in BN(0)S] were approximatively identical in BN(0) solid or liquid media.

3.6.3 Third Set

After 6–7 months of culture, some embryos evolved into transferable protocolonies or protoclones by secondary embryogenesis. The female (FP) and male (MP) protoclones were macroscopically visible, and their respective frequencies (protoclones or P/ml) and efficiencies (%) in the liquid or solid hormone-free media (Table 1) were estimated, respectively, at 26 ± 3 FP/ml and 0.31% ± 0.10% in N(0)L, 25 ± 9.2 FP/ml and 0.27% ± 0.10% in N(0)S, 39.6 ± 19.6 MP/ml and 0.42% ± 0.14% and 25 FP/ml and 0.26% in BN(0)L, then 34.7 ± 12 FP/ml and 0.37% ± 0.18% in BN(2) medium (with GR) (Table 1). At this time, about 20 colonies were transferred to solid medium supplemented with NAA and kin to assure rapid proliferation. BN(0), BN(2), and N(0) solid or liquid media showed a similar capacity to produce transferable embryo clusters through cyclic embryogenesis from gametophytic protoplast cultures (Fig. 2).

4 Conclusions

Two rapidly proliferating cell suspension cultures were established, one from embryo colonies derived from microspores and the other from a prothallus-derived callus, and both were maintained in the presence of growth regulators by regular transfers at 2-week intervals. Such cells, in spite of the influence of prolonged auxin treatment, were capable of dividing, of adhering to each other, and of producing cell aggregates or PEMs [45]. Morever, these suspensions showed a high capacity to liberate totipotent protoplasts that were capable of regenerating cell walls, dividing in growth regulator-free media, and giving rise to microclones. The latter developed into individual globular embryos evolving up to the heart, torpedo, or (exceptionally) cotyledonary stages (Fig. 1). The embryogenesis was estimated at 320 ± 143 E/ml (i.e., approximatively 504 E/g of fresh cells), depending on the media.

The absence of growth regulators was a prerequisite for protoplast development into embryos. After the removal of auxin, the protoplasts (i.e., PEM cells) were released to complete their developmental program into embryos, as reported in hardwood trees [49]. After the fourth month of culture, the embryo numbers decreased notably (Tables 2, 3). Indeed, the "embryo-defective" were from later maturation stages that had entered necrosis (Fig. 1). At the same time, a beneficial secondary embryogenesis was observed from 4- to 6-month-old embryos (at various stages) that were capable of forming embryo colonies and cell suspensions. PEM-derived protoplasts generated a cyclic embryogenesis similar to that observed in microspores (Fig. 1).

A repetitive protoplast-derived embryogenesis was achieved in the media devoid of all growth regulators or supplemented with auxin. Auxins affect zygotic morphogenesis [46]. This repetitive phenomenon might be induced by high endogenous auxin levels giving rise to a premature change in the embryo morphology (from globular embryo to heart stage) to generate defective embryos (devoid of thickness) and secondary embryogenesis leading to colonies of small globular embryos. The addition of inhibitors of auxin synthesis to the culture media might stimulate embryo development [50]. Theoretically, such a cyclic system would be perpetually totipotent [51]. The male and female protoclones will be submitted to assays of regeneration and, subsequently, their ginkgolide contents will be investigated.

From a recent unpublished study, we report a female cell line that was reactivated in March 1996. From this line a subpopulation of protoplasts (data not shown) was liberated and gave rise to globular and spindle-shaped structures (51 ± 6.6/ml), enclosed within a thickened and highly colored cell wall (flavonoids and polyphenols), and macroscopically visible after 1 week of culture (Fig. 2g). We speculate that this orange fibrous coating, turning brown, would be able to provide a biochemical *Ginkgo* embryogenesis marker. Indeed, it has been postulated that certain proteoglycans influence in vitro embryogenesis [52] by their effect on particular cell wall polymers [53].

Acknowledgments. The author thanks Professor Louise Jacquet-Hoag for critical reading of the English manuscript.

References

1. Hirase S (1896) Sur le spermatozoïde de *Ginkgo* (in Japanese). Bot Mag Tokyo 10:325–327
2. Favre-Duchartre M (1958) *Ginkgo*, an oviparous plant. Phytomorphology 8:377–390
3. Braquet P (1987) The ginkgolides: potent platelet activating factor antagonists isolated from *Ginkgo biloba* L: chemistry, pharmacology and clinical application. Drugs Future 12:643–699
4. Van Beek TA, Secheeren HA, Rantio T, Melger WC, Lelyveld GP (1991) Determination of ginkgolides and bilobalide in *Ginkgo biloba* leaves and phytopharmaceuticals. J Chromatogr 543:375–387
5. Barroso JG, Miguel MG, Figueirodo AC (1993) Production of secondary metabolites in plant cell cultures. In: VII international colloquy of medicinal plants and substances of natural origin, Angers, pp 70–82
6. Phillipson JD (1990) Possibilities of finding new products from plant cell cultures. In: Nigkamp HJJ, van der Plas LHW, van Aartryk J (eds) Progress in plant cellular and molecular biology. Kluwer, Dordrecht, pp 592–600
7. Verpoort R, van der Heijden R, Hoge JHC, Ten Hoopen HJG (1993) Plant cell cultures for the production of alkaloids: feasibility and prospects. In: VII international colloquy of medicinal plants and substances of natural origin, Angers, pp 83–97
8. Tulecke WR (1953) A tissue derived from the pollen of *Ginkgo biloba*. Science 117:599–600
9. Tulecke WR (1957) The pollen of *Ginkgo biloba*: in vitro culture and tissue formation. Amer J Bot 44:602–608
10. Tulecke WR (1964) A haploid tissue from the female gametophyte of *Ginkgo biloba*. Nature 203:94–95
11. Wang FH, Chen TK (1965) Experimental studies of young *Ginkgo* embryos. The effect of coconut milk on the embryos cultured in vitro. Acta Bot Sin 13:364–375
12. Wang FH, Lee SC (1966) Experimental studies on young *Ginkgo* embryos. The effect of casein hydrolysate, glutamine, asparagine, 2,4-D, and adenine sulphate on embryos grown in vitro. Kexue Tongbao 17:323–327
13. Nakanishi K, Habaguchi K (1971) Biosynthesis of ginkgolide B, its diterpenoid nature, and origin of the *tert*-butyl group. J Amer Chem Soc 93:3546–3547
14. Carrier J, Cosentino G, Neufeld R, Rho D, Weber M, Archambault J (1990) Nutritional and hormonal requirements of *Ginkgo biloba* embryo-derived callus and suspension cell culture. Plant Cell Rep 8:635–638
15. Jeon MH, Sung SH, Huch H, Kim YC (1995) Ginkgolide B production in cultured cells derived from *Ginkgo biloba* L. leaves. Plant Cell Rep 14:501–504
16. Laurain D, Trémouillaux-Guiller J, Chénieux JC (1993) Embryogenesis from microspores of *Ginkgo biloba* L., a medicinal woody species. Plant Cell Rep 12:501–505
17. Trémouillaux-Guiller J, Laurain D, Chénieux JC (1996) Microspore and protoplast culture in *Ginkgo biloba*. In: Jain SM, Sopory SK, Veilleux RE (eds) In vitro haploid production in higher plants Vol 3. Kluwer, Dordrecht, pp 275–295

18. Laurain D, Chénieux JC, Trémouillaux-Guiller J (1993) Direct embryogenesis from female haploid protoplasts of *Ginkgo biloba* L., a medicinal woody species. Plant Cell Rep 12:656–660
19. Trémouillaux-Guiller J, Laurain D, Chénieux JC (1996b) Direct embryogenesis in protoplasts of *Ginkgo biloba* (maidenhair tree). In: Bajaj YPS (ed) Plant protoplasts and genetic engineering. Biotechnology in agriculture and forestry, vol 38. Springer, Berlin Heidelberg New York, pp 33–47
20. Fontanel A, Serraf I, Pétiard V (1994) Essais de régénération chez le *Ginkgo biloba* L: aptitude à l'embryogenèse du mégagamétophyte selon le stade de développment. CR Acad Sci Paris Sci Vie 317:149–155
21. Yates WF (1986) Induction of embryogenesis in embryo-derived callus of *Ginkgo biloba* L. (abstract). In: VI international congress on plant tissue culture, University of Minnesota, Minneapolis
22. Laurain D, Chénieux JC, Trémouillaux-Guiller J (1996) Somatic embryogenesis from immature zygotic embryos of *Ginkgo biloba*. Plant Cell Tissue Organ Cult 44:19–24
23. Rohr R (1989) Maidenhair tree (*Ginkgo biloba* L.) In: YPS Bajaj (ed) Biotechnology in agriculture and forestry, vol 5. Trees. II. Springer, Berlin Heidelberg New York, pp 251–257
24. Carrier J, Chauret N, Neufeld R, Archambault J (1994) *Ginkgo biloba* L. (maidenhair tree): in vitro culture and the formation of gingolides. In: YPS Bajaj (ed) Biotechnology in agriculture and forestry, vol 26. Medicinal and aromatic plants. VII. Springer, Berlin Heidelberg New York, pp 136–145
25. Constabel FJ, Kambold S, Chatson KB, Kurz WGW, Kutney JP (1981) Alkaloid production in *Catharanthus roseus* (L.) G-Don. VI. Variation in alkaloid spectra of cell lines derived from one single leaf. Plant Cell Rep 1:3–5
26. Fujita Y, Takahashi S, Yamasa Y (1985) Selection of cell lines with high production of shikonin derivatives by protoplast cultures of *Lithospermum erythrorhizon* cells. Agric Biol Chem 49:1755–1759
27. Trémouillaux-Guiller J, Andreu F, Crèche J, Chénieux JC, Rideau M (1987) Variability in tissues cultures of *Choisya ternata*. Alkaloïd accumulation in protoclones and aggregate clones obtained from established strains. Plant Cell Rep 6:375–378
28. Bajaj YPS, Furmanowa M, Olszowska O (1988) Biotechnology of the micropropagation of medical and aromatic plants. In: Bajaj YPS (ed) Biotechnology in agriculture and foresty, vol 4. Medicinal and aromatic plants. I. Springer, Berlin Heidelberg New York, pp 58–94
29. Klimaszewska K (1989) Recovery of somatic embryos and plantlets from protoplast cultures of *Larix eurolepsis*. Plant Cell Rep 8:440–444
30. Gupta PK, Durzan DJ (1987) Somatic embryos from protoplasts of loblolly pine proembryonal cells. Bio/Technology 5:710–712
31. Attree SM, Bekkaoui F, Dunstan DI, Fowke LC (1987) Regeneration of somatic embryos from protoplasts isolated from an embryogenic suspension culture of white spruce (*Picea glauca*). Plant Cell Rep 6:480–483
32. Hartmann S, Lang H, Reuther G (1992) Differentiation of somatic embryos from protoplasts isolated from embryogenic suspension cultures of *Abies alba* L. Plant Cell Rep 11:554–557
33. Sangwan-Norreel BS, Sangwan RS, Pare J (1986) Haploïdie et embryogénèse provoquée in vitro. Bull Soc Bot Fr Actual Bot 133(4):7–39
34. Bourgin JP, Nitsch JP (1967) Obtention de *Nicotiana* haploïdes à partir d'étamines cultivées in vitro. Ann Physiol Veg (Paris) 9:377–382

35. Gamborg OL, Murashige T, Thorpe TA, Vasil IK (1976) Plant tissue culture media. In Vitro 12:473–478
36. Murashige T, Tucker DPH (1969) Growth factor requirements of *Citrus* tissues cultures. In: Chapman JV (ed) Proceedings, 1st international citrus symposuin, University of California at Riverside, 3, pp 1155–1161
37. Nitsch JP, Nitsch C (1969) Haploid plants from pollen grains. Science 163:85–87
38. Nadel BL, Altman A, Ziv M (1996) Somatic embryogenesis and synthetic seed in *Apium graveolens* (celery). In: Bajaj YPS (ed) Somatic embryogenesis and synthetic seed. II. Biotechnology in agriculture and forestry, vol 31. Springer, Berlin Heidelberg New York, pp 306–322
39. Lang H, Kohlenbach H (1989) Cell differentiation in protoplast cultures from embryogenic callus of *Abies alba* L. Plant Cell Rep 8:120–123
40. Scowcroft WR, Davey MR, Power JB (1973) Crown gall protoplasts isolation, culture and ultrastructure. Plant Sci Lett 1:451–456
41. Pedersen HC, Christiansen J, Wyndaele R (1983) Induction and in vitro culture of soybean crown gall tumors. Plant Cell Rep 2:201–204
42. Phillips R, Darrell NJ (1988) A simple technique for single-cell cloning of crown gall tumour tissue; *Petunia* protoplast regeneration without exogenous hormones. J Plant Physiol 133:447–451
43. Trémouillaux-Guiller J, Kodja H, Chénieux JC (1994) Single-cell cloning of a crown gall from protoplasts regenerated in hormone-free medium. Establishment of pure transformed cell lines of *Catharanthus roseus* G. Don. Plant Cell Tissue Organ Cult 37:25–30
44. Harms C, Potrykus I (1980) Hormone inhibition of *Citrus* protoplasts released by coculturing with *Nicotiana tabacum* protoplasts: its significance for somatic hybrid selection. Plant Sci Lett 19:125–301
45. Emons AMC (1994) Somatic embryogenesis: cell biological aspects. Acta Bot Neerl 43:1–14
46. Goldberg RB, de Paiva G, Yadegari R (1994) Plant embryogenesis: zygote to seed. Science 266:605–614
47. West MAL, Harada JJ (1993) Embryogenesis in higher plants: an overview. Plant Cell 5:1361–1369
48. Gerd J (1993) Axis formation in plant embryogenesis cues and lues. Cell 81:467–470
49. Merkle SA (1995) Strategies for dealing with limitations of somatic embryogenesis in hardwood trees. Plant Tissue Cult Biotechnol 1:113–121
50. Tulecke W (1987) Somatic embryogenesis in woody perennials. In: Bonga JM, Durzan DJ (eds) Cell and tissue culture in forestry, vol 2. Specific principles and methods: growth and developments. Nijhoff, Dordrecht, pp 61–91
51. Conger BV, Trigiano KN, Gray DJ, McDaniel JK (1995) Somatic embryogenesis in orchardgrass (*Dactylis glomerata* L.) In: Bajaj YPS (ed) Somatic embryogenesis and synthetic seed. II. Biotechnology in agriculture and forestry, vol 31. Springer, Berlin Heidelberg New York, pp 70–80
52. Kreuger M, van Holst GJ (1993) Arabinogalactan proteins of *Daucus carota* L. Planta 189:243–248
53. de Jong AJ, Schimdt EDL, de Vries SC (1993) Early events in higher plant embryogenesis. Plant Mol Biol 22:367–377

Tissue Culture Studies on *Ginkgo biloba*

WALTER TULECKE

1 Introduction

More than a half century of in vitro studies on *Ginkgo biloba* L. presents an opportunity for future research on one of the oldest, most respected trees on the planet. The unusualness of *Ginkgo* as an extant species with considerable heterozygosity and vigor makes it especially attractive for both fundamental and applied studies.

This chapter reviews the research that has been done in vitro and seeks to provide a framework and perspective that will be helpful for current and future work. Other contributors to this volume should be consulted for additional information on particular aspects, especially the chapters by Friedman and Gifford, Rohr, and Tremouillaux-Guiller.

2 Developmental Studies

Normal development of the male gametophyte of *Ginkgo* in vivo is reviewed elsewhere in this volume by Friedman and Gifford; the development of the female gametophyte is presented by Soma and the ultrastructural aspects of both the male and female gametophytes by Rohr. In vitro studies on the embryo and the male and female gametophytes of *Ginkgo* are covered here.

2.1 Embryo Culture

Among the earlier studies on *Ginkgo* embryos, Li [1–3] reported the accelerated growth of embryos that were removed from the female gametophyte nutritive tissue of the seed and cultured on a nutrient medium. Li emphasized the importance of adequate aeration and observed the rapid growth of the primary root

---

Department of Science, Antioch College, Yellow Springs, OH 45387, U.S.A.

and the greening of the cotyledons in light. Bulard and co-workers [4–6] used a defined medium with added growth substances such as gibberellic acid, which was found to inhibit growth even at low concentrations, and kinetin, which inhibited the root. The absorption of the nutrients was through the cotyledons, similar to that which occurs when the embryo is in the seed. Ball [7–9] extended these studies and used the simple media of Tukey or Knop. He also included anatomical observations on the origin of the root and the effects of longitudinally bisecting the hypocotyl; this produced two roots.

Very young embryos of *Ginkgo* within the archegonia of the female gametophyte were grown on a complex medium with yeast extract by Radforth [10]. Equally small embryos (1.0–1.6mm) were excised by Wang and co-workers [11–13] and placed on a variety of media containing coconut milk, casein hydrolysate, or glutamate, all of which promoted growth, or adenine sulfate, bee royal jelly, or 2,4-dichlorophenoxyacetic acid (2,4-D), which had no effect; asparagine inhibited growth. More recent studies by Webb et al. [14] confirmed the need for adequate aeration as the use of Parafilm closures inhibited embryo growth and promoted callus formation.

2.2 The Female Gametophyte

The excellent, detailed studies by Favre-Duchartre [15] on the cytology and the development of the reproductive tissues of *Ginkgo* were useful for many studies that followed. His collaborator, M.H. Sachet [15], cultured the female gametophyte of *Ginkgo* on a simple Knops solution with added glucose and thiamin and reported on the cytological changes over a 2-month period. Rohr [16–21] used female gametophytes that were 4–6 months post pollination. He used the medium of Bourgin and Nitsch, adding the trace elements of Murashige and Skoog, with the omission of potassium iodide and the addition of 1.0mg/l 2,4-D and 50–100ml of coconut milk. Rohr described the abnormal growth of cells around the archegonia and the growth of a white tissue from the apical end of the gametophyte. He also observed the complete development of the zygotic embryos into entire plants.

Green haploid tissue cultures from the female gametophyte of *Ginkgo* [22] were obtained by removing the tissue from surface-sterilized ovules at 11–13 weeks post pollination. Other earlier stages were tried but without success. The gametophyte was started on White's medium with 2,4-D and coconut water, which contains an array of amino acids, salts, and growth substances [23]. Later, this green tissue, an albino tissue derived from it, the white pollen tissue, and green tissue cultures derived from stem and leaf tissue of *Ginkgo* were analyzed and compared; their free amino acid pools were very different [24–26]. All these tissues were grown by subculturing continuously for periods of 2 to 15 years. The haploid green tissue from the female gametophyte contained resin canals, tracheary elements, and crystal druses of oxalate. In a few cultures some roots and abortive shoots were observed.

2.3 The Male Gametophyte

A clear picture of the development of the male gametophyte within the nucellar tissue of the developing ovule of *Ginkgo* is provided by the observations and results of Friedman [27] and Friedman and Gifford [28] and Gifford and Lin [29]. They reported an anticlinal annular ringlike division of the generative cell to give rise to the stalk and the spermatogenous cell. This discounted the protrusion theory, even though actual division figures were not observed. The three-dimensional representation of the male gametophyte was obtained by serial sections and computer graphic digitized models to give coherent pictures. The spermatogenous cell with its blepharoplast, astral rays, and osmiophilic granules was observed by both microscopic and ultrastructural means to show these developmental events.

Friedman [27] reported on extensive in vitro studies of the male gametophyte, using a medium similar to Rohr's [16]. Friedman confirmed that the level of sucrose did influence gametophyte development and the starch accumulation that occurs early in pollen germination; lower sucrose increased gametophyte branching. He reported that only "a small percentage of the gametophytes in

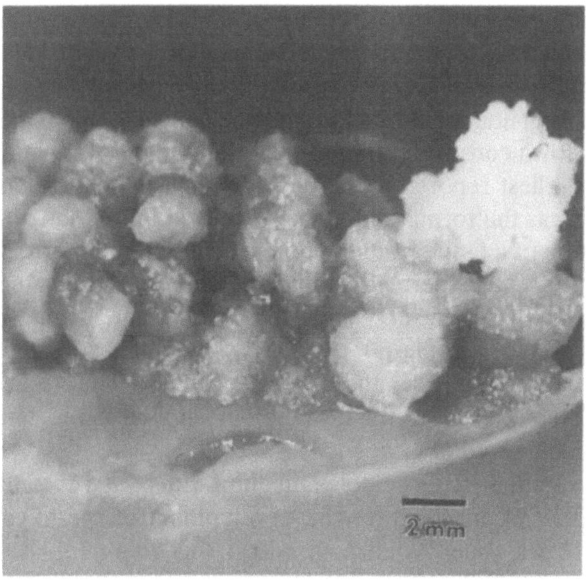

FIG. 1. The first tissue culture from a pollen source, *Ginkgo biloba*, observed growing from a sporangium of a strobilus that was placed in culture. The strobilus was surface sterilized and placed in culture May 3, 1951, on White's medium plus one-third tomato juice by volume. The tissue was observed Feb. 2, 1952, and subcultured as a continuously growing clone on White's plus 0.25% yeast extract plus 1.0 mg/l indole-3-acetic acid

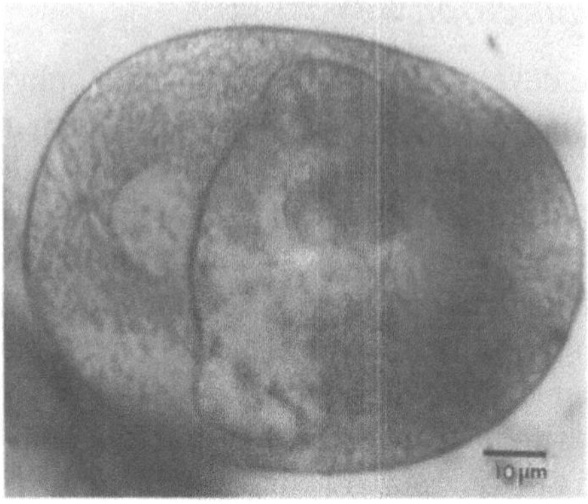

FIG. 2. The spermatogenous cell of a *Ginkgo* male gametophyte in culture before division to form two sperm cells. The astral rays, nucleus, nucleolus, and osmiophilic granules characteristic of a normally developing spermatogenous cell are present

culture initiate division of the generative cell." Late stages of development were not examined, and he did not report sperm cell formation.

An interesting and useful review by Rohr [21] covers many studies done in vitro on the male gametophyte of *Ginkgo*. Rohr used Bourgin and Nitsch medium without adenine for culturing the male gametophyte. He reported the reproductive cell (generative cell) gave rise to a basal cell and a gamete mother cell (the spermatogenous cell). He also noted the development of sperm cells on a medium of known constitutents, although the number was small.

Among the earliest reports on the development of the male gametophyte in culture [30–32] was the formation of sperm cells in more than 50 gametophytes and the formation of a tissue from the pollen (Fig. 1). The sperm cells developed from the large nucleated spermatogenous cell (Fig. 2) (designated the body cell by Favre-Duchartre [15], Lee [34], and Tulecke [32]) and included the astral rays (microtubules or blepharoplasts) and the osmiophilic granules (extranuclear body of Favre Duchartre, Lee, and Tulecke) (Fig. 3). The studies of Friedman and Gifford [28] have clarified the terminology applied to the various structures of the male gametophyte of *Ginkgo*.

In these early studies [30], a complex medium with added coconut milk or yeast extract was used and provided for the development of sperm cells (Fig. 3). In addition, "there was one instance in which the blepharoplasts of a pair of sperm cells differentiated to form spiral bands of two revolutions, but they lacked cilia" (Fig. 4). These were regarded as immature sperm cells because fully mature sperm cells developed in vitro would be observed swimming within the pollen tube or in an appropriate medium. The culture of male gametophytes in which the spiral bands were observed was done on May 2, 1952, and the sperm cells were seen July 5, 1952; this development occurred on yeast extract medium with inoculations of fresh pollen. Observations on 56 pairs of sperm cells

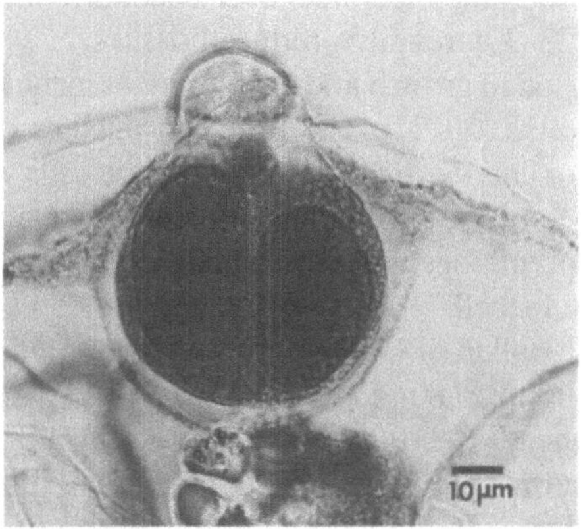

FIG. 3. A pair of immature sperm cells with the original exine of the pollen grain visible above them

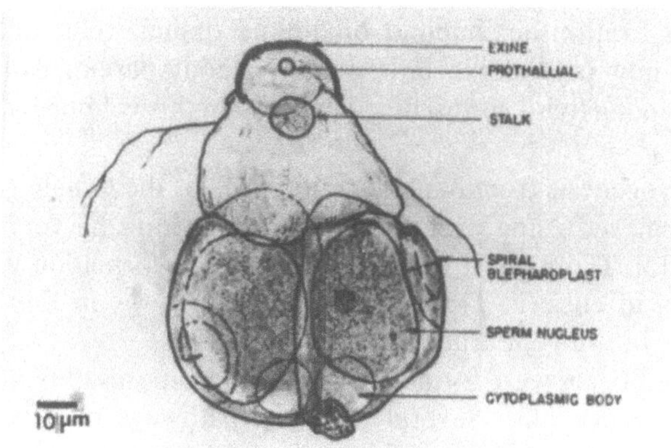

FIG. 4. The spiral blepharoplasts of two sperm cells within the spermatogenous cell. This was the most advanced stage of development of the male gametophyte observed in vitro

showed that the orientation of the two daughter cells to the stalk cells, prothallial cells, and the exine of the pollen grain was precisely what occurs in vivo.

3 Tissue Cultures of *Ginkgo*

The first tissue culture of a pollen was obtained from *Ginkgo* pollen by collecting mature strobili in May 1951, surface-sterilizing them, and placing them on a complex medium of White's plus tomato juice. The growth of a tissue was first observed in February 1952 (Fig. 1). Subsequent studies with fresh and preserved pollen grown on various media, some with coconut water [25] and others of known constituents [35–39], readily produced tissues.

Studies on normal pollen growth and development included work on the origin of the tissue from pollen [30–32], The tube cell nucleus often divided to form septate pollen tubes or coenocytic gametophyte tube cells with as many as 20–40 nuclei. The quiescence of the prothallial cell and the relatively infrequent divisions of the stalk or spermatogenous cell suggested that the tissues arose from the cell divisions of the tube cell.

The tissue from the pollen of *Ginkgo* grew well in liquid media [35–39], and it was possible to obtain up to 80g/l (fresh weight) of tissue at the peak phase of growth at 3 weeks when 7- to 10-l volumes were grown in a continuous culture system designed to permit the harvest of tissue and the replacement of an equal volume of fresh medium. The most successful culture, of 10-l, persisted for 76 days with periodic harvest of tissue and replacement of medium [37].

4 Embryogenesis

Embryogenesis from either haploid or diploid tissues, cells, or protoplasts of many plants is now possible, including many woody perennials [40]. *Ginkgo* is now included among these, as discussed in detail in this volume by Tremouillaux-Guiller.

Direct embryogenesis from haploid protoplasts of the female gametophyte of *Ginkgo* was achieved using protoplasts prepared from the very early stage of septation [41, 42]. The highest frequency of embryo formation was 1.5%–1.9% after 3 months in culture. This system of gynogenesis in *Ginkgo* offers the opportunity for future gene transfer.

Androgenic embryos were formed from the microspores of *Ginkgo* cultured at the uninucleate stage [43]. Several different pathways of development were suggested as the origin of these embryos. For success with this protocol, it was necessary to have a favorable uninucleate stage of microspore development, suitable cell densities, and appropriate culture media. Electrostimulation of some cultures seemed to be responsible for early embryo formation.

Embryogenesis was induced in female gametophyte tissues of *Ginkgo* ovules taken from trees in Argentina, Australia, and Chile [44]. The tissues were cultured on eight different media designed for the induction, expression, and development of embryos. The most embryogenic cultures were from ovules obtained from Chile; the highest percentage of ovules that produced embryos were from those cultured at 8–11 weeks post pollination, the "critical period" for culture. Meristematic activity leading to embryo formation occurred in the apical and marginal cells of the female gametophyte (Figs. 5 and 6). The embryos reported from this work were not fully developed or germinated to give plants. Other experiments designed to induce embryogenesis from leaf tissue or from zygotic embryos produced only callus or green excrescences.

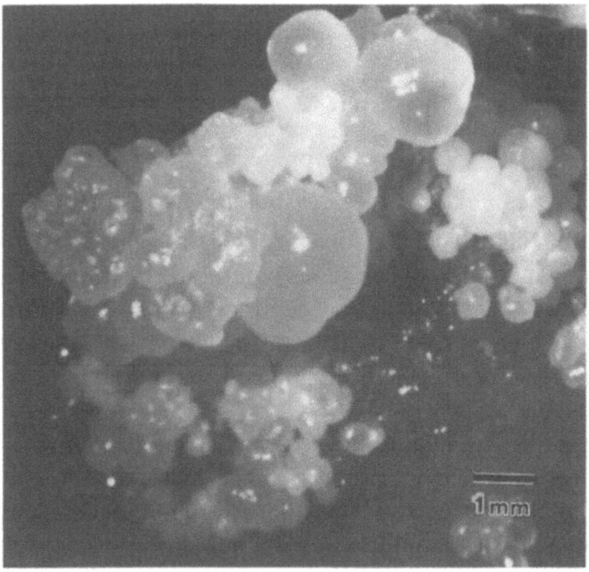

FIG. 5. Haploid embryos arising from the surface of a megagametophyte of *Ginkgo* collected 9 weeks after pollination and placed in culture. (From [44], with permission)

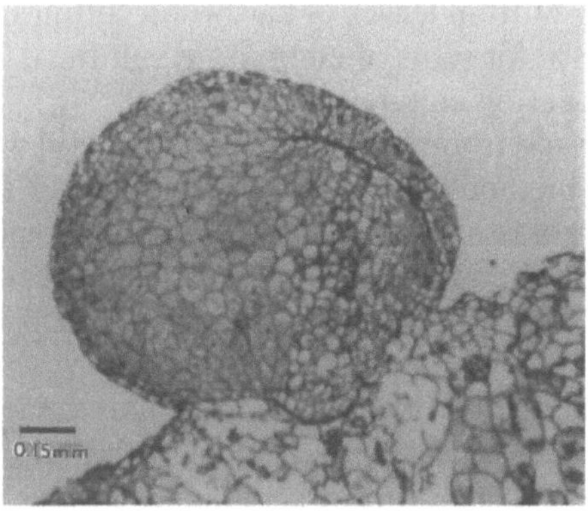

FIG. 6. Longitudinal section of a globular embryo induced on a megagametophyte after 3 weeks in the presence of 2,4-dichlorophenoxyacetic acid (2,4-D), 1 mg/l, followed by 2 months on a hormone-free medium. (From [44], with permission)

Immature zygotic embryos of *Ginkgo* (1–3.6 mm) at the globular, torpedo, or early cotyledonary stage were induced to form embryogenic callus with embryos forming on the cotyledons [45]. This occurred on a Murashige and Tucker medium with only benzyladenine as a growth regulator. This medium gave the highest frequency of embryogenesis (90%–95%) and more than nine somatic embryos per explant. A two-step treatment with media for induction and then development allowed embryos to be removed from the cultures and grown to the cotyledonary stage.

5 Biosynthesis

The early work of Nakanishi and Habaguchi [46] suggested that ginkgolides have a terpenoid origin, and their experiments with 11 different media produced some callus but no detectable ginkgolides. They then turned to culturing 5-month-old *Ginkgo* embryos, placing them on cotton wicks with labeled mevalonate in the liquid medium. Low levels of labeled ginkgolides in young roots were detected after 14 days in culture.

More recent studies by Chauret and co-workers [47] on *Ginkgo* cells produced in mass culture used gas chromatographic methods for the detection of ginkgolides. They confirmed the biosynthesis of ginkgolides A and B in 20g of lyophilized tissue, but the quantities were in the range of picomoles per gram (dry weight).

Huh and Staba [48] investigated tissue cultures of *Ginkgo* as one part of a larger study on the ontogenetic aspect of ginkgolide biosynthesis. In addition to seedlings of different ages, as well as leaves and roots, they analyzed root and callus cultures. Using liquid gas chromatography, they reported (as percent of dry weight) 0.007%–0.013% from leaves of hardwood cuttings, 0.026%–0.42% from roots of cuttings, 0.03% for roots of embryos in culture, 0.0026% for callus, and 0.0047% for cell suspension cultures.

Callus cultures derived from the leaves of *Ginkgo* [49] were analyzed by Jeon et al. [50], using the methods of Huh and Staba [43]. In studying the effect of ammonium, nitrate, potassium dihydrogen phosphate, and naphthalene acetic acid on ginkgolide production, Jeon et al. concluded that ammonium ion in a ratio to nitrate of 1:3 gave the best yield of ginkgolide, 0.0003% per gram (dry weight).

Normal seedling cotyledons of *Ginkgo* and a callus tissue derived from cotyledons were extracted and assayed for ginkgolide by thin-layer chromatography by Camper et al. [51]. Naphthalene acetic acid-induced callus was compared to 2,4-D-induced callus, using Murashige and Skoog medium and benzyladenine. The callus induced by 2,4-D contained more ginkgolide.

Catechin and proanthocyanidin production by callus cultures and cell suspensions derived from *Ginkgo* stem tissue was reported by Schrall and Becker [52]. They found that light was essential for production of these compounds on Murashige and Skoog medium with 1.0mg/l each of naphthalene acetic acid and kinetin. Thin-layer chromatographic co-chromatography with leaf and tissue culture extract showed equivalent concentrations of catechins and proanthocyanidins on an equal tissue weight basis.

In a series of papers, Stafford and her colleagues [53–55] used cell suspension cultures of *Ginkgo* to study the distribution and intracellular occurrence of condensed tannins. They found major amounts of flavan-3-ols in both the tissue cultures and in the leaves of *Ginkgo*, but these differed in their stereoisomers. The cultures produced greater amounts of proanthocyanidins than the leaves when compared on a dry weight basis. Extracts of callus tissue or cell suspensions derived from petioles contained enzymes that converted dihydromyricetin to flavan-3,4-diol or leucodelphinidin and then to (+)gallocatechin. Stafford [56] has summarized the importance of understanding the metabolism and regulation of phenolics in plants.

Phenolics such as anthocyanins and tannins are encountered in many plant tissue systems. Often a browning reaction occurs in older cultures under conditions of stress, dehydration, low temperature, or mineral deficiences. Sometime these red or brown pigments are related to regeneration, including embryogenesis.

Two haploid tissues of *Ginkgo*, one derived from the female gametophyte and one from *Ginkgo* pollen, were among several tissues tested for tetrapyrrole synthesis using δ-aminolevulinic acid as a precursor in the culture medium [57]. No tetrapyrroles were formed in the light, but in the dark they were shown to be localized in the amyloplasts by ultraviolet microscopy and identified as protoporphyrin and intermediates by paper chromatography. There was no greening of the tissues, but they turned a light copper color as they synthesized the tetrapyrroles.

6 Physiology

There have been a number of studies on *Ginkgo* tissues grown in the laboratory under experimentally defined conditions. One of the recent papers by Carrier et al. [58–60] reported the hydrolysis of sucrose in the medium before uptake by the cells of *Ginkgo*. The cells were derived from callus of *Ginkgo* embryos grown on Murashige and Skoog (MS) medium. Gamborg B5 medium was also used, but better growth was obtained from the MS medium, which had higher concentrations of sucrose and nitrogen salts plus 2.0 mg/l naphthalene-acetic acid and 1.0 mg/l kinetin. In suspension culture, the cells utilized the extracellular phosphate and ammonium by day 11 and the sucrose by day 14. A maximum yield of wet weight tissue after 15 days growth was 104 g/l.

It is well known that the seasonal aspects of plant growth can affect the response of some tissues placed in culture. Sussex and Clutter [61] tested the seasonal effect on secondary phloem explants from 11 species of woody perennials. For *Ginkgo*, the maximum growth of the phloem explants occurred in the spring when vegetative buds were beginning to open.

Lin and Mathes [62] found that a water extract of callus tissue derived from the cambium of *Ginkgo* would inhibit the growth of *Lemna minor*. This study was done on Linsmaier and Skoog medium with 1 mg/l 2,4-dichlorophenoxyacetic acid. The medium on which the *Ginkgo* tissues grew was also inhibitory to *Lemna* and to tomato seed germination. *Ginkgo* tissue secretions into the medium did not inhibit the growth of *Bacillus subtilis*, although the callus tissues of pokeberry, serviceberry, sweet potato, and aspen did inhibit bacterial growth to a distance of 2–2.9 cm from the callus.

Makino et al. [63] found that the callus cultures derived from the petiole of leaves of *Ginkgo* had a faster growth rate than callus tissue from *Sequoia*, *Cryptomeria*, or *Pinus*. All tissues were grown on Saitos medium with added amino acids, indole-butyric acid, and kinetin. The *Ginkgo* callus was the most friable, composed mostly of single cells, small clumps, and some nodules with tracheary elements.

The availability of fresh *Ginkgo* pollen for experimental use is limited to a brief period of 7–10 days of the yearly cycle. To overcome this, a method was used to surface sterilize the strobili and place them in sterile aluminum cups in micro-desiccators at 4°C [64]. Dehiscence of a strobilus in the cold released approximately 5000 pollen grains. This dry pollen in powder form was then available for experimentation year round; it maintained a 35%–45% viability for more than 2 years [57]. Studies with this pollen dusted on media in petri dishes led to the formation of haploid clones of tissue, presumably from single pollen grains. Some of these clones were maintained for more than 10 years [39].

In 1963, strobili of *Ginkgo* were brought into the laboratory and allowed to dehisce and release pollen, which was collected for analysis. Subsequently, the larvae of a fruit tree leaf roller, *Archips argyrospilus* (Walker), was found as a contaminant of a number of tissue cultures in the laboratory, especially in the albino and green tissue cultures derived from *Ginkgo* [65]. A search in the field revealed that both male and female *Ginkgo* trees had similar larval infestations, and this was reported as a new insect pest of *Ginkgo* [58]. The tissue cultures of *Ginkgo* were then used to rear the larvae originally found in the cultures. Feeding for 1–3 weeks led to pupation; adults emerged from the pupae after 2 weeks. The adults were mated and the females laid eggs in the manner typical for this species of leaf roller. This is the first report of the rearing of leaf rollers to functional adults on plant tissue cultures.

A modified White's medium with added casein hydrolysate or coconut milk and 0.1 mg/l 2,4-D was used by Bardinskaya and Moskaleva [66] to grow the callus of *Ginkgo* stems. They observed lignin formation in tracheary cells with bordered pits. Lignin biosynthesis from phenylalanine as a precursor was reported by Barnoud et al. [67], who used Knop's solution and stem sections of *Ginkgo*.

Ginkgo tissue cultures and the plant parts from which they were derived were compared using the pollen tissue culture and the pollen [24]. Citric and malic acids were high in both pollen and pollen tissue culture; the stored pollen contained mostly sucrose and the tissue contained principally fructose and glucose. The amino acid pool of the tissue resembled that of the germinated pollen. The abundance of shikimic and quinic acids in leaves was not found in either the pollen or the pollen tissue culture.

Later, a study was made of the amino acid composition of the medium and the cells of the *Ginkgo* pollen tissue during growth in a liquid medium containing arginine, grown in a 10-l carboy culture [35]. Arginine accumulated in the tissue during the lag phase of growth and was degraded by the tissue at a maximum rate during the second week. The maximum loss from the medium occurred in the third week.

The pollen-derived tissues of *Ginkgo* [35] were provided to several investigators for research projects related to pollen physiology. Yih [68] and Yih and Clark [69] tested the general boron requirement of pollens by growing the pollen tissue of *Ginkgo* in the absence of this trace element, using special media components and boron-free culture vessels. Without boron in the medium, the tissue ceased growth. At boron levels less than 1.0 mg/l (0.1–0.005 mg/l), the principal effect in these deficient media was that less carbohydrate was incorporated in the cell walls of the tissue. More galactose, arabinose, glucose, rhamnose, and xylose were found in cell wall hydrolysates when sufficient boron was present in the medium.

The arginine-requiring strain of *Ginkgo* pollen tissue [39] was used by Klein [70] to demonstrate ultraviolet light inhibition of tissue growth and the reversal of this inhibition by blue light. Klein and Edsall [71] further demonstrated a 40% reduction in tissue growth under cool white florescent lights compared to dark-grown tissue or tissue exposed to incandescent light. Ultraviolet and green light inhibition could be relieved by red light.

Lamport [72–74] also used the *Ginkgo* pollen tissue [39] to study the composition of primary cell wall components. The tissue, grown in liquid media, was friable and pipetteable. Lamport reported that the hydroxyproline content of the cell walls was low; the alkaline hydrolysate contained both hydroxypoline and hydroxyproline arabinosides.

In other studies on lignin, cellulose, and hemicellulose, Higuchi and Barnoud [75] used callus of *Ginkgo* and gymnosperms. They used 14 C labelled shikimic acid, l-phenylalanine and other compounds to study their incorporation into cell wall structures.

7 Discussion

Somatic embryogenesis is a natural process in some plants and the initiation of this process in vitro has been successful in a great variety of woody species and herbaceous plants. The success of Laurain et al. [41–43, 45] and Fontanel et al. [44] in obtaining somatic embryogenesis in *Ginkgo* from cells and protoplasts of the female gametophyte and from pollen is an excellent beginning. Eventually it should lead to the development of a system for the clonal propagation of *Ginkgo*. Moreover, the first successful introduction of genes into somatic embryos of *Juglans* (walnut) species by McGranahan et al. [76] suggests that this may be possible with *Ginkgo*, using other means of transformation. The apparent heterogeneity of the *Ginkgo* genome could also be another way of using somatic embryogenesis to increase the yield of desired chemicals in plantation forests of *Ginkgo* through a method of selective clonal propagation. Moreover, this might be a way to circumvent the problems of topophysis, the tendency for cuttings to propagate the growth proclivities of their lateral branch source; plants from somatic embryo would produce the normal excurrent growth form characteristic of *Ginkgo*. Of course, the somaclonal variation widely inherent in in vitro culture systems would need to be investigated to determine the reliability of any clonal propagation system derived from plant tissues, cells, or protoplasts grown in culture.

In the perhaps distant future, when DNA sequencing of cycads, conifers, and *Ginkgo* reveals interesting homologies and disparities of function, somatic embryogenesis could be used for wider gene transfer systems. This would include the unusual biochemical capabilities of *Ginkgo* that provide the basis for its relative resistance to diseases and pests. In this way, *Ginkgo* might be useful in reciprocal ways in studies on medicinal, agricultural, horticultural, and forest practice applications.

A recent report on the occurrence of the neurotoxic chemical 4′-*O*-methyl pyridoxine in the leaves and in the medicines derived from the leaves of *Ginkgo* [77] suggests that efforts will be made to select and propagate specific chemotype tissues of *Ginkgo* and then regenerate plants from those cells. Either through the selection of the source of cuttings or by the propagation of somatic embryos from such a source, it seems that clonal forests low in 4′-methyl pyridoxine will be developed at some point in the future.

There is a need for a protocol for going beyond the initiation of somatic embryos, a protocol for growing them to a viable form, followed by germination, the production of young seedlings, and plants in the field. It would also be useful to have a scheme for initiating somatic embryos from the leaves of *Ginkgo* and producing plants. The earlier studies by Wang et al. [12] on the growth of very small zygotic embryos of *Ginkgo* (1.0–1.6mm) reported success on complex media containing the liquid endosperm of coconut. The use of reported constituents of liquid endosperm [23], i.e., amino acids (190–757mg/l), sugars (9–21g/l), sorbitol (15g/l), and some growth substances and vitamins might be useful in designing a medium to nourish the somatic embryos of *Ginkgo*. The chemical constituents of the female gametophyte, the normal source of nutrients for zygotic embryos, would also be helpful.

The interest of the pharmaceutical and industrial sectors in the chemicals produced by *Ginkgo* has stimulated some support for work in vitro, as seen by reports from the laboratories of Biolega in Canada, Nestlé in Switzerland, and Beaufor in France. There may be further interest in the future if somatic embryogenic cloning and apical meristem culture and shoot propagation systems are perfected for *Ginkgo* [78, 79] or if other medicinal properties of *Ginkgo* are proven effective.

8 Conclusion

This chapter has reviewed the in vitro studies on *Ginkgo*, including research on the male and female gametophyte, embryo culture, somatic embryogenesis, and the growth, physiology, and biosynthetic capabilities of various tissues in culture.

References

1. Li TT (1934) The development of *Ginkgo biloba* in vitro. Sci Rep Natl Tsinghua Univ Ser B 2:41–52
2. Li TT, Chen SM (1934) Temperature and the development of the *Ginkgo biloba* embryo. Sci Rep Natl Tsinghua Univ Ser 2:37–39
3. Li TT, Shen T (1934) The effect of "pantothenic acid" on the growth of the yeast and on the growth of the radicle of *Ginkgo* embryo in artificial media. Sci Rep Natl Tsinghua Univ Ser 2:53–60
4. Bulard C (1967) Modifications provoquees par divers régulateurs de croissance sur des plantules de *Ginkgo biloba* L. cultivées in vitro. Bull Soc Bot Fr 114:119–127
5. Bulard C (1967) Un cas d'inhibition de croissance de l'épicotyl chez *Ginkgo biloba* L. obtenu sous l'influence de gibberellines. C R Acad Sci 265:1301–1304
6. Bulard C (1967) Culture aseptique d'embryons de *Ginkgo biloba* L. Role des cotyledons dans l'absorption du sucre et la croissance de la tige. C R Acad Sci 235:739–741
7. Ball E (1956) Growth of the embryo of *Ginkgo biloba* under experimental conditions. I. Origin of the first root of the seedling in vitro. Amer J Bot 43:488–495
8. Ball E (1956) Growth of the embryo of *Ginkgo biloba* under experimental conditions. II. Effects of a longitudinal split in the tip of the hypocotyl. Amer J Bot 43:802–810
9. Ball E (1959) Growth of the embryo of *Ginkgo biloba* under experimental conditions. III. Growth rate of the root and shoot upon media absorbed through the cotyledons. Amer J Bot 46:130–139
10. Radforth W (1936) The development in vitro of the proembryo of *Ginkgo*. R Can Inst Trans 21:87–94
11. Wang FH, Chen TK, Lee SC (1963) Experimental studies of young *Ginkgo* embryos. I. The effect of the bee royal jelly on the embryos. Acta Bot Sin 11:217–224
12. Wang FH, Chen TK (1965) Experimental studies of young *Ginkgo* embryos. The effect of coconut milk on the embryos cultured in vitro. Acta Bot Sin 13:224–239
13. Wang FH, Lee SC (1966) Experimental studies of young *Ginkgo* embryos. The effect of casein hydrolysate, glutamine, asparagine, 2,4-D, and adenine sulphate on embryos grown in vitro. Kexue Tongbao 17:323–327
14. Webb DT, Arias W, Hostos E (1986) Callus formation by *Ginkgo biloba* embryos on hormone-free media controlled by closures and media components. Phytomorphology 36:121–128
15. Favre-Duchartre M (1956) Contribution a l'étude de la reproduction chez le *Ginkgo biloba*. Rev Cytol Biol Veg 17:1–218
16. Rohr R (1980) Développment in vitro du pollen de *Ginkgo biloba*. Cytologia (Tokyo) 45:481–495
17. Rohr R (1977) Evolution en culture in vitro des prothalles femelles âgés chez le *Ginkgo biloba* L. Z Pflanzenphysiol 85:61–69
18. Rohr R (1978) Ultrastructure du tissu haploide du *Ginkgo biloba* L. Z Pflanzenphysiol 87:365–378
19. Rohr R (1978) In vitro culture of female gametophytes from *Ginkgo biloba*. Z Pflanzenphysiol 85:61–70
20. Rohr R (1980) De differentiation en culture in vitro des cellules a reserves dans l'endosperme de *Ginkgo biloba*. Z Pflanzenphysiol 96:423–434

21. Rohr R (1989) Maidenhair tree (*Ginkgo biloba* L.). In: Bajaj YPS (ed) Biotechnology in agriculture and forestry, vol 5. Trees. II. Springer, Berlin Heidelberg New York, pp 574–590
22. Tulecke W (1964) A haploid tissue culture from the female gametophyte of *Ginkgo biloba L.* Nature 203:94–95
23. Tulecke W, Weinstein L, Rutner A, Laurencot HJ (1961) The biochemical composition of coconut water (coconut milk) as related to its use in plant tissue culture. Contrib Boyce Thompson Inst 21:115–128
24. Tulecke W, Weinstein LH, Rutner A, Laurencot HJ (1963) Biochemical and physiological studies of tissue cultures and the plant parts from which they were derived. II. *Ginkgo biloba* L. Contrib Boyce Thompson Inst 21:153–163
25. Tulecke W (1967) Studies on tissue cultures derived from *Ginkgo biloba* L. Phytomorphology 17:381–386
26. Tulecke W (1965) Haploidy versus diploidy in the reproduction of cell type. In: Locke M (ed) Reproduction: molecular, subcellular, and cellular. Academic, New York, pp 217–241
27. Friedman WE (1987) Morphogenesis and experimental aspects of growth and development of the male gametophyte of *Ginkgo biloba* in vitro. Amer J Bot 74:1816–1830
28. Friedman WE, Gifford EM (1988) Division of the generative cell and late development in the male gametophyte of *Ginkgo biloba.* Amer J Bot 75:1434–1442
29. Gifford EM, Lin J (1975) Light microscope and ultrastructural studies of the male gametophyte in *Ginkgo biloba*: the spermatogenous cell. Amer J Bot 62:974–981
30. Tulecke W (1954) Studies in vitro on the pollen of *Ginkgo biloba.* Diss Abstr xiv: 1–103
31. Tulecke W (1953) A tissue from the pollen of *Ginkgo biloba.* Science 117:599–600
32. Tulecke W (1957) The pollen of *Ginkgo biloba*: in vitro culture and tissue formation. Amer J Bot 44:602–608
33. Lee CL (1953) Reproductive morphology of *Ginkgo biloba.* PhD dissertation, University of Illinois, Champaign
34. Lee CL (1955) Fertilization in *Ginkgo biloba.* Bot Gaz 117:79–100
35. Tulecke W, Rutner A (1965) Changes in the amino acid composition of medium and cells of a plant tissue culture during growth in a liquid medium containing arginine. In: White PR, Grove AR (eds) Proceedings, international conference on plant tissue culture at Pennsylvania State University. McCutchan, Berkeley, pp 103–116
36. Tulecke W, Nickell LG (1959) The production of plant tissue by submerged culture. Science 130:863–864
37. Tulecke W, Taggart R, Colavito L (1965) Continuous cultures of higher plant cells in liquid media. Contrib Boyce Thompson Inst 23:33–46
38. Tulecke W (1966) Continuous cultures of higher plant cells in liquid media: the advantages and potential use of a phytostat. Ann NY Acad Sci 139:162–175
39. Tulecke W (1960) Arginine-requiring strains of tissue obtained from *Ginkgo* pollen. Plant Physiol (Bethesda) 35:19–24
40. Tulecke W (1987) Somatic embryogenesis in woody perennials. In: Bonga JM, Durzan DJ (eds) Cell and tissue culture in Forestry, vol 2. Specific principles and methods: growth and development. Nijhoff, Dordrecht, pp 61–69
41. Laurain D, Chenieux JC, Tremouillaux-Guiller J (1993) Direct embryogenesis from female haploid protoplasts of *Ginkgo biloba* L., a medicinal woody species. Plant Cell Rep 12:656–660

42. Laurain D (1994) Etablissement de cultures cellulaires haploids, diploids et transformees de *Ginkgo biloba* par diverses strategies variabilisantes. PhD thesis, Department of Molecular and Cell Biophysics and Biology, University of Tours, Biophys Biol Mol Cell France, pp 1–138
43. Laurain D, Trémouillaux-Guiller J, Chenieux JC (1993) Embryogenesis from microspores of *Ginkgo biloba* L. Plant Cell Rep 12:501–505
44. Fontanel AI, Serraf I, Petiard V (1994) Regeneration in *Ginkgo biloba* L.: induction of embryogenesis from megagametophyte according to the development state. C R Acad Sci 317:149–155
45. Laurain D, Chenieux JC, Tremouillaux J (1996) Somatic embryogenesis from immature embryos of *Ginkgo biloba*. Plant Cell Tissue Organ Cult 44:19–24
46. Nakanishi K, Habaguchi K (1971) The biosynthesis of ginkgolide B, its diterpenoid nature, and origin of the tertiary butyl group. J Amer Chem Soc 93:3546–3547
47. Chauret N, Carrier J, Mancini M, Neufeld R, Weber M, Archambault J (1991) Gas chromatographic-mass spectrometric analysis of ginkgolides produced by *Ginkgo biloba* cell cultures. J Chromatogr 588:281–288
48. Huh H, Staba EJ (1993) Ontogenetic aspects of ginkgolide production in *Ginkgo biloba*. Planta Med 59:232–239
49. Jeon MH, Sung SH, Jeon S, Huh H, Kim J, Kim YC (1993) Culture of *Ginkgo biloba*. effect of nutritional and hormonal factors on the growth of cultured cells derived from *Ginkgo biloba*. Arch Pharmacal Res (Seoul) 16:244–250
50. Jeon MH, Sung SH, Jeon S, Huh H, Kim YC (1995) Ginkgolide B production in cultured cells derived from *Ginkgo biloba* L. leaves. Plant Cell Rep 14:501–504
51. Camper ND, Wedge DE, Keese RJ (1995) In vitro culture of *Ginkgo*: ginkgolide detection. In: Gustine DL, Flores HE (eds) Phytochemicals and health. American Society of Plant Physiologists, Rockville, MD, pp 33–34
52. Schrall R, Becker H (1977) Production of catechins and oligomeric proanthocyanidins in tissue and suspension cultures of *Crataegus monogyna, Crataegus oxyacantha* and *Ginkgo biloba*. Planta Med 32:297–307
53. Stafford HA, Lester HH (1985) Flavan-3-0.1 biosynthesis, the conversion of dextro dihydromyricetin to its flavan-3,4-diol leucodelphinidin and to dextro gallocatechin by reductases extracted from tissue cultures of *Ginkgo biloba* and *Pseudotsuga menziesii*. Plant Physiol (Bethesda) 78:791–794
54. Stafford HA, Lester WH, Weider RM (1987) Histochemical assay of proanthocyanidin heterogeneity in cell cultures. Plant Science (Shannon) 52:99–104
55. Stafford HA, Kreitlow KS, Lester HH (1986) Comparison of proanthocyanidins and related compounds in leaves and leaf-derived cell cultures of *Ginkgo biloba* L., *Pseudotsuga menziesii* Franco and *Ribes sanguineum* Pursh. Plant Physiol (Bethosda) 82:1132–1138
56. Stafford HA (1990) Flavonoid metabolism. CRC Press, Boca Raton, pp 1–298
57. Tulecke W (1967) Plastid function in plant tissue cultures. I. Porphyrin synthesis by haploid albino and diploid achlorophyllous cultures. Amer J Bot 54:797–804
58. Carrier DJ, Cosentino G, Neufeld R, Rho D, Weber M, Archambault J (1990) Nutritional and hormonal requirements of *Ginkgo biloba* embryo-derived callus and suspension cell culture. Plant Cell Rep 8:635–638
59. Carrier DJ, Chauret N, Mancini M, Coullombe P, Neufeld R, Weber M, Archambault J (1991) Detection of ginkgolide A in *Ginkgo biloba* cell cultures. Plant Cell Rep 10:256–259

60. Carrier J, Chauret N, Archambault J (1994) *Ginkgo biloba* L. (maidenhair tree): in vitro culture and the formation of ginkgolides. In: Bajaj YPS (ed) Biotechnology in agriculture and forestry, vol 26. Medicinal and aromatic plants. VII. Springer, Berlin Heidelberg New York, pp 136–145
61. Sussex IM, Clutter ME (1959) Seasonal growth periodicity of tissue explants from woody perennial plants in vitro. Science 129:836–837
62. Lin A, Mathes MC (1973) The in vitro secretion of growth regulators by isolated callus tissues. Amer J Bot 60:34–41
63. Makino RH, Kurola H, Shimaji K (1985) Morphological observation of the calli derived from 4 coniferous species in vitro. Wood Res 71:1–2
64. Tulecke W (1954) Preservation and germination of the pollen of *Ginkgo* under sterile conditions. Bull Torrey Bot Club 81:509–512
65. Tulecke W, Colavito L (1966) Fruit-tree leaf rollers on plant tissue cultures and on leaves of *Ginkgo biloba*. J Econ Entomol 59:1277–1278
66. Bardinskaya MS, Moskaleva VE (1961) Some data concerning the structure of tracheid cells in callus cultures of *Ginkgo biloba* L. Dokl Bot Sci 135:242–245
67. Barnoud F, Higuchi T, Josleau J, Mollard A (1964) A la biosynthése des lignines dans les tissus végétaux cultivés in vitro: rôle de précurseur de la phénylalanine. C R Acad Sci 259:4339–4341
68. Yih RY (1963) A search for an essential role of boron in plant metabolism. PhD dissertation, Rutgers State University, New Brunswick, NJ P 113
69. Yih RY, Clark HE (1963) The role of boron in carbohydrates and cell wall metabolism examined with pollen derived tissue culture of *Ginkgo biloba* L. Plant Physiol (Bcthesda) 38:lii
70. Klein RM (1963) Interaction of ultraviolet and visible radiation on the growth of cell aggregates of *Ginkgo* pollen tissue. Physiol Plant 16:73–81
71. Klein RM, Edsall PC (1967) Interference by near ultraviolet and green light with growth of animal and plant cell cultures. Photochem Photobiol 6:841–850
72. Lamport DTA (1965) The protein component of primary cell walls. Adv Bot Res 2:151–218
73. Lamport DTA, Miller DH (1971) Hydroxyproline arabinosides in the plant kingdom. Plant Physiol (Bethesda) 48:454–456
74. Lamport DTA (1964) Cell suspension cultures of higher plants: isolation and growth energetics. Exp Cell Res 33:195–206
75. Higuchi T, Barnoud F (1966) Biogenesis of lignins of the tissues and plants cultured in vitro. J Jpn Wood Res Soc 12:36–43
76. McGranahan GH, Leslie CA, Uratsu SL, Dandekar AM (1988) *Agrobacterium*-mediated transformation of walnut somatic embroyos and regeneration of transgenic plants. Biotechnology 6:800–804
77. Arenz A, Klein M, Fiehe K, Gross J, Drewke C, Hemscheidt T, Leistner E (1996) Occurrence of neurotoxic 4′-*O*-methyl pyridoxine in *Ginkgo biloba* leaves, *Ginkgo* medications, and Japanese *Ginkgo* food. Planta Medica 62:548–551
78. Wedge DE, Keese RJ, Depwa A (1995) In vitro culture of *Ginkgo* (abstract). In Vitro Cell Dev Biol 31:20
79. Yates WF (1986) Induction of embryogenesis in embryo-derived callus of *Ginkgo biloba* L. In: Abstracts, VI international congress on plant tissue culture. Univ Minnesota, p 43

Phylogeny and Fossil History

The Genetic Diversity of Isozymes and the Possible Dissemination of *Ginkgo biloba* in Ancient Times in Japan

YOSHIHIKO TSUMURA[1] and KIHACHIRO OHBA[2]

1 Introduction

Ginkgo biloba, a deciduous plant that was called a living fossil by Charles Darwin, is one of the most ancient gymnosperms in the world. Species of this genus were widely distributed throughout the world during the Jurassic period. Based on fossil records, it is estimated that at least 12 *Ginkgo* species existed. Most species of this genus and many other plant species perished during the glacial epochs of the Tertiary period. However, *G. biloba* is considered to have survived in southern China because of the warm climate. Natural *Ginkgo* trees are said to still exist in southern China. *G. biloba* has now spread by introduction to many countries worldwide as an ornamental tree.

It is thought that this species was introduced into Japan from China during the Kamakura and Muromachi eras (A.D. 1192–1573) because old documents from previous eras contain no record of *G. biloba*. In Japan, *G. biloba* has been commonly planted within the precincts of temples and shrines and it is used as an ornamental tree. The tree also is known for its edible fruit, called *ginnan* (in English, "ginkgo nut"), which has been commercially produced in orchards using several kinds of fecund clones in some prefectures. The genetic variation of these clones is not great, and most of them show a similar random amplified polymorphic DNA (RAPD) pattern (data not shown).

To investigate the genetic diversity of plant species for which we have no genetic information, isozyme analysis is one of the best methods because the inheritance mode and localization of many enzyme systems and most isozymes show codominant inheritance [1, 2]. The gymnosperm endosperm is also a maternal haploid tissue (megagametophyte), and the genotype of the female tree can be determined with open pollinated seeds by analyzing more than six megagame-

[1] Bio-Resources Technology Division, Forestry and Forest Products Research Institute, Kukizaki, Ibaraki 305, Japan
[2] Institute of Agriculture and Forestry, University of Tsukuba, 1-1-1 Tennodai, Tsukuba, Ibaraki 305, Japan

tophytes ($P = 1 - (1/2)^{n-1}$) where n is the number of megagametophytes analyzed and P is the probability of having at least one heterozygote. In analyzing more than six megagametophytes, P was >.96875. Using megagametophyte tissue, isozymes of many enzyme systems can be detected because this tissue has few phenolic and tannin-like compounds.

The purpose of this study is to clarify the genetic diversity of old *G. biloba* trees in Japan and calculate the possible dissemination routes from China and in Japan after analyzing their allozyme variation. As *G. biloba* was introduced from China, it may be possible to clarify what propagation methods were used and how many examples were brought to Japan.

2 Materials and Methods

2.1 Collection of Fruits

Ginnan fruits were collected from 98 old *Ginkgo biloba* trees grown in precincts of temples and shrines (locations are shown in Fig. 1). Estimates of the age of these trees by local people ranged from 32 to 1600 years, but most trees were more than 100 years old (Table 1). The seeds were extracted from the fruits, matured at room temperature for about 1 month, and then stored at 4°C until isozyme analysis.

FIG. 1. Locations *(circles)* of the 98 surveyed old or memorial trees of *Ginkgo biloba* and their geographical regions

TABLE 1. Location and characterization of surveyed *Ginkgo biloba* trees

No.	Tree	Location	Group	Putative age (yr)[a]	DBH (cm)	Height (m)
1	Tochigi 1	Kamimikawa	N. Kanto	350	162	22
2	Ibaraki 1	Mito	N. Kanto	700	—	—
3	Ibaraki 2	Daigo	N. Kanto	—	—	—
4	Ibaraki 3	Kashima	N. Kanto	—	—	—
5	Ibaraki 4	Tsukuba	N. Kanto	—	—	—
6	Ibaraki 5	Kasama	N. Kanto	—	—	—
7	Ibaraki 6	Kasama	N. Kanto	—	—	—
8	Ibaraki 7	Kitaura	N. Kanto	—	71	—
9	Ibaraki 8	Tsukuba	N. Kanto	—	—	—
10	Kyoto 1	Kameoka	Kansai	300	133	26
11	Kyoto 2	Ayabe	Kansai	140	99	24
12	Kyoto 3	Wachi	Kansai	85	91	20
13	Kyoto 4	Yakuno	Kansai	100	178	27
14	Kyoto 5	Kyoto	Kansai	200	130	25
15	Hyogo 1	Yamazaki	Kansai	—	78	14
16	Hyogo 2	Shingu	Kansai	—	150	25
17	Hyogo 3	Mikazuki	Kansai	—	75	18
18	Hyogo 4	Nankou	Kansai	70	60	17
19	Hyogo 5	Chourai	Kansai	—	70	20
20	Nara 1	Amakawa	Kansai	—	201	27
21	Nara 2	Soni	Kansai	—	92	17
22	Nara 3	Higashiyoshino	Kansai	—	72	17
23	Nara 4	Mitsue	Kansai	—	162	25
24	Nara 5	Goze	Kansai	—	60	15
25	Wakayama 1	Iwade	Kansai	500	154	19
26	Wakayama 2	Misato	Kansai	650	175	22
27	Wakayama 3	Minabe	Kansai	300	137	35
28	Wakayama 4	Kudoyama	Kansai	300	137	35
29	Tottori 1	Kooge	Kansai	123	100	30
30	Tottori 2	Nichinan	Kansai	400	160	35
31	Tottori 3	Tottori	Kansai	1000	320	40
32	Tottori 4	Shikano	Kansai	400	180	40
33	Tottori 5	Aoya	Kansai	100	120	25
34	Tottori 6	Yodoe	Kansai	100	90	25
35	Shimame 1	Shinji	Kansai	—	—	—
36	Shimane 2	Hikawa	Kansai	—	—	—
37	Shimane 3	Izumo	Kansai	—	—	—
38	Shimane 4	Hikawa	Kansai	—	—	—
39	Shimane 5	Hikimi	N. Kyushu	—	56	23
40	Okayama 1	Kumenan	Kansai	820	118	20
41	Okayama 2	Akasaka	Kansai	150	87	28
42	Okayama 3	Katsuyama	Kansai	700	280	35
43	Okayama 4	Kagaminno	Kansai	300	121	40
44	Okayama 5	Shouou	Kansai	80	56	23
45	Okayama 6	Shouou	Kansai	32	35	10
46	Yamaguchi 1	Yamaguchi	N. Kyushu	850	213	45
47	Yamaguchi 2	Mitou	N. Kyushu	200	143	23
48	Yamaguchi 3	Asahi	N. Kyushu	200	120	30
49	Yamaguchi 4	Yamaguchi	N. Kyushu	100	70	19
50	Yamaguchi 5	Hagi	N. Kyushu	150	100	30

TABLE 1. *Continued*

No.	Tree	Location	Group	Putative age (yr)[a]	DBH (cm)	Height (m)
51	Tokushima 1	Irida	Shikoku	100	61	14
52	Tokushima 2	Kamiyama	Shikoku	150	103	23
53	Tokushima 3	Naruto	Shikoku	120	81	15
54	Tokushima 4	Tokushima	Shikoku	96	54	15
55	Tokushima 5	Kiyadaira	Shikoku	200	170	25
56	Ehime 1	Niihama	Shikoku	1000	178	—
57	Ehime 2	Oda	Shikoku	1000	378	42
58	Fukuoka 1	Maebaru	N. Kyushu	250	100	18
59	Fukuoka 2	Dazaifu	N. Kyushu	300	130	28
60	Fukuoka 3	Chikushino	N. Kyushu	400	135	25
61	Fukuoka 4	Yoshii	N. Kyushu	—	95	20
62	Fukuoka 5	Ukiha	N. Kyushu	150	110	24
63	Fukuoka 6	Yame	N. Kyushu	300	120	—
64	Fukuoka 7	Hoshino	N. Kyushu	600	110	18
65	Fukuoka 8	Kuroki	N. Kyushu	150	100	15
66	Fukuoka 9	Kuroki	N. Kyushu	100	100	22
67	Fukuoka 10	Yabe	N. Kyushu	460	110	22
68	Fukuoka 11	Kawara	N. Kyushu	600	200	41
69	Fukuoka 12	Kitakyushu	N. Kyushu	—	270	37
70	Fukuoka 13	Kitakyushu	N. Kyushu	—	100	15
71	Fukuoka 14	Kitakyushu	N. Kyushu	—	110	16
72	Fukuoka 15	Nakama	N. Kyushu	250	115	14
73	Fukuoka 16	Mizumaki	N. Kyushu	40	73	15
74	Fukuoka 17	Genkai	N. Kyushu	—	75	—
75	Saga 1	Tosu	N. Kyushu	200	100	22
76	Saga 2	Nakabaru	N. Kyushu	300	100	25
77	Saga 3	Taku	N. Kyushu	500	100	24
78	Saga 4	Ariake	N. Kyushu	50	40	14
79	Saga 5	Takeo	N. Kyushu	510	290	25
80	Nagasaki 1	Takashima	N. Kyushu	—	139	21
81	Nagasaki 2	Toyotama	N. Kyushu	—	199	22
82	Nagasaki 3	Hirado	N. Kyushu	—	71	16
83	Nagasaki 4	Kawatana	N. Kyushu	—	103	19
84	Ooita 1	Innai	N. Kyushu	1600	357	36
85	Ooita 2	Nakatsu	N. Kyushu	100	76	18
86	Ooita 3	Kunimi	N. Kyushu	300	111	15
87	Ooita 4	Aki	N. Kyushu	300	108	25
88	Ooita 5	Ooita	N. Kyushu	1360	254	32
89	Ooita 6	Hita	N. Kyushu	300	102	25
90	Miyazaki 1	Saito	S. Kyushu	200	100	25
91	Miyazaki 2	Takaoka	S. Kyushu	700	50	38
92	Miyazaki 3	Miyazaki	S. Kyushu	200	70	20
93	Miyazaki 4	Yamado	S. Kyushu	600	40	50
94	Kagoshima 1	Iriki	S. Kyushu	—	—	—
95	Kagoshima 2	Ookuchi	S. Kyushu	—	—	—
96	Kagoshima 3	Kushikino	S. Kyushu	—	—	—
97	Kagoshima 4	Tsuruda	S. Kyushu	—	—	—
98	Kagoshima 5	Kamou	S. Kyushu	—	—	—

DBH, breast height diameter.
[a]Putative age was estimated by each temple and shrine in survey year (1988).

TABLE 2. Investigated enzyme systems

Enzyme system	Abbreviation	E. C. No.[a]	Reference
Alcohol dehydrogenase	ADH	1.1.1.1	11
Sorbitol dehydrogenase	SODH	1.1.1.14	11
Shikimate dehydrogenase	SDH	1.1.1.25	11
Malate dehydrogenase	MDH	1.1.1.37	1
6-Phosphogluconate dehydrogenase	6PG	1.1.1.44	1
Glucose-6-phosphate dehydrogenase	G6P	1.1.1.49	11
Glutamate dehydrogenase	GDH	1.4.1.2	1
Diaphorase	DIA	1.6.4.3	1
Glutamate oxaloacetate transaminase	GOT	2.6.1.1	11
Phosphoglucomutase	PGM	2.7.5.1	11
Acid phosphatase	ACP	3.1.3.2	11
Phosphoglucose isomerase	PGI	5.3.1.9	1

[a] Enzyme Commission number.

2.2 Isozyme Analysis

Seed coats were removed and the megagametophytes taken from each seed. Megagametophyte tissue (100 mg) was homogenized with 500 μl of extraction buffer [50 m*M* Tris-HCl, pH 7.5, 4 m*M* dithiothreitol, 2 m*M* nicotinamide adenine dinucleotide (NAD), 2 m*M* NADP]. These homogenates were centrifuged at 10000 × *g*, 0°C for 40 min; 10 μl supernatant was used as a sample for each enzyme detection. Electrophoresis was carried out using the polyacryamide vertical slab gel method at 4°C, 12.2 mA/cm^2 for 150 min. Twelve enzyme systems were analyzed in this study (Table 2); the staining methods for these enzyme systems were modified slightly [3].

2.3 Statistical Analysis

Statistical analysis for isozyme inheritance was made with the chi-square test. Megagametophytes of heterozygous mother trees were expected to be segregated in a 1:1 ratio because each megagametophyte is maternal haploid tissue. Between 14 and 55 seeds per tree were analyzed to determine the genetic control of isozymes. Seven megagametophytes per tree were analyzed to estimate the genotypes. To estimate allozyme variation, the following indices were used: the proportion of polymorphic loci *(Pl)* (95% level), the number of alleles per locus *(Na)*, the effective number of alleles per locus *(Ne)*, and the observed heterozygosity *(Ho)* [4]. For convenience, the 98 *Ginkgo* trees were grouped into five areas according to their location to estimate the genetic differences among these grouped trees (see Fig. 1). These five geographical regions have no inherent biological meaning, but it is thought that plant dissemination was closely related to trade and traffic routes of people in earlier times.

The following methods were used to evaluate the genetic difference between regions: (1) a heterogeneity test of genotype frequency in each locus using the chi-square test for 9 polymorphic loci, and (2) genetic identity between regions. The $I = J_{XY}/(J_X J_Y)1/2$ formula was used for calculation of genetic identity where J_X, J_Y, and J_{XY} are the arithmetic means of $j_X = \Sigma xi^2$, $J_Y = \Sigma yi^2$, and $J_{XY} = \Sigma xiyi$. The *xi* and *yi* are the frequencies of the *i*th alleles in populations *X* and *Y*, respectively [4].

Allozyme profiles of each tree were coded according to their genotypes following the method outlined by Smouse and Neel [5] and Yeh et al. [6]. Two multivariate procedures were used: (1) a principal component analysis and (2) a canonical discriminant analysis. The principal component analysis, based on a variance–covariance matrix, was applied to individual genotypes without regard to the regions. The canonical discriminant analysis was conducted using four local group individuals to define group membership. Two multivariate analyses were performed using SAS (Statistical Analysis System) Ver. 6.0 (SAS Institute, Cary, NC, USA) [7].

3 Results

3.1 *Inheritance of Isozymes*

We used 1 to 6 families for the segregation of allozymes in each enzyme, and in total 39 families, 1242 seeds, were used (Table 3). All families showed no segregation distortion except Ibaraki 2 family in *Acp-4* locus, which was significant at the 5% level. In this locus, 4 other families showed segregation with the same genotype as Ibaraki 2; therefore, we pooled them and checked the segregation distortion. The pooled data were fitted to a 1:1 segregation ratio. Totals of the segregation of 39 families also showed no segregation distortion; that chi-square value was 1.20 and *P* was .20.

Ten isozymes of 12 enzyme systems investigated showed segregation. The other 2 enzyme systems, alcohol dehydrogenase (ADH) and glucose-6-phosphate dehydrogenase (G6P), had no variation for two bands, Therefore, the genetic mode for these two enzyme systems could not be determined. The other 10 enzyme systems were found to be controlled by at least 14 loci [8, 9]. Sorbitol dehydrogenase, shikimate dehydrogenase, glutamate oxaloacetate transaminase, malate dehydrogenase, 6-phosphoglucomutase, glutamate dehydrogenase, and diaphorase were controlled by at least a single locus each. Phosphoglucomutase and phosphoglucose isomerase were coded by 2 loci each. Acid phosphatase was coded by at least 4 loci, but in the following study of genetic diversity in grouped trees, only the *Acp-4* locus was investigated.

3.2 *Allozyme Variation Within Species*

Twelve loci of ten enzyme systems were investigated to estimate the allozyme variation of *Ginkgo biloba* in Japan. The loci of ADH and G6P were not included in this estimation because the genetic mode of these enzymes was not decided at

TABLE 3. Segregation of allozyme loci in megagametophytes of *Ginkgo biloba*

Locus	Family	Genotype	*n*	Segregation *S*	Segregation *F*	Expected segregation *S*	Expected segregation *F*	Chi-square test χ	Chi-square test *P*
Sod	Ibaraki 4	*a/o*	24	12	12	12.0	12.0	0.00	1.00
	Ibaraki 5	*a/o*	14	6	8	7.0	7.0	0.29	0.59
		(Subtotal)	(38)	(18)	(20)	(19.0)	(19.0)	(0.11)	(0.75)
Shd-1	Ibaraki 3	*a/b*	46	29	17	23.0	23.0	3.13	0.08
	Ibaraki 5	*a/b*	14	9	5	7.0	7.0	1.14	0.29
	Ibaraki 7	*a/b*	34	16	18	17.0	17.0	0.12	0.73
	Ibaraki 8	*a/b*	15	5	10	7.5	7.5	1.67	0.20
		(Subtotal)	(109)	(59)	(50)	(54.5)	(54.5)	(0.74)	(0.39)
Got-1	Ibaraki 5	*a/b*	14	7	7	7.0	7.0	0.00	1.00
	Ibaraki 3	*a/c*	46	25	21	23.0	23.0	0.35	0.56
	Ibaraki 4	*a/c*	29	11	18	14.5	14.5	1.69	0.19
	Ibaraki 6	*a/c*	34	20	14	17.0	17.0	1.06	0.30
		(Subtotal)	(109)	(36)	(39)	(37.5)	(37.5)	(0.12)	(0.73)
Pgm-1	Ibaraki 1	*a/b*	19	6	13	9.5	9.5	2.58	0.11
	Ibaraki 3	*a/b*	31	17	14	15.5	15.5	0.29	0.59
	Ibaraki 4	*a/b*	21	11	10	10.5	10.5	0.05	0.83
	Ibaraki 5	*a/b*	14	9	5	7.0	7.0	1.14	0.29
	Ibaraki 7	*a/b*	19	11	8	9.5	9.5	0.47	0.49
		(Subtotal)	(104)	(54)	(50)	(52.0)	(52.0)	(0.15)	(0.69)
Pgm-2	Ibaraki 1	*a/b*	19	9	10	9.5	9.5	0.05	0.82
	Ibareki 8	*a/b*	15	6	9	7.5	7.5	0.60	0.44
		(Subtoal)	(34)	(15)	(19)	(17.0)	(17.0)	(0.47)	(0.49)
Acp-1	Tochigi 1	*a/o*	26	13	13	13.0	13.0	0.00	1.00
	Ibaraki 3	*a/o*	46	18	28	23.0	23.0	2.17	0.14
	Ibaraki 4	*a/o*	29	14	15	14.5	14.5	0.03	0.85
	Ibaraki 6	*a/o*	34	15	19	17.0	17.0	0.47	0.49
	Ibaraki 7	*a/o*	34	18	16	17.0	17.0	0.12	0.73
	Ibaraki 8	*a/o*	15	6	9	7.5	7.5	0.60	0.44
		(Subtotal)	(184)	(84)	(100)	(92.0)	(92.0)	(1.39)	(0.24)
Acp-2	Ibaraki 2	*a/o*	28	15	13	14.0	14.0	0.14	0.71
	Ibaraki 6	*a/o*	34	12	22	17.0	17.0	2.94	0.09
		(Subtotal)	(62)	(27)	(35)	(31.0)	(31.0)	(1.03)	(0.31)
Acp-4	Tochigi 1	*a/o*	26	12	14	13.0	13.0	0.15	0.69
	Ibaraki 1	*a/b*	34	16	18	17.0	17.0	0.12	0.73
	Ibaraki 2	*a/b*	28	8	20	14.0	14.0	5.14	0.02
	Ibaraki 4	*a/b*	29	18	11	14.5	14.5	1.69	0.19
	Ibaraki 7	*a/b*	34	18	16	17.0	17.0	0.12	0.73
	Ibaraki 8	*a/b*	15	10	5	7.5	7.5	1.67	0.20
		(Subtotal)	(140)	(70)	(70)	(70.0)	(70.0)	(0.00)	(1.00)
Mdh	Nagasaki 4	*a/b*	51	25	26	25.5	25.5	0.02	0.89
6Pg	Nagasaki 4	*a/b*	51	24	27	25.5	25.5	0.18	0.67
Gdh	Nagasaki 4	*a/b*	49	19	30	24.5	24.5	2.47	0.12
Dia	Tottori 4	*b/c*	55	33	22	27.5	27.5	2.20	0.14
	Oita 2	*a/c*	55	29	26	27.5	27.5	0.16	0.69
	Nagasaki 4	*c/o*	52	29	23	26.0	26.0	0.69	0.41
Pgi-1	Tottori 4	*a/b*	55	22	33	27.5	27.5	2.20	0.14
Pgi-2	Saga 3	*a/b*	54	22	32	27.0	27.0	1.85	0.17
		Total	1242	585	623	604.0	604.0	1.20	0.27

TABLE 4. Proportion of polymorphic loci *(Pl)*, mean allele per locus *(Na)*, unbiased mean alleles per locus *(Ne)*, and mean observed heterozygosity *(Ho)* of five local groups of *Ginkgo biloba*

Region	*n*	*Pl* (%)	*Na*	*Ne*	*Ho*
North Kanto	9	75.00	1.92	1.49	28.83
Kansai	35	75.00	2.25	1.40	25.00
Shikoku	7	58.30	1.75	1.35	19.10
North Kyushu	38	58.30	2.50	1.40	25.10
South Kyushu	9	75.00	1.92	1.41	20.40
Mean	19.6	68.32	2.07	1.41	23.69
Overall	98	75.00	2.50	1.41	23.69

the megagametophytic level. The following genetic parameters were used to evaluate the genetic diversity within species in Japan: (1) the proportion of polymorphic loci *(Pl)*, (2) the number of alleles per locus *(Na)*, (3) the effective number of alleles per locus, and (4) the observed heterozygosity *(Ho)*. These values were *Pl* = 75.0% (95% criterion), *Na* = 2.50, *Ne* = 1.41, and *Ho* = 0.237, respectively, for the 98 trees surveyed (Table 4). Hamrick et al. [10] compiled allozyme variability data for 89 gymnosperm species with the following values: *Pl* = 71.1%, *Na* = 2.38, *Ne* = 1.22, and *He* = 0.169. Polymorphic indices of *G. biloba* were mostly larger than those of Hamrick et al. [10]. In total, 91 multigenotype patterns were found in 98 surveyed trees; therefore, 85 mother trees of these have unique genotypes.

3.3 Genetic Difference Among Five Local Groups of Ginkgo biloba

The polymorphic index showed a larger genetic difference between five groups. The *Pl* ranged from 58.3% in Shikoku and North Kyushu to 75.0% in the other three groups. The *Na* values ranged from 1.75 in Shikoku to 2.50 in North Kyushu. The highest value of *Ne* was 1.49, in North Kanto, and the lowest was 1.35, in Shikoku. *Ho* varied between 19.1% in Shikoku and 25.1% in North Kyushu.

Allele frequencies between regions were different from each other (Table 5). A heterogeneity test of genotype frequency also showed a significant value at four of nine loci, which were *Shd-1*, *Pgm-1*, *Acp-4*, and *Pgi-1* (Table 6). The *Shd-1*, *Pgm-1*, and *Pgi-1* loci especially were highly significant at the 1% level (P = .01). Genetic identity varied from 0.845 to 0.997 (mean, 0.947), indicating that there are four clusters: North Kanto, Shikoku, Kansai, and two Kyushu groups (Table 7).

Genotypes of 12 loci in 98 mother trees were scored and analyzed by principal component analysis (Table 8). In the analysis based on the variance–covariance matrix, the first principal component accounted for 25.9% of the variation, and

TABLE 5. Allele frequencies of 12 allozyme loci in five local groups

Locus	Allele	Region				
		N. Kanto	Kansai	Shikoku	N. Kyushi	S. Kyushu
Sod	*a*	0.889	0.986	1.000	0.961	0.944
	b	0.111	0.014		0.039	0.056
Shd-1	*a*	0.722	0.800	1.000	0.987	1.000
	b	0.278	0.200		0.013	
Mdh	*a*	0.500	0.229	0.214	0.447	0.333
	b	0.500	0.771	0.786	0.553	0.667
6Pg	*a*		0.157	0.143	0.197	0.056
	b	1.000	0.843	0.857	0.803	0.944
Gdh	*a*	1.000	0.986	0.857	0.961	0.944
	b		0.014	0.143	0.039	0.056
Dia	*a*				0.013	
	b		0.400	0.357	0.263	0.333
	c	0.500	0.529	0.643	0.579	0.611
	d	0.500	0.071		0.145	0.056
Got	*a*	0.667	0.443	0.500	0.513	0.389
	b	0.056	0.300	0.214	0.079	0.333
	c	0.277	0.257	0.286	0.408	0.278
Pgm-1	*a*	0.833	0.286		0.316	0.444
	b	0.167	0.741	1.000	0.684	0.556
Pgm-2	*a*	0.889	0.929	1.000	0.974	1.000
	b	0.111	0.071			0.026
Acp-4	*a*	0.278	0.600	0.286	0.770	0.944
	b	0.612	0.300	0.643	0.122	0.056
	c	0.055	0.043		0.014	
	o	0.055	0.057	0.071	0.095	
Pgi-1	*a*				0.039	
	b	0.944	0.229	0.071	0.382	0.500
	c	0.056	0.771	0.929	0.579	0.500
Pgi-2	*a*				0.013	
	b	1.000	1.000	1.000	0.987	1.000

TABLE 6. Heterogeneity test of nine polymorhic loci in five local groups of *Ginkgo biloba*

Locus	Heterogeneity	D.F.
Shd-1	23.55**	4
Mdh	7.34	8
6Pg	7.03	4
Gdh	2.51	4
Dia	13.30	16
Got	17.00	16
Pgm-1	23.22**	8
Acp-4	23.44*	12
Pg-1	40.46**	12

D.F., degrees of freedom.
*, $P = .05$; **, $P = .01$.

TABLE 7. Genetic identity of five local groups of *Ginkgo biloba*

	(1) North Kanto	(2) Kansai	(3) Shikoku	(4) North Kyushu
(2) Kansai	0.9142			
(3) Shikoku	0.8453	0.9820		
(4) North Kyushu	0.9416	0.9816	0.9493	
(5) South Kyushu	0.9476	0.9825	0.9276	0.9972

TABLE 8. Eigenvectors of the first four principal components and their eigenvalues covariance matrix

Locus	Allele	PRIN1	PRIN2	PRIN3	PRIN4
Sod	*a*	−0.0094	−0.0298	0.0485	−0.1235
Shd-1	*a*	0.0880	−0.1474	−0.0204	0.0251
6Pg	*a*	−0.0017	−0.0129	0.0054	0.1369
Gdh	*a*	−0.0311	−0.0363	0.0220	−0.0103
Got	*a*	0.1677	0.5410	−0.5380	−0.4178
	b	−0.0715	−0.2987	0.2177	0.2307
Pgm-1	*a*	0.0441	0.4718	0.8004	−0.3106
Pgm-2	*a*	0.0253	−0.0559	−0.0265	0.1355
Acp-4	*a*	0.7185	−0.1507	0.1247	−0.0520
	b	−0.6442	0.1532	0.0213	−0.0177
	c	−0.0214	−0.0041	−0.0405	−0.0058
Pgi-1	*a*	−0.0044	−0.0198	−0.0052	−0.0066
	b	0.1517	0.5660	−0.0132	0.7871
Pgi-2	*a*	−0.0255	−0.0375	0.0255	0.0096
Eigenvalue		0.2788	0.2345	0.1620	0.1124
Difference		0.0443	0.0724	0.0497	0.0387
Proportion		0.2590	0.2178	0.1505	0.1044
Cumulative		0.2590	0.4769	0.6274	0.7318

the first 3 of 18 principal components accounted for 62.7% of the variance. Based on the variance–covariance matrix, a scatter diagram was made by using the first two principal components (Fig. 2). According to this diagram, the North Kanto group was the group most separated from the groups of other regions.

Allozyme data of 98 mother trees in the five regional groups were analyzed by canonical discriminant analysis (Table 9). Two canonical discriminant functions were significant ($P < .001$), accounting for 50.26% and 26.95% of the total variance in 30 alleles of 12 polymorphic loci, respectively. Approximately 46.8% of the variation in the first canonical discriminant function was explained by the geographical contributions of the groups. The value in the second canonical discriminant function was also 26.9%. This canonical discriminant function was dominated by a large loading from alleles *Pam-1*^{a} and *Pgi-1*^{b}. As illustrated by plotting individual data against the first two canonical axes (CAN1 and CAN2), the North Kanto group canonical variable means were located in the positive area for CAN1 (Fig. 3).

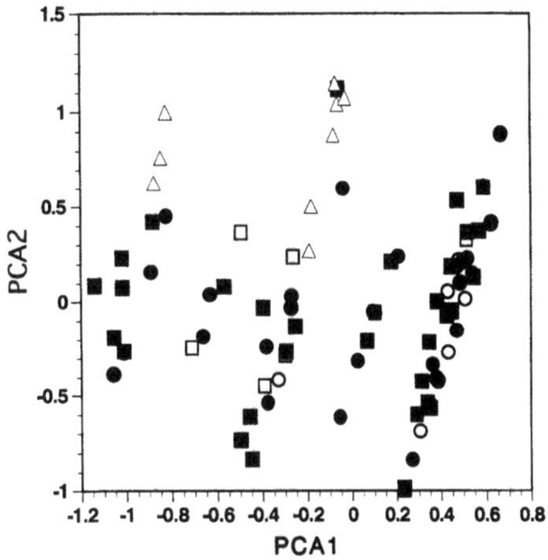

FIG. 2. Scatter diagram based on the variance–covariance matrix of the principal component analysis. PCA1 and PCA2 show the first and second components, respectively. *Solid squares*, Kansai; *open squares*, Shikoku; *solid circles*, North Kyushu; *open circles*, South Kyushu; *triangles*, North Kanto

TABLE 9. Correlation coefficients of the first two canonical variables for five local groups of *Ginkgo biloba*

Locus	Allele	Canonical variable 1	Canonical variable 2
Sod	*a*	−0.2435	−0.1744
Shd-1	*a*	−0.3584	0.5756
6Pg	*a*	−0.3226	−0.0199
Gdh	*a*	0.1102	−0.1107
Got	*a*	0.1775	0.1206
	b	−0.0893	−0.5619
Pgm-1	*a*	0.5297	0.0743
Pgm-2	*a*	−0.2241	0.2159
Pcp-4	*a*	−0.3475	0.0795
	b	0.3973	−0.1449
	c	0.1100	−0.1795
Pgi-1	*a*	−0.0722	0.2574
	b	0.6397	0.3351
Pgi-2	*a*	−0.0567	−0.1441
Eigenvalue		1.0097	0.5414
Proportion of variance		0.5026	0.2695
Canonical correlation		0.7088	0.5926
Probability		0.0001	0.0007

4 Discussion

Reference to *G. biloba* cannot be found in old documents from the Nara to the Heian eras (A.D. 710–1191) [11]. From this lack of historical evidence, it is considered highly likely that *Ginkgo biloba* was introduced from China during the

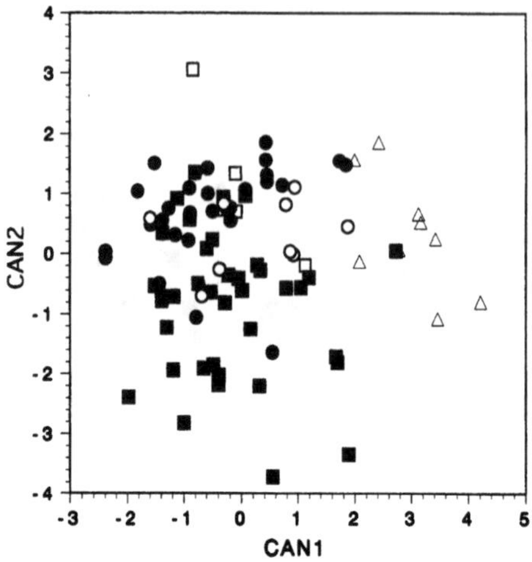

FIG. 3. Plot of canonical discriminant function of five geographical groups of *Ginkgo biloba* for the first two canonical variables, CAN1 and CAN2. *Solid squares*, Kansai; *open squares*, Shikoku; *solid circles*, North Kyushu; *open circles*, South Kyushu; *triangles*, North Kanto

Kamakura and Muromachi eras with some relation to Buddhism (A.D. 1192–1573) [11, 12].

If *G. biloba* was introduced from China, its genetic variation in Japan should have decreased relative to the original population because of the limited amount of seed, and therefore narrow genetic variation, brought to Japan. However, *G. biloba* in Japan has been shown to have a relatively larger isozyme variation than other gymnosperms [10]. Several reasons are thought responsible for this variation. First, the original *G. biloba* population is thought to have had a large variation. The mean heterozygosity *(Ho)* of 23.69% in 98 old *G. biloba* trees is thought to be high. Even if ADH and G6PD loci are included, the *Ho* value of this species might be decreased only a little. However, even that estimation (20.31%–17.77%) of *Ho* is still higher than the mean determined by Hamrick and Godt [13]. Second, judging from their larger isozyme genetic variation, it is inferred that many seeds or plant lots from different sites might have been introduced from China since the Kamakura and Muromachi eras. According to the multigenotype data, 87% of 98 mother trees have unique genotypes in the group. We found similar genotypes in 2 or 3 individuals but for only six multigenotype patterns. It was considered from these results that the propagation of old *G. biloba* trees in Japan had been mostly by seed. In modern times, *G. biloba* is usually propagated by cuttings and grafting, for ornamental or avenue trees (male) and fruit orchards (female). In ancient times, in the Muromachi or Kamakura era, it is considered that seeds of *G. biloba* were planted at temples and shrines, as the tree was related to Buddhism.

The result of the heterogeneity test showed that four of the nine polymorphic loci were significant, indicating a fairly large difference among the five local groups. The polymorphic index also indicated that the genetic components varied considerably between the five groups, especially the values of Shikoku, which were lower than the others. The genetic identity resulted in an average difference of 0.053, seen as four clusters. The results of canonical discriminant analysis followed a similar trend, clearly indicating larger differences between regions (see Fig. 3).

For convenience, we divided Japan into five geographical regions to estimate the genetic difference between local groups of the trees. If there were no human effect, the observed isozyme variation could not be explained merely by natural selection and chance distribution of the plant, especially among regions. While the scatter diagram generated by the principal component analysis showed that Shikoku and South Kyushu were separated, they were overlapped by the other regions. This result is thought to be closely related to later human transportation.

Li [12] reported the historical record of *Ginkgo* in China as follows. *Ginkgo* was unknown to the ancient Chinese; its record in the literature cannot be traced definitely beyond 1000 years ago. In the Sung dynasty of the early eleventh century, it first appeared in the literature as a plant native to eastern China. *Ginkgo* seems to have been carried directly to Japan from southern Kiangsu and Anhwei, its original habitat. Because the Japanese pronunciation of tree names is closer to the Wu dialect of southern Kiangsu than to northern Mandarin, the original population of this species is thought to have existed in eastern China, but this is conjecture.

From these isozyme analyses and the historical record of human activity, it might be possible to reason as follows. Within the past 1000 years, the seeds of *Ginkgo* may have been independently introduced sporadically from China to several regions in Japan. The trees of the South Kyushu region might be descendants of individuals from the North Kyushu group, and those of the Shikoku region might have come from individuals of the Kansai region. The sources of trees from the North Kanto region group might be different from those of other regional groups.

5 Conclusion

Inheritance of isozymes encoding 12 enzyme systems was investigated using megagametophyte tissues *(n)* of *Ginkgo biloba*. The allozyme variations of 98 old trees grown in precincts of shrines and temples in Japan were also investigated using 12 loci from 10 enzyme systems to clarify the genetic diversity and dissemination routes from China and in Japan. Fourteen loci of 11 enzyme systems were clarified by Mendelian segregation data (1:1). The genetic diversity of 98 mother trees was as follows: the proportion of polymorphic loci, the number of alleles per locus, the effective number of alleles per locus, and the observed heterozygosity were 75.0% (95% criterion), 2.5, 1.41, and 0.237, respectively. The variation of this species is relatively high in comparison with other gymnosperm species.

Genetic relationships between five groups of the trees clustered with their regions (North Kanto, Kansai, Shikoku, North Kyushu, and South Kyushu) were investigated by using two multivariate procedures. In both multivariate procedures, North Kanto was the most separated from the other regions. Genetic diversity of the other four region groups overlapped in some parts. The multigenotype data showed 87% of these were unique genotypes in the group. If *G. biloba* is an introduced tree, it is likely that seeds were brought to Japan at different times from China and planted in temples and shrines.

Acknowledgments. We are deeply grateful to the staff of the following Prefecture Forest Experiment Stations for their collection of the fruits: Kyoto, Nara, Wakayama, Hyogo, Okayama, Tottori, Shimane, Yamaguchi, Tokushima, Ehime, Fukuoka, Saga, Nagasaki, Oita, Miyazaki, and Kagoshima. We also thank H. Tsuburaya for collection of fruits in the North Kanto region, H. Motoike for isozyme analysis, and N. Tomaru and H. Yoshimaru for help with the statistical analysis.

References

1. Gottlieb LD (1981) Electrophoretic evidence and plant population. Prog Phytochem 7:1–46
2. Gottlieb LD (1982) Conservation and duplication of isozymes in plants. Science 216:373–380
3. Tsumura Y, Tomaru N, Suyama Y, Na'eim M, Ohba K (1990) Laboratory manual of isozyme analysis (in Japanese). Bull Tsukuba Univ For 6:63–95
4. Nei M (1987) Molecular evolutionary genetics. Columbia University Press, New York
5. Smouse PE, Neel JV (1977) Multivariate analysis of gametic disequilibrium in the Yanomama. Genetics 85:733–752
6. Yeh FC, Khalil MAK, El-Kassaby YA, Trust DC (1986) Allozyme variation in *Picea mariana* from Newfoundland: genetic differentiation. Can J For Res 16:713–720
7. SAS Institute (1986) SAS (Statistical Analysis System) user's guide: statistics. Ver. 6.0. SAS Institute, Inc., Cary, NC
8. Tsumura Y, Tsuburaya H, Ohba K (1987) Inheritance of isozyme variation of megagametophytes in *Ginkgo biloba*. J Jpn For Soc 69:386–390
9. Tsumura Y, Motoike H, Ohba K (1992) Allozyme variation of old *Ginkgo biloba* memorial trees in western Japan. Can J For Res 22:939–944
10. Hamrick JL, Godt MJW, Sherman-Broyles SL (1992) Factors influencing levels of genetic diversity in woody plant species. In: Adams WT, Strauss SH, Copes DL, Griffin AR (eds) Population genetics of forest trees. Kluwer, Dordrecht, pp 95–124
11. Ahida T (1989) Woody and herbaceous plants of the past and present (in Japanese). Soushisha, Tokyo
12. Li H-L (1956) A horticultural and botanical history of *Ginkgo*. Bull Morris Arbor 7:3–12
13. Hamrick JL, Godt MJW (1989) Allozyme diversity in plant species. In: Brown AHD, Clegg MT, Kahler AL, Weir BS (eds) Plant population genetics, breeding, and genetic resources. Sinauer, Sunderland, pp 43–63

Molecular Phylogeny of *Ginkgo biloba*: Close Relationship Between *Ginkgo biloba* and Cycads

MITSUYASU HASEBE

1 Introduction

Morphological characters have been widely used to infer phylogenetic relationships of organisms. Difficulty of distinction between plesiomorphic and apomorphic characters because of homoplasy or excess morphological diversity has often necessitated subjective character evaluation by authors, which has resulted in controversies among differently inferred phylogenetic trees. Recent progress in molecular biology has made possible the use of information on genome organization and macromolecule sequences in systematics, which is useful to evaluate previously proposed phylogenetic studies based on morphological characters. Especially in distantly related taxa like gymnosperms in which character evaluation is puzzling, molecular phylogeny is expected to be a powerful tool for inference of phylogenetic relationships.

Comparison of gene arrangements in the chloroplast genome and nucleotide or amino acid sequences of chloroplast-, mitochondria-, and nuclear-coded genes have given new insights into the phylogeny of *Ginkgo biloba*. In this chapter, the recent progress of molecular phylogenetic analyses is reviewed and future prospects are discussed.

2 Chloroplast Genome

The chloroplast has a highly conservative genome for its size, organization, and gene order (for review, see [1]). All angiosperm chloroplast DNA (cpDNA) are circular, and most range in size from 120 to 217 kb. The most characteristic organization is the large inverted repeat (IR), separated by a large and small single copy region. The size variation of IR is observed in several lineages by

Botanical Gardens, Faculty of Science, University of Tokyo, 3-7-1 Hakusan, Bunkyo-ku, Tokyo 112, Japan
(present address: National Institute for Basic Biology, 38 Nishigonaka, Myo-daiji-cho, Okazaki 444, Japan)

expansion or reduction of IR to single copy regions. The linear arrangement of genes is conservative in land plants, and occasional variations of the gene order are phylogenetically informative as synapomorphic characters. Palmer and Stein [2] studied *Ginkgo* cpDNA with restriction enzyme maps, gene mapping of ribosomal DNA and 17 protein-coding genes, and heterologous hybridization using cloned angiosperm chloroplast DNA fragments covering the entire genome. *Ginkgo* cpDNA is circular and has 158kb, with an inverted repeat of approximately 17kb. *Ginkgo* has the same gene organization as observed in the majority of seed plants [3, 4], and it has been confirmed that *Ginkgo* belongs to vascular plant lineages. In gymnosperms, conifer cpDNA has been extensively studied, and it has been shown that conifers including the Taxaceae have lost one side of the IR [3, 5, 6]. The results support the concept that the conifers examined belong to a monophyletic group in which *Ginkgo* is not included.

In addition to whole genome organization, comparison of loss or gain of genes in the genome is also useful information in systematics. Lack of *chlL* gene coding protochlorophyllide reductase related to chlorophyll synthesis under dark conditions has been reported for angiosperm cpDNA. A wide survey of the distribution of the gene in cpDNA in green plants including *Ginkgo* has shown that the lack of the gene is an autoapomorphic character in the angiosperms examined [7–9]. Other losses or gains of genes in cpDNA in comparison with tobacco have not been reported in *Ginkgo*. More detailed studies of gene organization in the chloroplast genome in cycads and Gnetales may give further information on the relationships among seed plants.

3 Comparison of Nucleotide or Amino Acid Sequences of Protein Genes Coded on the Chloroplast Genome

Nucleotide or amino acid sequences of genomes potentially give us a huge amount of information on phylogenetic reconstruction. The chloroplast and mitochondrion genomes are superior to the nuclear genome as sources of phylogenetic inferences. Most genes on these genomes exist as single copies per genome and do not form multigene families as is usually observed in nuclear-coded genes. In the case of multigene families, it is difficult to find homologous pairs of genes from different taxa instead of paralogous pairs, which may produce the wrong phylogenetic inferences. The chloroplast and mitochondrion genomes exist in multicopy in a cell, and this facilitates cloning and sequencing of the genes. The most extensively studied gene in plant molecular systematics is a gene encoding the large subunit of ribulose-1,5-bisphosphate carboxylase (*rbcL*). In last several years, most major lineages of land plants were examined with *rbcL* sequence data [10–14]. *RbcL* has a relatively slower evolutionary rate in chloroplast genes [15] and is suitable for phylogenetic studies above the family level. The first *rbcL* study using all classes of extant gymnosperms including *Ginkgo biloba* [16] demonstrated the existence of a monophyletic relationship of extant gymnosperms but could not infer relationships among the gymnosperms with statistical support because the amino acid sequences used had a few phylogenetically

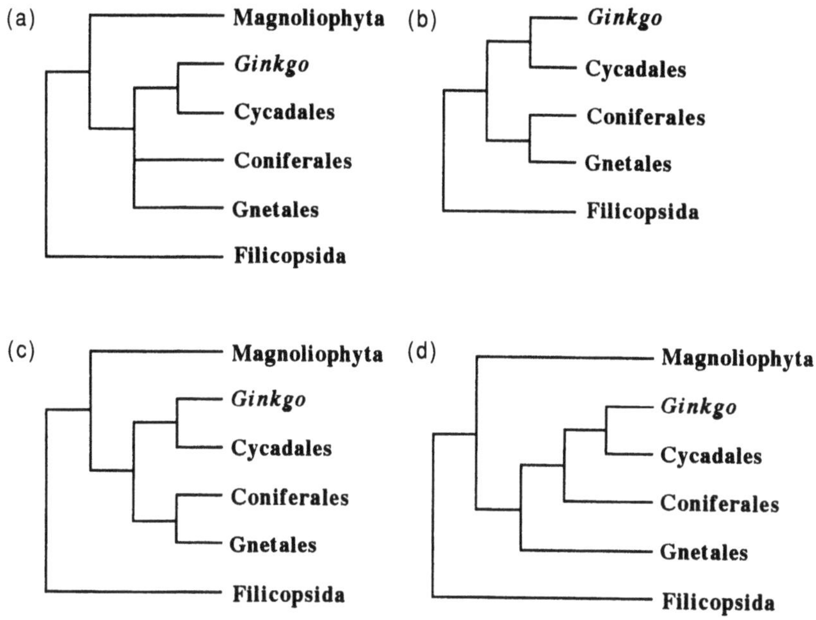

FIG. 1a–d. Phylogenetic position of *Ginkgo biloba* based on nucleotide sequences of (**a**) 1st and 2nd codon positions of *rbcL* [17], (**b**) nuclear 18*S* rRNA cited from [33], (**c**) chloroplast rRNA cluster from 3′-end of the 23*S* rRNA to 5′-end of the 5*S* rRNA, including the 4.5*S* rRNA and two intergenic transcribed spacer regions [17], and (**d**) combined data sets of (**a**) and (**b**)

informative sites among the gymnosperms. Analysis of nucleotide sequences using a larger number of extant angiosperms and gymnosperms [10] showed that cycads are most closely related to *Ginkgo*, although the study did not carry out statistical tests. Recent results using nucleotides of the 1st and 2nd codon positions of *rbcL* [17] showed that the clustering of cycads and *Ginkgo* was supported with more than 80% bootstrap probability (Fig. 1a). Organelle genomes are usually uniparently inherited, which may cause a bias for phylogenetic inferences. It is necessary to compare phylogenetic inferences obtained from different genomes.

4 Comparison of Nucleotide or Amino Acid Sequences of Protein Genes Coded on the Mitochondrion Genome

Another organelle genome in plants is in the mitochondrion. Mitochondrial genes have a slower evolutionary rate than chloroplast- and nuclear-coded genes [15] and should be suitable for phylogenetic inferences of distantly related taxa. Subunit III of the cytochrome *c* oxidase gene (*coxIII*) has been the most widely sequenced mitochondrial gene in plants. Comparison between genomic DNA sequence and cDNA sequences has revealed that 21 C in 381 nucleotides of the

Ginkgo genomic *coxIII* gene were edited as U in the mRNA [18]. The C → U type of editing has been observed widely in vascular plants, and the number of edited sites in *Ginkgo* was the most extensive in vascular plants examined [18]. A phylogenetic tree based on the *coxIII* genes covering 381 nucleotides supported the sister relationship between *Ginkgo* and cycads as inferred by chloroplast genes, although no statistical test was employed [19]. The sequence of subunit I of the cytochrome oxidase gene (*coxI*) has been submitted to a DNA database, but the detailed analysis has not been reported.

5 Comparison of Nucleotide Sequences of Nuclear-Coded Protein Genes

As seen in the preceding two sections, organelle genome-coded genes support the sister relationship between cycads and *Ginkgo*. Information on biparentally inherited nuclear gene sequences is important to confirm the phylogenetic relationships obtained using uniparently inherited organelle-coded genes, although few nuclear-coded protein genes were analyzed to cover all extant gymnosperms. Future studies on the following genes in gymnosperms will be useful to confirm the result of the organelle genes.

Nucleotide sequences of light-harvesting chlorophyll *a/b*-binding protein genes (*Lhcb1*, *-2*, and *-3*) [20, 21], the glycolysis-related glyceraldehyde-3-phosphate dehydrogenase gene (*GAPC*) [22], legumin-like seed storage protein genes [23], the 11*S* seed storage protein (ginnacin) gene [24], and phytochrome genes [25] were reported but phylogenetic analyses have not been well performed because of the meager sampling of the taxa. In all the genes except *GAPC* and ginnacin, duplicated genes are reported, suggesting that *Ginkgo* nuclear genes form a gene family like other vascular plants. Copia-like retrotransposons that are widely distributed in plants have also been found in the *Ginkgo* genome [26–28].

6 Comparison of Nucleotide Sequences of Ribosomal RNA Genes

Protein gene sequences coded on organelle genome support the close relationships between cycads and *Ginkgo*. This section considers the results of ribosomal RNA genes coded in both organelle and nuclear genomes.

Ribosomal RNAs (rRNAs) are coded on nuclear, chloroplast, and mitochondrial genomes [29]. Four nuclear rRNAs, i.e., 18*S*, 5.8*S*, 26*S*, and 5*S* rRNA, are known, and the first three genes are clustered on one or a few chromosomes as tandem repeats. Nuclear rRNA genes form a multigene family, and all the genes in the family show overall sequence homogeneity. The process that forms the congruity of the nucleotide sequences of the members is termed concerted evolution, which is caused by unequal crossing-over or gene conversion. Chloroplast-coded rRNAs, 16*S*, 23*S*, 4.5*S*, and 5*S* rRNA, are located in an inverted repeat, and the sequences of those rRNAs are highly conserved in chloroplast genes [15].

The first molecular phylogenetic analysis of rRNA including *Ginkgo* was done by Hori et al. [30] with nuclear 5*S* rRNA sequences. The results supported a sister relationship between *Ginkgo* and cycads, although it was not supported with statistical confidence because of the few informative sites resulting from both the conservative nature of the nucleotide substitution rates and the shortness (about 120 bp) of the gene. Nuclear-coded 18*S* rRNA has been widely sequenced in plants [31–33], and the analyses support the monophyletic relationship of *Ginkgo* and cycads with 76% bootstrap probability [33] (Fig. 1b). Doyle et al. [34] analyzed partial sequences obtained from 18*S* and 26*S* rRNA and showed that *Ginkgo* clustered with the conifer-cycad cluster with 88% bootstrap probability. This result should be reconsidered with revised data sets in future because their data sets include many ambiguous sites. Troitsky et al. [35] reported *Ginkgo* sequences of 4.5*S*, nuclear 5*S*, and chloroplast 5*S* rRNAs without any statistical tests, and the results support the clustering of *Ginkgo* to cycads.

Partial sequences of the chloroplast rRNA cluster from the 3′-end of the 23*S* rRNA to the 5′-end of the 5*S* rRNA, including the 4.5*S* rRNA, and two intergenic transcribed spacer regions (about 500 bp total) were reported [33]; the deduced phylogenetic tree supported monophyly of *Ginkgo* and cycads with 81% bootstrap probability (Fig. 1c). Combined data sets with nucleotides of the 1st and 2nd codon positions of *rbcL* supported the same result with 78% bootstrap probability (Fig. 1d). Sequences of mitochondrial rRNA have not been reported in *Ginkgo*.

7 Concluding Remarks

Molecular phylogenetic studies of *Ginkgo* have just begun and the available data are scarce. The *rbcL* and rRNA data suggest a close relationship of *Ginkgo* to cycads rather than to conifers or the *Gnetum* group (Fig. 1), although those results are based on relatively few data sets and the bootstrap values are not very high. Results based on cladistic analyses of morphological data [36–40] are controversial for the relationships among gymnosperms [reviewed in 41], but none of them support the molecular data. Molecular data have an advantage over traditional morphological data because more characters are obtainable and character evaluation is much easier than with morphological data. On the other hand, molecular data suffer the same difficult issue for phylogenetic inferences as morphological characters, that is, homoplasy. Therefore, there are no a priori criteria to decide which datum is better than another one and we should always compare the results on the basis of the different kinds of data sets. We should wait for future studies to resolve the phylogenetic position of *Ginkgo*.

What kind of studies are necessary? The following two approaches are thought to be useful.

One approach to reconcile the discrepancies among molecular and morphological data is by increasing the amount of molecular data. More sequence data will make it possible to infer the phylogenetic position of *Ginkgo* with high statistical confidence. Another approach is to review morphological characters with molecular biological aspects. Studies on the genetical basis of morphological characters have just started, but the progress has accelerated, especially in flower development in angiosperms (reviewed in [42, 43] and references therein). If we can reveal the developmental hierarchies of morphological characters, we will obtain more information on morphological characters that will be useful to recognize homologous organs, which are now very difficult to identify between distantly related taxa like the gymnosperms. For example, an homologous organ to the *Ginkgo* collar in other gymnosperms may be revealed by examining expression patterns of other gymnosperm homologues of genes related to collar development in *Ginkgo*. Homologues of *LEAFY* and MADS gene family related to floral organ development in angiosperms have been cloned from *Ginkgo* [42], conifers [44], and *Gnetum* (Shindo, et al., unpublished data), and should be good candidates for the studies. Other transcription factors related to morphogenesis in reproductive organs and vegetative organs have been reported [45], and studies of those gene homologues in *Ginkgo* will give new insights for the evolution of *Ginkgo biloba* in the plant kingdom.

References

1. Downie SR, Palmer JD (1992) Use of chloroplast DNA rearrangements in reconstructing plant phylogeny. In: Soltis PS, Soltis DE, Doyle JJ (eds) Molecular systematics of plants. Chapman and Hall, New York, pp 14–35
2. Palmer JD, Stein DB (1986) Conservation of chloroplast genome structure among vascular plants. Curr Genet 10:823–834
3. Raubeson LA, Jansen RK (1992) Chloroplast DNA evidence on the ancient evolutionary split in vascular land plants. Science 255:1697–1699
4. Shinozaki K, Ohme M, Tanaka M, Wakasugi T, Hayashida N, Matsubayashi T, Zaita N, Chunwongse J, Obokata J, Yamaguchi-Shinozaki K, Ohto C, Torazawa K, Meng B-Y, Sugita M, Deno H, Kamogashira T, Yamada K, Kusuda J, Takaiwa F, Kato A, Tohdoh N, Shimada H, Sugiura M (1986) The complete nucleotide sequence of the tobacco chloroplast genome: its gene organization and expression. EMBO J 5:2043–2049
5. Strauss SH, Palmer JD, Howe GT, Doerksen AH (1988) Chloroplast genomes of two conifers lack a large inverted repeat and are extensively rearranged. Proc Natl Acad Sci USA 85:3898–3902
6. Tsudzuki J, Nakashima K, Tsudzuki T, Hiratsuka J, Shibata M, Wakasugi T, Sugiura M (1992) Chloroplast DNA of black pine retains a residual inverted repeat lacking rRNA genes: nucleotide sequences of trnQ, trnK, psbA, trnI and trnH and the absence of rps16. Mol Gen Genet 232:206–214
7. Suzuki J, Bauer CE (1992) Light-independent chlorophyll biosynthesis: involvement of the chloroplast gene, *chlL*. Plant Cell 4:929–940

8. Yamada K, Matsuda M, Fujita Y, Matsubara H, Sugai M (1992) A *frxC* homologue exists in the chloroplast DNAs from various pteridophytes and in gymnosperms. Plant Cell Physiol 33:325–327
9. Burke DH, Raubeson LA, Alberti M, Hearst JE, Jordan ET, Kirch SA, Valinski AEC, Conant DS, Stein DB (1993) The *chlL* (*frxC*) gene: phylogenetic distribution in vascular plants and DNA sequence from *Polystichum acrostichoides* (Pteridophyta) and *Synechococcus* sp. 7002 (*Cyanobacteria*). Plant Syst Evol 187:89–102
10. Chase MW, Soltis DE, Olmstead RG, Morgan D, Les DH, Mishler BD, Duvall MR, Price RA, Hills HG, Qiu Y-L, Kron KA, Rettig JH, Conti E, Palmer JD, Manhart JR, Sytsma KJ, Michaels HJ, Kress WJ, Karol KG, Clark WD, Hedrén M, Gaut BS, Jansen RK, Kim K-J, Wimpee CF, Smith JF, Furnier GR, Strauss SH, Xiang Q-Y, Plunkett GM, Soltis PS, Swensen SM, Williams SE, Gadek PA, Quinn CJ, Equiarte LE, Golenberg E, Learn GH Jr, Graham SW, Barrett SCH, Dayanandan S, Albert VA (1993) Phylogenetics of seed plants: an analysis of nucleotide sequences from the plastid gene *rbc*L. Ann MO Bot Gard 80:528–580
11. Manhart JR (1994) Phylogenetic analysis of green plant *rbcL* sequences. Mol Phylogenet Evol 3:114–127
12. Mishler BD, Lewis LA, Buchheim MA, Renzaglia KS, Garbary DJ, Delwiche CF, Zechman FW, Kantz TS, Chapman RL (1994) Phylogenetic relationships of the green algae and bryophytes. Ann MO Bot Gard 81:451–483
13. Hasebe M, Omori T, Nakazawa M, Sano T, Kato M, Iwatsuki K (1994) *RbcL* gene sequences provide evidence for the evolutionary lineages of leptosporangiate ferns. Proc Natl Acad Sci USA 91:5730–5734
14. Hasebe M, Wolf PG, Pryer KM, Ueda K, Ito M, Sano R, Gastony GJ, Yokoyama J, Manhart JR, Murakami N, Crane EH, Haufler CH, Hauk WD (1995) Fern phylogeny based on *rbcL* nucleotide sequences. Amer Fern J 85:134–181
15. Wolfe KH, Gouy M, Yang Y-W, Shart PM, Li W-H (1989) Date of the monocot-dicot divergence estimated from chloroplast DNA sequence data. Proc Natl Acad Sci USA 86:6201–6205
16. Hasebe M, Kofuji R, Ito M, Kato M, Iwatsuki K, Ueda K (1992) Phylogeny of gymnosperms inferred from *rbcL* gene sequences. Bot Mag Tokyo 105:673–679
17. Goremykin V, Bobrova V, Pahnke J, Troitsky A, Antonov A, Martin W (1996) Noncoding sequences from the slowly evolving chloroplast inverted repeat in addition to *rbcL* do not support Gnetalean affinities of angiosperms. Mol Biol Evol 13:383–396
18. Hiesel R, Combettes B, Brennicke A (1994) Evidence for RNA editing in mitochondria of all major groups of land plants except the Bryophyta. Proc Natl Acad Sci USA 91:629–633
19. Hiesel R, von Haeseler A, Brennicke A (1994) Plant mitochondrial nucleic acid sequences as a tool for phylogenetic analysis. Proc Natl Acad Sci USA 91:634–638
20. Chinn E, Silverthorne J (1993) Light-dependent chloroplast development and expression of a light-harvesting chlorophyll *a/b*-binding protein gene in the gymnosperm *Ginkgo biloba*. Plant Physiol (Rockv) 103:727–732
21. Chinn R, Silverthorne J, Hohtola A (1995) Light-regulated and organ-specific expression of types 1, 2, and 3 light-harvesting complex b mRNAs in *Ginkgo biloba*. Plant Physiol (Rockv) 107:593–602
22. Jansson S, Meyer-Gauen G, Cerff R, Martin W (1994) Nucleotide distribution in gymnosperm nuclear sequences suggests a model for GC-content change in land-plant nuclear genomes. J Mol Evol 39:34–46

23. Häger K-P, Braun H, Czihal A, Müller B, Bäumlein H (1995) Evolution of seed storage protein genes: legumin genes of *Ginkgo biloba*. J Mol Evol 41:457–466
24. Arahira M, Fukazawa C (1994) Ginkgo 11*S* seed storage protein family mRNA: unusual Asn-Asn linkage as post-translational cleavage site. Plant Mol Biol 25:597–605
25. Mathews S, Lavin M, Sharrock RA (1995) Evolution of the phytochrome gene family and its utility for phylogenetic analyses of angiosperms. Ann MO Bot Gard 82:296–321
26. Voytas DF, Cummings MP, Konieczny A, Ausubel FA, Rodermel SR (1992) Copia-like retrotransposons are ubiquitous among plants. Proc Natl Acad Sci USA 89:7124–7128
27. Flavell A, Dunbar E, Anderson R, Pearch SR, Hartley R, Kumar A (1992) Ty1-copia group retrotransposons are ubiquitous and heterogeneous in higher plants. Nucleic Acids Res 20:3639–3644
28. Hirochika H, Hirochika R (1993) Ty1-copia group retrotransposons as ubiquitous components of plant genomes. Jpn J Genet 68:35–46
29. Hamby RK, Zimmer EA (1992) Ribosomal RNA as a phylogenetic tool in plant systematics. In: Soltis PS, Soltis DE, Doyle JJ (eds) Molecular systematics of plants. Chapman and Hall, New York, pp 50–91
30. Hori H, Lim B-L, Osawa S (1985) Evolution of green plants as deduced from 5S rRNA sequences. Proc Natl Acad Sci USA 82:820–823
31. Nickrent DL, Soltis DE (1995) A comparison of angiosperm phylogenies from nuclear 18*S* rDNA and *rbc*L sequences. Ann MO Bot Gard 82:208–234
32. Chaw S-M, Long H, Wang B-S, Zharkikh A, Li W-H (1993) The phylogenetic position of Taxaceae based on 18*S* rRNA sequences. J Mol Evol 37:624–630
33. Chaw S-M, Sung H-M, Long H, Zharkikh A, Li W-H (1995) The phylogenetic positions of the conifer genera *Amentotaxus*, *Phylloclasdus* and *Nageia* inferred from 18*S* rRNA sequences. J Mol Evol 41:224–230
34. Doyle JA, Donoghue MJ, Zimmer EA (1994) Integration of morphological and ribosomal RNA data on the origin of angiosperms. Ann MO Bot Gard 81:419–450
35. Troitsky AV, Melekhovets YF, Rakhimova GM, Bobrova VK, Valiejo-Roman KM, Antonov AS (1991) Angiosperm origin and early stages of seed plant evolution deduced from rRNA sequence comparisons. J Mol Evol 32:253–261
36. Grane PR (1985) Phylogenetic analysis of seed plants and the origin of angiosperms. Ann MO Bot Gard 72:716–793
37. Doyle JA, Donoghue MJ (1986) Seed plant phylogeny and the origin of angiosperms: an experimental cladistic approach. Bot Rev 52:321–431
38. Laconte H, Stevenson DW (1990) Cladistics of the Spermatophyta. Brittonia 42:197–211
39. Nixon KC, Crepet WL, Stevenson DW, Friis EM (1994) A reevaluation of seed plant phylogeny. Ann MO Bot Gard 81:484–533
40. Rothwell GW, Serbet R (1994) Lignophyte phylogeny and the evolution of spermatophytes: a numerical cladistic analysis. Syst Bot 19:443–482
41. Rothwell GW (1994) Phylogenetic relationships among ferns and gymnosperms; an overview. J Plant Res 107:411–416
42. Weigel D, Meyerowitz EM (1993) Genetic hierarchy controlling flower development. In: Bernfield M (ed) Molecular basis of morphogenesis. Wiley-Liss, New York, pp 93–107

43. Theißen G, Saedler H (1995) MADS-box genes in plant ontogeny and phylogeny: Haeckel's "bionenergetic law" revisited. Curr Opin Genet Dev 5:628–639
44. Purugganan MD, Rounsley SD, Schmidt RJ, Yanofsky MF (1995) Molecular evolution of flower development: diversification of the plant MADS-box regulatory gene family. Genetics 140:345–356
45. Ramachandran S, Hiratsuka K, Chua N-H (1994) Transcription factors in plant growth and development. Curr Opin Genet Dev 4:642–646

Mesozoic Ginkgoalean Megafossils: A Systematic Review

ZHIYAN ZHOU

1 Introduction

Numerous megafossils of *Ginkgo* and its allies have been recorded in the Mesozoic since the last century [1–3]. Most of these fossils, however, are isolated leaves and other vegetative organs. Fossils of reproductive organs believed to belong to ginkgoaleans are scarce, and they are either fragmentary or were only doubtfully attributed to Ginkgoales. For a long time, our knowledge of the history of *Ginkgo* and Ginkgoales was based almost entirely on vegetative organs [4–6]. A number of genera and many species have been established on the basis of leaf gross morphology. Cuticular analysis was later introduced to the studies of ginkgoalean leaves [7–14] and was found to be of great value in specific determination. As pointed out by Harris [9], the cuticles, on the whole, are rather more characteristic than any macroscopic characters. For classification of higher taxonomic units, however, cuticles are less important because of the consistency in general structure. Up to the present, gross morphology is still the main criterion for classification of Mesozoic ginkgoaleans above the specific level.

As is shown here, a given leaf type (or form genus) may belong to different genera or even families established on the basis of reproductive organs. Moreover, considerable variation in leaf form has been known in *Ginkgo biloba* and various fossil taxa that have been studied in some detail. It is obvious that a classification based on leaf gross morphology alone will be of little phylogenetic significance. One well-known example is that *Ginkgo*-like leaves (*Ginkgoites tigriensis*) were found in connection with a kind of ovulate organ called *Karkenia* by Archangelsky [15], which is very different from *Ginkgo*. In view of the unsatisfactory situation in classification of the Mesozoic ginkgoaleans, Zhou [16] proposed a preliminary scheme based on a phylogenetic analysis of some more completely known taxa (so-called whole plants) of which reproductive as well as vegetative organs are found. In this chapter I have tried to review at the generic

Nanjing Institute of Geology and Palaeontology, Academia Sinica, Chi-Ming-Ssu, Nanjing 210008, People's Republic of China

level all the Mesozoic ginkgoalean megafossils so far described, except those proved to belong to other plant groups (such as Czekanowskiales). Furthermore, a new phylogenetic analysis was made, and an amended classification of Mesozoic Ginkgoales is proposed in the light of new evidence.

2 Genera for Reproductive Organs and "Whole Plants"

Allicospermum Harris 1935 (Type species: *A. xystum* Harris)

The form genus *Allicospermum*, as originally defined, would include, among recent plants, the seeds of *Ginkgo*, cycads, and some conifers, and among fossils, the seeds of *Nilssonia, Lepidopteris*, and *Stachyotaxus* [9]. The type species and some other species are believed to be seeds of fossil ginkgoaleans, i.e., *A. xystum* Harris (belonging to *Ginkgoites taeniata* Braun sp.) [9], *A. ginkgoideum* Tralau (*Ginkgoites regnellii* Tralau), or *A. baiereanum* Tralau (*Baiera gracilis* Bean MS) [17]. More or less similar seeds proved to belong to *Ginkgo yimaensis* Zhou et Zhang [18], *G. huttonii* (Sternberg) [19], and *Yimaia recurva* Zhou et Zhang [20].

Antholithus (*Antholithes*) Brongniart 1822 (Type species: *A. liliacea* Brongniart)

According to Andrews [21], the type species is a small "bud-like" impression showing no fertile parts and of unknown affinity. Fossils of wide diversity have been assigned to this genus. Some species proved later to belong well to the Peltaspermaceae (*Antevsia* Harris) or to the Caytoniales (*Caytonanthus* Harris) [22]. Those considered to be related to Ginkgoales are *Antholithus wettsteinii* Krasser belonging to *Ginkgoites lunzensis* (Stur) Florin and the *Antholithus*-type male fructification of *Sphenobaiera furcata* (Heer) Florin [23].

Bernettia Gothan 1914 (Type species: *B. inopinata* Gothan)

Harris [9] described a new species of male fructification from East Greenland attributed to Gothan's genus as *B. phialophora* and believed it to be related to the associated *Sphenobaiera spectabilis* Nathorst because of similarity in cuticular structure. Recent investigations revealed that the type species of this genus (*B. inopinata*) [24] may represent the ovulate part of *Chlamydolepis*, possibly of Gnetales [25], while *B. phialophora* which yielded monocolpate pollen should belong to *Androstrobus* (as *A. phialophora*), an organ genus believed to have a cycadean affinity [26].

Ginkgo L. 1771

The generic name *Ginkgo* is adopted here for fossils which, as suggested by Zhou and Zhang [18], are essentially similar to *G. biloba* not only in leaf morphology but also in reproductive organs. Taking no account of the isolated seeds attributed to this genus by some authors [19], ovule-bearing organs of *Ginkgo yimaensis* Zhou and Zhang [18] and *G. huttonii* (Sternberg) [19, 27] are among the few Mesozoic records of female reproductive organs of which the generic determination is based on more convincing evidence. Less dubious pollen organs have so far only been found in relation to *Ginkgo huttoii* [19, 28, 29]. Drinnen and Chambers [30] described also some pollen organs closely associated with *Ginkgo australis* (McCoy) from the Lower Cretaceous of Victoria, but no in situ pollen grains of the *Ginkgo* type are found in the material.

Ginkgoanthus Nathorst 1899

The type species of this genus was not originally designated. *Ginkgoanthus* sp. from the Upper Jurassic of Franz-Josef-Land, described and illustrated by Nathorst [31], was considered a male flower of a ginkgo, but according to Seward [32], the preservation is too imperfect to admit of a satisfactory determination.

Ginkgospermum Nathorst 1878 (Type species: *G. globulare* Nathorst)

As commented by Krassilov [14], there is no distinction between this genus [33] and the noncommittal organ genus *Carpolithus* (or *Carpolithes*) Sternberg. Its attribution to Ginkgoales is uncertain.

Grenana Samylina 1990 (Type species: *G. angrenica* Samylina)

The genus includes ovule-bearing organs as well as vegetative leaves from the Jurassic deposits of Middle Asia [34]. The leaves are fragmentary, but are dichotomously divided into narrow segments. They are amphistomatic, with longitudinal files of epidermal cells and randomly oriented stomata. The seeds are platyspermic, singly and erectly seated in the paired cuplike cupules. Samylina referred this genus to Lyginopteridopsida. She pointed out the resemblance in the paired cupules and deeply divided leaves, respectively, to those of the extant *Ginkgo* and those of geologically older forms of this genus, but thought *Grenana* to be a relict pteridosperm belonging to one of the lineages of Lyginopteridopsida, which gave rise to Ginkgopsida in the Early Triassic (or Late Permian).

Karkenia Archangelsky 1965 (Type species: *K. incurva* Archangelsky)

This genus was founded on the ovule-bearing organ from the Lower Cretaceous of Argentina, which consists of a penduncle and as many as 100 small, densely packed, orthotropous but inverted ovules. The type species occurs in close association with dwarf shoots and leaves of the *Ginkgo* type (*Ginkgoites tigriensis* Archangelsky) [15]. Most known species of this genus are found to be related to leaves of the *Sphenobaiera* type (*K. asiatica* Krassilov and *K. cylindrica* Schweitzer et Kirchner) [14, 36], but *K. hauptmannii* Kirchner and Van

Konijnenburg-Van Cittert [35] is associated with both simple leaves and leaves once or twice divided.

Schmeissneria Kirchner et Van Konijnenburg-Van Cittert 1994 (Type species: *S. microstachys* (Presl) Kirchner and Van Konijnenburg-Van Cittert)

The ovule-bearing organs originally described as *Pinites microstachys* and regarded as a male fructification [37] were subsequently referred to *Stachyopitys* as *S. preslii* together with some other male fructifications by Schenk [2]. Wcislo-Luaniec [38] was the first to find that such specimens of *Stachyopitys preslii* do not show spreading pollen sacs but are cupule bearing. According to the recent study of Kirchner and Van Konijnenburg-Van Cittert [35], they belong to a new genus of Ginkgoales, *Schmeissneria*. The organs consist of a main axis and a number of sessile or pedunculate (pedicled) cupules. Each cupule bears one ovule, which is winged when ripe. The leaves, inserted apically on the dwarf shoots together with the ovule-bearing organs (*S. microstachys*), bear a resemblance to *Glossophyllum* and *Eretmophyllum* but have more than two veins at the base. The associated male fructification is of the *Stachyopitys*-type (see following).

Sphenobaieroanthus Yang (Type species: *S. sinensis* Yang)

The type species is a catkin-like organ with a long peduncle believed to be attached to the apex of a dwarf shoot and given off from the axil of a *Sphenobaiera*-type leaf (*Sphenobaierocladus* Yang; see Section 3). The fructification is 1.5 cm long and attains to a maximum width of 0.5 cm in the middle part. The detailed structure of the fructification is obscure, as shown in the figures [39], but according to the author it is an inflorescence with many staminate flowers, each consisting of a single filament and six or seven whorled anthers at the apical part. Neither cuticle nor pollen has been detected.

Stachyopitys Schenk 1867 (Type species: *S. preslii* Schenk)

The type species from the Lower Liassic of the Beyreuth area, southern Germany, has long been known as the male fructification of a ginkgoalean plant (*Baiera muensteriana*) [2, 40]. Gothan [24] also believed *Stachyopitys* sp. from the Nürenburg area to be related to *Baiera taeniata* (Braun). Recent work [35, 38] has revealed that Schenk included two different kinds of fructifications in the type species. While the female fructification proved to belong to a new genus *Schmeissneria* of Ginkgoales (see foregoing), the male fructifications may also possibly be related to it, although they are not found in direct connection. *Schmeissneria preslii*, as newly emended, comprises a striated longitudinal main rachis and irregularly arranged lateral branches that bear six (five to ten) spreading, probably longitudinally dehiscent elongate pollen sacs at their apex. No pollen grains have so far been detected [35].

Toretzia Stanislavsky (1971) 1973 (Type species: *T. angustifolia* Stanislavsky)

The type species was described from the Upper Triassic Novoraick Formation of the Donetz Basin, Ukraine [41, 42]. The ovule-bearing organs are believed to be given off from the axils of leaves on a dwarf shoot. The ovules are inverted and borne singly at the apex of the peduncle. The leaves attached to the long and dwarf shoots are linear to ribbonlike and with two (to four) indistinct veins.

Umaltolepis Krassilov (1970) 1972 (Type species: *U. vachrameevii* Krassilov)

This type of reproductive organ was first reported from the Upper Jurassic and Lower Cretaceous of the Bureja River Basin, Siberia [14, 43]. It consists of a bract and a stalk with scale leaves at the base. The bracts are elongate, entire or sometimes divided into two lobes. The ovules are believed to be borne singly on the abaxial side of the bract. The cuticles of the bract are similar to the associated leaves of the *Pseudotorellia* type.

Yimaia Zhou and Zhang 1992 (Type species: *Y. recurva* Zhou and Zhang)

This genus was established for pedunculate female reproductive organs bearing up to eight or nine terminal ovules that are sessile, contiguous, and orthotropus, but mostly recurved when mature, in association with *Baiera hallei* Sze from the Middle Jurassic Yima Formation, Henan, China [20]. Similar reproductive organs were found in connection with *Baiera gracilis* (Bunbury) (= *B. furcata* Lindley et Hutton sp. according to Harris and Millington [19]) in the Middle Jurassic Yorkshire flora [44]. The ovulate organs of *B. muensteriana* originally described from the Basal Liassic of Franconia [1, 2], as recently reexamined by Kirchner [45], are immature and bear a close resemblance to the young specimen of *Yimaia recurva* [20] in having very short pedicels.

Up to the present, few fossil reproductive organs so far recorded of Ginkgoales may be considered authenticated. *Bernettia, Ginkgoanthus*, and *Ginkgospermum* are either completely unrelated to Ginkgoales or were based on unsatisfactory material and dubiously defined. The pollen-bearing organs are only occasionally preserved in relation to *Ginkgo huttonii, Ginkgoites lunzensis*, and *Sphenobaiera furcata*, and *Stachyopitys preslii* is presumed to be related to *Schmeissneria*. It is doubted whether the catkin-like reproductive organ named *Sphenobaieroanthus* is attached to the dwarf shoot bearing the *Sphenobaiera*-type leaves. Moreover, its detailed structure is still unknown.

More female reproductive organs of Ginkgoales are known as compared with the male ones. Besides those essentially similar to the ovulate organs of the extant *Ginkgo biloba* (i.e., *Ginkgo yimaensis* and possibly *G. huttonii*), there are *Karkenia, Schmeissneria, Toretzia, Umaltolepis*, and *Yimaia*. In this writer's opinion, *Grenana* may also belong to Ginkgoales [20]. Although no dwarf shoots or resin bodies have ever been found in this genus, the similarities between its ovule-bearing organs and those of *Ginkgo*, and the associated bifurcated leaves and many ginkgoalean leaves are impressive. Some *Allicospermum*-type seeds ascribed to Ginkgoales are of little significance as the ovulate organs are not known.

In summary, only the following genera of Mesozoic reproductive organs may be regarded as useful:

Female reproductive organs: *Ginkgo, Grenana, Karkenia, Schmeissneria, Toretzia, Umaltolepis, Yimaia*, and *Allicospermum* (pro parte)

Male reproductive organs: *Ginkgo, Stachyopitys*, and *Antholithus* (pro parte)

3 Genera for Leaves or Vegetative Shoots

Baiera F. Braun 1843 (Homonym: *Bajera* Sternberg 1825; synonym: *Jeanpaulia* Unger 1845; type species: *Baiera muensteriana* (Presl in Sternberg 1833) Heer 1876 [1, 3, 46, 47]

Braun's original diagnosis of this genus [1] included characters of both leaves and reproductive organs. Although his description and interpretation were not totally correct, the ovulate organs he figured, according to the recent work of Kirchner [45], are trustworthy. The generic name *Baiera*, however, has been used later in a rather different sense. Florin [10], in a thorough investigation into the Mesozoic ginkgoaleans, emended this genus and separated it effectively from the then confusable leaf genera *Ginkgoites* (or *Ginkgo*) and *Sphenobaiera*. He did not, however, include the characters of reproductive organs in the diagnosis. Under his influence, more than 150 species, according to Jongmans and Dijikstra [48], have been described of this genus based solely on compression/impression of leaves. Before Kirchner [45] reexamined the type material of ovulate organs figured by Braun [1] and Schenk [2] from the Rhaeto-Liassic "Grenzschichten" of Franconia, *Baiera* had become, as generally accepted, a form (or organ) genus for those leaves that fall within the morphological category as defined by Florin [10].

Baierella Potonié 1933 (Type species: *B. bohnei* Potonié)

The type species from the Upper Jurassic of Iran [49] includes fanshaped, strongly divided leaves with a petiole that dichotomizes toward its apex; each branch then repeatedly bifurcates. The ultimate segments are narrow and linear, each with a single vein. The cuticular structure remains hitherto unknown. As pointed out by Florin [10], our knowledge of this genus is still too imperfect to discuss the systematic position with even any degree of certainty. Dorf [50] also considered this genus a doubtful ginkgophyte. Krassilov [14, 43], however, in studying ginkgophytes of the Bureja River Basin, transferred *Sphenobaiera uninervis* Samylina [12, 51] to Potonié's genus. According to Schweitzer and Kirchner [36], *B. bohnei* is a junior synonym of *Baiera muensteriana* (Presl).

Baierophyllites Jain et Delevoryas 1967 (Type species: *B. florinii* Jain and Delevoryas)

Baierophyllites is the only genus of Ginkgoaceae established for fossil leaves from the Southern Hemisphere. The type species *B. florinii* was described from the Middle Triassic of western Argentina [52]. In its general gross morphology it does not differ from *Sphenobaiera* Florin, but the venation is indistinct. It is also not clear whether the leaf has a clean-cut abscission base as that typical of *Sphenobaiera* [19]. Jain and Delevoryas [52] believed that this form is probably closer to *Ginkgoites* than to *Sphenobaiera* because the subsidiary cells bear no papillae and the terminal (polar) subsidiary cells are smaller. The material on which this genus was based is inadequate. No large pieces of cuticle were figured, and the authors did not describe the upper and lower cuticles separately or in detail. Krassilov [14] thought that the characters mentioned by Jain and Delevoryas cannot warrant a generic separation of *Baierophyllites* from *Sphenobaiera* because ginkgo leaves usually show considerable morphological variation.

Eretmoglossa Barale 1981 (Type species: *E. lacasae* Barale)

This rarely known genus was founded on specimens collected from the Lower Cretaceous of Montsech, Spain [53] and has not been reported from elsewhere. The leaves are found still attached to an axis believed to be of indeterminate (unlimited) growth. They are entire, sessile, and with resin bodies in the mesophyll. Irregularly oriented stomata occur in the stomatal strips alternately between the nonstomatal strips on both surfaces. The veins are parallel for the greater part of the leaf but dichotomize in the basal part and converge at the apex. Barale referred this genus to Glossophyllaceae of Ginkgoales.

Eretmophyllum Thomas 1913 (Type species: *E. pubescens* Thomas)

Thomas [7], in erecting this genus, gave the following diagnosis:

> Leaves oblanceolate to linear, tapering into a distinct petiole. Apices rounded or retuse. Veins distinct, dichotomizing near base of lamina and more or less parallel above, slightly convergent near apex. Epidermal cells more or less rectangular or polygonal. Stomata with group of angular subsidiary cells regularly arranged round and above the guard cells.

He described, besides *E. pubescens*, another species, *E. whitbiense* Thomas, which differs from the type species mainly in being amphistomatic instead of hypostomatic. According to Harris and Millington [19], however, stomata do occur on the upper cuticle in the basal part of lamina in *E. pubescens*. The emended generic diagnosis given by them added the words "leaf shed individually by abscission at base of petiole," "veins ending separately at apical margin," and "round to spindle-shaped resin bodies often present."

Euryspatha Prynada (1956) 1963 (Type species: *E. rarinervis* Prynada)

This generic name was first made public in *Osnovy Paleontologii* [54]. Only one species was described, based on leaves collected from the Jurassic and the Lower Cretaceous of the Bureja River Basin. According to Krassilov [14], who studied the ginkgophytes of the Bureja River Basin in detail, this genus was based on fragmentary leaves and cannot be separated from *Eretmophyllum*.

Furcifolium Kräusel 1943 (Type species: *F. longifolium* (Seward) Kräusel)

This generic name was proposed by Kräusel [55] for the Upper Jurassic specimens from Solenhofen, originally described as *Baiera* sp.? *longifolia* Heer and *Baiera* sp. by Salfeld [56] and later referred by Seward [32] to *Sewardia* Zeiller as *S. longifolia* (Salfeld). The leaves are deeply bifurcated and without a distinct petiole. They are borne in bundles on stout dwarf shoots attached to a long shoot. The segments are 3–6 mm wide and with two to four parallel veins. The cuticular structure of the leaves is not known. There is a thick backward pointed spinelike projection at the base of each dwarf shoot.

Ginkgodium Yokoyama 1889 (Type species: *G. nathorstii* Yokoyama)

This genus originally included both entire or bilobed ginkgoalean leaf impressions from the Middle Jurassic of Japan characterized by having a short petiole and thickened lateral margins near the leaf base. Veins are numerous, simple, and parallel. There are fine interstitial ribs [57]. Recent work has revealed the presence of this genus in the Lower Liassic and Lower Cretaceous [58, 59], but the generic delimitation remains uncertain. *Ginkgodium glandulosum* Samylina [51, 60] is the only species hitherto referred to this genus of which the leaf cuticular structure is known. According to Krassilov [14], who transferred this species to *Eretmophyllum* Thomas, *Ginkgodium* is indistinguishable from Thomas' genus. A similar opinion has been expressed by Kimura and Sekido [58] and by Kimura and Tsujii [59].

Ginkgoites Seward 1919 (Type species not designated, but according to Andrews (1955) *Ginkgoites obovata* (Nathorst) Seward)

Ginkgoites was not instituted on the basis of definite characters and does not necessary imply a generic identity with *Ginkgo*. It is a name proposed for fossil ginkgoalean leaves that it is believed belong either to plants generically identical with *Ginkgo* or to very closely allied types but which lack sufficient information [32]. There was much confusion as to the definition of this name later. Both Florin [10] and Tralau [5] intended to further define the genus based on morphological characters that can readily distinguish it from leaves of the extant *Ginkgo* (and also other ginkgoalean fossils). Harris and Millington [19] criticized it; because of the great morphological variation, the distinctions used for separating *Ginkgoites* are in practice hardly sustainable. They decided to drop Seward's generic name, referring all Yorkshire Jurassic *Ginkgo*-like fossil leaves directly to the extant genus. Zhou and Zhang [18] proposed to restrict the generic name *Ginkgo* to those fossil species that are essentially similar to *Ginkgo biloba* not only in leaf morphology but also in reproductive organs. The name *Ginkgoites* may still be used as an organ (or form) generic name for *Ginkgo*-like leaves that cannot be attributed to any natural genus with certainty. Similar views have been advanced by Krassilov [14], who gave an emended diagnosis of this genus.

Ginkgoitocladus Krassilov 1972 (Type species: *G. burejensis* Krassilov)

This is an organ genus proposed by Krassilov [14] for fossil ginkgoacean dwarf shoots, which are distinguished from *Pityocladus* Seward [32] by having double vascular bundle scars and by the form of secretory canals and epidermal characters. Incidentally, *Ginkgocladus* Ettingshausen [61] was based on an Eocene leaflike impression from New Zealand marked with a midrib. According to Seward [32], it is different from ginkgos and of little value as a botanical record.

Glossophyllum Kräusel 1943 (Type species: *G. florinii* Kräusel)

This genus was founded on well-preserved leaves from the Keuper of Lunz [23]. Some of them are still attached helically to the shoot. The leaves are coriaceous, entire, and nearly tongue shaped, straight, or sickle like, with a rounded apex and narrowing gradually to a petiole-like base with two vascular bundles. The veins are parallel, but branched in the lower part of the leaves. The stomata occur in the stomatal strips alternated with nonstomatal strips on both surfaces of leaves. No resin bodies have so far been found in leaves of this genus.

Kalantarium Dobruskina 1980 (Type species: *K. kraeuselii* Dobruskina)

In a comprehensive study on the Triassic plant-bearing strata of Eurasia, Dobruskina [62] instituted this genus on the basis of specimens from the Middle and Upper Triassic of the Pechora Basin. The leaves of the type species are elongate, parallel sided, lanceolate, or wedge shaped, narrowing toward the base into a petiole, with entire or dichotomously divided lamina. Veins are parallel or fan shaped. There are three to six veins in the 5 mm-wide leaves. The leaves are amphistomatic. Papillae are usually absent on the surface of epidermal cells. The guard cells are not sunken, and stomata are predominantly longitudinally oriented. No resin bodies have hitherto been found in the mesophyll. According to Dobruskina, this genus closely resembles *Glossophyllum* Kräusel in gross morphology and in epidermal structure and should belong to Glossophyllaceae of Tralau [5]. The differences between the two genera lie in the dimorphic leaves of *Kalantarium*, which, moreover, normally have more stomata and poorly developed vein courses and epidermal cell papillae. In *Glossophyllum*, only entire leaves are known. The other genus believed to be very close to *Kalantarium* is *Kirjamkenia* Prynada (see following), from the early Triassic of the Tongusk Basin, but the latter genus was based on insufficient material and only divided leaves of the *Sphenobaiera* type were known. There is a resemblance in general gross morphology between the entire and divided leaves of *Kalantarium* and some more well-defined ginkgoalean genera, such as *Sphenobaiera* Florin [10] and *Eretmophyllum* Thomas [7], respectively, but Dobruskina [62] considered *Kalantarium* to be different from these genera in cuticular structure.

Kirjamkenia Prynada 1970 (Type species: *K. lobata* Prynada)

This genus was originally regarded as a pteridosperm [63]. Dobruskina [62] believed that it is related to *Glossophyllum* Kräusel [23], and included *Ginkgoites lunzensis* (Stur) and *Sphenobaiera furcata* (Heer) Florin from the Middle and Upper Keuper of Lunz, Austria, and of Neue Welt near Basel, Sweitzerland, in this genus as *Kirjamkenia lunzensis* (Kräusel) Dobruskina comb. nov. and *K. furcata* (Kräusel) Dobruskina comb. nov. (note that she did not cite Stur and Heer as the authors of the two species). Dobruskina attributed *Kirjamkenia* together with *Kalantarium* to Glossophyllaceae, which she regarded as a special group of Ladinian to Carnian (and Permain?) ginkgophytes.

Leptotoma Kiritchkova and Samylina 1979 (Type species: *L. lenaensis* Kiritchkova and Samylina)

Kiritchkova and Samylina [64] erected this genus for ginkgophyte leaves that are petiolate and dichotomously divided into thick and narrow segments of quadrangular or trapezoid cross section. According to the Russian authors, the distinguishing features of the epidermal structure of this genus are the amphistomatic leaves with irregularly oriented stomata that spread over the lower and lateral surfaces but occur only in the marginal zones of the upper surface. This genus ranges from the Middle Jurassic to Lower Cretaceous. The type species occurs in the Upper Jurassic of northeastern Russia.

Mirovia Reymanówna 1985 (Type species: *M. szaferi* Reymanówna)

The generic name was proposed for detached leaves from the Jurassic of Kradów region, Poland, which bear a close resemblance to *Pseudotorellia* and were believed to belong to Ginkgoales [65]. According to Bose and Manum [66, 67], however, such leaves are not related to Ginkgoales but represent an extinct group of conifers, Miroviaceae.

Pseudotorellia Florin 1936 (Type species: *P. nordenskioldii* (Nathorst))

According to Florin [10], this genus is characterized by small, coriaceous, entire, sessile, linear to narrow tonguelike and more or less sickle-shaped leaves that are hypostomatic and with regularly longitudinally oriented haplocheilic stomata. Harris and Millington [19] thought that the leaves of this genus are normally shed individually by abscission at the leaf base. Krassilov [14, 43], however, figured a leaf of *P. angustifolia* still attached to the dwarf shoot. His work also proved the presence of resin bodies in the leaf of this genus as in many other ginkgoalean leaves.

Sphenobaiera Florin 1936 (Type species: *S. spectabilis* (Nathorst) Florin)

Florin [10] instituted this genus for leaves originally included in *Baiera* Braun that are distinguished from those of the type species *B. muensteriana* mainly by having no distinct petiole. Harris and Millington [19] emended the genus later by adding the words "leaves shed separately" in the diagnosis. This character had been neglected until it came to be considered of taxonomic importance. They proposed to separate leaves from this genus, which are believed to be persistent on caducous dwarf shoots, and to refer them to a distinct genus *Sphenarion* of Czekanowskiales. They supposed that *Sphenobaiera* as restricted is a natural genus and closely related to *Ginkgo*, although they pointed out that two different kinds of pollen organs have been attributed to *Sphenobaiera spectabilis* (Nathorst) and *S. furcata* (Heer), respectively. Kräusel [23] included pollen organs of *Antholithus* type also in the latter species.

Sphenobaierocladus Yang 1986 (Type species: *S. sinensis* Yang)

Yang [39] instituted this genus for a long shoot with several spirally arranged leafy dwarf shoots collected from the Upper Triassic Xujiahe Formation of Dazu, Sichuan Province. The shoot is almost indistinguishable from that of *Ginkgo biloba* in gross morphology, but the leaves found attached are of the *Sphenobaiera* type. The long shoot is smooth, while the dwarf shoots are covered with closely spirally arranged, transverse, oval leaf scars, in which there are scars of vascular bundles (more than two in number) in transverse arrangement.

3.1 Basic Morphological Types (Form Genera)

As discussed previously, a number of the 18 genera for leaves or vegetative shoots of Mesozoic ginkgoaleans are based on unsatisfactorily preserved material and are diagnostically poorly defined. It is often difficult to decide whether they are well separated from some other already established forms. For some, even their attribution to Ginkgoales may be doubted. *Mirovia*, for instance, has recently been transferred to Coniferales. The type species of *Baierella*, *Euryspatha*, and *Ginkgodium* are all based on very few impressions (only one in the second genus). Neither the range of morphological variation nor the cuticle is known. Opinions are divergent as to the later attribution of some other species to these genera, of which the cuticle is preserved, such as *Baierella uninervis* (Samylina) and *Ginkgodium glandulosum* Samylina (see Section 3). In gross morphology, there is a close resemblance between *Baierella* and segments of *Czekanowskia*, some *Baiera* (for instance, *B. leptophylla* Harris [9] and *B. muensteriana*, as pointed out by Schweitzer and Kirchner [36]), *Sphenobaiera* (*S.

furcata Heer; see Section 3), or even some *Stenopteris* Saporta [68] (= *Rhaphidopteris* [69]): *S. dinosaurensis* Harris [70]. These forms may all bear single-veined ultimate segments as *Baierella* and are hardly distinguishable from one another without knowledge of their cuticular structure or complete leaf morphology. The cases are more or less similar with *Euryspatha* and *Ginkgodium.* As already mentioned, both of these cannot be easily separated from *Eretmophyllum* in leaf form.

The separation of *Baierophyllites* and *Kirjamkenia* from *Sphenobaiera*, and of *Kalantarium* from *Eretmophyllum*, also need consideration. In cuticular structure, they seem to agree generally with each other. *Leptotoma* also does not differ from *Baiera* in any essential respect. They are essentially similar in cuticular structure. It is doubtful that the thickness of leaves alone could warrant a generic separation of the two forms.

If found detached, *Eretmoglossa* appears to be different from *Eretmophyllum* only slightly in leaf shape, and the leaf of *Sphenobaierocladus* and *Furcifolium* may well be referred to *Sphenobaiera.* The names *Eretmoglossa*, *Sphenobaierocladus*, and *Furcifolium* are valid in nomenclature, because they are based on specimens that are different from the diagnoses of the relevant genera. It is wondered, however, whether it is necessary to give independent generic names for each kind of leaves that differ in no essential respect from some others just because they are attached to the shoot. These kinds of names are superfluous and of little botanical value. Many palaeobotanists [19, 23, 71] still included *Sphenobaiera furcata* Heer in that genus, although its leaves are sometimes found preserved with shoots in original position, and Krassilov also referred a specimen of leafy shoot to *Pseudotorellia angustifolia* (see Section 3) [14]. Leaves of *Eretmophyllum tetoriensis* Kimura and Sekido [58] are also found attached to the apex of a shoot as in *Eretmoglossa.* While those leaf-bearing shoots may well be referred, respectively, to appropriate form (or organ) genera for ginkgoalean leaves, the generic name *Ginkgoitocladus* Krassilov is adopted here for ginkgoalean shoots without indication of their relation to any leaf genera.

In the present writer's opinion, the following eight more or less well-defined form (or organ) genera represent the basic morphological types of Mesozoic ginkgoalean leaves and shoots so far described: *Baiera* Braun, *Eretmophyllum* Thomas, *Ginkgodium* Yokoyama, *Ginkgoites* Seward, *Ginkgoitocladus* Krassilov, *Glossophyllum* Kräusel, *Pseudotorellia* Florin and *Sphenobaiera* Florin.

Quite a number of the established genera are not included in this list. It is possible that some of them represent useful morphological types, or even natural taxa of Mesozoic ginkgoaleans, but further collection and studies are needed. Besides, two rarely known types are found associated with reproductive organs of Mesozoic ginkgoaleans. The leaves of *Schmeissneria* Kirchner and Van Konijnenburg-Van Cittert [35] appear to be of the *Eretmophyllum* or *Glossophyllum* type, but were described as having more than two veins at the base. The leaves of *Toretzia* Stanislavsky [41, 42] are linear and bear two to four parallel veins. Unfortunately, the cuticles of the two genera have not been studied, and there is insufficient information concerning their detailed gross morphology and range of variation.

Although most of these morphological types may be fairly well separated from one another, there are intermediate forms that blur the boundary lines. *Eretmophyllum, Glossophyllum, Pseudotorellia*, and the Miocene *Torellia* Heer [10] may be easily distinguished from one another by the distribution and orientation of stomata and the presence and absence of resin bodies, but they are to some extent overlapping in shape and size, and it is rather difficult to identify them without knowing their cuticular structure. Some *Sphenobaiera* with deeply dissected lamina may be confused with *Baiera*, as it is sometimes difficult to decide whether they are petiolate or not, although those characterized by less divided, wedge-shaped leaves, and broad segments are well separated from the latter form. Under-developed leaves of *Sphenobaiera* are in fact indistinguishable from *Eretmophyllum* (and *Ginkgodium*). Well-developed leaves of the *Ginkgoites* type are easily distinguished from those of the *Eretmophyllum* and *Ginkgodium* types, or the *Sphenobaiera* type, but some under-developed leaves are not.

It is indeed very difficult to draw a line of demarcation between *Ginkgoites* and *Baiera*. Harris [9, 19] has repeatedly pointed out that the distinction is merely arbitrary. Florin [10] proposed to distinguish *Baiera* from *Ginkgoites* and *Ginkgo* by more deep and frequent division of laminae and fewer (two to four) veins per segment. Krassilov [14] used the following three characters: (1) laminae deeply divided right to the petiole, not only in the center but also at the sides; (2) stomata more regularly longitudinally oriented and arranged in files; and (3) trichomes less frequently encountered.

As mentioned by Harris [9], the cuticular structure cannot be used as a basis for separating the two genera. The extent of division and the number of veins per segment vary in *Ginkgo biloba* and in many Mesozoic ginkgoaleans. Some leaves of the *Ginkgoites* type (*G. longifolius, G. taeniata*, and *G. marginata*) [9, 13, 19] may be deeply divided and have few (two to four) veins in the segments. On the other hand, less divided leaves with up to five or six (to nine) veins in the segment are known in some species of *Baiera* (*B. furcata* in Harris and Millington's sense [19], *B. polymorpha* Samylina [51], and *B. hallei* [20]). In spite of the fact that different reproductive organs (*Ginkgo, Karkenia*, and *Yimaia*) have been correlated with leaves of the two types and that the majority of leaves are easily classified, morphological overlap between the two leaf forms is conspicuous.

3.2 Foliar Polymorphism and Heterogeneity of the Morphological Types

Quite a number of authors believed that at least some of these leaf types may correspond to certain natural taxa and are of taxonomic significance in Ginkgoales. The classification proposed by Tralau [5] was based entirely on leaf

TABLE 1. Selected Mesozoic ginkgoaleans bearing more than one type of leaves

Taxa	Geological age	Areas	Leaf types
**Ginkgo yimaensis* [18]	Middle Jurassic	China	GI, GD
Ginkgo insolita [79]	Middle Jurassic	Western Siberia	GI, E, S
Ginkgo longifolius [19]	Lower and Middle Jurassic	England	GI, B
Ginkgo dahllii [80]	Middle Jurassic	Norway	GI, GD, E, S
Ginkgoites jampolensis [14]	Upper Jurassic	Russia	GI, B
**Yimaia hallei* [20]	Middle Jurassic	China	B, GI
Baiera polymorpha [51]	Lower Cretaceous	Russia	B, GI
Baiera manchurica [14]	Upper Jurassic and Lower Cretaceous	Russia	B, GI
**Karkenia incurva* [15]	Lower Cretaceous	Argentina	GI
**Karkenia asiatica* [14]	Upper Cretaceous	Russia	S
**Karkenia hauptmannii* [35]	Lower Jurassic	Germany	S, E, or L
Sphenobaiera boeggildiana [9]	Lower Jurassic	Greenland	S, E, GD
Sphenobaiera gyron [19]	Middle Jurassic	England	S, E
Sphenobaiera nipponica [59]	Lower Jurassic	Japan	S, E, GD
**Grenana angrenica* [34]	Middle Jurassic	Middle Asia	B or S?

*Taxa with an asterisk are based on reproductive organs.
B, *Baiera*; E, *Eretmophyllum*; GD, *Ginkgodium*; GI, *Ginkgoites*; L, *Glossophyllum*; S, *Sphenobaiera*.

genera. Although Harris [9] had considered the division of the leaf genera to be arbitrary, he [19] eventually referred all the *Ginkgo*-type leaves of the Yorkshire Jurassic flora to the extant genus *Ginkgo* and believed *Sphenobaiera* to be a natural genus. Krassilov [14] also referred all the Bureja species of *Sphenobaiera* to Karkeniaceae. As revealed by recent studies [15, 18], at least two different kinds of ovulate organs are found in association with the *Ginkgo*-type leaves. It is quite possible for some species of *Sphenobaiera*, which are characterized by having less divided wedge-shaped leaves and broad segments, to be referred to Karkeniaceae (for instance, *S. umaltensis* proved to be related to *Karkenia asiatica*).

However, the suggestion that all species of this form genus bear the same type of ovulate organs is hardly convincing. Up to the present, only male reproductive organs were known in a few species of *Sphenobaiera* that have deeply divided leaves and narrow segments, such as *S. furcata*. Furthermore, leaves believed to be related to some very different ovulate organs, such as *Grenana* Samylina [34], are almost indistinguishable from the *Sphenobaiera*- or *Baiera*-type leaves. So far, no reproductive organs have been found in connection with the *Eretmophyllum*- and *Glossophyllum*-type leaves. Some species of *Baiera* and *Pseudotorellia* have proved to be related to certain reproductive organs (*Yimaia* and *Ulmatolepis*, respectively; see Section 2). This does not mean, however, that all leaves of the two types are related respectively to these two kinds of reproductive organs, because only few reproductive organs have so far been found of Mesozoic ginkgoaleans. The foliar polymorphism and heterogeneity of the leaf forms, as revealed by the available evidence, also do not support regarding them as natural taxa.

As described by many authors [32, 72–75], the extant *Ginkgo biloba* L. shows

TABLE 2. Distribution of different types of ginkgoalean leaves in natural taxa

Leaf types or form genera	Natural taxa (or taxa based on reproductive organs)	
	Genera	Families
Baiera	*Yimaia, Grenana*?	Ginkgoaceae?, Yimaiaceae
Eretmophyllum	*Karkenia, Ginkgo*?	Karkeniaceae, Ginkgoaceae?
Ginkgodium	*Ginkgo, Karkenia*?	Ginkgoaceae, Karkeniaceae?
Ginkgoites	*Ginkgo, Karkenia, Yimaia*	Ginkgoaceae, Karkeniaceae, Yimaiaceae
Glossophyllum	?	?
Pseudotorellia	*Umaltolepis*, ?	Umaltolepidaceae, ?
Sphenobaiera	*Karkenia, Grenana?, Ginkgo*?	Karkeniaceae, Ginkgoaceae?
Shoot		
Ginkgoitocladus		All known families of Mesozoic ginkgoaleans

a considerable variation in leaf form. Three main causes appear to be responsible for such variation: the heteroblastic development, age, and sexuality of trees, the nature of shoots (vegetative or reproductive, long or dwarf), and environmental conditions. Foliar polymorphism is also a remarkable phenomenon of Mesozoic ginkgoaleans (Table 1). Almost all species that have been studied in some detail exhibit a rather wide spectrum of leaf variation. Some *Ginkgo* (or *Ginkgoites*), such as *G. yimaensis*, *G. dahllii*, and *G. insolita*, include leaves of the *Ginkgodium*, *Eretmophyllum*, and *Sphenobaiera* types as well as the *Ginkgoites* type. Some others (*G. jampolensis* and *G. longifolius*) may match those of the *Baiera* type. *Yimaia* is dominated by the *Baiera*-type leaves, but some less divided leaves are hardly separated from those of the *Ginkgoites* type. At least three different types of leaves known to be related to *Karkenia*. Leaves of the *Sphenobaiera*, *Eretmophyllum*, and *Ginkgodium* types may sometimes occur in close association and show a transition in shape from one to another [35, 58, 59]. Although *Glossophyllum* appears to occur in a solitary state, according to Dobruskina [62], *Kirjamkenia*, including leaves of both the *Ginkgoites* and *Sphenobaiera* types, is allied to *Glossophyllum*, and so is another closely related genus *Kalantarium*, which includes leaves of both the *Glossophyllum* and *Sphenobaiera* types.

The heterogeneity of the leaf types of Mesozoic ginkgoaleans is remarkable as it is cognate with the foliar polymorphism. As shown in Table 2, the *Ginkgoites*-type leaves may be found in three different genera belonging to two families. The *Sphenobaiera*-type leaves may occur even in some Palaeozoic forms. Their relation to Ginkgoales is doubtful. Similar leaves are known in Czekanowskiales, such as *Sphenarion* Harris et Millington [18] and *Arctobaiera* Florin [10]. When the leaf base is not known, it is impossible to separate them from *Sphenobaiera*. It is also likely that leaves of the *Baiera* type are not homogeneous, and the same may be said of the *Eretmophyllum*-type leaves (see also Table 1). The *Pseudotorellia*- and *Glossophyllum*-type leaves appear to come from some special groups (or taxa), but it is premature to regard them as natural taxa. In addition to *Umaltolepis* Krassilov [14], another kind of ovulate organ similar to that of *Ginkgo* may also be related to *Pseudotorellia* [76, 77].

TABLE 3. Character analysis and data matrix of character states of ginkgoaleans

Characters	Character states 0	Character states 1
1 Ovulate shoots	Compound	Single
2 Dwarf shoots	Absent	Present
3 Resin bodies in ovule	Absent	Present
4 Ovules	Without a true wing	Winged
5 Position of ovules	Inverted	Erect
6 Number and size of ovules	Numerous and small	Few and large
7 Pedicels	Present or present and absent	Absent
8 Collar (or "cupule")	Absent	Present
9 Integument outer cuticle	Thick	Thin
10 Nucellus and integument	Separated	Largely fused
11 Bract	Not reduced	Reduced
12 Leaf	Divided or occasionally entire	Entire
13 Lamina	Without planation	Dorsiventral
14 Petiole	Absent or sometimes present	Present

Taxa	Characters 1	2	3	4	5	6	7	8	9	10	11	12	13	14
Trichopitys	1	0	?	0	0	0	0	0	?	?	0	0	0	0
Karkenia	1	1	1	0	0	0	0	0	1	0	0	0	1	0
Toretzia	1	1	?	0	0	1	1	0	?	?	0	1	1	0
Umaltolepis	1	1	?	?	?	1	1	0	?	?	1	1	1	0
Schmeissneria	1	1	?	1	1	1	0	1	?	?	0	1	1	0
Yimaia	1	1	1	0	0	1	1	0	0	1	0	0	1	1
Grenana	?	?	?	0	1	1	?	1	?	?	?	0	1	?
Ginkgo	1	1	1	0	1	1	0	1	0	1	0	0	1	1

4 Phylogenetic Analysis and Classification

A cladistic analysis was made using Hennig 86, version 1.5 [78]. A total of eight genera (taxa) that are among the most completely known ginkgoaleans so far recorded were adopted in this analysis. The Palaeozoic representative *Trichopitys* is used as an outgroup, as it has been considered to be an archetype or its analogue of post-Palaeozoic ginkgoaleans [16]. Altogether, 14 characters were considered (Table 3), all of which are binary. Those referred as "0" are considered primitive and "1" as advanced. The commands *ie-* (implicit enumeration, to identify one tree of minimal length) and *mh*;bb** (contrasting multiple initial cladograms by adding taxa in several different sequences, retaining the shortest cladogram of each; and "*branch breaker*", generating multiple equally parsimonious cladograms), were applied.

The cladogram shown in Fig. 1 was generated by implicit enumeration and of minimal length (*length*, 14; *ci*, consistency index, =85; *ri*, retention index, =80).

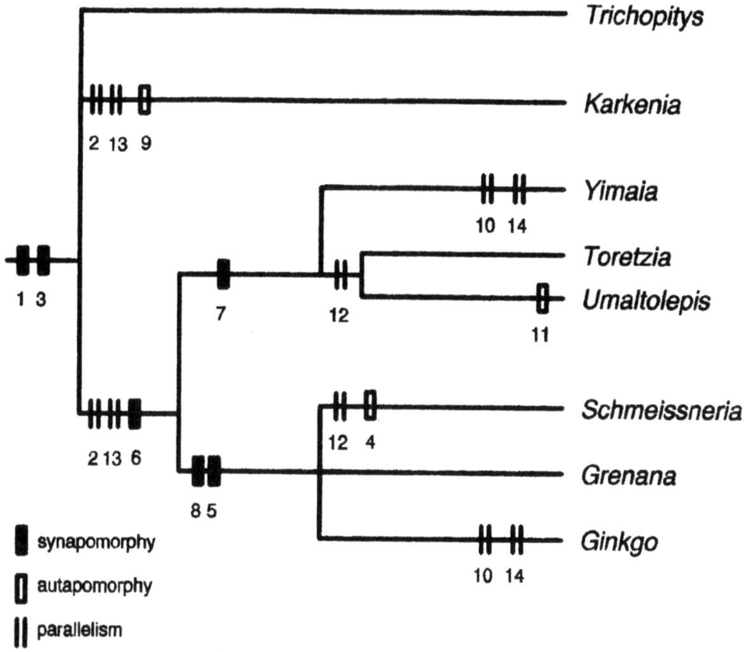

FIG. 1. The most parsimonious cladogram

Two equally parsimonious cladograms were obtained by using the options *mh*;bb**. One of them is exactly the same as that in Fig. 1, while the other (not figured) differs in having one more step for separating *Schmeissneria* from the *Ginkgo* and *Grenana* group. Because only a rather small number of characters were developed and some of them are found to be less or not at all phylogenetically informative, it is obvious that the analysis is still preliminary. Although there are two different groups of Mesozoic ginkgoaleans besides the unique genus *Karkenia* in the figured tree, the relationships among taxa in each group are rather poorly resolved.

On the whole, the figured cladogram is similar to the most parsimonious one obtained by analyzing a smaller number of taxa and characters [16]. In both, *Karkenia* appears to represent a distinct lineage among Mesozoic ginkgoaleans. It is much nearer to *Trichopitys* than all other Mesozoic ginkgoaleans. *Ulmatolepis* and *Toretzia* are more closely related to each other than either is to any other member. *Yimaia*, however, appears to be a sister taxon of the *Ulmatolepis-Toretzia* clade and not closely related to *Ginkgo*. *Grenana*, originally considered to be a member of Lyginopteridopsida, is found to be most close to *Ginkgo*. The new ginkgoalean recently described as *Schmeissneria* appears also to be related to the *Ginkgo* clade, but the exact relationships between the three taxa remain uncertain. In the light of new evidence and the current study, the classification of ginkgoaleans proposed by this writer [16] is amended as follows:

Order Ginkgoales
Family: Trichopityaceae
Genus: *Trichopitys* Saporta
Family: Karkeniaceae
Genus: *Karkenia* Archangelsky
(Leaves of the *Ginkgoites*, *Sphenobaiera*, and *Eretmophyllum* types)
Family: Umaltolepidiaceae
Genus: *Umaltolepis* Krassilov
(Leaves of the *Pseudotorellia* type)
Genus: *Toretzia* Stanislavsky
Family: Schmeissneriaceae (fam. nov.)
Genus: *Schmeissneria* Kirchner and Van Konijnenburg-Van Cittert
(Male fructification believed to be of the *Stachyopitys* type)
Family: Yimaiaceae (stat. nov.)
Genus: *Yimaia* Zhou and Zhang
(Leaves mainly of the *Baiera* type)
Family: Ginkgoaceae
Genus: *Ginkgo* L.
Genus: *Grenana* Samylina
(Leaves of the *Sphenobaiera* or *Baiera* type)

Appendices

1. Form (or organ) genera of Ginkgoales:
Baiera Braun, *Eretmophyllum* Thomas, *Ginkgodium* Yokoyama, *Ginkgoites* Seward, *Ginkgoitocladus* Krassilov, *Glossophyllum* Kräusel, *Pseudotorellia* Florin, *Sphenobaiera* Florin, *Stachypitys* Schenk, *Allicospermum* Harris (pro parte), *Antholithus* (*Antholithes*) Brongniart (pro parte), *Torellia* Heer [10]
2. Ginkgoaleans unsatisfactorily known or not clearly distinguished from already established taxa:
Baierella Potonié, *Baierophyllites* Jain and Delevoryas, *Eretmoglossa* Barale, *Euryspatha* Prynada, *Furcifolium* Kräusel, *Kalantarium* Dobruskina, *Kirjamkenia* Prynada, *Leptotoma* Kirtchkova and Samylina, *Sphenobaierocladus* Yang, *Sphenobaieroanthus* Yang

It is obvious that the proposed classification is still far from satisfactory. Firstly, it is based only on a few taxa, although attempts have been made to eliminate as far as possible the doubtful information and all the authenticated taxa so far known have been taken into consideration. The deficiency in evidence of male fructification has remained unchanged since Zhou [16] made a preliminary phylogenentic study in 1991. In the classification, each lineage is referred as a family of the Linnaean rank system, although *Trichopitys* together with *Karkenia* may represent a higher taxonomic unit as compared with the others.

To refer *Schmeissneria* to a distinct family is not compatible with the cladogram. This genus shares indeed a number of common characters with *Ginkgo*, but differs in some important characters, such as the entire and sessile leaves and the winged seeds. The possible male fructification (e.g., *Stachyopitys*) is also quite different from that of *G. biloba*. It seems, therefore, reasonable to separate it from Ginkgoaceae rather than to alter arbitrarily the definition of Ginkgoaceae so as to include the fossil taxon. *Grenana* is ascribed tentatively to Ginkgoaceae despite its imperfect preservation because it is so close to *G. biloba* in its female reproductive organs. For the great majority of taxa based on vegetative organs, their systematic positions can only be determined when they are found to be related to some reproductive organs. Nevertheless, these taxa (form or organ genera) are useful for data accumulation and for biostratigraphical, ecological, and palaeophytogeographical studies.

5 Evolutionary Trends and Other Considerations

The general evolutionary trend of ginkgoaleans is reduction of both reproductive and vegetative organs [16]. As compared with the late Palaeozoic *Trichopitys*, all the Mesozoic members have reduced (shortened) shoots and more or less reduced ovulate organs; their leaves (and bracts) tend to be laminated (planted) and petiolate (complete leaf), and the deeply divided laminae become fused and webbed. Except for those in Karkeniaceae, ovules decrease in number and increase in size, and pedicels become reduced and eventually lost. Both vegetative and reproductive organs appear to evolve more or less in parallel in different lineages.

As in many other plant groups, the heterobathmy of characters is one of the interesting features in Mesozoic ginkgoaleans. *Karkenia* represents an independent lineage that is characterized by having rather primitive ovulate organs, while the integumental cuticle is thin, as in many bennettitaleans and angiosperms [79]. Umaltolepidaceae, including *Toretzia* and *Umaltolepis* (*Pseudotorellia*), is distinguished by sessile leaves, but their ovule-bearing organs have been greatly reduced. In Schmeissneriaceae the ovulate organs are specialized in having collars and winged seeds. The leaves, however, are still incomplete, without petiole. Both *Yimaia* and *Grenana* have ovules directly implanted at the apex of the peduncle (or pedicels disappearing in the mature stage), but their leaves are less fused and webbed.

Until the present, little information about the ecology and evolutionary mechanisms of Mesozoic ginkgoaleans had been obtained. It is believed that the Mesozoic ginkgos might have survived in environmental conditions where *k* selection predominated and the *G. biloba*-type ovulate organs arose through peramorphosis (most likely hypermorphosis) from the *G. yimaensis* type found in the Middle Jurassic [80]. It means that the ovulate organ of *G. biloba* represents an additional developmental stage beyond that of the ancestral adult. When ovules grew in size, they decreased in number and the pedicels disappeared to balance the weight of the ovules with the strength of the supporting organs. A comparative study of the ultrastructure seems also to reveal a similar heterochronic origin of the megaspore membrane of *G. biloba* [81].

6 Conclusion

So far, a total of 18 genera for vegetative organs of Mesozoic ginkgoaleans have been established. A critical review shows that only 8 of these may be recognized as basic morphological types (form genera). Most of the others were based on unsatisfactorily preserved material or were poorly defined, and some were doubtfully attributed to Ginkgoales. Some leaf genera are distinguished from already established ones only in being attached to the shoot. Even the 8 more or less well-defined genera overlap morphologically, and cuticular characters sometimes offer little help in generic identification. Almost all species studied in some detail, moreover, show a considerable variation in leaf morphology. There are 11 genera described for ginkgoalean reproductive organs that may be regarded as useful. Most of these were found to be correlated with more than one type of leaf. An amended classification is proposed for Mesozoic ginkgoaleans on the basis of more completely known taxa, including female reproductive organs as well as vegetative organs to eliminate as far as possible the disadvantages caused by the foliar polymorphism and heterogeneity of the vegetative organs.

References

1. Braun CFW (1943) Beiträge zur Urgeschichte der Pflanzen. In: Münster GG (ed) Beiträge zur Petrefactenkunde, vol 6. F C Birmer, Bayreuth, pp 1–25
2. Schenk A (1867) Die fossile Flora der Grenzschichten des Keupers und Lias Frankens. Kreidel Verlag, Wiesbaden
3. Heer O (1876) Beiträge zur Jura-Flora Ostsibiriens und des Amurlandes. Mém Acad Imp Sci St Pétersbg Seventh Ser 22:1–222
4. Tralau H (1967) The phytogeographic evolution of the genus *Ginkgo* L. Bot Not 120:409–422
5. Tralau H (1968) Evolutionary trends in the genus *Ginkgo*. Lethaia 1:63–101
6. Samylina VA (1967) On the final stages of the history of the genus *Ginkgo* L. in Eurasia (in Russian). Bot Zh (Leningr) 41:303–316
7. Thomas HH (1913) On some new and rare Jurassic plants from Yorkshire: *Eretmophyllum*, a new type of Ginkgoalean leaf. Proc Cambr Philos Soc 17:256–262
8. Ôishi S (1933) A study on the cuticles of some Mesozoic gymnospermous plants from China and Manchuria. Sci Rep Tohuku Imp Univ Second Ser 12:239–252
9. Harris TM (1935) The fossil flora of Scoresby Sound, East Greenland, vol 4. Medd Grønland 112:1–176
10. Florin R (1936) Die fossilien Ginkgophyten von Franz-Joseph-Land nebst Erörterungen über vermeintliche Cordatailes mesozoischen Alters. I. Palaeontogr Abt B Palaeophytol 81:71–173
11. Florin R (1936) Die fossilien Ginkgophyten von Franz-Joseph-Land nebst Erörterungen über vermeintliche Cordaitales mesozoischen Alters. II. Palaeontogr Abt B Palaeophytol 82:1–72

12. Samylina VA (1956) Epidermal structure in leaves of the genus *Sphenobaiera* (in Russian). Dokl Akad Nauk SSSR 106:537–539
13. Lundblad AB (1959) Studies in the Rhaetic-Liassic floras of Sweden. II. 1. Ginkgophyta from the mining district of NW Scania. K Sven Vetensk Akad Handl Fourth Ser 6:1–38
14. Krassilov VA (1972) Mesozoic flora from the Bureja River (Ginkgoales and Czekanowskiales) (in Russian). Nauka, Moscow
15. Archangelsky SA (1956) Fossil Ginkgoales from the Ticò flora, Santa Cruz Province, Argentina. Bull Br Mus (Nat Hist) Geol 10:121–137
16. Zhou Z (1991) Phylogeny and evolutionary trends of Mesozoic ginkgoaleans—a preliminary assessment. Rev Palaeobot Palynol 68:203–216
17. Tralau H (1966) Botanical investigations in the fossil flora of Eriksdal in Fyledalen, Scania. Sver Geol Unders Arbs Ser C 579:1–36
18. Zhou Z, Zhang B (1989) A Middle Jurassic *Ginkgo* with ovule-bearing organs from Henan, China. Palaeontogr Abt B Palaeophytol 211:113–133
19. Harris TM, Millington W (1974) The Yorkshire Jurassic flora. Br Mus (Nat Hist) Publ 6:2–78
20. Zhou Z, Zhang B (1992) *Baiera hallei* Sze and associated ovule-bearing organs from the Middle Jurassic of Henan, China. Palaeontogr Abt B Palaeophytol 224:151–169
21. Andrews HN Jr (1955) Index of generic names of fossil plants, 1820–1950. US Geol Surv Bull 1013:1–262
22. Harris TM (1937) Fossil flora of Scoresby Sound, East Greenland, vol 5. Medd Grønland 112:1–114
23. Kräusel R (1943) Die Ginkgophyten der Trias von Lunz in Nieder Österreich und von Neuewelt bei Basel. Palaeontogr Abt B Palaeophytol 87:59–93
24. Gothan W (1914) Die unterliassische (rätische) Flora der Umgegend von Nürenberg. Abh Nathist Ges Nuerenberg 19:91–186
25. Van Konijnenburg-Van Cittert JHA (1992) An enigmatic Liassic microsporophyll, yielding *Ephedripites* pollen. Rev Palaeobot Palynol 71:239–254
26. Van Konijnenburg-Van Cittert JHA (1993) *Androstrobus phialophora* (Harris) nov. comb., a new name for *Bernettia phialophora* from Liassic of Greenland. Rev Palaeobot Palynol 77:301–304
27. Harris TM (1976) The Mesozoic gymnosperms. Rev Palaeobot Palynol 21:119–134
28. Van Konijnenburg-Van Cittert JHA (1971) *In situ* gymnosperm pollen from the Middle Jurassic of Yorkshire. Acta Bot Neerl 20:1–96
29. Van Konijnenburg-Van Cittert JHA (1972) Some additional notes on male gymnosperm fructifications from the Jurassic flora of Yorkshire. Acta Bot Neerl 21:95–98
30. Drinnen AN, Chambers TC (1987) Flora of the Lower Cretaceous Koonwarra Fossil Bed (Korumburra Group). In: Jell PA, Roberts J (eds) Plants and invertebrates from the Lower Cretaceous Koonwarra Fossil Bed, South Gippsland, Victoria. Association of Australasian Palaeontologists, Sydney, pp 1–78
31. Nathorst AG (1899) Fossil plants from Franz Josef Land. In: Nansen F (ed) The Norwegian North Polar Expidition 1893–96, Scientific Results. III. Christiania, London, pp 1–26
32. Seward AC (1919) Fossil plants, vol IV. Cambridge University Press, Cambridge
33. Nathorst AG (1878) Om floran Skånes Kolförande Bildningar. I. Sver Geol Unders Arsb Ser C Avh Uppsats 27:1–52
34. Samylina VA (1990) *Grenana*—a new genus of seed ferns from the Jurassic deposits of Middle Asia (in Russian). Bot Zh (Leningr) 75:846–850

35. Kirchner M, Van Konijnenburg-Van Cittert JHA (1994) *Schmeissneria microstachys* (Presl, 1833) Kirchner et Van Konijnenburg-Van Cittert, sp. nov., plants with ginkgoalean affinities from the Liassic of Germany. Rev Palaeobot Palynol 83:199–215
36. Schweitzer HJ, Kirchner M (1995) Die rhäto-jurassischen Floren des Iran und Afghanistans, 8 Ginkgophyten. Palaeontogr Abt B Palaeophytol 237:1–58
37. Von Sternberg C (1833) Versuch einer geognostisch-botanischen Darstellung der Flora der Vorwelt, 5, 6. Kommission im Deutschen Museum, 1–80, Leipizig
38. Wcislo-Luaniec E (1992) A fructification of *Stachyopitys preslii* Schenk from the Lower Jurassic of Poland. Cour Forschungsinst Senckenb 147:247–253
39. Yang X (1986) *Sphenobaierocladus*—a new Ginkgophytes genus (Sphenobaieraceae n. fam.) and its affinities (in Chinese with English summary). Bull Chengdu Inst Geol Miner Chengdu 7:49–58
40. Schenk A (1890) Paläophytologie. In: Zittel KA (ed) Handbuch der Paläontologie, vol II. 9. München, R Oldenboug, pp 765–958
41. Stanislavsky FA (1971) Fossil flora and stratigraphy of Upper Triassic sediments in the Donetz Basin (Rhaetian flora from Raick) (in Russian). Naukova Dumka, Kiev
42. Stanislavsky FA (1973) The new genus *Toretzia* from the Upper Triassic of the Donetz Basin and its relation to the genera of the order Ginkgoales (in Russian). Paleont Zh 1:88–96
43. Krassilov VA (1970) Approach to the classification of Mesozoic "ginkgoalean" plants from Siberia. Palaeobotanist (Lucknow) 18:12–19
44. Black M (1929) Drifted plant-beds of the Upper Estaurine Series of Yorkshire. Q J Geol Soc Lond 85:389–437
45. Kirchner M (1992) Untersuchungen an einigen Gymnospermen der fränkischen Rhät-Lias-Grenzschichten. Palaeontogr Abt B Palaeophytol 224:17–61
46. Von Sternberg K (1825) Versuch einer geognostisch-botanischen Darstellung des Flora des Vorwelt, vol 4. Tentamen, Leipzig
47. Unger F (1845) Beiträge zur Flora der Vorwelt vols 6, 7. Wilhelm Engelmann, Leipizig
48. Jongmans WJ, Dijikstra SJ (1971) Fossilium catalogus. II. Plantae, vol 80, Gymnospermae, I-II. Uitgeverij Dr Junk NV, Gravenhage
49. Potonié R (1933) Über einige Pflanzenreste aus dem Jura Persiens. Inst Paläobot Petrogr Brennsteine Arbor 3:247–250
50. Dorf E (1958) The geological distribution of the *Ginkgo* family. Bull Wagner Free Inst Sci 33:1–10
51. Samylina VA (1963) The Mesozoic flora of the lower course of the Aldan River (in Russian with English summary). Palaeobotanica 4:59–139
52. Jain RK, Delevoryas T (1967) A Middle Triassic flora from the Cackeuta Formation, Minas de Petroleo, Argentina. Palaeontology (Lond) 10:557–589
53. Barale G (1981) *Eretmoglossa* nouveau genre de Ginkgophytes dans les calcaires lithographiques du Crétacé inférieur de la Sierra du Montsech (Espagne). Ilerda 42:51–61
54. Takhtajan AL, Vakhrammev VA, Radtschenko GP (1963) Gymnosperms and angiosperms. In: Orlov A (ed) Osnovy paleontologii, vol 15 (in Russian). Gosgeoltekhizdat, Moscow, pp 1–743
55. Kräusel R (1943) *Furcifolium longifolium* (Seward) n comb, eine Ginkgophyte aus dem Solenhofener Jura. Senckenbergiana 26:426–433
56. Salfeld H (1907) Fossile Landpflanzen der Rhät- und Juraformation Südwestdeutschlands. Palaeontographica 54:163–204

57. Yokoyama M (1889) Jurassic plants from Kaga, Hida and Echizen. J Coll Sci Imp Univ Tokyo 3:1–66
58. Kimura T, Sekido S (1965) Some interesting Ginkgoalean leaves from the Itoshiro-Subgroup, the Tetori Group, Central Honshu, Japan. Mem Mejiro Gakuen Women's Jr Coll 2:1–4
59. Kimura T, Tsujii M (1984) Early Jurassic plants in Japan, vol 6. Trans Proc Palaeont Soc Jpn New Ser 133:265–287
60. Samylina VA (1956) Two new ginkgoalean species from lower Cretaceous deposits. Bot Zh (Leningr) 41:1525–1527
61. Ettingshausen C (1887) Beiträge zur Kenntnis der fossilen Flora Neuseelands. K Akad Wiss Denkschr 53:143–192
62. Dobruskina IA (1980) Stratigraphic position of Triassic plant-bearing beds of Eurasia (in Russian). Trudy Geol Inta AN SSSR 346:1–163
63. Prynada VD (1970) Fossil flora of the Corvunchansk Suite, Lower Tunguska River Basin (in Russian). Nauka, Moscow
64. Kiritchkova AI, Samylina VA (1979) On the peculiarities of leaves of some Mesozoic Ginkgoales and Czekanowskiales (in Russian). Bot Zh (Leningr) 64:1529–1538
65. Reymanówna M (1985) *Mirovia szaferi* gen. et sp. nov. (Ginkgoales) from the Jurassic of the Kraków region, Poland. Acta Palaeobot 25:3–12
66. Bose MN, Manum SB (1990) Mesozoic conifer leaves with "*Sciadopitys*-like" stomatal distribution. A re-evaluation based on fossils from Spitsbergen, Greenland and Baffin Island. Nor Polarinst Skr 192:1–81
67. Bose MN, Manum SB (1991) Additions to the family Miroviaceae (Coniferae) from the Lower Cretaceous of West Greenland (*Mirovia groenlandica* n. sp., *Tritaenia crassa* (Seward) comb. nov.) and Germany (*Tritaenia linkii* Mägdefrau et Rudolph emend.). Polar Res 9:9–20
68. Saporta G (1873) Paléontologie francaise ou description des fossils de la France. 2. Végétaux, Plantes Jurassiques, vol I. Libraire de l'academie de Medicine, Paris
69. Barale G (1972) *Rhaphidopteris* nouveau nom de genre de feuillage filicoide mesozoique. C R Acad Sci Paris 274:1011–1014
70. Harris TM (1932) The fossil flora of Scoresby Sound, East Greenland. 2. Medd Grønland 85:1–112
71. Sze HC (1956) Older Mesozoic plants from the Yenchang Formation, Northern Shensi. Palaeontol Sin Ser A 139(5):1–217
72. Seward AC, Gowan J (1900) The maidenhair tree (*Ginkgo biloba* L.). Ann Bot (Lond) 14:109–154
73. Kräusel R (1917) Über die variation von *Ginkgo biloba* L. und ihre Bedeutung für die Paläobotanik. Centralbl Miner 1917:63–68
74. Sahni B (1933) On some abnormal leaves of *Ginkgo*. J Ind Bot Soc 12(1):50
75. Porterfield WM (1924) Sexual dimorphism and leaf variation in *Ginkgo biloba* L. China J Sci Arts 2:255–265
76. Reymanówna M (1980) A new ginkgoalean plant from the Jurassic of Kraków, Poland. In: Abstracts, first international palaeobatany conference, July 7–13, Reading, pp 1–64
77. Reymanówna M (1984) *Pseudotorellia*, a Mesozoic genus close to *Ginkgo*? In: Abstracts, second international organization for palaeobotany conference, August 19–26, Edmonton
78. Farris JS (1988) Hennig 86 version 1.5 (computer programme and reference).
79. Harris TM (1954) Mesozoic seed cuticles. Sven Bot Tidskr 48:281–291

80. Zhou Z (1994) Heterochronic origin of *Ginkgo biloba*-type ovule-bearing organs (in Chinese with English summary). Acta Palaeontol Sin 33:131–139
81. Zhou Z (1992) Comparative ultrastructure of fossil and living ginkgoalean megaspore membranes. Rev Palaeobot Palynol 78:167–182

Cenozoic History of *Ginkgo* in East Asia

KAZUHIKO UEMURA

1 Introduction

The ginkgo or maidenhair tree, *Ginkgo biloba* L., is one of the best known trees, with a long geological history of fossil antecedents [1, 2]. During the Mesozoic era, ginkgophytes were common and diverse in both the Northern and Southern hemispheres, culminating during the Late Jurassic and Early Cretaceous periods. The genus *Ginkgo*, one of the major groups in ginkgophytes, appeared in the Early Jurassic [3]. It was distributed widely by the Middle Jurassic and attained a high-latitude distribution during the Cretaceous in the Northern Hemisphere [3, 4]. The Cenozoic pattern of distribution of *Ginkgo* is typical of the Northern Hemisphere, excepting the early Eocene record in Patagonia in South America ([5, 6]; geologic age after [7]).

Evolutionary trends and the phytogeographic history of *Ginkgo* in the Northern Hemisphere have been summarized by Tralau [3, 4], based on a thorough bibliographicical survey and examination of the cuticular morphology of *Ginkgo* leaves. This review is, therefore, based primarily on the work by Tralau, with additions that include new fossil finds and a revised geochronology of the plant-bearing beds in East Asia. Because East Asia is one of the centers where many Tertiary remnants still survive, a brief Cenozoic phytogeographic history is discussed in relation to *Ginkgo* in the Northern Hemisphere.

2 Fossil Records

Fossil *Ginkgo* is recognized by its leaves, reproductive organs, and wood. Leaf remains are the most common, while reproductive organs, such as seeds and ovulate fructifications, and woods are not numerous [see 8, 9, for examples]. Dispersed pollen grains, although useful to trace the existence of many fossil

Department of Geology, National Science Museum, 3-23-1 Hyakunin-cho, Shinjuku-ku, Tokyo 169, Japan

plants, cannot be incorporated into the fossil record of *Ginkgo* because they are difficult to distinguish from the pollen of certain cycadophytes [9].

In the earlier literature, the genus name *Ginkgoites* Seward [10] was often used even for leaves indistinguishable from *Ginkgo* or even from the extant *Ginkgo biloba*. This system of Seward has been criticized by many authors [4, 11, 12], who have redefined the genus *Ginkgoites* as those leaves similar to *Ginkgo* but distinguished from it in morphological and anatomical characteristics (cuticular features in most cases).

Although some nomenclatural problems remain, the recognition of fossil leaves as *Ginkgo* is easy in such features as the gross morphology showing a fan-shaped lamina with distinct petiole, fernlike dichotomous venation originating from two vascular bundles in the petiole, resin bodies, a chemically resistant leaf texture, and the features of the leaf cuticles.

Leaves that are indistinguishable in their gross morphology from those of the extant *Ginkgo biloba* are known to occur from Early Cretaceous rocks [4] and become common in the Late Cretaceous (Fig. 1a). Such a leaf type has often been described from Cenozoic rocks under the name of *Ginkgo adiantoides* (Unger) Heer or some other species on the basis of cuticular features. However, some species were established on the basis of minor differences in their cuticular features, and many leaves, mostly impressions, are lumped into the single species

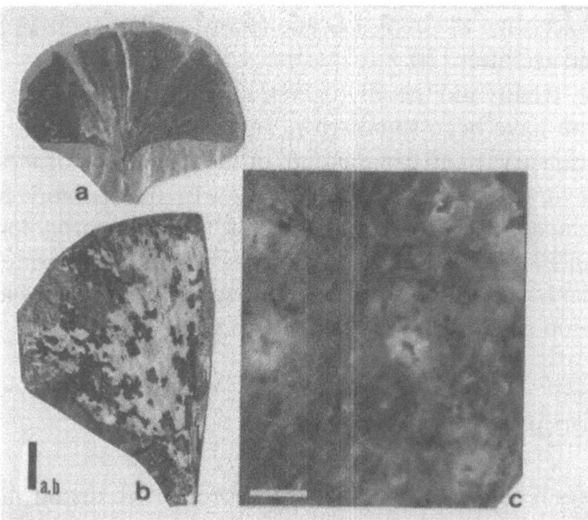

FIG. 1a–c. Cretaceous and early Eocene leaves from Japan. **a** *Ginkgo* sp., Upper Cretaceous from Yubari, Hokkaido; NSM-PP 10659. **b,c** *Ginkgo tzagajanica* Samylina, lower Eocene (Kuji Group) from Hanzaki, Kuji, Iwate Prefecture; NSM-PP 10663. **c** Epifluorescence photomicrograph of lower epidermis of the same specimen as in **b**. Note overarching papillae on subsidiary cells and papillate ordinary cells. **a**,**b**, *Bar* = 10 mm; **c**, *bar* = 50 μm

of *G. adiantoides*. Although taxonomic revisions are necessary to completely elucidate the phytogeographic history of *Ginkgo* on the basis of the evolutionary framework, this is beyond the scope of the present review (for taxonomic works on Cenozoic *Ginkgo*, see [13–16]).

3 Cenozoic Records in East Asia

Cenozoic floras in East Asia are numerous in both stratigraphic and geographic distribution [17]. The geographic term East Asia in this chapter includes Japan, the Korean Peninsula, China, and their surrounding areas such as Sakhalin, Primorye, and Amur. In Figs. 2 and 3, fossil records of *Ginkgo* in the north of East Asia are also included to demonstrate its distribution into eastern Eurasia.

All the Cenozoic records in East Asia are leaf remains: many are the impression type, although some are compression type in preservation. *Ginkgo* leaves are common in the Paleogene (Paleocene, Eocene, and Oligocene). On the other hand, Neogene (Miocene and Pliocene) leaves are rare, excepting the Pliocene records in southwestern Japan. Unfortunately, there are no Neogene or Quaternary records from China.

3.1 Paleocene

Although Paleocene floras in East Asia are not numerous [17], the early Paleocene Tsagajan flora in the Bureja basin has been studied in detail by many Russian authors (see [18]). The Tsagajan flora commonly contains *Ginkgo* leaves; "*G.* ex gr. *adiantoides* (Unger) Heer," *G. tzagajanica* Samylina, or *G. spitsbergensis* Manum [13, 18]. The flora is one of the representative early Paleocene floras in eastern Eurasia and is characterized by *Trochodendroides*, "*Platanus*," and many deciduous broad-leaved trees, together with ferns and conifers. Both the floral composition and leaf physiognomic features of dicots indicate a moderately warm and moist climate.

3.2 Eocene

Eocene records of *Ginkgo* are widely known in East Asia, extending north into Kamchatka. *Ginkgo*-bearing assemblages of probable early Eocene age are known from the Minato Formation in Kuji, Iwate Prefecture, Japan [15, 19] (Fig. 1b,c). These assemblages were assigned as Oligocene in age by Oishi [19] and cited by Tralau [3, 4], but recent palynological evidence indicates ages of early to middle Eocene (F. Akiba and A. Obuse, 1993, unpublishea manuscript) or late Paleocene to early Eocene [20]. Plant megafossil assemblages from Kuji are similar to the Tsagajan flora in their floral composition and components, although they contain more thermophilic plants.

Oishi [19] described the cuticles of his "*Ginkgoites* cf. *adiantoides*" from the Minato Formation, to which Samylina and Chelebaeva [14] gave a new name

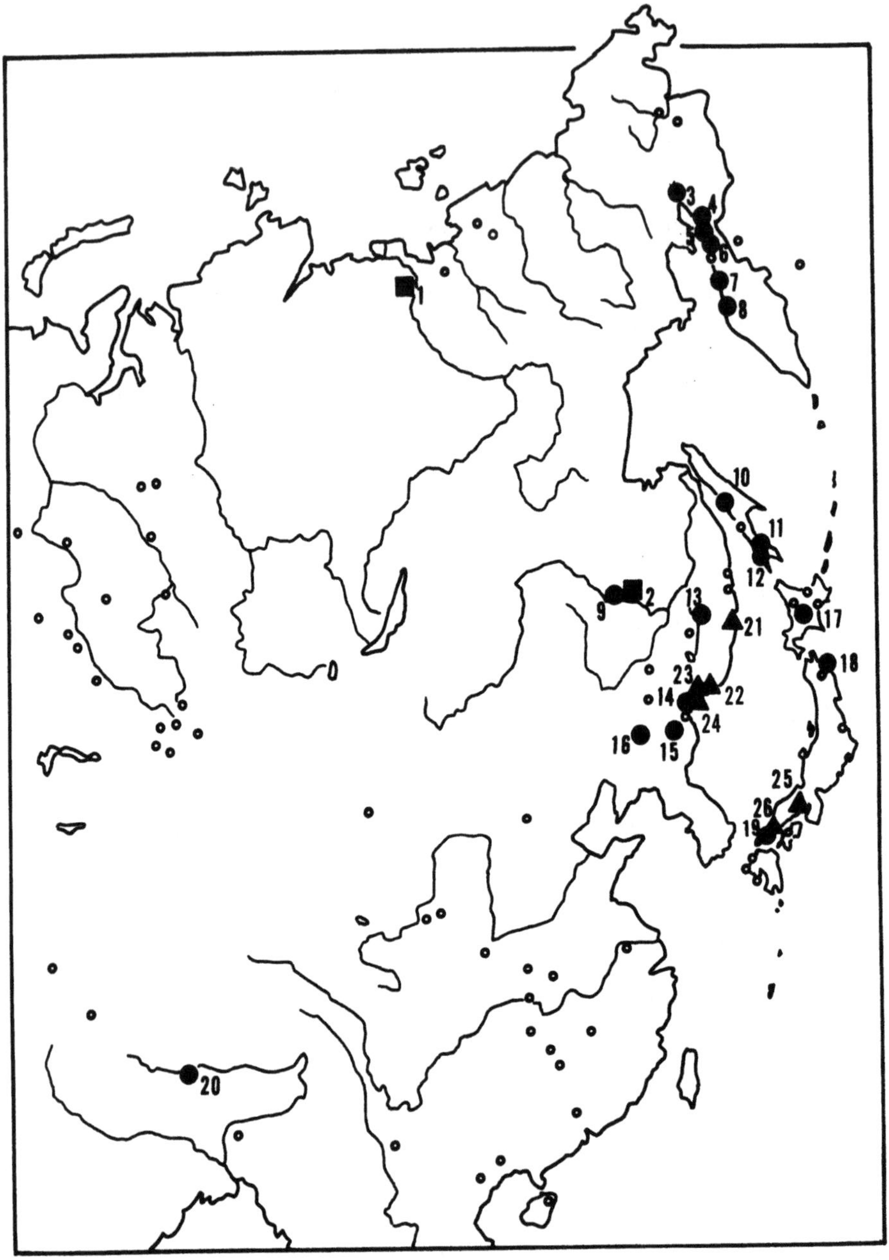

FIG. 2. Paleogene occurrence of *Ginkgo* in eastern Eurasia. *Small open circles* indicate plant megafossil localities (simplified from Tanai [17]) without *Ginkgo* remains. *1*, Kharaulakh; *2*, Tsagajan; *3*, Shestakov; *4*, Astronomicheskii; *5*, Getkilnin; *6*, Riv. Podkagernaya; *7*, Anadyrka; *8*, Levaya Napana; *9*, Raichikhi; *10*, Ogorodnoy; *11*, Riv. Naiba; *12*, Kril'on; *13*, Bikin; *14*, Rechnoi; *15*, Yusen; *16*, Fushun; *17*, Yubari; *18*, Kuji; *19*, Ube; *20*, Lhaze; *21*, Velikaya Kema; *22*, Bolotonoi, *23*, Rechinoi (Ust-Davidovskaja Formation); *24*, Kraskino and Shakhtersk; *25*, Kobe; *26*, Kiwa. *Square symbols*, Paleocene; *solid circles*, Eocene; *triangles*, Oligocene

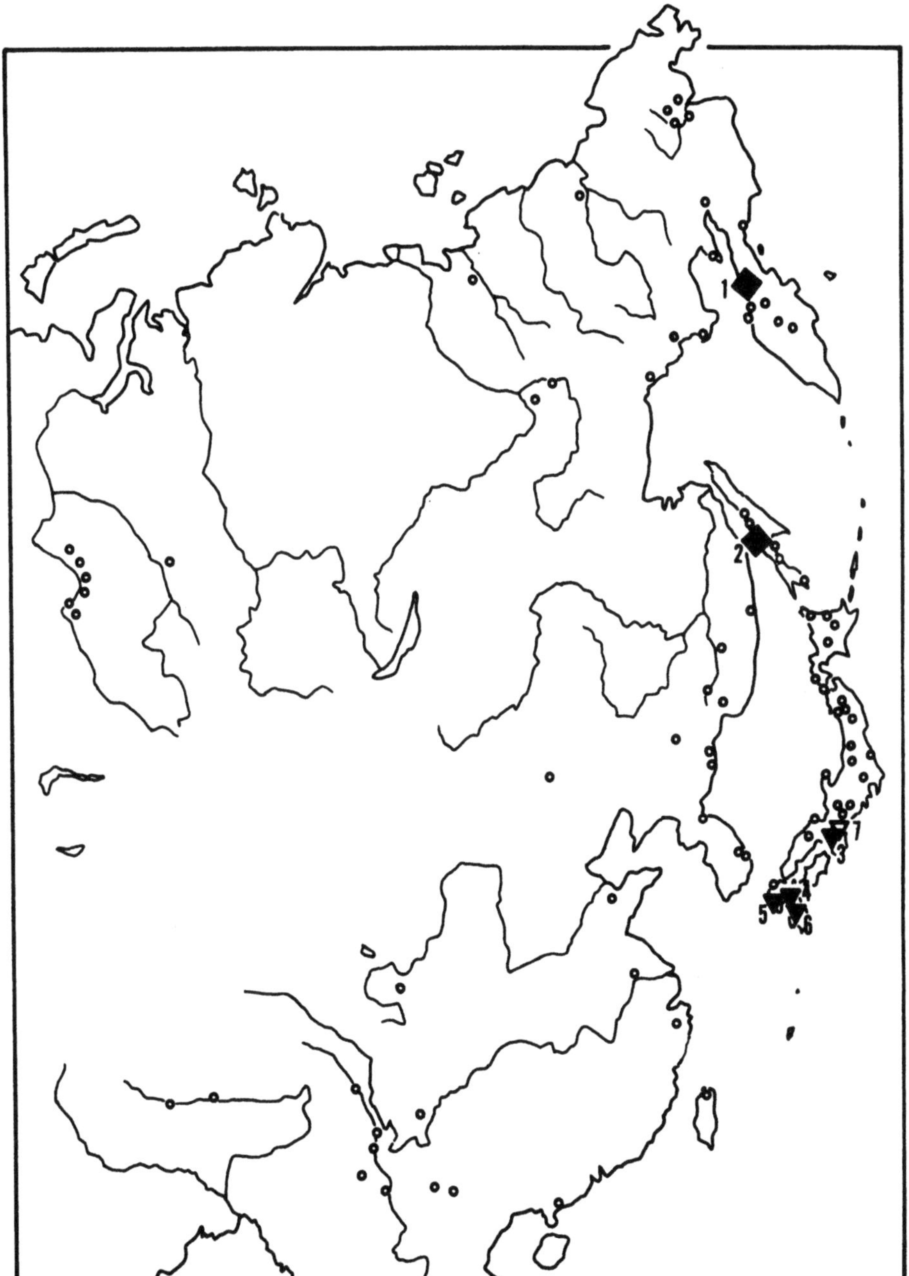

FIG. 3. Neogene and Quaternary occurrence of *Ginkgo* in eastern Eurasia. *Small open circles* indicate plant megafossil localities (simplified from Tanai [17]) without *Ginkgo* remains. *1*, Nepropusk; *2*, Shakhtersk; *3*, Sennan; *4*, Hoshiwara; *5*, Hiradoguchi; *6*, Daiwa; *7*, Seta. *Diamonds*, Miocene; *solid triangles*, Pliocene; *open triangles*, Pleistocene

under the genus *Ginkgo*, that is, *G. oishii* Samylina and Chelebaeva. Horiuchi and Kimura [15] also studied *Ginkgo* leaves from the same formation and referred them to *G. tzagajanica*, being characterized by the overarching papillae of subsidiary cells, papillate ordinary cells on the lower surface, trichomes, and the sinuosity of ordinary cell walls. Although Horiuchi and Kimura placed Oishi's specimens into synonymy with their *G. tzagajanica*, reexamination of the original specimens may be required before this can be confirmed.

Plant megafossil assemblages of middle to late Eocene ages are known from various localities in the Russian Far East, Sakhalin, Japan, the Korean Peninsula, and China [17]. *Ginkgo* leaves, most of which are referred to as *G. adiantoides* or "*G.* ex gr. *adiantoides*," were reported widely from northern areas of East Asia: Bikin and Raichikhi in Amur [13, 21, 22], Rechnoi (Uglovskaya and possibly Nadezhdinskaya formations) in Primorye [23], Ogorodnoy, Riv. Naiba, and Kril'on in Sakhalin [24, 25], Yubari in Hokkaido, Japan [26, 27] (Fig. 4a–e), Yusen in North Korea [28], and Fushun in China [29, 30]. Plant megafossil

FIG. 4a–f. Middle Eocene and early Oligocene leaves, *Ginkgo adiantoides* (Unger) Heer, from Japan. **a–d**, Middle Eocene (Ikushunbetsu Formation) from Yubari, Hokkaido; NSM-PP 10651-10654. **e**, Middle Eocene (Yubari Formation) from Yubari, Hokkaido; NSM-PP 10664. **f**, Lower Oligocene (Kiwa Formation) from Kiwa, Yamaguchi Prefecture; NSM-PP 10660. *Bar* = 10 mm

assemblages in these areas, although not uniform, are composed mainly of broad-leaved deciduous trees and taxodiaceous conifers, with many thermophilic elements. They reveal overall a moist, warm-temperate climate.

Rare occurrences of *Ginkgo* in the southern part of East Asia are known from two localities: Ube in southeastern Japan [31] and Lhaze in southwestern China [32]. Plant megafossil assemblages from these localities contain diverse broad-leaved evergreen trees and reveal an aspect of subtropical (paratropical) or tropical rain forests.

The cuticular features of middle to late Eocene *Ginkgo* leaves were studied by Samylina [13], Medjulianov [25], and Polyshchuk [23], describing *G. orientalis* Samylina and *G. samylinae* Krassilov et Medjulianov. These species differ from the Paleocene and early Eocene *G. tzagajanica* in their cuticular features and are more similar to the extant *G. biloba*. Furthermore, it is noted that the middle to late Eocene population shows generally small leaves without conspicuous lobes (Fig. 4a–e).

3.3 Oligocene

Oligocene records of *Ginkgo* have been reported from Kobe and Kiwa in southeastern Honshu, Japan [33, 34] (Fig. 4f). *Ginkgo* was also reported from Kraskino, Shakhtersk, Bolotonoi, Rechnoi (Ust-Davidovskaja Formation), and Velikaya Kema in Primorye [23, 35, 36]. Although the age of these plant-bearing beds of Primorye, whether Miocene or Oligocene, is controversial, they are usually referred to as Oligocene (see [37]). These *Ginkgo*-bearing assemblages contain diverse broad-leaved deciduous and evergreen trees and conifers of moist, warm-temperate vegetations.

3.4 Miocene

Miocene plant megafossil assemblages are widely known from the circum–Japan Sea areas and China and have been well studied compared with those in other geologic ages [17, 38]. These assemblages show diverse moist vegetations during the Miocene: broad-leaved deciduous forests mixed with conifers in the early early Miocene, mixed broad-leaved evergreen and deciduous forests in the late early to earliest middle Miocene, and broad-leaved deciduous forests occasionally accompanied by broad-leaved evergreen and coniferous trees in the majority of the middle Miocene and in the late Miocene. Nevertheless, the absence of *Ginkgo* in these assemblages should be pointed out. Some fossil records from the "Miocene" rocks in Primorye are excluded from the Miocene occurrence of *Ginkgo*, as noted previously. The leaf described as *G. adiantoides* from the late early Miocene coal-bearing rocks in Shakhtersk, Sakhalin, by Fotjanova [39] is the only record in East Asia. To the north of East Asia, Fotjanova [40] reported *Ginkgo* from the late Miocene plant megafossil assemblages in Kamchatka.

3.5 Pliocene and Pleistocene

Pliocene and Pleistocene occurrence of *Ginkgo* in East Asia is confined to southwestern Japan and has been reported from five areas under the extant species name, *G. biloba* L. As these fossil records represent the final stage of the last known fossil *Ginkgo* records, occurrence at each area is reviewed in some detail here.

Hikita [41] first confirmed the existence of *Ginkgo* in East Asian Pliocene rocks through his extensive carpological work on Plio-Pleistocene floras of the Sennan area, southern Osaka Prefecture, Japan (Fig. 5a). He found *Ginkgo* leaves at three localities in the Sennan area and later added one more locality [42]. All the specimens are leaf compressions, and their cuticular features were studied. The Sennan and its adjacent areas have been known as one of the well-established areas in the floral succession from the Pliocene to middle Pleistocene [42–44]. These *Ginkgo*-bearing beds are stratigraphically confined to the lowest part of the Osaka Group. More strictly they are below the Tuchimaru-II Tuff and

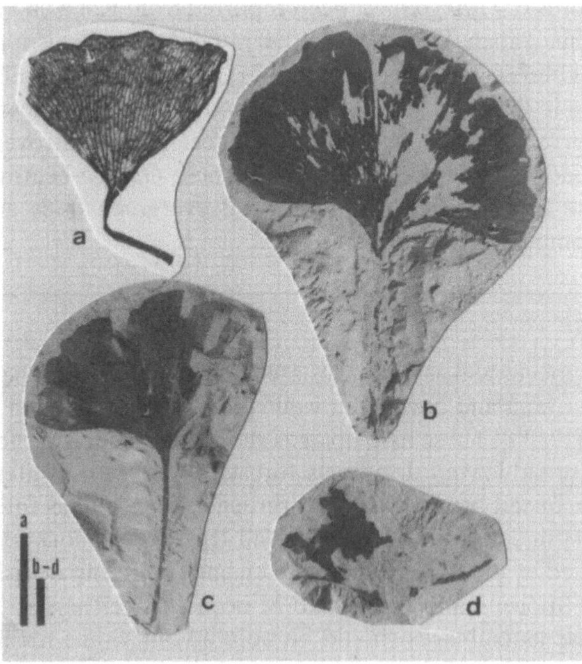

FIG. 5a–d. Late Pliocene leaves, *Ginkgo biloba* L., from southwestern Japan. **a** Lower part of the Osaka Group in Sennan, Osaka Prefecture (Fig. 4b in Hikita [41]); note *samll dots* (resin bodies) between veins. **b** Pliocene bed of Hiradoguchi, Nagasaki Prefecture (Pl. 1, Fig, 1 in Kamada et al. [45]). **c**, Hoshiwara Formation from Hoshiwara, Kumamoto Prefecture (Pl. 1, Fig. 1 in Iwao [46]). **d** Koriyama Formation in Daiwa, Kagoshima Prefecture; KSMP-3001b. *Bar* = 10 mm (with permission from [41, 45, 46])

above the Misaki Tuff. Evidence of fission track ages for both tuffs, paleomagnetic polarity, and lithostratigraphy indicates the *Ginkgo*-bearing beds are approximately 3 Ma, that is, early late Pliocene [44]. Hikita [41] listed 35 associated plants in which many elements not now indigenous to Japan are contained: *Keteleeria*, *Pseudolarix*, *Cunninghamia*, *Glyptostrobus*, *Metasequoia*, *Liquidambar*, and *Reevesia*. From the floral list of Hikita [41] and sedimentary facies in the lower Osaka Group, it is assumed that the *Ginkgo*-bearing assemblages were derived mainly from near-sea lowland forests under a moist, warm-temperate climate.

Ginkgo leaves have also been reported from three areas in Kyushu, Japan: Hiradoguchi in Nagasaki Prefecture [45] (Fig. 5b), Hoshiwara in Kumamoto Prefecture [46] (Fig. 5c), and Daiwa in Kagoshima Prefecture [47] (Figs. 5d; 6a,b).

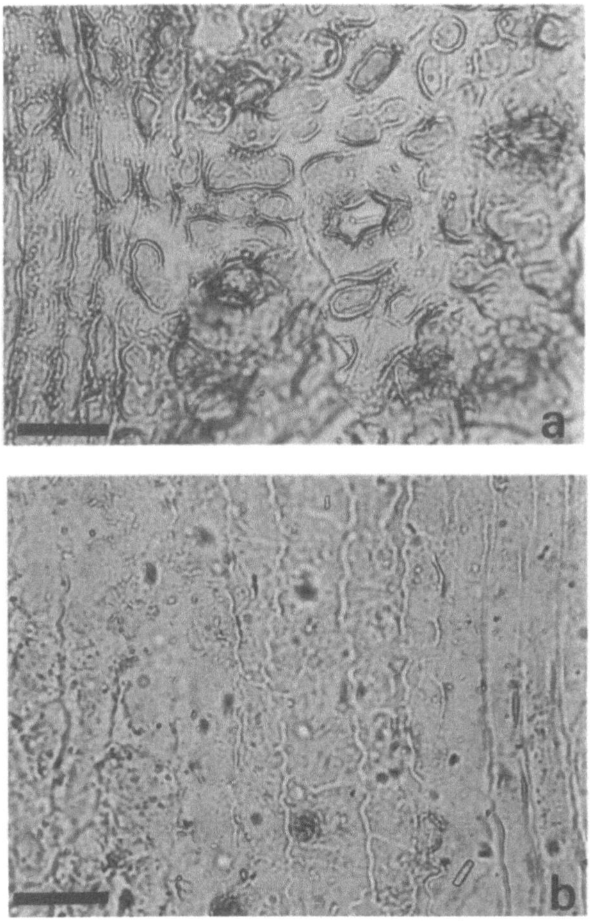

FIG. 6a,b. Cuticular features of late Pliocene *Ginkgo biloba* L. from Daiwa, Japan (see Fig. 5d; KSMP-3001b). **a** Lower cuticle; **b** upper cuticle. *Bar* = 50 μm

Ginkgo in the latter two areas are preserved in lacustrine beds of late Pliocene age [46, 47]. A fission track age of approximately 2.6 Ma for the beds in Daiwa [47] also supports the age assignment given earlier. Plant megafossil assemblages of these two areas are similar to one another, as indicated by the common occurrence of *Ginkgo*, *Pseudolarix*, *Metasequoia*, *Ficus* (evergreen), *Zelkova*, *Acer*, and others.

The leaves of *Ginkgo* from Hiradoguchi were found because of a river improvement program. Kamada et al. [45] suggested a Pleistocene age for these *Ginkgo*-bearing sediments, the Hiradoguchi Plant Bed, based principally on the floral composition. However, the assemblage contains *Pseudolarix* and *Carya*, which are unknown from the Pleistocene assemblages in Japan. In this review, *Ginkgo* leaves from Hiradoguchi are treated as a Pliocene record and are excluded from a reliable Pleistocene occurrence. All these *Ginkgo*-bearing assemblages in Kyushu show a composition of mixed broad-leaved deciduous and evergreen forests of a moist, warm-temperate climate.

The Kobiwako Research Group [48] reported *Ginkgo biloba* from the Kobiwako Group in the southwestern hilly area of Biwa-ko (Lake Biwa), Japan. The Kobiwako Group, excepting its basal and upper parts, is correlative to the Osaka Group mentioned previously. Unfortunately, this occurrence of *Ginkgo* is one listed without description or illustration, and the fossil specimens are unavailable for further examination at present. Nevertheless, the stratigraphic position of the fossil in question is worthy of note. *Ginkgo* occurs approximately 50 m above the well-dated and regionally traceable volcanic ash bed; correlative to the Fukuda Tuff bed, which is placed as just above the Pliocene–Pleistocene boundary or the Olduvai Normal Polarity subchron [49]. The record of *Ginkgo* from the Kobiwako Group is evidently from the Pleistocene and is the youngest fossil record of the genus in the world. Associated plants include *Metasequoia*, *Distylopsis*, *Fagus*, *Cyclobalanopsis*, *Quercus*, *Styrax*, and *Prunus* [48].

4 Discussion

During the Paleocene and Eocene of globally warmer ages, including the Cenozoic thermal maximum in the early Eocene, *Ginkgo* shows circumarctic distribution [3, 4]. This distributional pattern is similar to that seen in the Late Cretaceous. Figure 2 shows the Paleocene and Eocene records of *Ginkgo* added from records in the north of East Asia [50–52]. *Ginkgo*-bearing assemblages are evidently rich in broad-leaved deciduous vegetation developed under a moderately warm and moist climate.

Eocene records of *Ginkgo* are most abundant in East Asia or eastern Eurasia. Many are from assemblages characterized by broad-leaved deciduous and taxodiaceous conifers with common thermophilic elements of middle and high latitudes, and some records are from those characterized by subtropical to tropical rain-forest elements in the southern part of the middle latitude. All these records are from the moist assemblages of a summer rain type. In the Eocene floras of China, Guo [53] distinguished three provinces: North, Central, and South China. The floras of the central China province are characterized by the extinct genus of *Palibinia* and some other thick, coriaceous, small-leaved dicotyledons. As pointed out by Guo, these floras, together with the presence of evaporites and red-colored beds, might reflect an arid subtropical climate. It is worthy to note that no *Ginkgo* has been found in this region, contrasting with its occurrences in the North and South China provinces.

Marked reduction in the distribution of *Ginkgo* is evident in the Oligocene and succeeding Miocene, as is the case in other parts of the Northern Hemisphere [3, 4]. *Ginkgo* was confined mostly to the middle latitudes in three areas: East Asia, Europe, and western North America. This fact can probably be ascribed to climatic deterioration during and since the late middle Eocene and early Oligocene [54].

The paucity of *Ginkgo* in the Miocene floras in East Asia is difficult to explain, especially as the knowledge of Miocene floras in East Asia is more complete than that in other ages. In relation to this, its absence or paucity in the Oligocene and early Miocene floras in Europe and in the Oligocene floras of North America may be noted. These facts seem to reflect that *Ginkgo* was a plant of rare occurrence during this time interval.

Ginkgo existed in the Miocene of North America. Many localities have been known from the Pacific Northwest and its neighboring areas, such as Latah, Succor Creek, Vantage, Eagle Creek, and '49 Camp [4, 55], and even in Alaska [56]. These *Ginkgo*-bearing assemblages are actually confined in age to the early Miocene and early middle Miocene, or at least to the early Neogene (>13 Ma) assemblages of Wolfe [57]. Taking the succeeding floral change into consideration, the absence of *Ginkgo* in younger assemblages is probably related to the decrease in temperature after the middle Neogene thermal maximum around the early to middle Miocene boundary and initiation of a seasonally dry climate.

In Europe, *Ginkgo* has been recorded from many Miocene and Pliocene localities [3, 4]. Kovar-Eder et al. [58] recently summarized the geographic and stratigraphic distribution of *Ginkgo* in central Europe. In this region, *Ginkgo* reappeared in the late middle Miocene and attained a wide distribution during the late Miocene. According to these authors, *Ginkgo* persisted at least into the late Pliocene as an accessory element in warm deciduous forest communities. Subsequently, many plants in these communities, including *Ginkgo* and several East Asiatic elements, have been eliminated from Europe because of lower temperatures and a gradual shift in rainfall regimes from summer rain to dry summers.

Ginkgo reappeared in the Pliocene of Japan and survived at least into the early part of the Pleistocene. Because the early Pleistocene climate was still moist and warm in the lowlands of southwestern Japan, *Ginkgo* might have inhabited such lowlands, together with *Metasequoia*, *Pseudolarix*, and *Cunninghamia*. In floral composition these *Ginkgo*-bearing assemblages show the closest resemblance to the modern mixed mesophytic forest [59] in central China. Many Chinese authors state that *Ginkgo* grows naturally in some stands of the mixed mesophytic forest

[59, 60]. Unless the modern *Ginkgo* is spontaneous, it is probable that the areas in this mixed mesophytic forest today are one of the refugia of *Ginkgo* as well as other relict plants, such as *Metasequoia*, *Pseudolarix*, and others, that once flourished in the Tertiary of the Northern Hemisphere.

Cenozoic fossil records show that *Ginkgo* prevailed under moist and moderately warm climates. The same tendency is also recognized in Mesozoic *Ginkgo* since the genus attained a marked increase in frequency and also a wide occurrence during this era [4, 61]. At least, *Ginkgo* is not recorded from areas of prevailing arid or semiarid climates, which are indicated by evaporites and so-called red beds in addition to climate-sensitive plants. In other words, *Ginkgo* had only a tolerance to moist environments from a very early stage in its evolution.

5 Conclusion

Fossil *Ginkgo* was more common through the Paleogene in East Asia and rare in the Miocene. Some Pliocene and early Pleistocene occurrences are known in southwestern Japan, and the latter occurrence is the youngest record of the genus in the world. Comparing these East Asian fossil records with other areas in the Northern Hemisphere, the phytogeographic history of the genus is consistent with known climatic changes. *Ginkgo* has prevailed in moist and moderately warm climates since the Mesozoic. In this respect, *Ginkgo* has been conservative in its ecological tolerance.

Acknowledgments. I thank Dr. T. Tanai, Professor J.A. Wolfe, Professor G.W. Rothwell, Dr. J. Horiuchi, and Dr. P. Davies for their helpful discussion and improvement of this paper. For the loan of fossil specimens I thank Drs. Y. Hase, S. Hikita, Y. Iwao, Y. Kamada, G. Naito, and M. Tsukagoshi. Dr. T. Tanai kindly allowed the use of his fossil specimens from the Eocene of Yubari. Thanks are also extended to Drs. Y. Hase, A. Momohara, and A. Obuse for the geological and chronological discussion of the *Ginkgo*-bearing assemblages.

References

1. Seward AC (1938) The story of the maidenhair tree. Sci Prog 32:420–440
2. Stewart WN, Rothwell GW (1993) Paleobotany and the evolution of plants, 2nd edn. Cambridge University Press, Cambridge
3. Tralau H (1968) Evolutionary trends in the genus *Ginkgo*. Lethaia 1:63–101
4. Tralau H (1967) The phytogeographic evolution of the genus *Ginkgo* L. Bot Not 120:409–422
5. Berry EW (1935) A Tertiary *Ginkgo* from Patagonia. Torreya 35:11–13
6. Berry EW (1938) Tertiary flora from the Rio Pichileufu, Argentina. Geol Soc Am Spec Pap 12:1–149

7. Romero EJ (1986) Paleogene phytogeography and climatology of South America. Ann MO Bot Gard 73:449–461
8. Crane PR, Manchester SR, Dilcher DL (1990) A preliminary survey of fossil leaves and well-preserved reproductive structures from the Sentinel Butte Formation (Paleocene) near Almont, North Dakota. Fieldiana Geol 20:1–63
9. Scott RA, Barghoorn ES, Prakashi U (1962) Wood of *Ginkgo* in the Tertiary of western North America. Am J Bot 49:1095–1101
10. Seward AC (1919) Fossil plants, vol 4. Cambridge University Press, Cambridge
11. Harris TM (1935) The fossil flora of Scoresby Sound, east Greenland, vol 4. Medd Grønl 112:1–176
12. Florin R (1936) Die fossilen Ginkgophyten von Franz-Joseph-Land nebst Eröterungen über vermeintlische Cordaitales mesozoischen Alters. II. Allgemeiner Teil. Palaeontogr Abt B Palaeophytol 82:1–68
13. Samylina VA (1967) On the final stages of the history of the genus *Ginkgo* L. in Eurasia (in Russian). Bot Zh (Moscow) 52:303–316
14. Samylina VA, Chelebaeva AI (1986) New date on Tertiary species of *Ginkgo* from eastern Asia (in Russian). Paleontol Zh (Moscow) 1986–2:97–102
15. Horiuchi J, Kimura T (1986) *Ginkgo tzagajanica* Samylina from the Palaeogene Noda Group, Northeast Japan, with special reference to its external morphology and cuticular features. Trans Proc Palaeontol Soc Jpn New Ser 142:341–353
16. Boulter MC, Kvacek Z (1989) The Palaeocene flora of the Isle of Mull. Spec Pap Palaeontol 42:1–149
17. Tanai T (1992) Tertiary vegetational history of East Asia (in Japanese). Bull Mizunami Fos Mus 19:125–163
18. Krassilov VA (1976) Tsagajanskaya flora of Amurskoy Oblasti (in Russian). Izdatel'stro "Nauka," Moscow
19. Oishi S (1938) On the cuticles of Tertiary *Ginkgoites* leaves from Kuji, Iwate Prefecture. J Fac Sci Hokkaido Univ Ser IV Geol Mineral 4:103–106
20. Horiuchi J (1995) Pollen and spore composition of the Paleogene Noda Group in Iwate Prefecture (in Japanese). In: abstracts, annual meeting of the Palaeonto logical Society of Japan (February 2–4, 1995, Nagoya), p 52
21. Koshman MN (1964) Tertiary flora of the Bikin browncoal field (in Russian). Bot Zh (Moscow) 49:265–271
22. Fedotov VV (1983) Eocene flora of Raichikhi in Amur Oblasti, part 2 (in Russian). Akademiya Nauk SSSR, Komarov Botanical Institute, Leningrad
23. Polyshchuk LV (1975) Variation of epidermal characters of *Ginkgo* from the Paleogene and Neogene of Primorye in respect to climatic changes. In: Krassilov VA (ed) Fossil floras of the Far East (in Russian). Akademiya Nauk SSSR, Vladivostok, pp 76–83
24. Borsuk MO (1956) Paleogene flora of Sakhalin (in Russian). Tr All-Union Prosp Invest Geol Inst 12:1–131
25. Medjulianov VA (1969) *Ginkgoites* ex gr. *adiantoides* from the Lower Paleogene of South Sakhalin. In: Gramm MN, Krassilov VA (eds) Fossil fauna and flora of the Far East, vol I (in Russian). Far East Geol Inst Komarov Far East Filial Siberian Br Akademiya Nauk USSR, pp 139–152
26. Huzioka K (1951) *Ginkgo adiantoides* (Unger) Heer (in Japanese). Cenozoic Res 11:187–188
27. Huzioka K, Kobayashi M (1961) The mode of occurrence of fossil plants in the Yubari Formation at the Shimizusawa coal-mine in the Ishikari coal-field, Hokkaido (in Japanese). Mining Geol 11:245–252

28. Huzioka K (1972) The Tertiary floras of Korea. J Mining Coll Akita Univ A 5:1–83
29. Endo S (1942) On the fossil flora from the Shulan coal-field, Kirin Province and the Fushun coal-field, Fengtien Province (in Japanese). Bull Natl Mus Manchoukuo 3:33–43
30. Academia Sinica (1978) Cenozoic plants of China (Fossil plants of China, vol 3) (in Chinese). Science Press, Beijing
31. Huzioka K, Takahasi E (1970) The Eocene flora of the Ube coal-field, Southwest Honshu, Japan. J Mining Coll Akita Univ A 4:1–88
32. Tao JR (1988) Plant fossils from Lepuqu Formation in Lhaze County, Xizang and their paleoclimatological significances (in Chinese). Bull Inst Geol Acad Sin 3:223–238
33. Hori J (1987) Plant fossils from the Miocene Kobe flora (in Japanese). Hyogoken Seibutsu Gakkai, Kobe
34. Matsushita H, Matsuo S, Ishida S (1994) The Shimokurata flora of Ube City, Yamaguchi Prefecture, Southwest Honshu, Japan (in Japanese). Bull Mine City Mus 10:1–49
35. Akhmetiev MA, Bratzeva BM (1973) Fossil remains of the genus *Engelhardtia* from Cenozoic deposits of Sikhote-Alin and southern Primorye. Rev Palaeobot Palynol 16:123–132
36. Ablaev AG (1978) Geology and fossil flora of Japan Sea coast (late Cretaceous and Tertiary period) (in Russian). Akademiya Nauk SSSR, Moscow
37. Tanai T, Uemura K (1994) Lobed oak leaves from the Tertiary of East Asia, with reference to the oak phytogeography of the Northern Hemisphere. Trans Proc Palaeontol Soc Jpn New Ser 173:343–365
38. Tanai T (1961) Neogene floral change in Japan. J Fac Sci Hokkaido Univ Ser IV Geol Mineral 11:119–398
39. Fotjanova LI (1988) Flora of the Far East on the Paleogene and Neogene boundary (in Russian). Tr Paleontol Inst Akad Nauk SSSR 231:1–189
40. Fotjanova LI (1976) The floristic complex of the Erman suite and its position among similar by age complexes of the Northern Pacific. In: Menner VV (ed) The Mio-Pliocene of the West Kamtschatka (in Russian). Akademiya Nauk SSSR, Moscow, pp 67–108
41. Hikita S (1949) On the new localities and habitat of *Ginkgo biloba* L. in Pliocene, Japan (in Japanese). Kobutsu to Chishitsu (Miner Geol) 12:1–7
42. Hikita S (1954) A study on the floral remains around Osaka Bay in Japan since the Pliocene (in Japanese). Furitsu-gakko Kyoin Kenkyu-houkokusho (Teahcer's Res Rep Osaka Pref School) 1:1–17
43. Miki S (1948) Floral remains in Kinki and adjacent districts since the Pliocene with description of 8 new species (in Japanese). Sci Rep Osaka Second Teachers Coll 2:105–144
44. Momohara A (1994) Floral and paleoenvironmental history from the late Pliocene to middle Pleistocene in and around central Japan. Palaeogeogr Palaeoclimatol Palaeoecol 108:281–293
45. Kamada Y, Iwao Y, Miyaki M, Okazawa A (1981) On the Hiradoguchi Plant Bed discovered from Tabira-machi, Kitamatsuura-gun, Nagasaki Prefecture (in Japanese). Nagasaki-ken Chigaku Kaishi 33–34:11–28
46. Iwao Y (1978) Late Cenozoic *Ginkgo biloba* L. from the Hoshiwara Formation in Kumamoto Prefecture, Kyushu, Japan. Rep Fac Sci Eng Saga Univ 6:45–50

47. Hase Y, Yamamoto M, Nagamine S, Nomura S (1987) Volcano-stratigraphy of Yaeyama area in Kagoshima Prefecture, Japan (in Japanese). Assoc Geol Collab Jpn Monogr 33:189–206
48. Kobiwako Research Group (1980) The Kobiwako Group in the Seta-Ishibe area, southern part of Shiga Prefecture, Japan (in Japanese). Earth Sci 35:26–40
49. Yoshikawa S, Satoguchi Y, Nagahashi Y (1996) A widespread volcanic ash bed in the horizon close to the Pliocene-Pleistocene boundary: Fukuda-Tsujimatagawa-Kd38 volcanic ash bed occurring in central Japan (in Japanese). J Geol Soc Jpn 102:258–270
50. Chelebaeva AI (1971) Problems on stratigraphy of the continental Cainozoic in Kamchatka (in Russian). Akademiya Nauk SSSR, Leningrad
51. Budantsev LY (1983) History of the Arctic flora of the early Cainophytic period (in Russian). Akademiya Nauk SSSR, Leningrad
52. Ablaev AG (1985) Flora of Koryak-Kamchatka region and stratigraphic problems of the continental Cenozoic (in Russian). Akademiya Nauk SSSR, Vladivostok
53. Guo SX (1985) Preliminary interpretation of Tertiary climate by using megafossil floras in China. Palaeontol Cathayana 2:169–175
54. Wolfe JA (1985) Distribution of major vegetational types during the Tertiary. Geophys Monogr 32:357–375
55. Chaney RW, Axelrod DI (1959) Miocene floras of the Columbia Plateau. Carnegie Inst Washington Publ 617:1–237
56. Wolfe JA, Tanai T (1980) The Miocene Seldovia Point flora from the Kenai Group, Alaska. US Geol Surv Prof Pap 1105:1–51
57. Wolfe JA (1987) An overview of the origins of the modern vegetation and flora of the northern Rocky Mountains. Ann MO Bot Gard 74:785–803
58. Kovar-Eder J, Givulescu L, Hably L, Kvaček Z, Mihajlovic D, Teslenko J, Walther H, Zastawniak E (1994) Floristic changes in the areas surrounding the Paratethys during Neogene time. In: Boulter MC, Fisher HC (eds) Cenozoic plants and climates of the Arctic. Springer, Berlin Heidelberg New York pp 347–369
59. Wang CW (1961) The forest of China with a survey of grassland and desert vegetation. Maria Moors Cabot Found Publ 5:1–333
60. Cheng WC, Fu LK (1978) Ginkgopsida. In: Flora Reipublicae Popularis Sinicae, vol 7 (Gymnospermae) (in Chinese). Academia Sinica, Beijing, pp 17–23
61. Vakharameev VA (1988) Jurassic and Cretaceous floras and climates of the Earth. Cambridge University Press, Cambridge

Fossils and Phenology in the Evolution of *Ginkgo biloba*

Gar W. Rothwell and Ben Holt

1 Introduction

Among the unique and intriguing features that make ginkgophytes some of the most fascinating of seed plants is the rich fossil record for a clade that is monotypic in the modern flora, and for which there may be no natural populations in the wild [1]. Numerous apparently primitive features contribute to the systematic intrigue of ginkgophytes and to their apparently crucial role in seed plant evolution. Meyen [2] considered *Ginkgo* to represent one of three major groups of seed plants, Florin [3] interpreted *Ginkgo* to have evolved in parallel to conifers. Emberger [4] interpreted *Ginkgo* to be little changed in its reproductive biology from the most ancient seed plants, and modern cladistic analyses place *Ginkgo* below all or nearly all other living species on the seed–plant tree [e.g., 5–7].

The fossil record yields plants with *Ginkgo*-like leaves from the Carboniferous to the recent, but many species are known only as isolated compressions and their affinities often remain in question [8]. Most of the earliest leaf fossils from Permo-Carboniferous strata are highly dissected [e.g., 3, 9], but some species from as early as the Permian are remarkably similar to the living *Ginkgo biloba* [10]. Specimens of these types are most confidently assigned to ginkgophytes when they display a combination of overall morphology, venation, and the appropriate epidermal patterns [11].

Ginkgophytes appear to have reached the peak of their systematic diversity and phytogeographical distribution during the late Mesozoic and early Tertiary, where they achieved a circumpolar distribution in the Northern Hemisphere and extended into several regions of the Southern Hemisphere [10, 12, 13]. This was followed by a Tertiary decline that became evident by the Oligocene and appears to have accelerated from the Miocene to the Quaternary [14]. Estimates of systematic diversity during the Tertiary are complicated by the extreme variability of leaf form that is known to occur in *G. biloba* [11, 15], so that other types of

Department of Environmental and Plant Biology, Ohio University, Athens, Ohio 45701, U.S.A.

evidence are needed to confirm the taxonomic diversity and systematic diversity of the clade.

2 Whole-Plant Concepts

As has been the case with other major groups of vascular plants, whole-plant concepts for a number of the extinct species are playing a crucial role in clarifying these facets of ginkgophyte evolution. There are several candidates for developing such concepts, and excellent progress has already been made with some species (see the chapter by Zhou, this volume). Permian fossils assignable to *Trichopitys heteromorpha* [3], *Polyspermophyllum sergii* [9], and the plant that produced *Ginkgoites acosmia* shoots [9] have a great deal of promise for clarifying the origin of the ginkgophyte clade. Concept development for a variety of more recent fossils such as *Ginkgoites tiarensis*/*Karkenia incurva* [17], *Ginkgo huttonii* [18], and *G. yimaensis* [19] is well underway, and several additional assemblages of fossils are currently available for this purpose.

One excellent example is an assemblage preserved in association with abundant dinosaur remains from Upper Cretaceous (Maastrichtian) deposits of Alberta, Canada [20]. These fossils occur at several localities and represent a broad spectrum of organs that conform to those of *Ginkgo*. Leaves are most commonly preserved in a sandstone matrix. Specimens have a narrow petiole up to 2cm long and a fan-shaped blade that is up to 6cm long and 6cm wide (Figs. 1a and 1b). The distal margin is variable. Some specimens have a nearly entire margin, while in others the distal margin is dissected by a variable number of sinuses that may be as deep as one third the length of the blade (Fig. 1a). The intervening lobes are typically irregularly dentate (Fig. 1b). In some specimens, one or, more rarely, two of the lobes are longer than the rest (Fig. 1b). As in the living *G. biloba*, the venation is dichotomous. No anastomoses, like those that sometimes are present near the distal margin of the extant species [21], have been observed. Cuticle and epidermal features are not preserved in these specimens.

There are two types of flattened seeds present in the Upper Cretaceous deposits of Alberta. One has a thin sarcotesta, rounded apex, and somewhat cordate chalaza. The other (Fig. 1c, right) lacks sarcotesta, has two prominent sclerotestal ribs, and tapers to a somewhat pointed apex. Seeds of this type are smaller than those of the extant *G. biloba*, measuring approximately 1cm long, but are similar in shape to the living species (Fig. 1c).

One specimen of a *Ginkgo* pollen cone is present in the Maastrichtian Alberta assemblage. The specimen has helically arranged sporophylls that diverge from an axis that measures about 4mm in diameter (Fig. 1d). Attached sporophylls are terete and bear two terminal pollen sacs that are approximately 6mm long. As in the living species of *Ginkgo*, the pollen sacs are attached terminally and bend toward the base of the cone (Fig. 1d), and dehiscence is by a longitudinally oriented slit. The cone is senescent, but some pollen remains within a dehisced pollen sac (Fig. 1e). As in the living species (Fig. 1f), the fossil grain is ellipsoidal

FIG. 1a–f. Fossil remains of *Ginkgo* from Upper Cretaceous deposits of Central Alberta, Canada. **a,b** Compressed leaves showing variation in dissection of distal margin. *Scale bars*, 1 cm. **c** Seeds of *Ginkgo* with sarcotesta removed to show outer surface of sclerotesta. Extant *G. biloba* at *left*, and fossil seed at *right*; both seeds oriented with the micropyle upward and the primary axis parallel to the surface of the photograph. *Scale bar*, 1 cm. **d** Fossil pollen cone showing helical arrangement of sporophyll bases, and one intact sporophyll with two terminal pollen sacs. *Scale bar*, 5 mm. **e** Pollen grain from interior of dehisced pollen sac from cone in Fig. **d**, with distal sulcus oriented outward. *Scale bar*, 10 μm. **f** Pollen grain of extant *G. biloba* for comparison with fossil grain in **e**. Orientation of grain comparable to the fossil. *Scale bar*, 10 μm

with more or less pointed ends, and is prolate with a prominent distal sulcus (cf. Figs. 1e and 1f).

3 Phenology and Germination Biology

Ginkgo biloba is widely believed to be a "préphanérogame", (i.e., seed plants, both living and fossil, whose ovules develop embryos only after release from the parent sporophyte [22, 23]). Favre-Duchartre [24] used the term "oviparous" in reference to the primitive features of the reproductive biology of *G. biloba*. According to that author, the length of time from ovule initiation to seed germination is 12–14 months, with fertilization taking place either immediately before or closely following abscission from the parent tree in September [25]. Germination typically occurs the following spring or early summer after full development of the embryo on the ground. This developmental scenario separates *Ginkgo* from all other living seed plants with the possible exception of some members of the Cycadales. Of particular importance, relationships between *Ginkgo* and the Paleozoic gymnosperms have been suggested based on this developmental feature, an important contributor to the perception that *Ginkgo* is among the most primitive of seed plants [26].

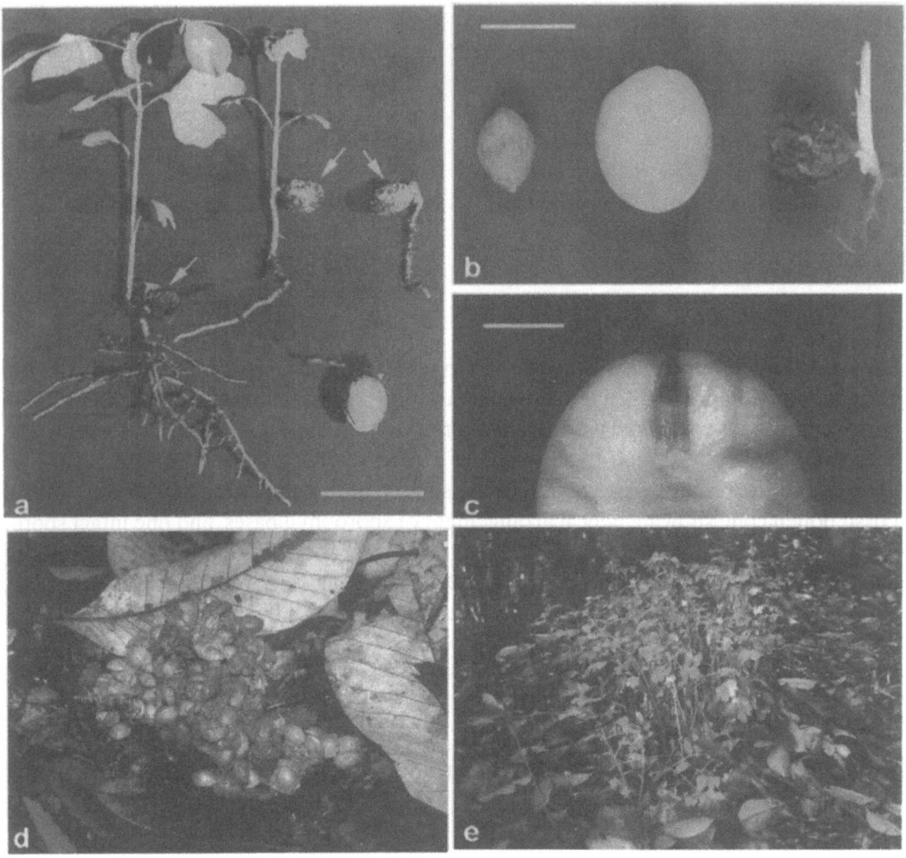

FIG. 2a–e. Seedlings, seeds, and embryo of extant *Ginkgo biloba*. **a** Seeds in various stages of germination to produce seedlings. Note that specimens with shrivelled sarcotesta (*arrows*) have germinated much more rapidly and have produced more mature seedlings than the seed with the undegraded sarcotesta (at lower right). *Scale bar*, 3 cm. **b** Seeds with differing dispositions of the sarcotesta. One seed with sarcotesta removed (*left*), one with undegraded sarcotesta (*center*), and one with degraded and shrivelled sarcotesta (*right*) showing emergence of seedling. *Scale bar*, 1 cm. **c** Megagametophyte of extant *G. biloba* that has been split open to show embryo at the time of seed dispersal from the tree. Specimen oriented with the micropylar end upward and cotyledons of embryo pointing downward. *Scale bar*, 5 mm. **d** Badger droppings that contain a large number of scarified *Ginkgo* seeds with the sarcotesta removed. **e** Clump of *Ginkgo* seedlings that have grown from the droppings of a badger

In a recent study [27], we compared the germination of *Ginkgo* seeds with intact and removed sarcotesta in both cold room and greenhouse conditions (Figs. 2a and 2b). Much to our surprise, and contrary to the accepted literature, we found that seeds removed directly from the tree at the time of abscission contained well developed embryos (Fig. 2c). These experiments [28] also confirmed that the presence of a sarcotesta significantly reduces the total germination percent when compared to seeds with removed sarcotesta [28]. Cold stratification is not necessary for germination although it improves the total germination percent [28]. Based on this information, it is reasonable to hypothesize that sarcotesta enforces dormancy before its decomposition and, therefore, that the sarcotesta reduces autumn germination [28]. While these results are interesting to the reproductive biology of *G. biloba*, their potential significance to the evolution of ginkgophytes becomes much more apparent when viewed in light of the fossil record.

4 Discussion

Although the sarcotesta of *Ginkgo* seeds clearly limits germination, it seems unlikely that such a characteristic would have evolved and persisted through time unless it were part of a successful life history pattern. This mechanism could have evolved to prevent autumn germination, but the Cretaceous and Paleogene were apparently more or less frost free in the Northern Hemisphere where ginkgophytes were most common [29, 30]. Under these conditions, it seems unlikely that the dormancy mechanism of *Ginkgo* would have been required as a response to a warm/cold seasonal cycle.

An alternative is that Mesozoic fossil species of the Ginkgoales lost their sarcotesta as a consequence of animal-mediated seed dispersal [30]. Tiffney [30] suggested that the odoriferous sarcotesta of *Ginkgo* would have been well adapted to attracting herbivorous reptiles, and that the loss of these dispersal vectors may have played a key role in the demise of the ginkgophytes during the Tertiary. According to this scenario, the pungent sarcotesta of *Ginkgo* would have both depressed germination and attracted animal dispersers. Following removal of the sarcotesta in the digestive tract of the disperser, seeds would have been deposited away from the parental tree and capable of almost immediate germination. Our discovery of well-developed embryos in seeds that were still attached to the parental trees demonstrates that the phenology of *Ginkgo* is, indeed, compatible with such a dispersal/germination scenario.

The mode of territorial maintenance and expansion of ginkgophytes in the Cretaceous flora [14], resulting partly from interactions with dispersal agents, is further suggested by comparing their seed size and ecological role to extant species with similar seed sizes and ecological roles [30–32]. According to Tiffney [30], larger seeds would improve the chance of establishment in closed canopies, but would also reduce the seed dispersal range. Therefore, the coevolution of large seeds and appropriate biotic dispersal agents is essential to the success of this life history pattern [30]. If correct, his hypothesis also accounts for the decline of *Ginkgo* following the extinction of dinosaurs and large reptiles at the end of the Mesozoic. Given our current understanding of evolutionary community ecology [33, 34], the loss of an effective dispersal mechanism may not have led to an immediate reduction in the geographic range of established *Ginkgo* communities. However, it would have severely limited the ability of *Ginkgo* to recolonize areas after the disruption of community structure [35]. As a result, an erratic, long-term reduction in the diversity and geographical distribution of *Ginkgo* would be expected. The fossil record suggests that just such an erratic reduction did occur throughout the Tertiary [14].

We find Tiffney's hypothesis [30, 31] that the seed characteristics associated with reptile herbivory in extant *Ginkgo* have "become a relict syndrome" to be an attractive explanation for evolution of the germination mechanism in *Ginkgo*. This hypothesis accounts both for the improved total germination of *Ginkgo* seeds following removal of the sarcotesta (as presumably would have occurred during scarification in the digestive tract of a dinosaur or large reptile), and for the contraction in both range and taxonomic diversity throughout the Tertiary [14] that brought *Ginkgo* to the brink of extinction.

In a recent study of *Ginkgo* in Japan, Hori [36] documents that a badger (*Nyctereutes procyonoides*, locally referred to as a "hondo-tanuki") eats *Ginkgo* seeds for the nourishment provided by the sarcotesta, and a similar feeding strategy by the cat-like animal *Paguma larvata* also has been reported [37]. The scarified seeds are subsequently deposited in droppings (Fig. 2d), and the following year they produce a clump of *Ginkgo* seedlings (Fig. 2e). Extinction of the putative dinosaurian and reptilian dispersers of *Ginkgo* at the end of the Cretaceous precludes a direct experimental test of the Tiffney hypothesis. Nevertheless, this modern corollary does provide a posteriori evidence for the plausibility of the scenario, and similar feeding strategies by Tertiary mammals also have been hypothesized [38]. Regardless of the role that plant–animal interactions may have played in the decline of ginkgophytes, the novel features of this ancient group provide vital evidence for interpreting the evolution of spermatophyte reproduction. In this regard, it is becoming increasingly clear that a combined paleontological/neontological approach to these questions has great promise for developing answers to questions that cannot be addressed using either approach alone.

Acknowledgments. The authors thank Rudolph Serbet for collecting assistance and for preparing the Cretaceous specimens of fossil *Ginkgo*. Dr. Terumitsu Hori, University of Tsukuba, contributed valuable discussions of *Ginkgo* herbivory and provided photographs for Figures 2d and 2e. Dr. Ruth Stockey provided laboratory facilities at the University of Alberta. This work was supported in part by a grant from the National Science Foundation (DEB9527920).

References

1. Li HL (1956) A horticultural and botanical history of *Ginkgo*. Bull Morris Arboretum 7:3–12
2. Meyen SV (1987) Fundamentals of palaeobotany. Chapman and Hall, London
3. Florin R (1949) The morphology of *Trichopitys heteromorpha* Saporta, a seed-plant of Palaeozoic age, and the evolution of the female flowers of the Ginkgoinae. Acta Horti Bergiani 15:79–109
4. Emberger L (1944) Les plantes fossiles dans leurs rapports avec les végétaux vivants. Masson, Paris
5. Loconte H, Stevenson DW (1990) Cladistics of the Spermatophyta. Brittonia 42:197–211
6. Nixon KC, Crepet WL, Stevenson D, Friis EM (1994) A reevaluation of seed plant phylogeny. Ann Missouri Bot Gard 81:484–533
7. Rothwell GW, Serbet R (1994) Lignophyte phylogeny and the evolution of spermatophytes: a numerical cladistic analysis. Syst Bot 19:443–482
8. Taylor TN, Taylor EL (1993) The biology and evolution of fossil plants. Prentice-Hall, Englewood Cliffs
9. Archangelsky S, Cùneo R (1990) *Polyspermophyllum*, a new Permian gymnosperm from Argentina, with considerations about the Dicranophyllales. Rev Palaeobot Palynol 63:117–135
10. Cùneo R (1987) Sobre la presencia de probables Ginkgoales en la Pérmico inferior de Chubut; Argentina. Actas VII Simposio Argentino Paleobot Palynol p 47–49
11. Stewart WN, Rothwell GW (1993) Paleobotany and the evolution of Plants, 2nd edn. Cambridge University Press, New York
12. Harris TM (1974) The Fossil flora of Scoresby Sound, East Greenland 4: Ginkgoales, Coniferales, Lycopodiales, and isolated fructifications. Meddelser om Grønland 112:1–176
13. Harris TM (1974) The Yorkshire Jurassic flora. IV. Ginkgoales and Czekanowskiales. British Museum (Nat Hist), London
14. Tralau H (1968) Evolutionary trends in the genus *Ginkgo*. Lethaia 1:63–101
15. Krassilov VA (1970) Approach to the classification of Mesozoic "Ginkgoalean" plants from Siberia. Palaeobotanist 18:12–19
17. Archangelsky S (1965) Fossil Ginkgoales from the Ticó flora, Santa Cruz Province, Argentina. Bull Brit Mus (Nat Hist) Geol 13:119–137
18. Harris TM, Millington W (1974) The Yorkshire Jurassic flora IV, 2. Czekanowskiales. British Museum (Nat Hist), London
19. Zhou Z, Zhang B (1989) A Middle Jurassic *Ginkgo* with ovule-bearing organs from Henan, China. Palaeontographica 211B:133–143
20. Serbet R (1996) A diverse assemblage of morphologically and anatomically preserved fossil plants from the Upper Cretaceous (Maastrichtian) of Alberta, Canada. International Organization of Palaeobotanists Conference V, Abstracts, Santa Barbara, California, p 89
21. Arnott HJ (1959) Anastomoses in the venation of *Ginkgo biloba*. Amer J Bot 46:405–411
22. Emberger L (1944) Les plantes fossiles dons leur rapports avec les vegetuax vivants. Masson, Paris
23. Stein WE, Beck CB (1987) Paraphyletic groups in phylogenetic analysis:

Progymnospermopsida and Préphanérogames in alternative views of seed plant relationships. Bull Soc Bot France 134, Actualités Botaniques 2:107–119
24. Favre-Duchartre M (1958) *Ginkgo*, an oviparous plant. Phytomorphology 8:377–390
25. Favre-Duchartre M (1956) Contribution à l'être de la reproduction chez le *Ginkgo biloba*. Rev Cytol Biol Vêg 17:1–218
26. Gifford EM, Foster AS (1989) Comparative morphology of vascular plants, 3rd edn. WH Freeman, San Francisco
27. Holt B, Rothwell GW (1995) Phenology and germination biology of *Ginkgo biloba*. Amer J Bot 82:20
28. Holt B, Rothwell GW (1997) Is *Ginkgo biloba* really an oviparous plant? Amer J Bot (in press)
29. Krassilov VA (1975) Climatic changes in eastern Asia as indicated by fossil floras II. Late Cretaceous and Danian. Paleogeog Palaeoclimat Palaeoecol 17:157–172
30. Tiffney BH (1984) Seed size, dispersal syndromes, and the rise of angiosperms: evidence and hypothesis. Ann Missouri Bot Gard 71:551–576
31. Tiffney BH (1986) Evolution of seed dispersal syndromes according to the fossil record. In: Murray DH (ed) Seed dispersal. Academic, Sidney, p 273
32. Tiffney BH (1992) The role of vertebrate herbivory in the evolution of land plants. Paleobotanist 41:87–97
33. DiMichele WA, Phillips TL, Peppers RA (1985) The influence of climate and depositional environment on the distribution and evolution of Pennsylvanian coal-swamp plants. In: Tiffney BH (ed) Geological factors in the evolution of plants. Yale University Press, New Haven, p 223
34. DiMichele WA, Phillips TL, Olmstead RG (1987) Opportunistic evolution: Abiotic environmental stress and the fossil record of plants. Rev Palaeobot Palynol 50:151–178
35. DiMichele WA, Aronson RB (1992) The Pennsylvanian-Permian vegetational transition: A terrestrial analogue to the onshore–offshore hypothesis. Evolution 46:807–824
36. Hori T (1996) *Ginkgo* to the Japanese people (in Japanese). Microscopia 13:184–185
37. Del Tredici P, Ling H, Yang G (1992) The Ginkgos of Tian Mu Shan. Conserv Biol 6:202–209
38. Del Tredici P (1989) Ginkgos and multituberculates: evolutionary interactions in the Tertiary. BioSystems 22:327–339

Environmental Aspects

Tolerance and Susceptibility of *Ginkgo* to Air Pollution

YOON SOO KIM[1], JAE KEE LEE[1], and GAP CHAE CHUNG[2]

1 Introduction

Trees face a variety of environmental stresses and must endure conditions unfavorable for their growth. The stress induced by numerous anthropogenic stress factors can, however, be calculated to a critical threshold value by morphological and physiological adaptations of woody plants. If the sum of the various stresses exceeds this critical value, trees start to develop symptoms in the plant organs [1].

Air pollution is one of the most serious anthropogenic stresses imposed upon forest ecosystems. Much of the world's forest vegetation has been declining, and air pollutions have been shown to be the main causal agents among a complex of inciting factors [2–5]. Some plant species can tolerate high concentrations of pollutants while other species are more sensitive to geochemical materials. Clearly, the tolerance of trees for air pollution is species specific and is considered to be an inherited characteristic [1, 6]. *Ginkgo* has been said to be the species of tree most resistant to air pollution [7], even though little is known about the biochemical mechanisms for this tolerance. This chapter provides a review of the current knowledge of the tolerance and susceptibility of *Ginkgo* to air pollution in an attempt to understand the response of *Ginkgo* to various air pollutants.

2 Response of *Ginkgo* to Air Pollutants

2.1 Forms of Air Pollutants

The deposition of air pollutants occurs through three processes: dry deposition of both particles and gases, wet deposition via precipitation, and cloud water

[1] Department of Forest Product and Technology
[2] Department of Horticulture, Chonnam National University, Yongbong-Dong 300, Kwangju 500-757, Korea

deposition. Dry and wet deposition occur in nearly all landscapes while cloud water deposition is limited mainly to high-elevation forests and some seacoast regions [8]. Deposition is highly dependent on the morphological or biochemical characteristics of the leaf surface as well as of the leaf interior. Dry deposition on vegetation takes place primarily by gaseous absorption processes on the leaves, including adsorption on the surface and absorption through stomata or cuticles [9]. Assessing damage to trees by air pollution has usually been accomplished by examining visible injuries appearing on the leaf surface because that is the first contact between plants and air pollutants [10, 11]. In particular, the cuticle covering the outermost leaf surface is of primary importance for the initial interaction with airborne pollutants because the characteristics of the cuticle determine to a considerable degree the wettability, retention, and leaching of some inorganic ions from the leaves [12–14]. Structural and chemical changes in the cuticle are thus prime indicators in the assessment of anthropogenic influences on plants [15–18].

From a biological or ecological perspective, the most important deposition is that of the gas-phase pollutants such as sulfur dioxide, the oxides of nitrogen, and tropospheric ozone, while wet deposition is acid precipitation. The present work focuses on the tolerance and susceptibility of *Ginkgo* to these deposits.

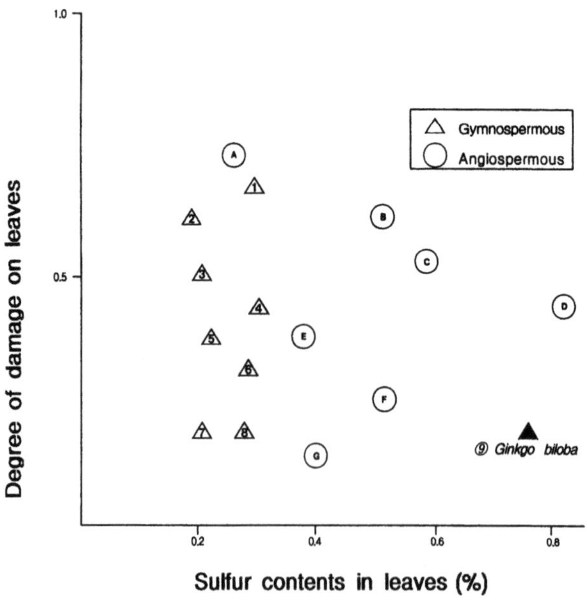

FIG. 1. Degree of damage and sulfur contents in the leaves of tree species. *1, Pinus koraiensis; 2, Pinus rigida; 3, Abies holophylla; 4, Pinus densiflora; 5, Cryptomeria japonica; 6, Chamaecyparis pisifera; 7, Chamaecyparis obtusa; 8, Juniperus chinensis; 9, Ginkgo biloba; A, Quercus serrata; B, Betula platyphylla; C, Populus occidentalis; D, Populus euramericana; E, Prunus jedoensis; F, Populus alba × Populus glandulosa; G, Paulownia coreana* (from [20] with permission)

2.2 *Relative Tolerance of* Ginkgo *to Gaseous Pollutants*

Many previous investigations have shown that *Ginkgo* exhibits a considerable tolerance for gaseous air pollutants but is susceptible to damage from simulated acid rain [19–21]. Oh et al. [20] investigated the degree of damage induced by air pollutants in 17 woody plants. *Ginkgo* was one of the tree species most tolerant to air pollution even though *Ginkgo* has a higher sulfur content than other tree species (Fig. 1). Experimental results on the relative sensitivity of *Ginkgo* trees to various air pollutants show tolerance for sulfur dioxide, nitrogen oxides, and ozone but intermediate tolerance for fluorides, indicating that *Ginkgo* trees are generally tolerant of gaseous air pollutants (Table 1).

2.3 *Resistance of* Ginkgo *Under Ambient Air Pollution*

Because most air pollution in the terrestrial ecosystem combines the action of various pollutants in the damage it inflicts, the foliage of *Ginkgo* trees grown under ambient air pollution conditions were investigated by Kim and Lee [26]. The foliage of *Ginkgo* trees grown in streets with heavy traffic were collected in October for morphological and chemical analysis. For comparison, 2-year-old conifer needles were collected from *Pinus densiflora* Sieb. et Zucc (Japanese red pine), *Pinus koraiensis* Sieb. et Zucc (Korean pine), and *Picea abies* (L.) Karst (Norway spruce) growing under different environmental conditions.

The leaf analyses showed that *Ginkgo* growing in heavy traffic areas contained five times more sulfur than *Ginkgo* trees growing in traffic-free areas (1.81% vs.

TABLE 1. Relative susceptibility of *Ginkgo* trees to various air pollutants

Sulfur dioxide[a]	Fluorides[b]	Nitrogen oxides[c]	Ozone[d]
Tolerant	Intermediate	Tolerant	Tolerant

[a] Davis and Gerhold [22].
[b] Weinstein [23].
[c] Davis and Wilhour [24].
[d] Karnosky [25].

TABLE 2. Sulfur contents in *Ginkgo* and conifer needles (%)

Species	Pollution-exposed areas (A)	Pollution-free areas (B)	(A)/(B)
Ginkgo biloba	1.810	0.360	5.03
Pinus densiflora	0.248	0.107	2.32
Pinus koraiensis	0.388	0.145	2.68
Picea abies	0.420	0.260	1.62

The needles of *Pinus koraiensis* and *P. abies* were collected in Seoul, the capital of Korea, and those of *P. densiflora* from Seoul and from industrial complexes in Onsan and Yeochun. (from Kim and Lee [26] with permission)

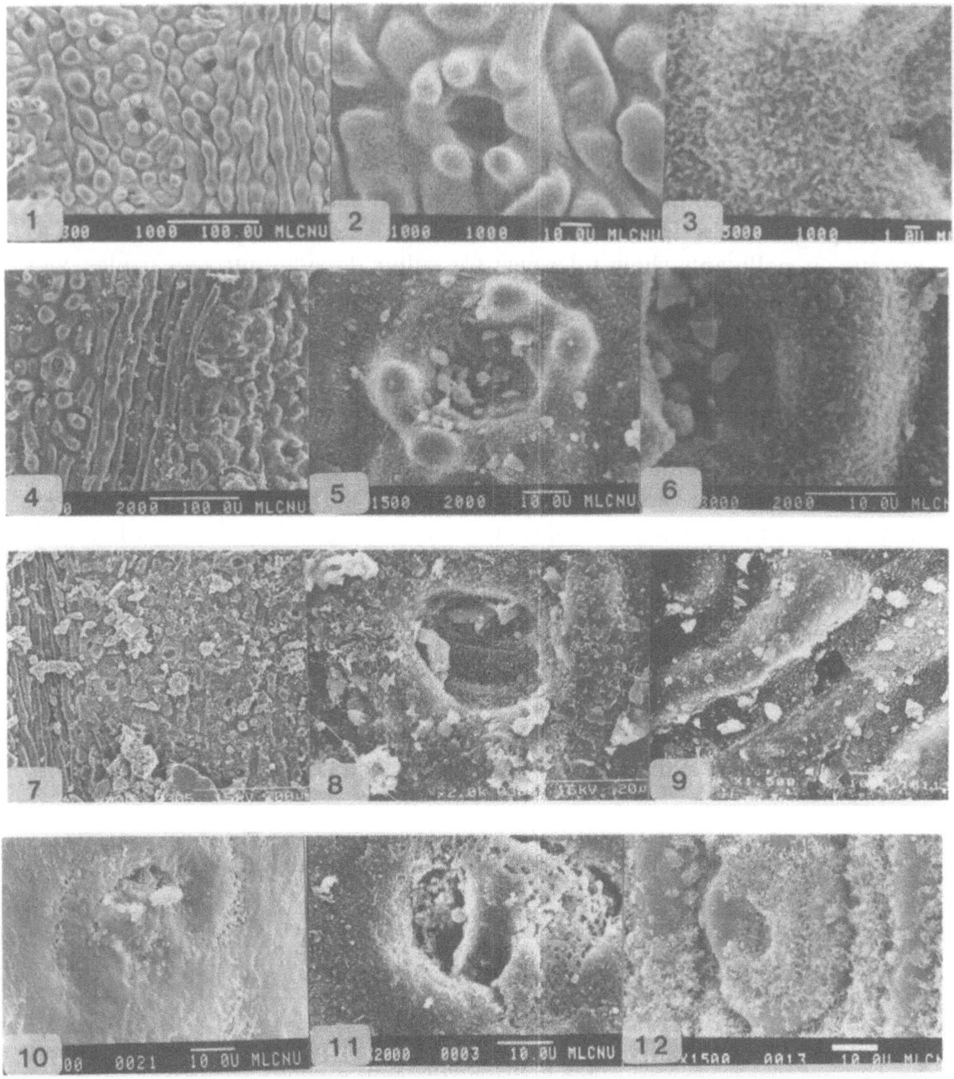

FIG. 2. Scanning electron microscope (SEM) photographs of *Ginkgo* leaves. *1–3*, Collected from traffic-free areas: *1*, general view of the abaxial surface of *Ginkgo* leaves; *2*, stomata of *Ginkgo* (note the granular wax crystalloids); *3*, high magnification of granular wax crystalloides distributed evenly on the surface. *4–6*, *Ginkgo* leaves from heavy traffic areas collected in 1987: *4*, slight accumulation of particles on the surface; *5*, slight erosion of wax structures and accumulation of particles in stomata; *6*, enlarged view of stomata shows relatively intact state of wax structures. *7–9*, *Ginkgo* leaves from heavy traffic area collected from 1995 from the same site as in 1987: *7*, heavy accumulation of particulates on the surface; *8*, partial degradation of wax structures in the stomata; *9*, partial erosion of epicuticular wax layer. *10–12*, Degradation mode of conifer needles collected from industrial complexes in 1987: *10*, Degradation of *Pinus densiflora* (note the total degradation of stomata and wax crystalloides); *11*, erosion of stomatal wax structures in *Picea abies*; *12*, partial degradation of stomatal wax structures in stomata of *Pinus koraiensis*

0.36%) (Table 2). Exhaust gases from automobiles and other particulates may have contributed to the higher accumulation of sulfur in the traffic-area specimens. The conifer needles collected from polluted areas generally contained only about twice as much sulfur than those collected from pollution-free areas. However, no visible leaf injuries were detected in the *Ginkgo*.

Scanning electron microscopy (SEM) (Fig. 2) showed that *Ginkgo* leaves collected from areas with heavy road traffic did not suffer any significant degradation of cuticular waxes except for the accumulation of unidentified particles (Fig. 2-4, 2-5) when compared to the controls (Fig. 2-1, 2-2, 2-3). Some erosion of the granular wax crystalloides distributed evenly on the leaf surface was observed at higher magnifications (Fig. 2-6). However, the degree of erosion of the wax crystalloids was not as severe as on the conifer needles. Most of the granular wax crystalloids on the *Ginkgo* remained intact, in contrast with the severe degradation of wax structures on Korean pine, Norway spruce, and Japanese red pine as revealed by SEM (Fig. 2-7, 2-8, 2-9). It is quite interesting that the foliage of *Ginkgo*, containing at least ten times more sulfur than the pine needles, did not show any marked degradation of wax structures, demonstrating the resistance of *Ginkgo* to air pollution and explaining its popularity as a roadside tree in Korea. In October 1995, 8 years after our initial investigations, we reexamined the morphological changes in *Ginkgo* leaves growing at the same sites studied in October 1987. The later *Ginkgo* samples showed heavy accumulation of particulates on the leaf surface (Fig. 2-10, 2-11), and partial erosion of the epicuticular wax (Fig. 2-12); these changes had not appeared in prior samples, indicating the worsening air quality at these sites. These more recent SEM observations demonstrate that no plant is immune to the damaging effects of air pollution.

2.4 *Susceptibility of* Ginkgo *to Acid Precipitation*

Although *Ginkgo* is resistant to gaseous dry deposition, this tree has shown susceptibility to damage from acid rain. Oh et al. [19, 27] sprayed 15 tree species with simulated acid rain at various pH levels. In angiosperms visible injuries occurred at pH 2.0, but no injuries appeared in most gymnospermous trees at the same pH levels (Table 3). Interestingly, *Ginkgo* showed the most severe leaf injury among the gymnosperms examined. The rate of leaf injury was 100% in *Ginkgo*, while the rates for *Abies holophylla* and *Pinus thungergii* were 76% and 12%, respectively, under the same pH conditions. Furthermore, simulated acid rain affected seed germination and growth of seedlings and decreased the chlorophyll content [21]. Stomatal resistance and contact angles in *Ginkgo* leaves increased remarkably after exposure to simulated acid rain [28], and leaf injuries increased as pH levels decreased [27]. The most frequently reported response to simulated acid rain in forest vegetation is the appearance of necrosis in the upper epidermis of leaves caused by the destruction of epicuticular waxes, distortion of epidermal cells on the upper leaf surface, and premature abscission of leaves [29–32].

TABLE 3. Leaf injury after exposure to simulated acid rain at pH 2.0

Species	Injury index	Injured leaves (%)	Outbreak time of symptoms (days)/number of sprays
Ginkgo biloba	2.99	100	3/1
Larix leptolepis	2.20	86.8	1/1
Crytomeria japonica	2.00	87.3	5/2
Chamaedaphne obtusa	1.66	85.0	5/2
Pinus virginiana	1.60	81.5	7/1
Abies holophylla	1.56	75.6	7/1
Pinus rigiteada	1.52	78.9	5/2
Pinus rigida	1.35	74.5	7/1
Pinus densiflora	1.26	73.6	5/2
Pinus thunbergii	0.16	11.8	3/14
Fraxinus spp.	2.98	100	1/1
Platanus spp.	2.95	100	1/1
Castanea crenata	2.91	100	1/1
Populus canadiensis	2.84	100	3/1
Populus nigra	2.78	100	3/1
Populus Koraiensis × *nigra*	2.78	100	3/1
Zelkova serrata	2.78	100	1/1
Quercus aliena	2.61	100	3/1
Betula platyphylla	2.51	100	1/1

Maximal value of injury index is 3, and minimal value is 0. (from [19] with permission)

The reason why *Ginkgo* leaves are so susceptible to simulated acid rain has not yet been clarified. Susceptibility to acid rain depends on many factors, including leaf age, integrity of the cuticle, anatomy and morphology of the leaf surface, and environmental factors such as temperature, light, and humidity. Turkey [33] reported that leaves which are relatively large, flat surfaced, juvenile, and easily wetted are more easily leached by acid precipitation. On the other hand, leaves that have a smooth, waxy surface and which are wetted with difficulty are less susceptible to leaching by aqueous solution.

Ginkgo is a broad-leaved tree, unlike most other gymnosperms. Broad-leaved trees are much more susceptible to acid precipitation than gymnospermous trees (see Table 3). Rinallo et al. [34, 35] observed alterations in the wax structures of the needles of silver fir and Norway spruce at pH 3.5. In contrast, acid treatment of angiospermous tree species at pH 2.0 caused necrosis after only 24 h exposure [36]. Similar results in *Eucalyptus globulus* were detected at pH 2.2 [37].

Anatomically, the epidermis of *Ginkgo* has relatively thin walls, and hypodermal sclerenchyma is absent. The stomata of *Ginkgo* trees occur almost exclusively on the abaxial surfaces (lower epidermis) of the leaves (i.e., they are hypostomatous), while almost all the conifers bear numerous stomata on both sides of their needles (amphistomatous) [38, 39]. In addition, the cuticle is not well developed in *Ginkgo* (Fig. 3). Considering the fact that the epicuticular wax

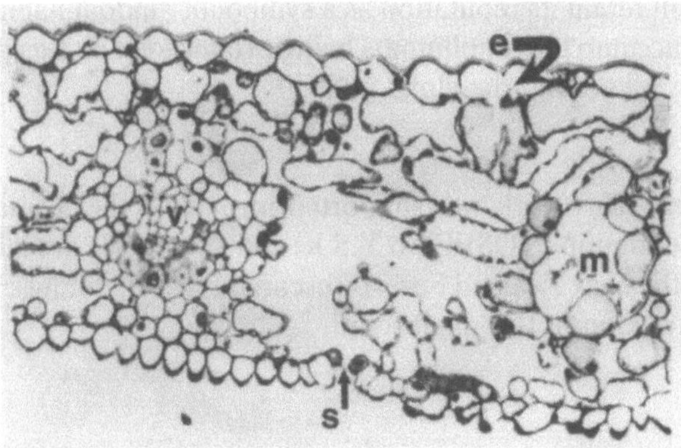

FIG. 3. Cross section of *Ginkgo bilŏba* leaf. *e*, epidermis; *m*, mucilage duct; *s*, stomata; *v*, vascular bundle

has many protective functions on plant surfaces and this function is of crucial importance in conifers for the longevity and evergreen form of their needles, the structural characteristics of the epicuticular wax and the different chemical compositions of wax [12, 40] would determine the sensitivity of *Ginkgo* to simulated acid rain. In addition, anatomical, morphological, and genetic factors such as stomatal density and stomata morphology [41] are likely to be other factors that explain the sensitivity of *Ginkgo* to pollution.

3 Conclusions

Forest plants have evolved progressively in response to environmental stresses during the development of life forms on Earth. The long survival of *Ginkgo* in a Chinese province demonstrated the tremendous adaptability of *Ginkgo* in the face of environmental influences as well as its resistance to pathogenic fungi and insects [42]. *Ginkgo* is the sole living survivor of a genus that has changed little for more than 80 million years, holding the secret to its immeasurable durability. When it became the first plant to bud at the central area of the Hiroshima bomb blast, the *Ginkgo* tree also became a symbol of indominability and hope [43].

Although the lack of published data introduces a considerable uncertainty into the estimates of the sensitivity of *Ginkgo* to air pollution, this work suggests that *Ginkgo* can tolerate gaseous air pollutants but is susceptible to acid precipitation. Further research is required to clarify its inheritance of resistance and susceptibility to air pollution and the extent of changes in the chemical composition of its leaves that are caused by pollution.

Finally, changes in forest vegetation caused by air pollution are slow and are often overlooked unless the effects are dramatic and severe. So, one must wonder if the *Ginkgo* will retain its reputation as a symbol of "indominability and hope" or if it too will succumb to air pollutants being produced at an exceptional rate by human activities all around the globe [44].

Acknowledgment. This work was supported in part by the Korea Science and Engineering Foundation (KOSEF) to Y.S.K. and G.C.C. We are indebted also to Dr. A. Singh in the New Zealand Forest Research Institute for his critical reading and English correction of this manuscript.

References

1. Fowden L, Mansfield T, Stoddart J (eds) (1993) Plant adaptation to environmental stress. Chapman and Hall, London
2. Braun HJ, Sauter JJ (1983) Unterschiedliche Symptome des Waldsterbens im Schwarzwald, moegliche Kausalketten und Basis in Ursache. Allg Forst Ztg 38:656–660
3. Jensch UE, Jager HJ (1983) Zum Nachweis der Wirkung von Luftverunreinigungen auf die Vegetation durch physiologische und biochemische Reaktionen von Pflanzen. Angew Bot 57:157–171
4. Kozlowsik TT, Constantinidou HA (1986) Responses of woody plants to environmental pollution. For Abstr 47:1–51
5. Kozlowski TT, Constantinidou HA (1986) Environmental pollution and tree growth. For Abstr 47:105–132
6. Baker EA, Hunt GM (1986) Erosion of waxes from leaf surfaces by simulated rain. New Phytol 102:161–173
7. Raven PH, Evert RF, Eichhorn SL (1986) Biology of plants, 4th edn. Worth, New York
8. Smith WH (1981) Air pollution and forests. Springer-Verlag, New York
9. Reineret RA (1984) Plant response to air pollutant mixtures. Annu Rev Phytopathol 22:421–442
10. Karhu M, Huttunen S (1986) Erosion effects of air pollution on needle surfaces. Water Air Soil Pollut 31:417–423
11. Hasemann G, Jung G, Wild A (1990) The loss of structural integrity in damaged spruce needles from locations exposed to air pollution. II. Epidermis and stomata (dermal tissue). J Phytopathol (Berl) 128:33–45
12. Schutt P, Schuck HJ (1972) Zusammenhaenge zwishchen Rauchhaerte und Cuticularwachsen bei Koniferen. Mitt Forstl Bundesversuchanst Wien 97:399–417
13. Percy KE, Baker EA (1987) Effects of simulated acid rain on production, morphology and composition of epicuticular wax and on cuticular membrane development. New Phytol 107:577–589
14. Percy KE, Baker EA (1988) Effects of simulated acid rain on leaf wettability, rain retention and uptake of some inorganic ions. New Phytol 108:75–82
15. Grill D, Pfeifhofer H, Halbwachs G, Waltinger H (1987) Investigations on epicuticular waxes of differently damaged spruce needles. Eur J For Pathol 17:246–255
16. Huttunen S (1994) Effects of air pollutants on epicuticular wax structures. In: Percy KE et al. (eds) Air pollutants and the leaf cuticle. Springer, Berlin Heidelberg New York, pp 82–96 (NATO ASI Series, vol G 36)

17. Kim YS (1985) REM Beobachtungen immissionsgeschaedigter Fichtennadeln. Centralbl Gesamte Forstwes 102:96–105
18. Swiecke TT, Endress AG, Taylor OC (1982) The role of surface wax in susceptibility of plants to air pollutant injury. Can J Bot 60:316–319
19. Oh JW, Kim YG, Chae JS (1987) Studies on the periodical changes of air pollution at several forest stands. In: Effects of air pollution and acid precipitation on the forest ecosystem (in Korean, summary in English). Ministry of Science and Technology, Seoul, pp 49–69
20. Oh JW, Chae JS, Yi CK (1983) Effects of atmospheric sulfur dioxide concentration on the growth of some tree species (in Korean, abstract in English). Bull Korean Forest Res Inst 30:243–258
21. Kim GT (1987) Effects of simulated acid rain on growth and physiological characteristics of *Ginkgo biloba* L. seedlings and on chemical properties of the tested soil. I. Seed germination and growth. II. Leaf surface area, visible leaf injury, leaf chlorophyll content and photosynthetic ability of the leaf tissue. J Korean For Sci 76:99–108, 230–240
22. Davis DD, Gerhold HD (1976) Selection of trees for tolerance of air pollutants. In: Santamour FS, Gerhold HD, Little S (eds) Better trees for metropolitan landscapes. Gen Tech Rep NE For Exp Stn USDA For Serv NE-22:61–66
23. Weinstein LH (1977) Fluoride and plant life. J Occup Med 19:49–78
24. Davis DD, Wilhour RG (1976) Susceptibility of woody plants to sulfur dioxide and photochemical oxidants. EPA-600/3-76-102. U.S. Environmental Protection Agency, Corvallis, OR
25. Karnosky DF (1978) Selection and testing program for developing air pollution tolerant trees for urban areas. Proceedings, IUFRO air pollution meeting, Institute of Forest and Wood Economy Ljubljana, pp 233–242
26. Kim YS, Lee JK (1990) Chemical and structural characteristics of conifer needles exposed to ambient air pollution. Eur J For Pathol 20:193–200
27. Oh JW (1986) Effects of simulated acid rain on the growth of trees and soil. Master's thesis, Kyung Hee University, Seoul
28. Lee KH, Chung GC, Lee JS (1993) Effects of simulated acid rain on stomatal resistance, wettability and anatomical changes in *Quercus acutissima* and *Ginkgo biloba* seedlings. J Korean For Soc 82:328–336
29. Adams CM, Dengler NG, Hutchingson TC (1984) Acid rain effects on foliar histology of Artemisia tilesii. Can J Bot 62:463–474
30. Chung GC, Kim YS, Lee SH (1994) Physiological and morphological responses of perilla and cucumber to simulated acid rain. J Korean Soc Hortic Sci 35:587–592
31. Novick NJ, Klein TM, Alexander M (1984) Effect of simulated acid precipitation on nitrogen mineralization and nitrification in forest soils. Water Air Soil Pollut 23:317–330
32. Rather TM, Frink CR (1984) Simulated acid rain: effects on leaf quality and yield of broad leaf tobacco. Water Air Soil Pollut 22:389–394
33. Turkey HB Jr (1980) Some effects of rain and mist on plants with implications for acid precipitation. In: Hutchinson TC, Havas M (eds) Effects of acid precipitation on terrestrial ecosystem. Plenum, New York, pp 141–150
34. Rinallo C, Raddi P, Gellini R, Di Lonardo V (1986) Effects of simulated acid deposition on the surface structure of Norway spruce and silver fir needles. Eur J For Pathol 16:440–446

35. Rinallo C, Raddi P (1989) Effects of simulated acid rain and ABS on leaf surfaces of some broadleaf seedlings. Eur J For Pathol 19:151–160
36. Neufeld HS, Jernstedt JA, Haines BL (1985) Direct foliar effects of simulated acid rain. I. Damage, growth and gas exchange. New Phytol 99:389–495
37. Valentini R, Scarascia-Mugnozza G, De Angelis P, Monaco R (1989) Short-term effects of simulated acid mist on gas exchange of Eucalyptus globulus. Eur J For Pathol 19:200–205
38. Esau K (1977) Anatomy of seed plants, 2nd edn. Wiley, New York, pp 321–372
39. Neidner H, Mansfield TA (1968) Physiology of stomata. McGraw-Hill, London
40. Schmidt U, Rutze M, Liese W (1987) REM Untersuchungen an Stomata von Fichtenund Tannenadeln nach Begasung und saurer Beregnung. Eur J For Pathol 17:118–124
41. Raddi P, Monicca S, Paoletti E (1994) Effects of acid rain and surfactant pollution on the foliar structure of some tree species. In: Percy KE et al. (eds) Air pollutants and the leaf cuticle. Springer, Berlin Heidelberg New York, pp 205–216 (NATO ASI Series vol G 36)
42. Dietrich H (1994) *Ginkgo* biloba—ein Überlebensstratege im Pflanzenreich. In: Schmid M, Schmol GED (eds) *Ginkgo* Ur-Barum und Arzneipflanze Mythos, Dichtung und Kunst. Wissenschaftl Verlagsgesell, Stuttgart, pp 15–22
43. Beuchert M (1995) Symbolik der Pflanzen. Insel Verlag, Frankfurt, pp 110–115
44. Evans LS (1984) Acidic precipitation effects on terrestrial vegetation. Annu Rev Phytopathol 22:393–420

Ginkgo and Insects

HIROSHI HONDA

1 Introduction

The ginkgo, or maidenhair tree, *Ginkgo biloba* L., is the only living species of the family Ginkgoaceae, which were gymnosperms that thrived about 175 to 200 Million years ago. Native to southern China, *Ginkgo* is now distributed widely. *Ginkgo* was introduced to Japan about 900 years ago and has been used locally for firewood and handcrafts, and the seed roasted as a delicacy. It has been exploited mainly as a landscape tree rather than as a source of timber. For centuries, because of its high desirability as an ornamental tree with handsome foliage and autumn colors, its resistance to insect pests and plant diseases and its adaptability to urban conditions, planting of *Ginkgo* has increased along roadsides, in public parks, and in private gardens. However, I will note in this chapter some insect pests that occur *Ginkgo* trees and the relationships between insects and *Ginkgo*.

2 Pest Insects of the Ginkgo in East Asia

The fact that there are extremely few insect pests on *Ginkgo* trees worldwide should be noteworthy. For example, of 344 pest species found on 40 species of landscape tree in Japan, only 6 species (2 leipdopterous, 4 homopterous) were found on the *Ginkgo* tree [1]. However, no species-specific lepidopterous defoliators for *Ginkgo* trees have been recorded, and all pests feed on the *Ginkgo* as an alternative host in Japan. The giant silk moth, *Dictyoploca japonica* Butler (Saturniidae), is well known as a serious pest of the *Ginkgo* tree [2]. But as shown by the Japanese common name of *kususan*, where *kusu* means the camphor tree and *san* the silk moth, this insect is not a ginkgo specialist and infests various kinds of trees [3–5]. Sporadic infestations of *Ginkgo* by *D. japonica* larvae have been observed in unexpected outbreaks [5–12].

Institute of Agriculture and Forestry, University of Tsukuba, 1-1-1 Tennodai, Tsukuba, Ibaraki 305, Japan

Larvae of the Oriental tea tortorix, *Homona magnanima* Diakonoff and *Spilarctia subcarnea* (Walker), are leaf-mining hervibores on *Gingko* leaves [1, 13]. The fall webworm, *Hyphantria cunea* Drury, native to the United States, was introduced to Japan after World War II, and since then terrible damage has been observed in 337 tree species but not in *Ginkgo* trees [14]. Yoneyama [15] compared larval development and fecundity of *H. cunea* reared on six species of plants including *G. biloba*, and observed the lowest levels for *Ginkgo* leaves.

Although dead or decayed *Ginkgo* trees on the ground may be infested with many bark beetles or scavengers, coleopterous pests on live trees are extremely few compared with those on other species of forest and garden trees. Go [16] found larvae of *Acalolepta sejuncta* (Bates) on *Ginkgo* twigs cut in Tokyo, but this longicorn beetle is also a typical decaying-wood borer. After this finding, Kojima and Nakamura [17] and the Japanese Society of Applied Entomology and Zoology [13] also reported the *Ginkgo* tree as a host plant for *A. sejuncta*. Recently, Onagamitsu and Kaneko [18] and Onagamitsu [19] observed serious damage in Kyushu (Japan) by two species of cerambycids in a ginkgo orchard grown for production of dry seeds. One of these cerambycid pests was *A. sejuncta* and the other which feeds on live bark, was closely related to *A. kusamai* Hayashi. The latter species was, however, revised and identified as *A. gingovora* Makihara (*Icho-hige-biroudo-kamikiri*) [20, 21].

The last group of insect pests on *Ginkgo* trees in Japan is the homopterous insect group [1, 9, 22–28]. The most common species is the Comstock mealbug, *Pseudococcus comstoki* (Kuwana), which infests the leaves and twigs of *Ginkgo*. This infestation is accompanied by sooty mold on honey dew excreted by the mealbug [1]. Of three other minor scale pests, *Ceroplastes ceriferus* Anderson mainly feeds on twigs but not on leaves or trunk, and *Parlatoreopsis pyri* (Marlatt) and *Lepidosaphes kuwacola* Kuwayama occur on trunk, twigs, or leaves [1]. Watanabe [11] added *C. rubens* Maskell to the above four species. Severe infestations by these scale insects on the *Ginkgo* tree were clearly associated with environmental pollution in areas where they were planted; many more scale pests occurred at the more polluted urban sites [1, 23, 27].

The *Ginkgo* tree is also an important tree in making the country green and is a major source of dry fruit in China. Zheng [29] showed that young *Ginkgo* trees are damaged by beetles, longicorn beetles, termites, and caterpillars of Saturniidae, but gave no detailed species names. However, *Phassus excrescens* Butler [30, 31], *Holcocerus vicarius* Walker [31, 32], *Setora postornata* (Hampson) [33], *Ocneria disper* L. (Lymantridae) [31, 34], *Cryptothelea variegata* Snellen (Psychidae) [31], *Cnidocampa flavecens* Walker (Limzcodidae) [32], and *Vapocheima robiniae* Chu (Geometridae) [34] were recorded as the most common lepidopterous pests [31]. The two saturniid moths, *D. japonica* Butler and *Eriogyna pyretorum* Westwood are also known as pests of *Ginkgo* tree [35, 36]. The former is named *ying xin da can e* (in Chinese), meaning a giant silk moth fed on the ginkgo tree, and the latter is *zhange can e* (in Chinese), which means a giant silk moth fed on the camphor tree. However, the giant silk moth, *kususan*, which has been recorded as a pest of the *Ginkgo* tree and named in Japan, is *D. japonica* Butler but not *E. pyretorum* Westwood. In China, the latter species is, however, described as *Kususan* in Chinese letters [35].

Recently, Zhang and Li [37] reported that more than 10 years ago, serious damage to buds, twigs, and seeds of *Ginkgo* tree was caused by an unknown tortorix pest, *Pammene* sp. Liu [38] reported detailed morphological characteristics of this pest and identified it as a new species, *P. ginkgoicola*, which is closely allied to *P. ochsenheimeriana* (Zeller). Although not an insect, the broad mite, *Polyphagotarsonemus latus* (Bank) (Tarsonemidae), is widely distributed throughout temperate and tropical regions, and infests various kinds of orchard crops, field vegetable crops, and garden flowers in these areas [39]. Damage to *Ginkgo* trees by this polyphagous mite was also observed in China [40]. In Taiwan, ginkgo plantations for economic production and for ornamental or shading agents are very few, so there have been no reports of any pest insects of *Ginkgo* trees (Y.I. Chu 1996, personal communication).

Pest insect fauna on *G. biloba* in Korea seems to be similar to that of Japan or China. As lepidopterous pests, *D. japonica* Butler (saturniid moths), *Eurneta minuscula* Butler, *E. japonica* Heylaerts (the bagworm moths), *H. cunea* (Drury), *Spilosoma subcarmeum* Walker (tiger moths), and *Archips fuscocupreanus* Walsingham (leaf-roller moths) have been recorded from the ginkgo tree [41, 42]. Recently, Lee et al. [43] revised *E. minuscula* Butler and identified it as *Mahasena aurea* Butler. These larvae feed on flower buds and soft bark of the tree before new foliage has expanded. A closely related species to *A. fuscocupreanus*, *A. argyrospilus* (Walker), has been found as a contaminant of pollen-derived tissue culture of *G. biloba*; however, this was a record from the United States [44]. Other species of insect pests in Korea are scale insects (Homoptera). *Iceria purchasi* Maskell, cottony cushion scale, *Cerostegia japonicus* (Green), Japanese wax scale, *Pseudococcus comstocki* (Kuwana), Comstock mealybug, and *Pseudaulacaspis pentagona* (Targioni), mulberry scale, have been recorded [41, 42]. *Criconemoides morgensis* (Hofmanmer and Menzel) Taylor is the only nematode that attacks *Ginkgo* tree [42].

3 Insecticidal Property of Ginkgo

Dry *Ginkgo* leaves have been used traditionally as bookmarks in Japan because of their unique shape and beautiful yellow color in autumn. People believed that when the bookmark was used, there was less damage by booklice (Pscoptera) or silverfishes (Thysanura), which feed on the starch in book bindings or labels [45, 46]. The wood has been used to make insect-proof cabinets [47]. The efficacy of ginkgo leaves as an agent to control agricultural pest insects was known in Japan 250 years ago, when *Fukihouzoki*, a manual on pest control during the Edo (Tokugawa) era, was used [45, 48]. This idea is thought to have been introduced from China. In fact, Zheng [29] described practical application of extracts of *Ginkgo* leaves for controlling pests in paddy fields. Yagi [49] also recommends this method in organic or natural agriculture with reduced agrochemicals. These facts strongly suggest the existence of some chemical barriers against pest insects in *Ginkgo* trees.

Hartzell and Wilcoxon [50] and Heal *et al.* [51] found that acetone-water extracts of *Ginkgo* leaves had no lethal effect on mosquito larvae, cockroaches, or carpet beetles. However, Beck [52] showed that alcohol extracts of ground whole roots of the *Ginkgo* tree strongly inhibited growth of the European corn borer, *Pyrausta (Ostrinia) nubilalis*, but similar extracts of stems did not. This inhibitory effect might be due to acidity from oxalic acid and malic acid [47]. European corn borer larvae were repelled olfactorially by 2-hexenal [53]. This green leaf aldehyde is produced when *Ginkgo* leaves are damaged [54]. These two facts indicate the *Ginkgo* leaves may be protected from severe herbivorous insect infestation by producing repellents such as 2-hexenal.

The Japanese beetle, *Popilia japonica*, which has been recorded to attack 350 species of plants from more than 24 families, eating foliage, flowers and fruits, berries, garden crops, and ornamental shrubs and trees [55], never feeds on *Ginkgo* leaves. However, Metzger and Grant [56] showed that extracts of fresh *Ginkgo* leaves applied to apple or peach tree leaves did not repel Japanese beetles, *P. japonica*. When *Ginkgo* leaves were also treated with eugenol or n-valeric acid from the favorite host plants of the beetle, they never refused to feed on them. This indicates that *Ginkgo* leaves lack feeding stimulants or attractants for inducing beetle feeding [47, 53, 57].

As mentioned above, lack of stimulants or attractants in food plants strongly affects the survival of phytophagous insects, while some secondary plant chemicals such as toxins or feeding deterrents in plants also affect host plant preferences of herbivorous insects [58]. Since knowledge of natural pest-specific toxins and feeding deterrents has increased rapidly in recent years, these secondary plant chemicals have been focussed on for developing environmentally safe types of insecticides [59].

G. biloba was reported up as a source of insecticidal natural chemicals by Grainger et al. [51, 60]. Methanol extracts of the leaves deterred feeding by *Pieris rapae crucivora* on cabbage leaf disks treated with the extract. The active components were 6-penta-decylsalicylic acid, 6-hepta-decenylsalicylic acid and sesquiterpene, bilobalide [61]. Kwon et al. [62] tested extracts of 43 species of plants including *G. biloba* leaves for their insecticidal and acaricidal activities against six species of insects and one mite species. They found that the methanol extracts of the leaves were selectively toxic against the brown plant hopper, *Nilaparvata lugens* (Stal), which is a serious pest of rice plants in tropical and temperate Asia. The toxic component or components have not been identified yet but were soluble in ethylacetate when the crude methanol extract was fractionated with several kinds of organic solvents, and were highly toxic to the plant hopper at concentrations of 0.001 μg/female; this concentration produced no mortality with Diazinon, Carbofuran, or Fenobucarb [63]. From the studies reported above it is clear that *G. biloba* contains some chemicals that are toxic to at least some insects. The high resistance promised by sophisticated chemical defense to insect herbivores may be one of the reasons why *G. biloba* could survive as a "living fossil." Therefore, we may be able to develop new and environmentally safe agents to control insect pests on further study of the chemical defense strategies of *G. biloba*.

4 Conclusion

The earlist fossil insects were from the Devonian (about 400 million years ago) and the Carboniferous saw the rise of various insects and all modern plant divisions except the angiosperms. After 200 million years Ginkgopsida including *G. biloba* arose and diversified during the Mesozoic. This paleontological fact suggests that an ecological interaction between *Ginkgo* trees and insects might have been established, because without doubt the main gymnosperm herbivores at these times were insects [64]. However, as reviewed in the first half of this chapter, the number of insect species feeding on *Ginkgo* trees is extremely small compared with those attacking other species of gymnosperms. Also, some phytochemicals found in *Ginkgo* trees have been shown to be effective defense factors against insect herbivores. Although the highly developed resistance of *G. biloba* to phytophagous insects cannot be explained by chemical defense alone, such unusually high resistance accounts for the longevity of the tree and its long reproductive period. Understanding evolutionary interactions between *G. biloba* and herbivorous insects will provide fruitful information on coevolutionary interaction between modern angiosperms and insects.

Acknowledgments. I thank Prof. M. Tamura and S. Kawai (Tokyo University of Agriculture, Japan), Prof. H. Dai (Nanjing Agricultural University, China), Dr. K.Y. Seol (National Institute of Sericultre and Insects, Korea), Prof. K. Furuta (University of Tokyo, Japan), Prof. Y.I. Chu (National Taiwan University, Taiwan), Mr. Y. Kusakabe (Yokohama, Japan), and Dr. W. Tulecke (Antioch University, USA) for kindly providing various information and references. I also thank Prof. D. Taylor of our institute for his critical reading of the manuscript.

References

1. Kawai S (1977) The insect pests occurring on trees lining a street and the real status of their damages. In: Report of Studies on Protection of Trees Lining Streets (in Japanese). Tokyo Agr Exp Stat pp 277–309
2. Sasaki C (1902) Giant silk moth. In: Pest insects of Japanese cultivated trees (in Japanese). Sanrakusya, Tokyo, pp 89–93
3. Esaki T, Issiki, Mutuura A, Ioue H, Ogata M, Okagaki H, Kuroko H (1958) Icones Hetrocerorum Japonicorum In Coloribus Naturalibus, Vol. II (in Japanese). Hoikusha, Osaka, p 303

4. Inoue H, Okano M, Shirozu T, Sugi S, Yamamoto H (1959) Iconographia Insectorum Japonicorum Colore naturali edita, Vol. I (in Japanese). Hokuryukan, Tokyo, p 284
5. Furukawa H (1970) The giant silk moth. In: Encyclopedia of entomology. Tokyo-do, Tokyo, p 97
6. Matsusita M (1943) Giant silk moth. In: Forest entomology (in Japanese). Toyamabo, Tokyo, pp 268–271
7. Saito K (1957) Giant silk moth. In: Forest entomology (in Japanese). Asakura, Tokyo, pp 115–116
8. Aino H, Ito K (1958) Giant silk moth. In: Diseases and pests of plants in colour (in Japanese). Hokuryukann, Tokyo, p 153
9. Kobayashi F (1977) Insect pests of ginkgo and Japanese sago palm. In: Pest diseases and insects of landscape trees (II), Insect pests and their control (in Japanese). Jpn For Tech Assoc, Tokyo, pp 69–73
10. Okuno T, Tanaka H, Kimura Y (1977) Insect pests of ginkgo tree. In: Diseases and pests of cultivated trees and shrubs in colour (in Japanese). Hoikusya, Osaka, p 285
11. Watanabe F (1992) Ginkgo tree. In: list of insect pests on trees and shrubs. Ariake, Tokyo, p 198
12. Kobayashi F (1994) Giant silk moth. In: Kobayashi F, Taketani A (eds) Forest entomology (in Japanese) Yokendo, Tokyo, p 345
13. Japanese Society of Applied Entomology and Zoology (1987) Major insect and other pests of economic plants in Japan (in Japanese). Jpn Plant Prot Assoc, Tokyo, p 379
14. Tanaka M (1995) Control of the fall webworm, Hyphantria cunea Drury in Japan (in Japanese). Hist Jpn Plant Prot 9:93
15. Yoneyama M (1968) An experiment on the fall webworm (in Japanese). Bull Yamanashi Sericul Exp Stat 51:71–75
16. Go N (1975) Record on *Acalolepta sejuncta* from ginkgo tree (in Japanese). Gekkan-mushi 52:27
17. Kojima K, Nakamura S (1986) List of host plants for Japanese Cerambycidae (in Japanese). Hiba Sci Edu Prom p 336
18. Onagamitsu J, Kaneko S (1983) Ecology and control of some longicorn beetles on maidenhair tree. 1. Species and their infestation (in Japanese). Proc Assoc Plant Prot Kyushu 29:138–140
19. Onagamitsu J (1991) Unknown longicorn beetle as a pest of ginkgo tree. In: Japan Forest Technical Association (ed) 100 Wonder stories on forest insects (in Japanese). Jpn For Tech Assoc, pp 198–199
20. Makihara H. (1992) Revisional study of the genus *Acalolepta* of Japan—Studies on Cerambycidae (Coleoptera) of Japan (8). Acta Coleopterol Japon 2:45–72
21. Makihara H (1992) Genus Acaloleota In: Obayashi N, Sato M, Kojima K (eds) An illustrated guide to identification of longicorn beetles of Japan (in Japanese). Tokai, Tokyo, pp 584–591
22. Nagasawa M, Kawai S (1971) Important pest insects on garden trees (in Japanese). Plant Protection 25:347–352
23. Kawai S (1972) Diagnostic notes and biology of the Coccid-species occurring on cultivated or wild trees and shrubs in Japan (Homoptera: Coccoidae) (in Japanese). Bull Tokyo Agric Exp Stat 6:1–54.
24. Kawai S (1972) Scale insects on garden trees (in Japanese). Noyaku Kenkyu, 18: 10–13

25. Kawai S (1973) Studies on coccid-species under urban environments. In: Environmental Agency of Japan (ed) Studies on indicator plants and animals for human life under urban environments (in Japanese). pp 18–57
26. Kawai S (1973) How to know the scale insects infesting trees (2) (in Japanese). For Prot News Letter 22:249–253
27. Kawai S, Motohashi S (1974) Scale insects occurring on trees in city parks. In: Report on production mechanisms of smog and its damage on plants in Tokyo (in Japanese). Institute of Environmental Sciences of Tokyo, Tokyo, pp 556–561
28. Kawai S (1980) Scale insects of Japan in color (in Japanese). Zenkoku Nouson Kyoiku Kyokai, Tokyo p 455
29. Zheng W (1983) *Ginkgo* L. In: Comm Sylva Sinica (ed) Sylva Sinica (in Chinese). Peking, pp 154–158
30. Li Y, Ling J (1983) The swift moth. In: Xiao G (ed) The forest insects of China (2nd edn) (in Chinese). China For Pub House, Peking, pp 471–473
31. Liang L (1988) Major insect pests and control. In: Chinese ginkgo tree (in Chinese). Shandong Sci Tech Pub, Shandong, pp 222–240
32. Qi R (1983) The willow holcocerus moth. In: Xiao G (ed) The forest insects of China (1st edn) (in Chinese). China For Pub House, Peking, pp 535–536
33. Ying H (1983) The brown slug caterpillar moth. In: Xiao G (ed) The forest insects of China (1st edn) (in Chinese). China For Pub House, Peking, pp 556–558
34. Zhan Y (1983) The acacia geometer moth. In: Xiao G (ed) The forest insects of China (1st edn) (in Chinese). China For Pub House, Peking, pp 670–672
35. Fang H, Lian Y (1983) The giant silk moth. In: Xiao G (ed) The forest insects of China (1st edn) (in Chinese). China For Pub House, Peking, pp 754–755
36. Tan Z (1983) The ginkgo giant silkmoth. In: Xiao G (ed) The forest insects of China (1st edn) (in Chinese). China For Pub House, Peking, pp 752–754
37. Zhang G, Li Z (1981) Preliminarly study on *Pammene sp* (in Chinese with English summary). J Nanjing Tech Coll For Prod 4:83–89
38. Liu Y (1992) Two new species of *Laspeyresiini* damaged cones and seed of forest (Lepidoptera: Tortricidae) (in Chinese with English summary). Sinozoologia 9:249–252
39. Ehara S, Shinkaji N (1996) Principles of plant acarology (in Japanese). Zenkoku Nouson Kyouiku Kyokai, Tokyo, p 419
40. Kuang H (1983) The broad mite. In: Xiao G (ed) The forest insects of China (1st edn) (in Chinese). China For Pub House, Peking, pp 937–938
41. Jeong SJ, Lee BY, Byeon BH (1995) Insect pests and diseases of trees and shrubs (in Korean). Korean Forest Administration Forestry Research Institute, Seoul, p 350
42. Korean Plant Protection Society (1986) *Gingko biloba* In: Major pest insects, Diseases of economic plants and weeds in Korea (in Korean) Kor Plant Prot Soc, Seoul, p 304
43. Lee S, Oh Y, Byun B (1996) Biological character of a bagworm sp. (Lepidoptera: psychidae) on *Ginkgo biloba* L (in Korean with English summary). Proc Ann Meet Kor Soc For pp 52–53
44. Tulecke W, Colavito L (1966) Fruit-tree leaf rollers on plant tissue cultures and on leaves of *Ginkgo biloba*. J Econ Entomol 59:1277–1278.
45. Konishi M (1996) Knowledge for practical use of plants in Edo era. In: Yagi A (ed) Control of disease and insect pests with plant extracts (in Japanese). Nobunkyo, Tokyo, pp 20–24

46. Jacobson M (1990) Family Ginkgoaceae. In: Glossary of plant-derived insect deterrents. CRC, Boca Raton, p 59
47. Major RT (1967) The Ginkgo, the most ancient living tree. Science 157:1270–1273
48. Honda H (1996) Practical application of plant chemicals to pest control (in Japanese). Agric Hort 71:548–554
49. Yagi A (1996) Control of disease and insect pests with plant extracts (in Japanese). Nobunkyo, Tokyo, p 137
50. Hartzell A, Wilcoxon F (1941) A survey of plant products for insecticidal purposes. Boyce Thompson Inst Cont 12:131
51. Heal RH, Rogers EF, Wallance RT, Starnes O (1950) A survey of plants for insecticidal activity. Lloydia 13:89–164
52. Beck SD (1960) The European corn borer, Pyrausta nubilalis (HUBN), and its principal host plant VII. Larval feeding behavior and host plant resistance. Ann Entomol Soc Amer 53:206–212
53. Major RT, Tietz HJ (1962) Modification of the resistance of *Ginkgo biloba* leaves to attack by Japanese beetles. J Econ Entomol 55:272
54. Bevan CW, Birch AJ, Casewell H (1961) An insect repellant from black cocktail ants. J Chem Soc 1961:488
55. Metcalf RL, Metcalf ER (1992) Plant kairomones in insect ecology and control. Chapman & Hall, New York, p 168
56. Metzger FW, Grant DH (1932) Repellency to the Japanese beetle of extracts made from plants immune to attack. USDA Tech Bull p 299
57. Hirano T (1971) Japanese beetle and ginkgo. In: Host selection in phytophagous insects (in Japanese). Kyoritsu, Tokyo, pp 32–34
58. Dethier VG (1947) Chemical insect attractants and repellents. Blakiston, Toronto, p 289
59. Hedin PA (1991) Naturally occurring pest bioregulators, ACS Symp Ser 449. Am Chem Soc, Washington, p 456
60. Grainger M, Ahamed S, Mitchell W, Hylin JW (1985) Plant species reportedly possessing pest-control properties—An EWC/UH Database. East-West Resource Systems Institute, Honolulu, p 249
61. Matsumoto T, Sei T (1987) Antifeedant activities of *Ginkgo biloba.* L. components against the larvae of *Pieris rapae crucivora.* Agr Biol Chem 51:249–250
62. Kwon M, Lee SB, Ahn YJ, Park NJ, Cho KY (1994) insecticidal and acaricidal activities of plant extracts. Agr Chem Biotech 37:492–497
63. Kwon M, Ahn YJ, Yoo JK, Choi BR (1996) Potent insecticidal activity of extracts from *Ginkgo biloba* leaves against *Nilapalvata lugens* (Homoptera: Delphacidae). Appl Ent Zool 31:162–165
64. Swain T (1978) Plant-animal coevolution: a synoptic view of the Paleozoic and Mesozoic. In: JB Harborn (ed) Biochemical aspects of plant and animal coevolution. Academic, London, pp 3–19

Fungal Association with *Ginkgo biloba*

Takayuki Aoki

1 Introduction

In the Gymnospermae, coniferous substrata have been well examined for associating fungal species, e.g., in phytopathological studies, in floral investigation including edible or toxic mushrooms, and in the research of fungal succession phenomena following decomposition of leaf litter or wood [1, 2]. Fungi in relation to *Ginkgo biloba* have not been well studied, however. Because of its less common distribution in the wild and because of the limited industrial use of *Ginkgo* wood, interests of many mycologists until now have been mainly confined to fungi pathogenic to *Ginkgo* street trees.

2 Literary Records: Symbionts, Parasites, and Pathogens

Some fungal species have been recorded as the symbionts in vesicular-arbuscular mycorrhiza of *Ginkgo* in Italy, i.e., *Glomus* spp. [3, 4], and as parasites, i.e., *Ceuthospora melaleuca* Ferd. and Winge in England [5], *Fomes meliae* (Underw.) Murrill, *Glomerella cingulata* (Stoneman) Spauld. and Schrenk, *Irpex lacteus* (Fr.: Fr.) Fr. (= *Irpex tulipiferae* (Schwein.) Schwein.), *Trametes hirsuta* (Wulfen: Fr.) Pilát (= *Polyporus hirsutus* (Wulfen: Fr.) Fr.), *Trametes versicolor* (L.: Fr.) Pilát (= *Polyporus versicolor* (L.: Fr.) Fr.) in North America [6], *Pestalotia sinensis* Shen in China [7], and others . Tree pathogens have been more intensively studied, and approximately ten pathogenic fungi have been reported from Japan, i.e., *Botryotinia fuckeliana* (de Bary) Whetzel (gray mold) [8], *Ceratobasidium anceps* (Bres. and Syd.) H.S. Jacks. (= *Corticium koleroga* (Cooke) Höhn. = *Pellicularia koleroga* Cooke; thread blight) [9–15], *Erythricium salmonicolor* (Berk. and Broome) Burds. (= *Corticium salmonicolor* Berk. and Broome; pink disease) [16], *Fusarium* spp. (canker) [12–14, 17, 18], *Gonatobotryum apiculatum*

Department of Genetic Resources I, National Institute of Agrobiological Resources, Ministry of Agriculture, Forestry and Fisheries, Kannondai, Tsukuba, Ibaraki 305, Japan

(Peck) S. Hughes [13, 14, 19, 20], *Helicobasidium mompa* N. Tanaka (violet root rot, also in Taiwan) [13, 14, 21, 22], *Hendersonia ginkgonis* Naito (brown leaf-spot disease) [14, 23, 24], *Pestalotia ginkgo* Hori (pestalotia disease) [7, 13, 14, 25–27], *Rosellinia necatrix* Prill. (white root rot) [13, 14, 28], *Thanatephorus cucumeris* (A.B. Frank) Donk (web blight) [13, 14], and others.

These are important studies for fungal species associated with *Ginkgo*, but overall mycological work on *Ginkgo* is very limited. The above list is a compilation of occasional records, and these examinations mostly focused on a single species of fungi as a pathogen, and even then the substrate of *Ginkgo*, e.g., a single leaf, had not also been thoroughly examined for other inhabiting fungi. In addition, the abundance of each fungal species was not evaluated, even as a relative value.

3 Fungi Inhabiting Decomposing Leaves: Saprophytes

A preliminary examination was made of microfungi inhabiting the decomposing leaves of *G. biloba* during a winter in Japan.

3.1 Materials and Methods

Leaf litter of *G. biloba* planted on the campus of the National Institute of Agrobiological Resources (NIAR), Ministry of Agriculture, Forestry and Fisheries (MAFF), Tsukuba City, Ibaraki Prefecture, Japan, was collected on 22 January 1996. Collected leaves were classified into three different stages of decay: L-type leaves, golden brown to clay to light brown, elastic; L/F-type leaves, light brown to brown, partly discolored to grayish brown or brownish gray, fairly thin but still elastic; and F-type leaves, dark brown to grayish brown to brownish gray, thin and fragile. The apparent succession of leaf decomposition might progress from L- through L/F- to F-type leaves. By using a 6-mm cork borer, leaf discs were cut randomly from each type of leaf. Ten leaf discs from each stage of decay were put in a sterile test tube. Test tubes were then filled with 10 ml of a sterilized detergent, 0.005% aerosol OT solution (di-*iso*-octyl sodium sulfosuccinate) and were settled on a Vortex genie mixer (Scientific Industries, INC., Bohemia, NY). They were washed 10 times using the procedure given by Tokumasu [29] as a modified washing method. Washed leaf discs were then dried on sterilized filter paper in a petri dish for 1 night at room temperature. Each leaf disc was then placed on a half-strength cornmeal-agar plate (Difco, Detroit, MI) and incubated at room temperature. Observations of fungal species that appeared on and around the leaf discs were made three times: after 1 week, 2 weeks, and 4 weeks of incubation. Fungal species were isolated into pure cultures and then identified. Frequencies of occurrence of individual fungi were calculated for each stage of decay by the following equation [1, 2]: The percentage frequency of a given fungus from 6-mm leaf discs of a certain stage of decay (%) = [number

TABLE 1. Vertical distribution of fungi in the leaf litter[a] of *Ginkgo biloba* L. in Tsukuba (6-mm leaf discs: L-, L/F-, and F-type leaves[b])

Species	Frequency (%)		
	L	L/F	F
Pestalotiopsis negrecta (Thuem.) Steyaert	10		
Ordus tribrachiatus K. Ando and Tubaki	10		
Tripospermum sp.	10		
Dark sterile hyphae	10		
Microdochium bolleyi (R. Sprague) de Hoog and Herm.-Nijh.	10	10	
Yeasts (budding)	10		10
Pestalotiopsis gracilis (Kleb.) Steyaert	40	40	10
Aureobasidium pullulans (de Bary) Arnaud	40	50	20
Alternaria alternata (Fr.) Keissl.	40	90	90
Cladosporium oxysporum Berk. and M.A. Curtis	40	60	70
Cladosporium cladosporioides (Fresen.) G.A. de Vries	50	80	100
Phoma sp.	30	50	70
Dark Coelomycete	10	40	20
Pithomyces chartarum (Berk. and M.A. Curtis) M.B. Ellis		30	
Epicoccum nigrum Link		10	
Penicillium sp.		10	
Nigrospora oryzae (Berk. and Broome) Petch		10	
Acremonium fusidioides (Nicot) W. Gams		10	
Rhizoctonia sp.		20	20
Hyalodendron cf. *album* (Dowson) Diddens		10	10
Scolecobasidium humicola G.L. Barron and L.V. Busch		10	10
Scolecobasidium constrictum E.V. Abbott		10	30
Dactylaria humicola G.C. Bhatt and W.B. Kendr.		50	80
Mirandina corticola Arnaud ex Matsush.			60
Mucor hiemalis Wehmer f. *hiemalis*			20
Fusarium sacchari (E.J. Butler) W. Gams var. *sacchari*			10
Fusarium proliferatum (Matsush.) Nirenberg			10
Fusarium graminearum Schwabe			10
Fusarium sporotrichioides Sherb.			10
Fusarium tricinctum (Corda) Sacc.			10
Fusarium semitectum Berk. and Ravenel			10
Gliocladium catenulatum J.C. Gilman and E.V. Abbott			10
Acremonium strictum W. Gams			10
Dactylaria cf. *fusiformis* Shearer and J.L. Crane			10
Colletotrichum gloeosporioides Penz.			10
Geotrichum sp.			10

L-type leaves, golden brown to clay to light brown, elastic; L/F-type leaves, light brown to brown, partly discolored to grayish brown or brownish gray, fairly thin but still elastic; F-type leaves, dark brown to grayish brown to brownish gray, thin and fragile.

[a] collected at the campus of National Institute of Agrobiological Resources (NIAR), Ministry of Agriculture, Forestry, and Fisheries, Kannondai, Tsukuba, 22 Jan 1996.

[b] a washing method modified by Tokumasu [29] was performed.

of leaf discs from which the fungus was detected / 10 (total number of leaf discs for each stage of decay)] × 100.

3.2 Results

Vertical distribution of microfungi on decomposing leaves of *G. biloba* is shown in Table 1, in which the calculated occurrence frequencies are also indicated. Some fungal species occurred at all stages of leaf decay: *Pestalotiopsis gracilis*, *Aureobasidium pullulans*, *Alternaria alternata*, two species of *Cladosporium*, *Phoma* sp., and an unidentified coelomycetous species. Other species showed partial distribution and were isolated from particular stages of decay: *Pestalotiopsis negrecta* and *Ordus tribrachiatus* were isolated from the L-type leaf of *Ginkgo*, as was a species of *Tripospermum*. *Microdochium bolleyi* was found with L and L/F leaves. Among the dematiaceous hyphomycetes isolated, most species appeared with L/F- (to F-) type leaves, e.g., *Pithomyces chartarum*, *Epicoccum nigrum*, *Nigrospora oryzae*, two species of *Scolecobasidium*, and *Dactylaria humicola*. A zygomycete, *Mucor hiemalis*, and species of *Fusarium*, *Gliocladium*, and *Acremonium*, which are known as so-called soil fungi, were mostly restricted to F-type leaves. *Mirandina corticola* occurred at fairly high frequency with the F-type leaves.

3.3 Discussion

A fungal flora in the leaf litter of *G. biloba* during a winter in Tsukuba was recorded. Thirty-six species of microfungi were recorded from three different types of fallen leaves. From the vertical distribution of the fungi recorded, it was shown that leaves at a certain stage of decay had their own subfloras.

Some species isolated from the *Ginkgo* leaves collected in winter in Tsukuba are known as leaf-surface or phylloplane fungi, e.g., *Tripospermum* sp. from the L-type leaves [30]. So-called primary and secondary saprophytes [1, 31] also appeared with the *Ginkgo* leaves, e.g., species of *Cladosporium*, *Alternaria alternata*, *Mucor hiemalis*, and *Penicillium* sp. Soil fungi such as species of *Fusarium* were isolated mainly from the F-type leaves. Saprophytes on the *Ginkgo* leaves detected in this pilot study were found to be quite similar to those on fir needles in Tsukuba [1], although their occurrence patterns and frequencies showed differences. Species characteristic of the *Ginkgo* substratum are not clear yet.

As listed previously, fungal pathogens have been recorded for *G. biloba*. These pathogenic species reported were, however, totally different from species recorded in the present study, although some species of *Fusarium* and *Pestalotia* (or *Pestalotiopsis*) are known to be pathogenic. It is possible that some parasitic or pathogenic species of fungi could not be isolated, because only fallen and dead leaves were examined in the present study. It may be that these species could be found if living leaves were studied in the same way.

It has been pointed out that monthly mycoflora on fir needles is not stable and its alternation should correspond to the actual decay of leaf litter [1, 2]. Some fungi in the previous stages of decay may remain in the successive stages and others may disappear. In the later stages of decay, a new fungal colonization may occur. Such alternation may also be repeated until the end of total decomposition of the substratum. Winter and summer mycofloras could be quite different. Further, geographical differences may affect the fungal flora on similar substrata, although this may be limited to microfungi, as shown by comparison of fir needles in Japan and Germany [1, 2].

In the present study, only three types of fallen leaves were studied for fungi. Other parts of the *Ginkgo* plant, such as living leaves, and soil of the basal ground were not examined. It could be suggested that fungal species on the list (Table 1) were also only a transection figure for microfungi colonizing *Ginkgo* leaves during winter in Tsukuba. It is expected that more fungal species can be recorded when similar examinations are made in other seasons and other geographical areas and for other plant bodies of *Ginkgo*.

4 Conclusion

- Fungal species associated with *G. biloba* have not been well examined, except for pathogenic species.
- From a preliminary survey performed during winter in Tsukuba, Japan, a clear vertical distribution of microfungi (mostly saprophytes) was illustrated with only three different stages of leaf decomposition.
- Individual stages of decomposing *Ginkgo* leaves have their own submycofloras.
- More fungal species are expected when more intensive studies are made of *Ginkgo* substrata in different areas.

As a footnote, it may be added that a jelly fungus, *Exidia glandulosa* Fr., has been observed to colonize dead branches of *G. biloba* (T. Aoki, pers obs).

References

1. Aoki T, Tokumasu S, Tubaki K (1990) Fungal succession on momi fir needles. Trans Mycol Soc Jpn 31:355–374
2. Aoki T, Tokumasu S, Oberwinkler F (1992) Fungal succession on fir needles in Germany. Trans Mycol Soc Jpn 33:359–374
3. Bonfante-Fasolo P, Fontana A (1985) VAM fungi in *Ginkgo biloba* roots: their interaction at cellular level. Symbiosis 1:53–67
4. Fontana A (1985) Vesicular-arbuscular mycorrhizas of *Ginkgo biloba* L. in natural and controlled condition. New Phytol 99:441–447
5. Sutton BC (1980) The Coelomycetes, Fungi Imperfecti with Pycnidia, Acervuli and Stromata. CMI, Kew, pp 471–473, 666

6. Seymour AB (1929) Host index of the fungi of North America. Harvard University, Cambridge, p 28
7. Guba EF (1961) Monograph of *Monochaetia* and *Pestalotia*. Harvard University, Cambridge, p 136
8. Takano K (1992) *Botryotinia* Whetzel. In: Kobayashi T, Katumoto K, Abiko K, Abe Y, Kakishima M (eds) Illustrated genera of plant pathogenic fungi in Japan (in Japanese). Zenkoku Noson Kyoiku Kyokai, Tokyo, pp 148–149
9. Ito T (1954) Thread blight of *Ginkgo biloba* by *Corticium koleroga* (Abstract, in Japanese) Ann Phytopathol Soc Jpn 19:65
10. Ito T (1957) On *Pellicularia koleroga* Cooke causing thread blight (in Japanese). J Jpn For Soc 39:483–485
11. Ito T (1958) On *Pellicularia koleroga* Cooke causing thread blight of *Ginkgo biloba* (in Japanese). Bull Gov For Exp Sta (Japan) 105:11-18
12. Ito K (1974) Pathology of forest trees, vol 3 (in Japanese). Norin Shuppan, Tokyo, pp 100–102, 307–308
13. Kobayashi T (1977) Ryokuka-jumoku no Byo-chugai (Diseases and pests of green trees), vol 1, Byogai to sono bojo (Diseases and protection) (in Japanese). Nihon Ringyo Gijutsu Kyokai, Tokyo, pp 79–84
14. The Phytopathological Society of Japan (1983) Common names of economic plant diseases in Japan, vol 4, Conifers and bamboos, 2nd edn (in Japanese). Japan Plant Protection Association, Tokyo, pp 2–3
15. Oniki M (1992) *Ceratobasidium* Rogers. In: Kobayashi T, Katumoto K, Abiko K, Abe Y, Kakishima M (eds) Illustrated genera of plant pathogenic fungi in Japan (in Japanese). Zenkoku Noson Kyoiku Kyokai, Tokyo, pp 306–307
16. Abe Y (1992) *Corticium* Persoon. In: Kobayashi T, Katumoto K, Abiko K, Abe Y, Kakishima M (eds) Illustrated genera of plant pathogenic fungi in Japan (in Japanese). Zenkoku Noson Kyoiku Kyokai, Tokyo, pp 316–317
17. Takai S (1965) A canker of *Ginkgo biloba*, a tentative name (in Japanese). Forest Pests 14:206–208
18. Kobayashi T (1992) *Fusarium*. In: Kobayashi T, Katumoto K, Abiko K, Abe Y, Kakishima M (eds) Illustrated genera of plant pathogenic fungi in Japan (in Japanese). Zenkoku Noson Kyoiku Kyokai, Tokyo, pp 537–538
19. Horie H, Kobayashi T, Sugata S, Abe Z (1975) Tree diseases at Metropolitan Jindai Botanical Garden (in Japanese). Forest Pests 24:71–75
20. Horie H (1992) *Gonatobotryum* Saccardo. In: Kobayashi T, Katumoto K, Abiko K, Abe Y, Kakishima M (eds) Illustrated genera of plant pathogenic fungi in Japan (in Japanese). Zenkoku Noson Kyoiku Kyokai, Tokyo, pp 448–449
21. Naubu N (1915) Investigation of diseases of flowers and bonsais (potted plants) (in Japanese). J Plant Prot 2:1092–1093
22. Sawada K (1917) Treatment for the violet root rot of mulberry trees in Taiwan and for the future (in Japanese). Rep Dept Agr Gov Res Inst Formosa 129:609–619
23. Naito T (1952) The mycoflora of southern Kiusiu IV. Sci Rept Kagoshima Univ 1:71–81
24. Katumoto K (1992) *Hendersonia* Berkeley. In: Kobayashi T, Katumoto K, Abiko K, Abe Y, Kakishima M (eds) Illustrated genera of plant pathogenic fungi in Japan (in Japanese). Zenkoku Noson Kyoiku Kyokai, Tokyo, p 559
25. Nanbu N (1915) Specimens collected in June (in Japanese). J Plant Prot 2:743
26. Shirai K, Hara K (1927) A list of Japanese fungi hitherto known, 3rd edn (in Japanese). Yokendo, Tokyo, p 252

27. Tokunaga Y (1984) Plant Pathogenic Mycology (in Japanese). Hakuyu-sha, Tokyo, p 292
28. Kobayashi T, Kusunoki M, Hayashi H (1981) Note on outbreaks of soil-borne diseases of green trees (in Japanese). Forest Pests 30:153–155
29. Tokumasu S (1980) Observations on fungal flora on pine leaf litter. In: Biseibutsu no seitai (Ecology of microorganisms) (in Japanese). Gakkai Shuppan Center, Tokyo, vol 7, pp 129–144
30. Tubaki K, Ando K, Tokumasu S, Tsurumi Y (1984) List of Hyphomycetes in Tsukuba area (I) *Tripospermum* (in Japanese). Tsukuba Environmental Studies, University of Tsukuba 8:54–59
31. Hudson HJ (1968) The ecology of fungi on plant remains above the soil. New Phytol 67:837–874

Ginkgo Landscapes

Mariko Handa[1], Yasuo Iizuka[2], and Nobuo Fujiwara[3]

1 Background to *Ginkgo* Landscapes

1.1 Ginkgo biloba *and Japan*

Since ancient times, *Ginkgo* has stimulated people's imaginative powers and permitted culture of many kinds to bloom. Among many distinctive features, the shape of its leaves especially has stimulated the creative powers of artists and served as the theme of poetry. Johann Wolfgang Goethe (1749–1832) sang as follows in the poem titled "Gingo Biloba" in the *West-Eastern Divan*: "Is it one living thing that has become divided within itself? Are these two who have chosen each other, so that we know them as one?" (translated by David Luke) [1]. *Ginkgo* personifies a world of harmony with oneself and is also called the tree of love. It is a mysterious tree that reminds us of how the occident and the orient have met and influenced each other. This chapter considers the *Ginkgo* in Japan, a nation of the orient.

There are various, still unproven, theories concerning the era when *Ginkgo* arrived in Japan. Some say the *Ginkgo* tree grew in Japan during the Mesozoic era, but later died out. Others argue that it arrived from China along with Buddhism in the sixth century, but this has not been proven.

Ginkgo is now loved by Japanese people and deeply rooted in their daily lives. There are many reasons why *Ginkgo* is so common and so widespread in Japan. The first reason is its contribution to landscapes. It is beautifully shaped, and its colors vary widely throughout the seasons. The second is its contribution to

[1] Environment Department and [2] Planning Division, Planning and Research Administration Department, Public Works Research Institute, Ministry of Construction, 1 Asahi, Tsukuba, Ibaraki 305, Japan
[3] Michinoku Lakewood National Government Park Works Office, Tohoku Regional Construction Bureau, Ministry of Construction, 53-9 Nihonmatsu, Ono, Kawasaki Cho, Shibata Gun, Miyagi 989-15, Japan

culture. *Ginkgo* is frequently established as a symbol of a region or as an object of worship, and it plays roles in many legends. The third is environmental. *Ginkgo* matures rapidly, is easily transported, can be cared for with ease, withstands the urban environment, and is hardy against fire and cold weather. The fourth is its merit as a source of useful products, for example, foods and boxes.

It is assumed that these spiritual and physical characteristics of *Ginkgo* with which people have endowed this tree have combined to stimulate its popularity in Japan. Today we can see landscapes including *Ginkgo* trees (called here *Ginkgo* landscapes) throughout Japan, along roadsides and alleys, in parks, gardens, and school yards, on shrine grounds, and close to private houses. These facts testify eloquently to the extent to which the Japanese people have accepted *Ginkgo* in ways suited to the natural features and climates of their homeland, and how they have represented *Ginkgo* trees in their environment.

SPATIAL FORM	ROLE FORM	IMAGE	UTILIZATION FORM (EXAMPLE)
Point	Main		Single Tree Giant Tree
Line	Main	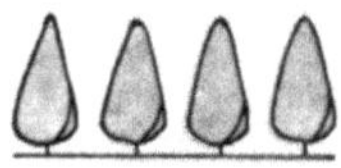	Roadside Trees Alleys
	Not Main		
Area	Main	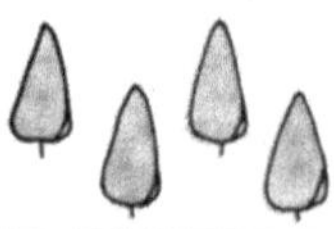	Forest
	Not Main	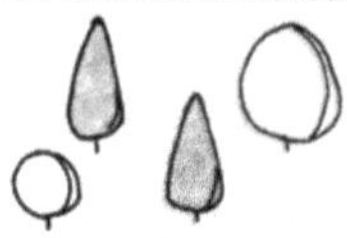	

FIG. 1. Applicatione forms of *Ginkgo biloba* (provided by Public Works Research Institute, Ministry of Construction)

1.2 Landscaping Methods Using Ginkgo *Trees*

The key point in any discussion of landscapes is the method by which the landscapes are formed and the materials used to form them. Thus, when *Ginkgo* trees are to be incorporated into a landscape, landscape architects must use methods to maximize the advantages of their special characteristics and value as a landscaping resource. Because the methods of using *Ginkgo* trees in landscaping vary widely according to local conditions and desires of the citizens related to them, we classify these methods from the point of view of so-called application forms, an issue that must be considered under all conditions. These application forms are usually classified as follow: (1) spatial form—points, lines, or areas; (2) role form—used as the main element (leading role) or not (nonleading role); and (3) utilization form—used or mixed with other varieties of trees (Fig. 1).

More detailed explanations of this classification are as follows. Examples of a "point" include a single *Ginkgo* tree as the leading player in a scene: a single tree, a giant tree growing on the grounds of a shrine, a giant tree grown as a natural commemorative object, or a symbol tree at a major intersection, etc. Examples of a "line" include a series of *Ginkgo* trees planted as roadside trees or as alleys. In many cases, *Ginkgo* trees play the leading role in these arrangements, but they are also combined with medium-sized or small trees. There are also cases where several rows of trees are planted, with one row consisting of *Ginkgo* trees, another of camphor trees, and so on, without any one tree variety playing the leading role. A typical example of an "area" is represented by many trees scattered over a reforested field. As is often seen in urban parks, *Ginkgo* trees are usually mixed with other varieties of trees. It is rare to see a broad area occupied solely by *Ginkgo* trees, but it seems to be worth trying.

This chapter is focused on points and lines, in which *Ginkgo* trees play the leading role. To make it easier to grasp the situation throughout Japan, we first describe lines along networks of roads all over Japan; this is followed by a description of points.

2 Linear Landscapes

2.1 Landscapes of Roadside Trees

This section outlines the history and change of roadside trees, describes their present state, and clarifies the positioning of *Ginkgo* trees in a linear landscape.

2.1.1 History of Roadside Trees

Some theories hold that roadside trees in Japan date back to the middle of the eighth century, when fruit trees were planted along roads for the benefit of travelers [2]. Rather than going back that far, this section provides an outline of events since the latter half of the nineteenth century, when Japan opened its borders and began the process of modernization and roadside trees became an element in modern urban landscaping.

The forerunners of modern roadside trees were first seen in Yokohama and the capital, Tokyo. Willow and pine trees were planted along the street called Bashamichi in Yokohama in 1867. In Tokyo in 1873, black pine, cherry, maple, and other trees were planted along the newly adorned Ginza Street modernized by the construction of European-style brick buildings. The year 1875 saw the first use of imported trees to line city streets, when *Robinia pseudoacacia* trees were planted in Tokyo. They were grown from seeds brought back to Japan from the International Exposition held in Vienna in 1873. In addition to *R. pseudoacacia, Ailantus altissia*, maple, Chinese bottle tree, and other varieties of trees were introduced at that time, but because of either inappropriate planting methods or poor care, they died, and many of them were replaced with willow trees because willows can withstand both dry and moist conditions, heavy rain, or strong winds. But insufficient care was taken to protect these willow trees, resulting in a deterioration in the appearance of the city.

A plan introduced in 1907 to restore the appearance of the trees along the streets in Tokyo signaled the beginning of work to nurture seedlings of ten selected varieties of trees: *Ginkgo*, *Platanus*, tulip tree, Chinese bottle tree, marronniere, trident maple, *Sophora japonica*, dogwood, ash, and *Mallotus japonicus*. It was the debut of *Ginkgo* trees along streets in earnest. These varieties are largely fast-growing trees capable of withstanding the urban environment, and their selection gives us a glimpse of the atmosphere in this age of rapid growth aimed at the creation of modern cities. In line with this plan, 33 *Ginkgo* trees were planted along the street in front of the Tokyo City Hall [3].

The fires following the Great Kanto Earthquake of 1923 burned down more than half of the roadside trees in Tokyo, leaving only about 10000 still standing. As part of the restoration work, however, the national government planted 16280 trees and the city of Tokyo planted another 4681 trees along the streets. The principal varieties planted during these projects were *Ginkgo*, *Platanus*, *R. pseudoacacia*, and others, indicating that many fast-maturing trees were planted.

The bombing of Japan during World War II destroyed a total of 121162 roadside trees, or about 45% of the total of 271168 formerly standing along Japan's roads. The War Recovery Agency, a national agency for recovering from war damage, appointed a committee to study roadside trees. For this roadside tree restoration project, the committee suggested selecting local trees that could take advantage of the environmental characteristics of each region. The intention toward diversification of trees was already indicated in this suggestion.

2.1.2 Change of Varieties of Roadside Trees

Table 1 shows the numbers of each of the top ten trees by variety planted along roads and the distribution ratios of each variety, as revealed by past surveys. This

TABLE 1. Numbers and distribution ratios of the top ten trees (provided by Public Works Research Institute, Ministry of Construction)

Year of survey	Scope of survey	Source	Rank	1	2	3	4	5			
1904	1 city	[4]	Species	Willow	Cherry	Japanese pine	Maple	Chinese bottle tree	Others	Total of 10 species	Total
			Number	3424	2336	216	99	45	328	6448	6448
			Ratio	53.1	36.2	3.3	1.5	0.7	5.1	100	
1922	1 city	[5]	Species	*Platanus*	***Ginkgo***	Chinese bottle tree	Cherry	Willow	Others	Total of 10 species	Total
			Number	5826	5461	2914	2358	1866	6423	23 077	24 848
			Ratio	23.4	22.0	11.7	9.5	7.5	25.9	92.9	
1936	5 cities	[6]	Species	*Platanus*	***Ginkgo***	Cherry	*Robinia pseudoacacia*	Chinese bottle tree	Others	Total of 10 species	Total
			Number	58 890	28 933	13 976	12 647	5528	19 474	135 435	139 448
			Ratio	42.2	20.7	10.0	9.1	4.0	14.0	97.1	
1938	6 cities	[6]	Species	*Platanus*	***Ginkgo***	Cherry	*Robinia pseudoacacia*	Chinese bottle tree	Others	Total of 10 species	Total
			Number	83 800	37 417	17 300	12 210	9289	32 158	181 991	192 174
			Ratio	43.6	19.5	9.0	6.4	4.8	16.7	94.7	
1954	111 cities	[7]	Species	*Platanus*	***Ginkgo***	Weeping willow	Yoshino cherry	*Robinia pseudoacacia*	Others	Total of 10 species	Total
			Number	88 930	59 854	21 745	21 724	21 354	71 013	253 212	284 620
			Ratio	31.3	21.0	7.6	7.6	7.5	25.0	89.0	
1967	160 cities	[7]	Species	*Platanus*	***Ginkgo***	Weeping willow	*Robinia pseudoacacia*	Carolina poplar	Others	Total of 10 species	Total
			Number	107 839	105 927	54 721	40 461	25 207	92 079	418 937	426 234
			Ratio	25.3	24.9	12.8	9.5	5.9	21.6	98.3	
1982	229 cities	[8]	Species	***Ginkgo***	*Platanus*	Trident maple	Weeping willow	Japanese zelkova	Others	Total of 10 species	Total
			Number	301 938	199 830	142 037	90 620	78 966	964 658	1 123 798	1 778 049
			Ratio	17.0	11.2	8.0	5.1	4.4	54.3	63.2	
1986	Nationwide	[9]	Species	***Ginkgo***	Cherry	*Platanus*	Trident maple	Japanese zelkova	Others	Total of 10 species	Total
			Number	485 577	260 195	258 462	238 143	191 393	2 274 362	2 015 862	3 708 132
			Ratio	13.1	7.0	7.0	6.4	5.2	61.3	54.4	
1991	Nationwide	[10]	Species	***Ginkgo***	Cherry	Japanese zelkova	Trident maple	*Platanus*	Others	Total of 10 species	Total
			Number	552 407	345 563	305 007	281 427	251 399	3 048 899	2 463 085	4 784 702
			Ratio	11.5	7.2	6.4	5.9	5.3	63.7	51.5	

table, however, was prepared from documents of various kinds, and for reasons related to survey methods, the cities included in surveys conducted in different periods differ to some degree. The table indicates that during the Meiji Period (1868–1912), the period when planting of roadside trees began, many willow and cherry trees were planted. Surveys conducted between 1922 and 1967 show that the tree planted most often was *Platanus*, and *Ginkgo* took second place. The reason why many *Platanus* trees were planted was that they were suited to the urgent conditions which prevailed at that time; the need to recover quickly from the earthquake and, later, from the devastation left by the war. But *Platanus*, an overwhelming leader according to a 1936 survey which revealed that its distribution ratio was 42.2% of all roadside trees, later began to decline, and was forced into second place by *Ginkgo* in 1982. Since then *Ginkgo* has been the leader.

The factors of these changes are considered to be as follow. The relative importance of *Platanus* is believed to have declined because its characteristic fast growth, excellent sprouting capacity, and relatively big leaves, which were positive factors during the recovery of the city, meant that as the city matured these trees were, inversely, troublesome to maintain. *Ginkgo* trees, on the other hand, were used more and more in recognition of their merits. They grow well in the urban environment, their shape remains regular, and their changing appearance contributes to the clear marking of the progress of the

FIG. 2. Midosuji Avenue (Osaka City)

FIG. 3. Otemae Avenue (Himeji City) (photo courtesy of Himeji City)

seasons. Similar factors also explain the popularity of Japanese zelkova and trident maple. The growing use of the evergreen camphor tree was a result of a desire to maintain greenery throughout the year. Many cherry trees were planted despite the fact that they are difficult to take care of because of many diseases which threaten them. This reflects the sentiments of the Japanese people and the special concern they have for these trees, which are the most popular blossom-viewing trees. The distribution ratio of the top ten varieties had fallen to 51.5% in 1991, indicating a trend to diversification of the tree varieties used for roadside reforestation projects.

As typical *Ginkgo* landscapes along ordinary roads, two examples are illustrated here: one example is the configuration of a modern, symbolic landscape along the main avenue of a city, (Midosuji Avenue in Osaka) (Fig. 2), and another example is combined with a historical landscape and planted in combination with camphor trees, (Otemae Avenue in Himeji City) (Fig. 3).

2.1.3 Present State of Roadside Trees

The present state of roadside trees is discussed from the point of view of (1) roadside greenery ratio and (2) numbers and varieties of roadside trees.

2.1.3.1 Roadside Greenery Ratio

The results of calculations of the roadside greenery ratio (length of all roads lined by trees/total length of all roads), prepared as part of national road status surveys conducted by the Road Bureau of the Ministry of Construction, revealed that at the time of a 1977 survey only 2.8% of roadsides were lined by trees, but that this increased to 6.4% in 1990, indicating steady growth in this index.

2.1.3.2 Numbers and Varieties of Trees

A survey of the state of roadside trees conducted by the Public Works Research Institute of the Ministry of Construction throughout Japan was completed by March 31, 1992 [10]. In this survey, the term "roadside trees" is defined as follows: "Trees which have a height of 3m or more (includes trees less than 3m at the time of planting, but which are expected to exceed 3m in the future under care.)" The following roads were included in the survey. "Roads managed by the Ministry of Construction, prefectural governments, municipalities (cities, towns, etc.), regional road public corporations (below called 'ordinary roads'), and roads managed by other road-related public road corporations (Japan Highway Public Corporation, Tokyo Expressway Public Corporation, Hanshin Superhighway Corporation, Honshu-Shikoku Bridge Authority) (below called 'public corporation-managed roads')." The aggregation and classification of the survey results were done separately for ordinary roads and for public corporation-managed roads.

Ordinary Roads

Number of trees. Concerning the number of trees planted along ordinary roads in Japan, there were a total of 4784702 trees nationwide on March 31, 1992. This means that there are 4 trees for every kilometer of ordinary road in Japan. It also represents 39 trees/1000 people and 13 trees/km^2 of land in Japan.

Varieties of trees. Approximately 500 varieties of trees were lining ordinary roads in Japan on March 31, 1992. Figure 4 shows the numbers and distribution ratios of the top 10 tree varieties. The most common tree used throughout Japan is *Ginkgo*; 552000 *Ginkgo* trees account for 11.5% of all trees standing along ordinary roads. Various kinds of cherry stand in second place at 346000 (7.2%), followed by 305000 Japanese zelkova (6.4%), 281000 trident maple (5.9%), and 251000 *Platanus* (5.3%). The total number of the 10 most common tree varieties account for 51.5% of all trees planted along ordinary roads, indicating that more than half of all these trees are among the top 10 varieties.

Conclusion. These results show both that *Ginkgo* trees are the most common and that many varieties of trees are planted along ordinary roads. *Ginkgo* and

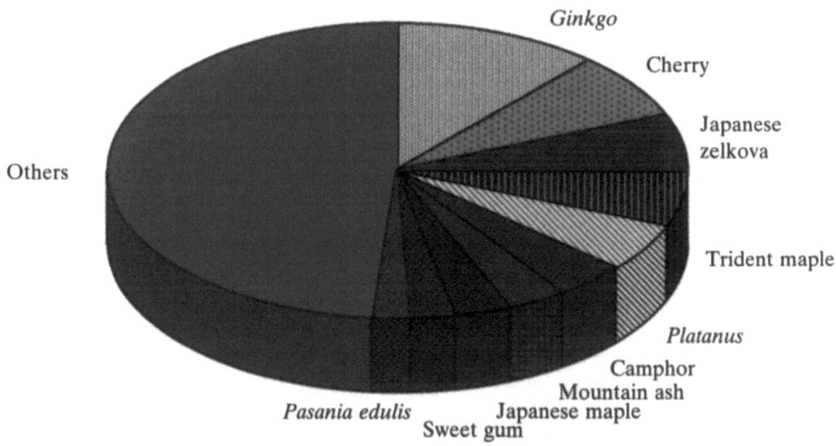

Ranking	Tree	Number	Ratio
1	*Ginkgo*	552 407	11.5
2	Cherry	345 563	7.2
3	Japanese zelkova	305 007	6.4
4	Trident maple	281 427	5.9
5	*Platanus*	251 399	5.3
6	Camphor	193 060	4.0
7	Mountain ash	162 067	3.4
8	Japanese maple	139 881	2.9
9	Sweet gum	119 852	2.5
10	*Pasania edulis*	112 422	2.3
	Others	2 321 617	48.5
	Total	4 784 702	

FIG. 4. Top ten roadside tree varieties in Japan (ordinary roads) (provided by Public Works Research Institute, Ministry of Construction)

other popular varieties are selected because they can be relied on to thrive beside roads used by motor vehicles, environments that are not good for plants and trees to grow.

An examination of the most popular varieties in different prefectures or regions of Japan showed that *Ginkgo* is the front-runner in and around the Kanto region, while cherry varieties are most popular nationwide and camphor trees are preferred south of the Kansai Region, in the Chugoku Region, and in the Kyushu Region. *Ginkgo* trees are the leader in 16 of 47 prefectures; among these 16, *Ginkgo* trees account for more than 20% of all roadside trees in Yamanashi, Yamagata, Niigata, and Gunma prefectures (Fig. 5).

The prefectures where *Ginkgo* trees are among the total number of 552 000 *Ginkgo* trees along ordinary roads include Tokyo at 11.6%, followed by Hokkaido at 9.3%, Kanagawa at 7.9%, Osaka at 7.8%, and Aichi at 7.3%.

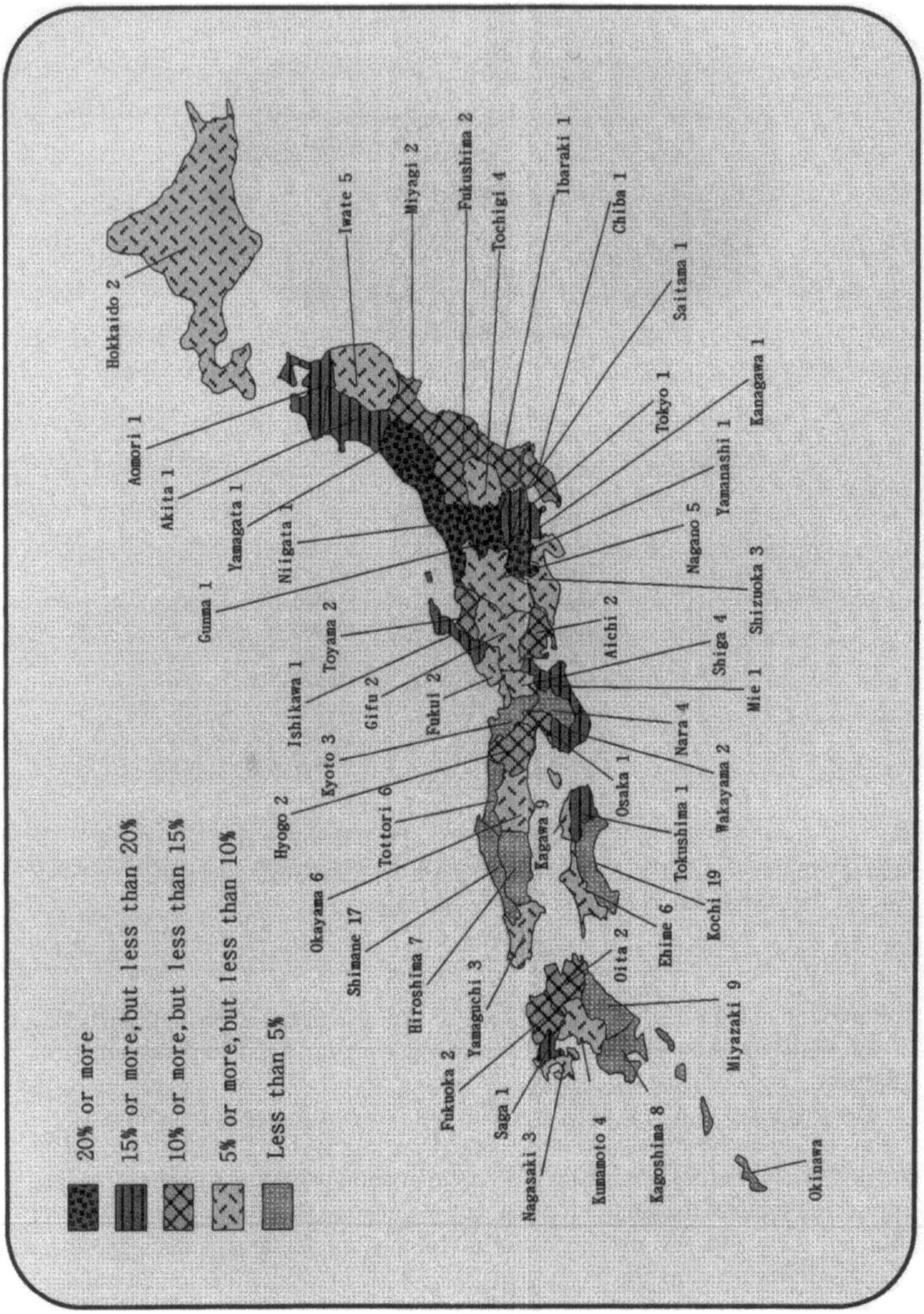
20% or more
15% or more,but less than 20%
10% or more,but less than 15%
5% or more,but less than 10%
Less than 5%
Hokkaido 2
Aomori 1
Akita 1
Yamagata 1
Gumma 1
Niigata 1
Iwate 5
Miyagi 2
Fukushima 2
Tochigi 4
Ibaraki 1
Chiba 1
Saitama 1
Tokyo 1
Kanagawa 1
Yamanashi 1
Nagano 5
Shizuoka 3
Aichi 2
Shiga 4
Mie 1
Nara 4
Wakayama 2
Osaka 1
Toyama 2
Ishikawa 1
Gifu 2
Fukui 2
Kyoto 3
Hyogo 2
Tottori 6
Okayama 6
Shimane 17
Hiroshima 7
Yamaguchi 3
Kagawa 9
Tokushima 1
Kochi 19
Ehime 6
Oita 2
Miyazaki 9
Fukuoka 2
Saga 1
Nagasaki 3
Kumamoto 4
Kagoshima 8
Okinawa

The *Ginkgo* tree is designated as the prefectural tree in Tokyo, Kanagawa, and Osaka prefectures, indicating that it is used preferentially in these prefectures.

Public Corporation-Managed Roads

Number of trees. A total of 1 732 721 trees are planted along public corporation-managed roads (March 31, 1992).

Varieties of trees. About 210 varieties of trees are planted along public corporation-managed roads throughout Japan (March 31, 1992). Figure 6 shows the numbers and distribution ratios of the top 10 tree varieties planted along these roads. Only 1844 (0.1%) *Ginkgo* trees are included. The trees most commonly used are red pine and black pine, at 307 000 (17.7%). Theses are followed by cedar at 250 000 (14.4%), cherry varieties at 145 000 (8.3%), *Picea abies* at 93 000 (5.4%), and Japanese varieties of maple at 65 000 (3.8%). The top 10 species account for 64% of all trees.

Conclusion. The foregoing results reveal that, along public corporation-managed roads, few *Ginkgo* trees are used while evergreens such as red pine, black pine, cedar, and *Picea abies*, which are rarely seen along ordinary roads, are planted in profusion. Large numbers of these evergreens are planted because it is necessary to provide a wide environment separating an expressway managed by a public corporation from surrounding residential areas, and trees along expressways have an important role to play in protecting environmental conditions in the regions they pass through by reducing traffic noise produced on the expressway.

FIG. 5. Percentage of *Ginkgo* trees in each prefecture and *Ginkgo* tree rank in each prefecture (provided by Public Works Research Institute, Ministry of Construction).
1. This map concerns the trees along ordinary roads in Japan.
2. "Percentage of *Ginkgo* trees in each prefecture" represents the percentage of *Ginkgo* trees planted along ordinary roads in each prefecture.
3. "*Ginkgo* tree rank in each prefecture" represents the rank of *Ginkgo* trees in a list of all varieties used along ordinary roads, arranged from the most common variety.
4. The following prefectures (in parentheses) belong to each region of Japan:
   a. Tohoku Region (Aomori, Iwate, Miyagi, Akita, Yamagata, and Fukushima)
   b. Kanto Region (Ibaraki, Tochigi, Gunma, Saitama, Chiba, Tokyo, Kanagawa, Yamanashi, and Nagano)
   c. Hokuriku Region (Niigata, Toyama, and Ishikawa)
   d. Chubu Region (Gifu, Shizuoka, Aichi, and Mie)
   e. Kinki Region (Shiga, Kyoto, Osaka, Hyogo, Nara, Wakayama, and Fukui)
   f. Chugoku Region (Tottori, Shimane, Okayama, Hiroshima, and Yamaguchi)
   g. Shikoku Region (Tokushima, Kagawa, Ehime, and Kochi)
   h. Kyushu Region (Fukuoka, Saga, Nagasaki, Kumamoto, Oita, Miyazaki, and Kagoshima)

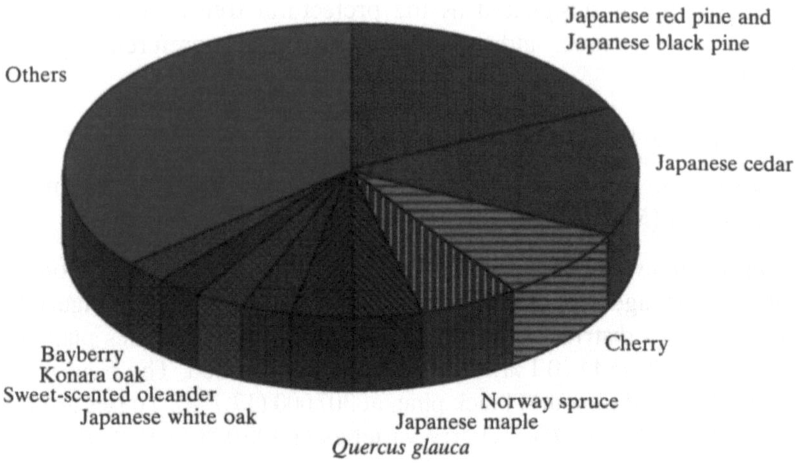

Ranking	Tree	Number	Ratio
1	Japanese red pine and Japanese black pine	307 206	17.7
2	Japanese cedar	250 151	14.4
3	Cherry	144 588	8.3
4	Norway spruce	93 316	5.4
5	Japanese maple	65 419	3.8
6	*Quercus glauca*	61 967	3.6
7	Japanese white oak	51 438	3.0
8	Sweet-scented oleander	46 865	2.7
9	Konara oak	44 421	2.6
10	Bayberry	43 457	2.5
	Others	623 893	36.0
	Total	1 732 721	

FIG. 6. Top ten roadside tree varieties in Japan (Public Cooperation Managed Roads) (provided by Public Works Research Institute, Ministry of Construction)

There is one more factor behind the selection of trees planted along expressways of the Japan Highway Public Corporation. Outside the large cities, these expressways pass through many seminatural areas such as mountain foothills, and therefore the trees planted have been selected to achieve harmony and continuity with the surrounding natural vegetation. Thus, *Ginkgo* trees are limited to service areas and similar places.

2.2 Landscapes of Alleys

Among linear landscapes of alleys with *Ginkgo* trees, there are examples applying Western landscaping methods. This section examines examples of alley landscapes based on the perspective representation method and examples of the formal garden style.

FIG. 7. *Ginkgo* tree alley at Meiji Jingu Gaien, 1927 (photo courtesy of Meiji Jingu Gaien)

FIG. 8. *Ginkgo* tree alley at Meiji Jingu Gaien, 1988 (photo courtesy of Meiji Jingu Gaien)

2.2.1 Alleys Based on the Perspective Representation Method

The alley at the Meiji Jingu Gaien in Tokyo (Figs. 7 and 8) is an example of a landscape featuring the artificial and natural beauty of a tree line configured by planting *Ginkgo* trees. The area of the Meiji Jingu Gaien was completed in 1926, while the *Ginkgo* trees of the alley were planted previously at their present location in 1923 [11].

Two special features of this alley are the method of creating space and the materials used. To create the space, the alley trees were planted along the street extending at right angles from Aoyama Street to the Meiji Memorial Gallery; to provide a perspective when the gallery is viewed from Aoyama Street with the tree lines on both sides, the trees were planted so that their heights decline from Aoyama Street to the gallery. The perspective is even more effective because the ground falls slightly from Aoyama Street toward the gallery. The alley trees, all of which convey a feeling of massive foliage, guide the viewers' line of sight sharply down the center to focus on the gallery. The four lines of trees include 146 *Ginkgo* trees; this number includes two other short tree lines that branch to the left. When they were planted, the *Ginkgo* trees were about 6m in height, but now the largest is about 24m tall with a trunk diameter of about 2.80cm; the shortest is 17m tall with a trunk diameter of about 1.80cm.

The second feature, the materials used, refers to the fact that these are *Ginkgo* trees, which were rarely used for alleys in Europe at that time. Dr. Yoshinobu Orishimo, one of the pioneers of modern landscape architecture in Japan, sowed and carefully nurtured the seedlings used to form the alley. Dr. Orishimo worked at the Imperial Household Agency, where he was in charge of the management of the Shinjuku Gyoen Garden and the Yoyogi Estate. In 1908, he obtained seeds from a *Ginkgo* tree in the Shinjuku Gyoen Garden and sowed these in a nursery in the Yoyogi Estate. Later, he was assigned responsibility for the management of the Meiji Jingu Gaien. As the design work advanced, it was necessary to decide what kind of tree to plant for the alley. While many suggestions were considered, Dr. Orishimo showed the *Ginkgo* trees to others involved in the project and finally obtained approval of their use [12]. Today, the *Ginkgo* trees are trained carefully, pruned once every 4 years to maintain their beautiful shape.

2.2.2 Alleys in the Formal Style

The next example described is a *Ginkgo* tree alley in the formal style. The National Showa Memorial Park (in Tachikawa City and Akishima City in Tokyo), a national government park covering 180 hectares, was created by the Ministry of Construction in commemoration of the fiftieth anniversary of the accession of the Emperor Showa (the Showa Period, 1926–1989). Formerly the site of the U.S. Tachikawa Base, it was flat land occupied by 600 buildings. A large project was implemented to return the land to its natural condition; the buildings were removed, ponds excavated, hills constructed, trees planted, and grasses sown to configure forests and fields.

FIG. 9. *Ginkgo* alley in National Showa Memorial Park (photo courtesy of National Showa Memorial Park Works Office)

When visitors enter the Tachikawa Gate of the National Showa Memorial Park, they are greeted by a stately and impressive sight: a water canal reaching straight down the middle of the space, sculptures and fountains at the opposite end of the canal, and lines of *Ginkgo* trees on both sides (Fig. 9). The canal is 230 m long and 108 *Ginkgo* trees form four lines on both sides. The ground around their trunks is covered with *Hedera canariensis*. The promenade surface is formed by 91 varieties of stones from 25 countries around the world. Overall, the atmosphere is tranquil. The stone materials, which consist of scraps left over after the removal of the important parts of the stones, form various shapes both large and small and provide the scene with marvelous coloring.

The *Ginkgo* trees used to form this alley formerly stood at other places inside the site, but they were moved to form the alley in 1982. At the time of this work, their trunk diameters ranged from 0.75 to 1.5 m, but now in 1996, they have grown to between 0.8 and 1.6 m. They have been trimmed constantly to keep their heights to 7 m to conform with height restrictions required by the proximity of the Self-Defense Force Air Base to the north of the site.

2.3 Cultural Change of Landscapes

As for Western landscapes, there have been almost no plans to use *Ginkgo* trees along roads or alleys, although the perspective representation method or the formal garden style exists in Western culture. The landscape architects in Japan have tried to use *Ginkgo* trees along roads or in alleys, and as a result original landscapes suited for the mentality and the climate of users have been configured. In the case of the Meiji Jingu Gaien, there is sufficient reason to conclude that Dr. Orishimo wanted to create the alley applying *Ginkgo* trees to the perspective representation method because he was also an expert on Western landscapes.

In the case of the National Showa Memorial Park, the formal style formed by using straight lines to the left and to the right makes the space very impressive. When people hear the term formal style, most think of geometric French-style formal gardens such as the garden in Versailles. Although the space with the canal and *Ginkgo* alley in the National Showa Memorial Park is stylistically similar, the atmosphere in this space is different, and as a result of the idea of using *Ginkgo* trees to form alleys, it is also original. In the summer, their leafy branches provide cooling shade, and in the autumn their yellow leaves flutter to the ground. Families with children are a common sight in this park. The landscaping throughout this space is filled with both warmth and quiet, and harmony with nature suited to the natural features of the area and the sensibilities of the inhabitants of the country. This way of combining a style (a vessel) from the West with materials (contents) from the East to devise an innovative method of creating unique landscapes is a fascinating example of the process of the change of culture and the transmission of culture in Japan.

3 Point Landscapes

3.1 Giant Ginkgo *Trees*

This section, which is focused on points or in other words, single *Ginkgo* trees, considers the status of *Ginkgo* among giant trees found throughout Japan.

3.1.1 Location of Giant *Ginkgo* Trees

The source used to clarify the location of *Ginkgo* trees throughout Japan is the report by the Environment Agency, "Giant Trees and Giant Groves in Japan" (here after called "the report") [13]. The report is a summary of the results of surveys of giant trees and giant trees in groves or in tree lines conducted by the 47 prefectures in 1988 as part of the Fourth Basic Natural Environment Conservation Survey carried out by the Environment Agency. The report mentions giant trees of various kinds, but this chapter considers only those giant trees that are *Ginkgo* (here called "giant *Ginkgo* trees").

A giant tree in the report is defined as follows: "A tree with a trunk diameter of 300 cm or more at a height of approximately 130 cm above the ground. In a case where a trunk is divided into two or more sections at a height of 130 cm above the ground, the total of the diameters of all the sections is at least 300 cm, and the diameter of the main trunk is at least 200 cm." According to the report, the share of giant *Ginkgo* trees among all Japanese giant trees is as follows. Of the total of 55 798 giant trees of all kinds in Japan, there are 4318 giant *Ginkgo* trees, which represent 7.7% of the total. They are the fourth most common after cedar, Japanese zelkova, and camphor trees.

3.1.2 Distribution of Giant *Ginkgo* Trees

The Public Works Research Institute of the Ministry of Construction has selected 47 giant *Ginkgo* trees according to certain conditions (here called "valuable giant *Ginkgo* trees"), based on the report "Giant Trees and Groves in Japan" by the Environment Agency and the book *New · 100 Valuable Trees in Japan* by the *Yomiuri Shinbun* (newspaper) [14], and has prepared a map, the "Valuable Giant *Ginkgo* Map," showing the distribution of these trees (Fig. 10).

3.1.3 Shape of Giant *Ginkgo* Trees

When *Ginkgo* trees grow naturally to immense size, some of them develop distinctive and unique shapes. (1) In the Senbon (one-thousand)-*Ginkgo*, the central trunk is surrounded by many secondary trunks, forming a single large tree shape. It is called One-thousand-*Ginkgo* because it looks like a collection of many separate trees. (2) In the Chichi (breast)-*Ginkgo*, a number of aerial roots droop down from the thick branches and trunk of the tree and appear to have grown backward. These resemble pendulous breasts and are called "breasts" or "breast columns"; *Ginkgo* trees with many of these forms are called breast *Ginkgo*, which in Japanese is "chichi ichou." They have become sacred trees because many women began to pray to them for the ability to nurse their babies. These forms appear on both male and female *Ginkgo* trees. (3) Sakasa (upside-down) *Ginkgo* are so called because their branches, especially the lower branches, sag down so that they appear to be upside down. These trees are also called pendulous. (4) The names meoto (husband-wife) *Ginkgo* and oyako (parent-child) *Ginkgo* refer to a number of *Ginkgo* trees growing close together that appear to be related as husband and wife or as parent and child. (5) Ohatsuki *Ginkgo* is a name used when female and male flowers bloom at the margins of leaves because the Japanese word "ohatsuki" means "stuck to leaves."

3.2 Single Tree

This section introduces typical examples of single *Ginkgo* trees associated with worship, legends, and episodes to analyze how people direct their feelings to *Ginkgo* trees and the significance of the *Ginkgo* landscapes.

3.2.1 Worship

There are many single *Ginkgo* trees associated with worship, such as the *Ginkgo* tree in the Jonichiji Temple (Toyama Prefecture), the Ohatsuki *Ginkgo* trees in the Hongokuji Temple and in the Jotakuji Temple (Yamanashi Prefecture), the Chichi-*Ginkgo* trees in Hiraishi (Kochi Prefecture) and in the Nyubo Shrine (Tokushima Prefecture), and so on.

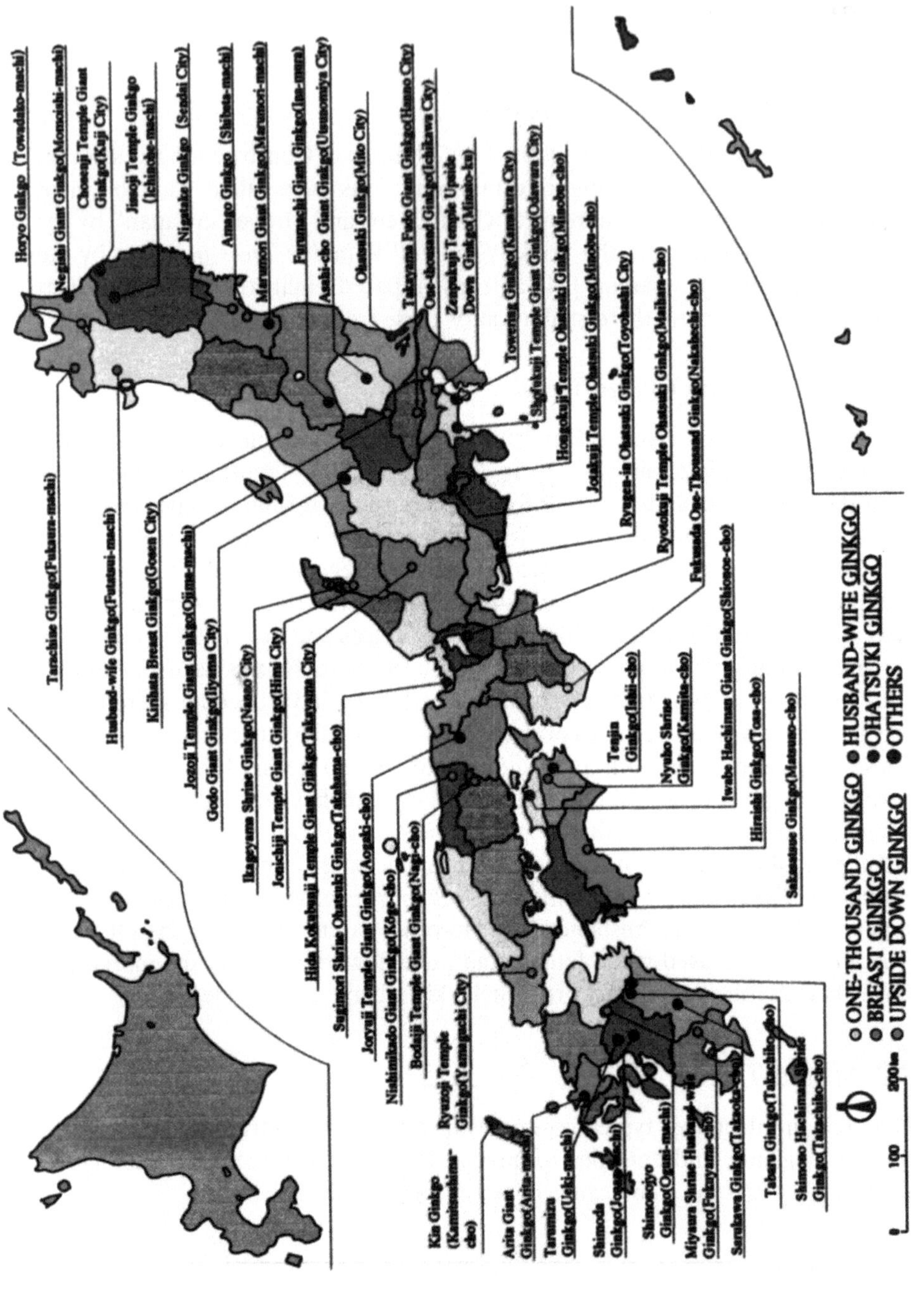
ONE-THOUSAND GINKGO
HUSBAND-WIFE GINKGO
BREAST GINKGO
OHATSUKI GINKGO
UPSIDE DOWN GINKGO
OTHERS
100
200

Among these *Ginkgo* trees, the Nigatake-*Ginkgo* (at Ichou Machi, Sendai City, Miyagi Prefecture) (Fig. 11) is an example. It is a female tree about 35 m in height, with a trunk diameter of about 2.5 m at chest height, and is said to be about 1000 years old. It has many "breast columns" (aerial roots like breasts), and the largest of these has a diameter of about 1.6 m. It is said that a woman named Byakkouni, who was a wet nurse of the Emperor Shoumu (reign, A.D. 724–749), expressed her dying wish and according to it this *Ginkgo* tree was planted on the mound where her corpse was buried. This tree is called "Chichi-*Ginkgo*" or "wet-nurse-*Ginkgo*," and it is said that the god is enshrined at the foot of this tree. Women who cannot produce their own milk often worship here.

FIG. 11. Nigatake-*Ginkgo* at Ichou Machi, Sendai City

FIG. 10. Valuable Giant *Ginkgo* Map (provided by Public Works Research Institute, Ministry of Construction). (From [13] and [14], with permission)

3.2.2 Reverence for an Old Tree

A famous old *Ginkgo* tree called "Mizufuki-*Ginkgo*" (water-spray *Ginkgo*) stands in front of the Goeidou at the Nishi Hongwanji Temple in Kyoto (Fig. 12). Its age is estimated to be between 400 and 500 years. A legend reports that as the flames of fire that swept through Kyoto in 788 threatened to spread to the Goeidou, this large *Ginkgo* sprayed a column of water saving the Goeidou from the fire. This legend is the origin of the name of this tree. It is a male *Ginkgo* tree with a height of 11.8m, a trunk circumference at the roots of 8.92m, and an overall branch width of 26m. However, it lost its former vigor; as more and more of its branches began to dry up, the size of its leaves was reduced and their density declined. A tree doctor in Kyoto, Mr. Shoji Yamada, treated this *Ginkgo* tree in hopes of restoring its vigor [15].

The treatment was carried out from January to March 1994 (Fig. 13). Before beginning the treatment, Mr. Yamada and others, who were involved in the effort, not only studied the tree externally, but also excavated the ground at its roots to survey its root system, and devised restoration measures. Their survey demonstrated that the main causes of the tree's loss of vigor were (1) rot and hollowing advancing from already rotten parts of its trunk and (2) pressure on its root system caused by trampling of the ground.

During study of the root system, trial excavations 1.2m deep and 1.0m wide were made at three locations under the tips of the branches to investigate the soil layers and the distribution of the root system. The results revealed that the soil was severely compacted by foot pressure to a depth between 10 and 15cm from the surface of the ground, a condition that caused asphyxiation of the roots, hampering the growth of the feeder roots.

The following measures were taken to help the tree recover its vigor. (1) While taking care not to harm the natural beautiful shape of the tree, dead parts of the trunk and large branches were cut off to remove the rot. Then, urethane resin was injected into the tree and putty was applied to prevent rainwater from penetrating this repair work. Finally, an antibacterial agent was applied. (2) To protect the root system from trampling pressure, a fence and curbstones were placed around the edge of the area under the branches to keep people away from the tree. (3) Inside the newly installed fence and curbstones, the ground was manually excavated to a depth of approximately 50cm, then filled with soils, a ground improvement agent, and fertilizer, to improve its condition. The excavation was performed with great care to avoid damaging the fine roots. (4) A shallow embankment was formed in the area, which was then covered with ground-cover plants to protect the root system from drying up in the summer and from freezing in the winter [16].

In the middle of April of the year in which this treatment was done, young fresh buds had already formed. The *Ginkgo* tree has recovered its vigor and appears to be in good condition. When trimming and pruning the tree, the workers are extremely careful about the locations where the cutting is done and take special pains to conserve the beautiful shape of the tree.

FIG. 12. Mizufuki-*Ginkgo* at Nishi Hongwanji Temple, Kyoto, 1996

FIG. 13. Mizufuki-*Ginkgo* at Nishi Hongwanji Temple during treatment, 1994 (photo courtesy of Shoji Yamada)

3.2.3 Prayer for Peace

During World War II, an atomic bomb was dropped on Hiroshima on August 6, 1945. In spite of the disasters, some trees have survived, and remain in relative healthy condition today in the temple named Housenbou located about 1 km away from the blast center of the atomic bombing. At that time a *Ginkgo* tree was standing close to the main building of the temple. The building was instantly destroyed by the force of the blast. In spite of the blast, this *Ginkgo* tree has survived; fresh young buds appeared soon afterward, and new branches were formed.

The temple followers were eager to rebuild the main building, but they had to decide what to do about the *Ginkgo* tree. Because the temple site was smaller after the war than before, there was not even space to move the tree. Also, it would have been dangerous to try to transplant the large tree, estimated to be 150 years old. Cutting it down was proposed, but in response to the dominant view that it should be conserved, it was decided to conserve the *Ginkgo* tree where it

FIG. 14. *Ginkgo* Tree in Housenbou, Hiroshima City

stood that day. The old main building had big slanting roofs, but if the *Ginkgo* tree was to remain standing, there would be no space left for the same design. Therefore, the new main building was designed following the style of the Tsukiji Hongwanji Temple in Tokyo, and the outlook reminds us of an Indian atmosphere [17]. Today we see the main building with the stairs in the front divided into left- and right-hand sides, shaped like an inverted letter "U" that protects the *Ginkgo* tree inside it (Fig. 14). It is an appealing design with a powerful impact on all who see it.

Other special measures taken to conserve the *Ginkgo* tree include an opening under the stone steps that allows air to flow past the tree. The design of the main building expresses the intensity of the feelings that people hold for this *Ginkgo* tree. The *Ginkgo* tree of Housenbou is living today as a precious witness of the disaster: "No more Hiroshima." It is engraved with the people's prayers for peace.

4 Conclusions

Ginkgo landscapes in Japan have been configured according to the atmosphere and the needs of the time. The *Ginkgo* trees in Japan have grown in people's surroundings in various forms. Since ancient times, they have lived in landscapes showing natural forms, as giant trees in shrines and temples. On the other hand, in concert with modernization, urban *Ginkgo* landscapes have been configured showing artificial-natural forms as roadside trees and alleys. In Europe also the use of *Ginkgo* trees along roads is beginning in earnest, for example, in Val de Marne, France [18]. Further in the future, various kinds of *Ginkgo* landscapes will be configured in various places of the world. By using *Ginkgo* trees, we will endeavor to develop diverse landscape cultures.

Acknowledgments. We wish to express our gratitude to the officials of Kanagawa and Hiroshima Prefectures, Hiroshima City, and National Showa Memorial Park, Meiji Jingu Gaien, Housenbou, Ms. Chitose Nagano, Dr. Cheng Xiaohua, Mr. Yasuo Bansho, and our colleagues for their assistance for this manuscript.

References

1. Luke D (1964) Goethe—selected verse. Penguin, Harmondsworth, Middlesex, England, p 249
2. Fujisaki K (1990) The history of roadside trees. In: The Japanese history of trees (in Japanese). Shin Jinbutsu Ohrai Sha, Tokyo, pp 58–65
3. Parks and Green Spaces Division, Construction Bureau, Tokyo Metropolitan Government (1988) Second report by the Roadside Trees Committee of Construction Bureau (in Japanese). Construction Bureau, Tokyo Metropolitan Government, Tokyo
4. Tokyo City (1939) History of roads in Tokyo (in Japanese). Tokyo City, p 455
5. Mizutani S (1939) Alleys (in Japanese). Parks and Open Space, 3:54

6. Sato A (1969) Road and landscape—alley and planting (in Japanese). City Planning Research Institute, Tokyo
7. Honma A (1968) Cities and roadside trees in our country (in Japanese). Parks and Open Space 29:1–9
8. Planting Division, Public Works Research Institute, Ministry of Construction (1986) Roadside trees in our country. Technical memorandom of Public Works Research Institute no. 2318 (in Japanese). Public Works Research Institute, Ministry of Construction, Tsukuba
9. Planting Division, Public Works Research Institute, Ministry of Construction (1988) Roadside trees in our country. II. Technical memorandom of Public Works Research Institute no. 2693 (in Japanese). Public Works Research Institute, Ministry of Construction, Tsukuba
10. Landscape and Ecology Division, Public Works Research Institute, Ministry of Construction (1994) Roadside trees in our country. III. Technical memorandom of Public Works Research Institute no. 3239 (in Japanese). Public Works Research Institute, Ministry of Construction, Tsukuba
11. Meiji Jingu Gaien (1991) Alley of *Ginkgo* trees (pamphlet in Japanese). Meiji Jingu Gaien, Tokyo
12. Ichou Kai (1992) Episodes concerning "*Ginkgo* tree alley" at Meiji Jingu Gaien, vol 1, vol 2 (*Ginkgo* News special edition 1992, no. 7, no. 8). In: Heisei *Ginkgo* One Hundred Stories (in Japanese). Ichou Kai (Japan Green Wave Ltd), Tokyo
13. Environment Agency (1991) Giant trees and giant groves in Japan. Fourth basic natural environment conservation survey. Giant trees and giant groves survey report (in Japanese). Printing Bureau, Ministry of Finance, Tokyo
14. Yomiuri Shinbun (1990) New·100 Valuable Trees in Japan (in Japanese). Yomiuri Shinbun, Tokyo
15. Yamada S (1993) The way to tree doctor. (in Japanese). Green age 20:36–40
16. Yamada S (1994) Fair of flowers and greens in Kyoto and treatment of mizufuki Ginko in Nishi Hongwanji (in Japanese). Green age 21:28–32
17. Okawa E (1995) Trees surviving the atomic bomb in Hiroshima (in Japanese). Takanoshobou, Kyoto, pp 25–29
18. Direction des Espaces Verts, Department du Val de Marne (1991) La Bievre, une Riviere du Val de Marne. Direction des Espaces Verts, Department du Val de Marne

Bibliography

Committee of Education, Miyagi Prefecture (1993) Cultural properties in Miyagi prefecture (in Japanese). Association for Conservation of Cultural Properties in Miyagi Prefecture, Sendai

Construction Committee for Reconstruction of Main Building of Housenbou (1995) Housenbou (in Japanese). Construction Committee for Reconstruction of Main Building of Housenbou, Hiroshima

Encyclopedia of Miyagi prefecture (in Japanese) (1982) Kahokushinposha, Sendai

Handa M (1995) Landscapes of roads in Japan. In: Views & visions—Conference papers IFLA. Organizing Committee, IFLA Eastern Regional Conference, Christchurch, pp 82–102

Hayashi Y, Furusato K, Nakamura T (1985) Illustrated trees in colour (in Japanese). Hokuryukan, Tokyo, p 749

Kanai T (1981) Old trees and valuable trees in Tokyo (in Japanese). In: Tokyo park libraries. Kyogakusha, Tokyo
Karizumi N (1979) Illustrations of tree roots (in Japanese). Seibundoshinkousha, Tokyo, pp 521–522
Kato M, Numata M, Watanabe K, Hata M (1995) Natural monuments in Japan (in Japanese). Kodansha, Tokyo, pp 472–487
Kitamura F, Koshimizu H, Nakamura T, Fujita N (1982) Illustrated trees in cities (in Japanese). Kodansha, Tokyo, pp 15, 17
Kitamura S, Murata G (1984) Colored illustrations of woody plants of Japan, vol II. In: Colored illustrations by Hoikusha 50 (in Japanese). Hoikusha, Tokyo, pp 454–455
Koshimizu H, Handa M, Akiyama C, Arimura T, Iizuka Y, Kozawa T, Sato T, Fujisaki K, Maeda H, Yamagishi Y, Yoshinaga Y (1992) Trees used for roads (in Japanese). Japan Highway Landscape Association, Tokyo, pp 99, 371–378
Kouyousonshou Shujin (1916) Guide of Miyagi Prefecture, 1st edn (in Japanese). Possession of Nagano family
Section for Conservation of Cultural Properties, Educational Agency, Kanagawa Prefecture (1987) 100 Valuable Trees of Kanagawa (in Japanese). Committee of Education, Kanagawa Prefecture, Yokohama

Chemistry and Clinical Application of *Ginkgo* Substances

Flavonoids of *Ginkgo biloba*

KUNIJIRO YOSHITAMA

Introduction

The leaves of *Ginkgo biloba* have been used from ancient times to protect books from blight by inserting them between pages. Because of this insecticidal use and its pharmacological usefulness, many reports of the chemical constituents in the plants have been published. These chemical investigations have led to the isolation and characterization of such unique compounds as biflavones and lactonic terpenes (ginkgolide and bilobalide).

Of the great variety of flavonoid components isolated from leaves, biflavones are the best-known compounds, and many studies of the constituents have been undertaken since 1932. Ginkgetin and isoginkgetin were the first biflavones whose structures were elucidated and confirmed by complete syntheses (Table 1).

The resolving of the structure of *Ginkgo* biflavones was summarized in detail by Baker and Ollis in 1961 [1]. They described how two Japanese scientists, S. Furukawa and K. Nakazawa, made a great contribution to the isolation of *Ginkgo* biflavones and the resolution of the structure.

Studies of the other flavonoid compounds began later. Reports of flavonol glycosides have been published since the 1960s, and those of flavones, flavan-3-ols, and proanthocyanidins a little later. Until now, there have been no reports of the presence of isoflavone and anthocyanin. Therefore, in this chapter, the biflavones, flavonols, flavones, flavan-3-ols, and proanthocyanidins so far detected in the plant are presented.

1 Biflavones

Ginkgo leaves contain a biflavone mixture of the amentoflavone type, the individual compounds differing from each other in the number and position of methoxyl groups (Table 1).

Department of Biological Science, Faculty of Science, Kumamoto University, Kumamoto 860, Japan

TABLE 1. Amentoflavone-type biflavones present in the leaves of *Ginkgo biloba*

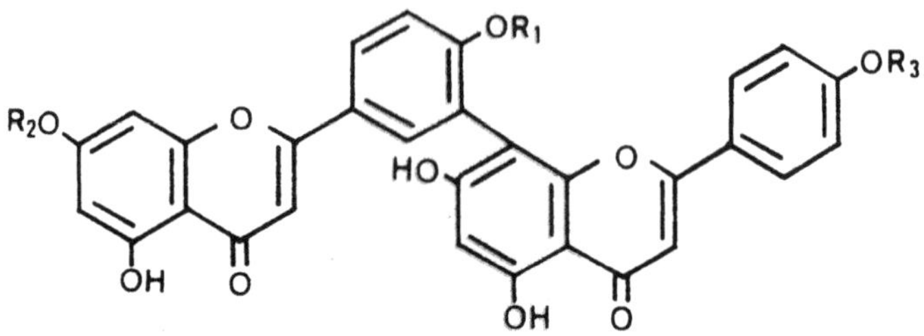

No.	Compounds	R_1	R_2	R_3
I	Amentoflavone	H	H	H
II	Bilobetin	CH_3	H	H
III	Isoginkgetin	CH_3	H	CH_3
IV	Ginkgetin	CH_3	CH_3	H
V	Sciadopitysin	CH_3	CH_3	CH_3

The flavonoid constituents of the leaves were studied for the first time by Furukawa with a view to obtaining some insecticidal compounds and some yellow coloring compounds [2, 3]. He separated five crystalline substances A–E from an alcohol extract of air-dried yellow autumnal leaves [2]. Only substance B was assumed to be a coloring compound of the flavone series, and the formula 5,8-dihydroxy-4′-methoxyflavone was given to substance B from the results of various analyses such as methylation, acetylation, alkali hydrolysis, and absorption spectrum [3] (Fig. 1). Later, Baker et al. isolated a similar compound and called the flavone-like compounds "ginkgetin." The knowledge of the structure of ginkgetin was considerably extended by the studies of Nakazawa [4, 5]. He isolated pure ginkgetin from autumnal fallen leaves through its sparingly water-soluble potassium salt, and suggested that ginkgetin was possibly derived from two molecules of apigenin (5,7,4′-trihydroxyflavone) type flavone; the molecular formula $C_{30}H_{16}O_8(OCH_3)_2$ was proposed for ginkgetin. By carrying out detailed analyses of the degradation products of the compound by alkaline hydrolyses [4, 5] (Fig. 2), he made the tentative proposal that ginkgetin contained two genkwanin (5,4′-dihydroxy-7-methoxyflavone) type flavones linked to each other by 3,8″-carbon-carbon linkage. Although Baker et al. had also actively studied the structure of ginkgetin, communication between Baker's group and Nakazawa was hindered until 1950 because of World War II. Baker et al. synthesized 5, 8-dioxy-4′-methoxyflavone deduced by Furukawa to be the flavone-like compound (substance B) in the extract of ginkgo leaves, but the synthesized compound was not identical to ginkgetin [6, 7, 8]. By examinating the oxidation, UV and IR absorptions, and other biochemical aspects, Baker et al. also proposed a new structure for ginkgetin coupling two flavone units between 3′- and 8-positions [9, 10]. The proposed structure was finally supported by Nakazawa's syntheses of ginkgetin tetramethyl ether [11] and ginkgetin [12]. In addition,

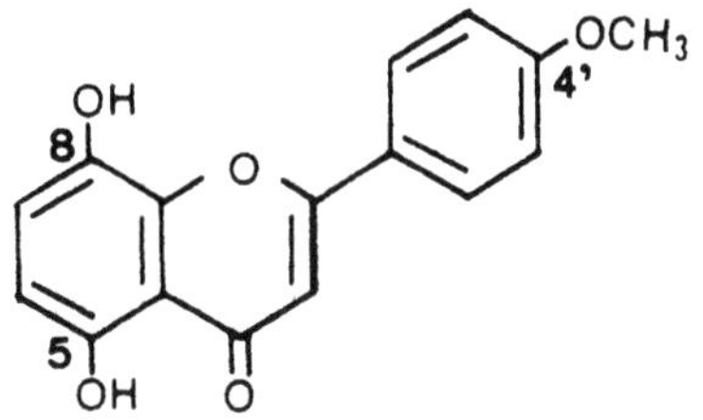

FIG. 1. Structure of "Substance B" proposed by Furukawa [3]

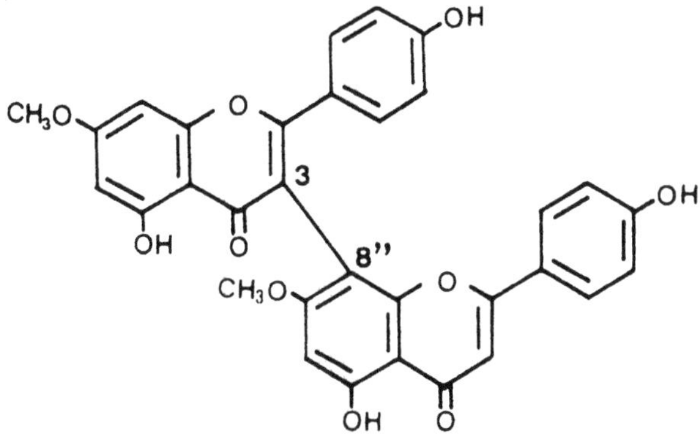

FIG. 2. Structure of ginkgetin proposed by Nakazawa [4]

Baker et al. separated two new biflavones by lengthy countercurrent distribution between ethylmethyl ketone and borate buffer. The two biflavones were characterized as isoginkgetin (III) and bilobetin (II); the content ratio from the dried leaves was 0.03%, 0.04%, and 0.016% in ginkgetin, isoginkgetin, and bilobetin, respectively [10].

In addition to these three isoflavones, sciadopitysin (V), which was identified for the first time in the leaves of *Sciadopitys verticillata* by Kariyone and Kawano [1], was also separated by Miura et al. from ginkgo leaves [13]. Moreover, a novel biflavone 5′-methoxybilobetin (VI) in the leaves was identified by Joly et al. [14] (Fig. 3).

Using high-performance liquid chromatography (HPLC), rapid and reliable methods for the identification and the quantitative determination of biflavones in ginkgo-leaf extracts were developed by Briançon-Scheid et al. [15, 16] and Pietta et al. [17, 18]. The best HPLC separation was achieved by Briançon-Scheid et al. using normal-phase chromatography with gradient elution systems [15], and a quantitative determination was also done using a similar elution system [15]. The presence of amentoflavone (I) in ginkgo-leaf extract was shown for the first time by Briançon-Scheid et al. using the HPLC system [15]. They also found seasonal variation in biflavone content with the highest levels in autumn leaves [16]. For example, the amount of sciadopitysin (V), the major biflavone in ginkgo leaves,

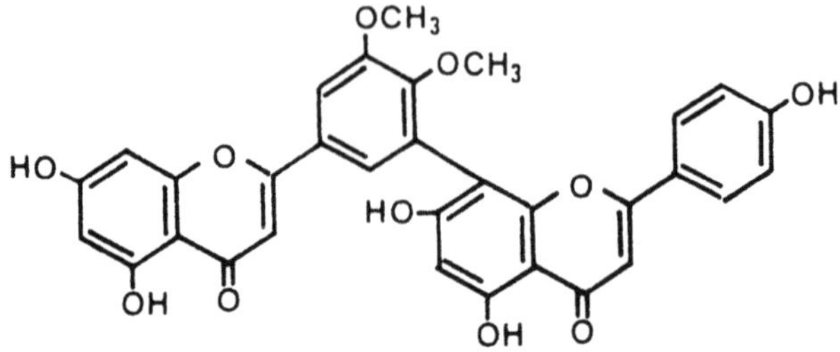

FIG. 3. 5′-Methoxybilobetin (VI) isolated from the leaves by Joly et al. [14]

was 2.9 mg and 2.5 mg per gram dry extract in spring and summer leaves, respectively; however, the amounts in green or yellow autumn leaves increased by three times (ca. 10 mg per gram dry extract) [16]. The procedure for isolation of biflavones was also described by Xiao et al. [19]. Recently, the presence of biflavones in the testa was described by Pan et al. [20].

2 Flavonol Glycosides

Flavonol glycosides are also major compounds in ginkgo leaves. However, studies of flavonol constituents are more recent because interest in ginkgo flavonoid constituents had been mostly directed toward biflavones until the 1960s. From the latter half of the 1980s, more complex acylflavonol glycosides have been identified using HPLC equipped on-line photodiode-array detectors and nuclear magnetic resonance (NMR)- and mass-spectrometry. Of more than 25 flavonol glycosides identified, kaempferol, quercetin, and isorhamnetin glycosides are the main ones.

In simple flavonol glycosides, kaempferol 3-*O*-rutinoside (VII), quercetin 3-*O*-rutinoside (rutin, VIII), isorhamnetin 3-*O*-rutinoside (IX) and 3′-*O*-methylmyricetin 3-*O*-rutinoside (X) were isolated from autum leaves [21, 22] (Table 2). In addition to (VII) ~ (X), kaempferol 7-*O*-glucoside (XI) was identified in the leaves; quercetin 3-*O*-rhamnoside (XII) and quercetin 3-*O*-glucoside (XIII) were isolated as minor components [23] (Table 2).

Moreover, unusual acylated flavonol glycosides, kaempferol/quercetin 3-*O*-α-(6‴-*p*-coumaroylglucosyl-β-1,4-rhamnoside) (XV and XVI) were isolated by Nasr et al. from the leaves [24–26] (Fig. 4). The two glycosides are unique because there is a sequence of sugar in which rhamnose, not glucose, is directly linked to the C-3 hydroxyl of the aglycone. Subsequently, kaempferol 3-*O*-β-glucoside (astragalin, XIV) and two novel flavonol triglycosides, kaempferol/quercetin 3-*O*-[α-rhamnosyl-(1 → 2)-α-rhamnosyl-(1 → 6)]-β-glucoside (XVII and XVIII) were isolated by Vanhaelen and Vanhaelen-Fastré [27] (Fig. 5); the content ratio per dry leaf of the both triglucosides was 0.07% [27].

Recently, five new flavonol glycosides were isolated from the leaves by Hasler et al., and their structures were determined by spectral and hydrolysis experi-

TABLE 2. Flavonol glycosides present in the leaves of *Ginkgo biloba*

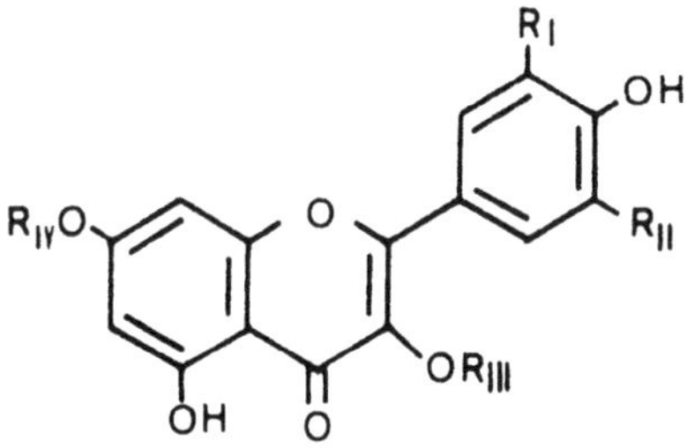

Flavonol glycosides	R_I	R_{II}	R_{III}	R_{IV}
VII	H	H	rutinose	H
VIII	OH	H	rutinose	H
IX	OCH_3	H	rutinose	H
X	OCH_3	OH	rutinose	H
XI	H	H	H	glucose
XII	OH	H	rhamnose	H
XIII	OH	H	glucose	H
XIV	H	H	glucose	H

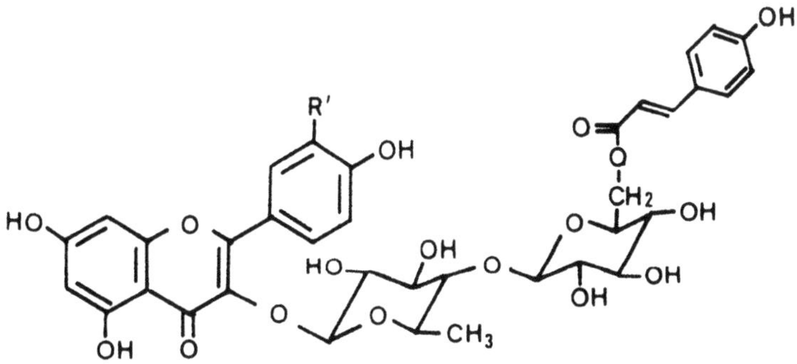

FIG. 4. Acylflavonol diglycosides identified in the leaves by Nasr et al. [24–26]. XV, R′ = H; XVI, R′ = OH

ments to be kaempferol/quercetin 3-[2″-glucosyl] rhamnoside (XIX and XX), kaempferol/quercetin 3-[2″-6‴-{*p*-(7⁗-glucosyl) coumaroyl} glucosyl] rhamnoside (XXI and XXII), and quercetin 3-[2″-(6‴-*p*-coumaroyl) glucosyl] rhamnosyl-7-glucoside (XXIII) [28] (Table 3).

For the isolation and the purification of seven predominant flavonol glycosides from ginkgo leaves, a procedure that combined countercurrent chromatography (CCC) and HPLC was developed by Vanhaelen and Vanhaelen-Fastré [29]. Verotta et al. developed an improved method for the separation of two acylflavonol glycosides using gel filtration and droplet CCC chromatography [30].

TABLE 3. Five flavonol glycosides isolated from the leaves by Hasler et al. [28]

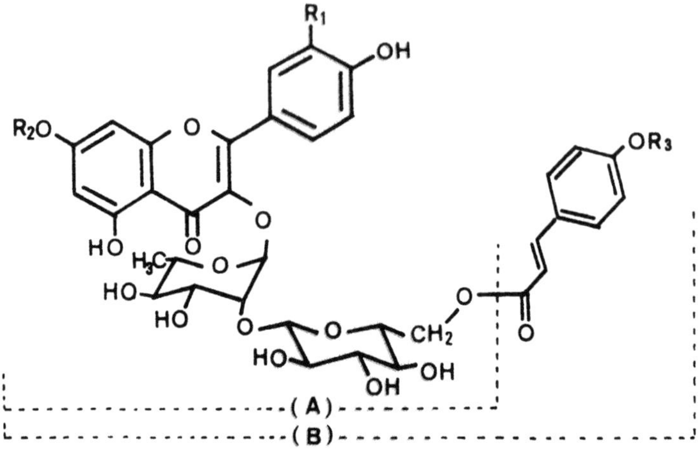

	Flavonol compounds	R_1	R_2	R_3
(A)	XIX	H	H	—
	XX	OH	H	—
(B)	XXI	H	H	glucose
	XXII	OH	H	glucose
	XXIII	OH	glucose	H

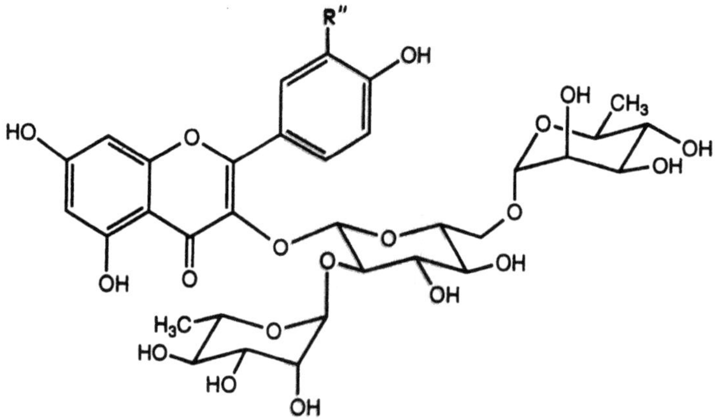

FIG. 5. Flavonol triglycosides isolated from the leaves by Vanhaelen and Vanhaelen-Fastré [27]. XVII, R″ = H; XVIII, R″ = OH

The separation by HPLC of flavonol aglycones after hydrolysis of ginkgo extract was also carried out by Hasler et al. [31].

Lobstein et al. chose a successful and reproducible separation system of HPLC, and used it to analyze the seasonal variations of flavonoid content in the leaves

[32]. Their results showed that apart from acyltypes the highest levels of total flavonol glycosides and acylflavonol glycosides were observed in spring leaves (1221 mg per 100 g dried leaves in April) and in buds (666 mg per 100 g dried leaves in March), respectively. The levels of total flavonol glycosides increased conspicuously from March to April and gradually decreased from April to October, following by an increase of the levels in fallen leaves; the levels of total acylflavonol glycosides gradually decreased from March to November [32]. Although biflavone was not found in buds, the levels in the leaves gradually increased from spring to autumn. The highest levels of total biflavones were observed in fallen leaves in November (295 mg per 100 g dried leaves) [32].

Good separation of ginkgo flavonoids by HPLC equipped with a diode-array UV detector was also carried out by Pietta using a different eluent [33]. Recently, some flavonol glycosides in the leaves were analyzed by micellar electrokinetic capillary chromatography [34] and thermospray liquid chromatography mass spectrometry [35]. The method of fingerprint analysis by HPLC was elaborated by Hasler et al. [36]. The method was satisfactory for the determination of great varieties of genuine flavonoids in ginkgo leaves and for the standardization of the leaf extract as a phytomedicine [36]. Their results showed that the dried leaves used commercially contained a flavonol aglycone content of 0.2%–0.4% (w/w), corresponding to a calculated total flavonol glycoside content of 0.5%–1% (w/w), and also showed that kaempferol was the main component in the leaves after hydrolysis, followed by quercetin and isorhamnetin. The highest level of flavonol aglycones after hydrolysis was observed in green leaves collected in May, the results being identical to those described by Lobstein et al. [32]; the level of total flavonol aglycones was nearly constant from June to November [36]. By the fingerprint procedure, the 33 flavonoids separated within 30 min in the dry extract of the leaves were identified by comparing with reference compounds [36].

3 Flavones

Monomeric flavones are minor leaf components compared with biflavones. Recently, apigenin (XXIV), luteolin (XXV), apigenin 7-β-D-glucoside (XXVI) and luteolin 3′-β-D-glucoside (XXVII) have been detected by a fingerprint HPLC separation method (Fig. 6) [36].

4 Flavan-3-ols and Proanthocyanidins

Catechins and oligomeric proanthocyanidins (Fig. 7) were identified by Schrall and Becker in tissue and suspension culture of *Ginkgo biloba* [37].

Proanthocyanidins and flavan-3-ols (Fig. 8) in *Ginkgo* leaves and the leaf-derived callus and cell suspension cultures have been also isolated and analyzed

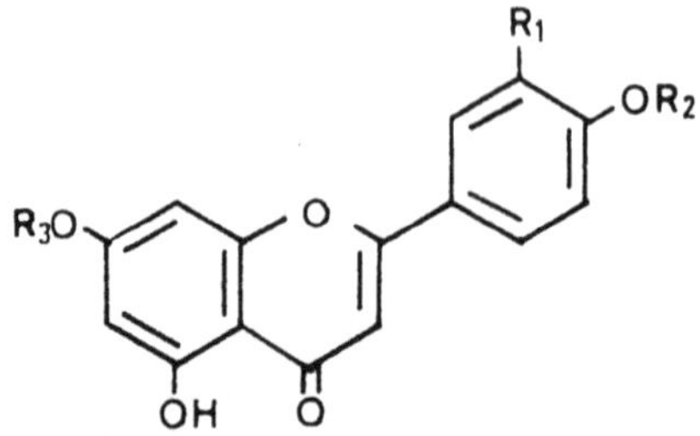

FIG. 6. Flavones detected in the leaves by Hasler et al. [36]. XXIV (apigenin), R_1 = H, R_2 = H, R_3 = H; XXV (luteolin), R_1 = OH, R_2 = H, R_3 = H; XXVI, R_1 = H, R_2 = H, R_3 = glucose; XXVII, R_1 = OH, R_2 = glucose, R_3 = H

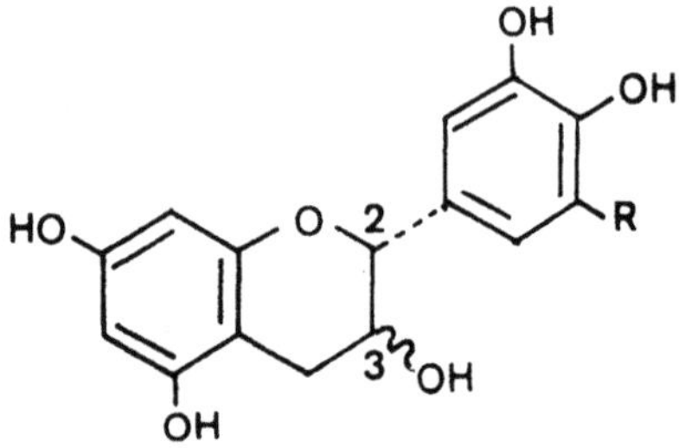

FIG. 7. Flavan-3-ols detected in the leaves and callus and in suspension cultures derived from ginkgo leaves [37, 38]. Catechin (XXVIII), 2,3-trans, R = H; gallocatechin (XXIX), 2,3-trans, R = OH; epicatechin, 2,3-cis, R = H; epigallocatechin, 2,3-cis, R = OH

chromatographically by Stafford et al. [38]. They showed that the cultures derived from *Ginkgo* leaves produced a much greater amount of proanthocyanidin than the intact leaves. The amount of total proanthocyanidins in the intact leaves was about 70 μg per mg dry weight, but the amount increased to about 529 μg and 565 μg per mg dry weight in callus and cell suspension cultures derived from ginkgo leaves, respectively [38]. In the cultures, the amounts of flavan-3-ols and dimeric proanthocyanidins increased: the amounts of catechin (XXVIII) and gallocatechin (XXIX) of flavan-3-ols increased 20–80 times compared with those of the intact leaves [38].

The amount of prodelphinidin relative to procyanidin was much less in both callus and cell suspension cultures. Although ginkgo leaves contained major amounts of flavan-3-ols and dimers with the 2,3-*cis*-stereo, the culture tends to synthesize 2,3-*trans*-isomers instead [38].

Stafford and Lester studied the biosynthesis of flavan-3-ol using the callus or suspension cultures from petioles of *G. biloba*, and showed the conversion of (+)-dihydromyricetin to (+)-gallocatechin via a leucodelphinidin intermediate by nicotinamide adenine dinucleotide phosphate, reduced (NADPH) dependent reductases [39]. They also studied microscopically the distribution patterns and intracellular localization of proanthocyanidins in cell suspension culture derived

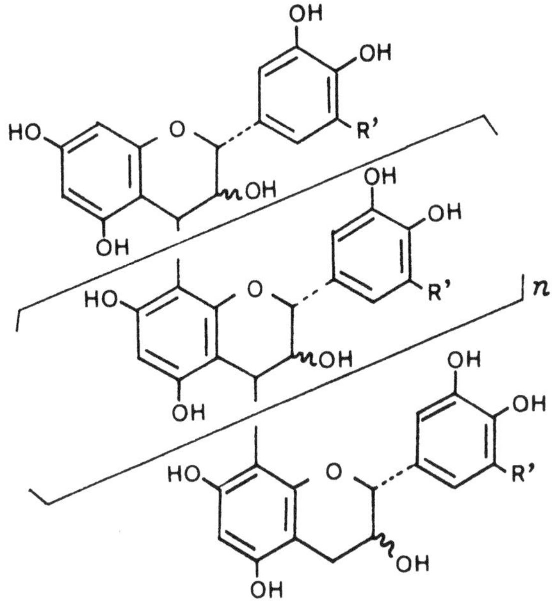

FIG. 8. Proanthocyanidin olygomers detected in the leaves and callus and in suspension cultures derived from *Ginkgo* leaves [37, 38]. R′ = H, procyanidin; R′ = OH, prodelphinidin. n = 0 ~ 8

TABLE 4. List of the flavonoids in the leaves of *Ginkgo biloba*

No.	Flavonoid class	Flavonoids identified		CN	Ref.
1	Biflavones	Amentoflavone		I	[1]
2		Bilobetin		II	[1]
3		Isoginkgetin		III	[1]
4		Ginkgetin		IV	[1]
5		Sciadopitysin		V	[13]
6		5′-Methoxybilobetin		VI	[14]
	Flavonol	[Kaempferol type]			
7	glycosides	Kaempferol		—	[36]
8		Kaempferol 3-*O*-glucoside	(astragalin[a])	XIV	[29]
9		Kaempferol 3-*O*-rhamnoside	(afzelin[a])	—	[36]
10		Kaempferol 3-*O*-rutinoside		VII	[23]
11		Kaempferol 3-[2″-glucosyl]rhamnoside		XIX	[28]
12		Kaempferol 3-*O*-α-(6′″-*p*-coumaroylglucosyl-β-1,4-rhamnoside)		XV	[24]
13		Kaempferol 3-[2″-6‴ {*p*-7‴-glucosyl)coumaroyl}-glucosyl]rhamnoside		XXI	[28]
14		Kaempferol 3-*O*-[α-rhamnosyl-(1→2)-α-rhamnosyl(1→6)]-β-glucoside		XVII	[27]
15		Kaempferol 3-*O*-α-(6″-*p*-coumarylglucosyl)-(1→2)-rhamnoside		—	[36]
16		Kaempferol 7-*O*-glucoside	(populnin[a])	XI	[23]
		[Quercet in type]			
17		Quercetin		—	[36]

TABLE 4. *Continued*

No.	Flavonoid class	Flavonoids identified		CN	Ref.
18		Quercetin 3-*O*-glucoside	(isoquercitrin[a])	XIII	[23]
19		Quercetin 3-*O*-rhamnoside	(quercitrin[a])	XII	[23]
20		Quercetin 3-*O*-rutinoside	(rutin[a])	VIII	[21]
21		Quercetin 3-[2″-glucosyl]rhamnoside		XX	[28]
22		Quercetin 3-*O*-[α-rhamnosyl-(1→2)-α-rhamnosyl (1→6)]-β-glucoside		XVIII	[27]
23		Quercetin 3-*O*-α-(6′″-*p*-coumaroylglucosyl-β-1,4-rhamnoside)		XVI	[25]
24		Quercetin 3-[2″-6‴-{*p*-(7‴-glucosyl)coumaroyl}-glucosyl]rhamnoside		XXII	[28]
25		Quercetin 3-*O*-α-(6″-*p*-coumarylglucosyl)-(1→2)-rhamnoside-7-glucoside		XXIII	[28]
26		Quercetin 3-*O*-α-(6″-*p*-coumarylglucosyl)-(1→2)-rhamnoside		—	[36]
		[Isorhamnetin type]			
27		Isorhamnetin		—	[36]
28		Isorhamnetin 3-*O*-glucoside		—	[36]
29		Isorhamnetin 3-*O*-rhamnoside		—	[36]
30		Isorhamnetin 3-*O*-rutinoside	(narcissin[a])	IX	[22]
31		Isorhamnetin 3-rhamnosyl-(1→2)-rhamnosyl(1→6)-glucoside		—	[36]
		[Myricetin type]			
32		Myricetin		—	[22]
33		Myricetin 3-*O*-rutinoside		—	[36]
		[3′-*O*-Methymyricetin]			
34		3′-*O*-Methymyricetin 3-*O*-rutinoside		X	[22]
35	Flavone	Apigenin		XXIV	[36]
36	glycoside	Apigenin 7-*O*-glucoside		XXVI	[36]
37		Luteolin		XXV	[36]
38		Luteolin 3′-*O*-glucoside		XXVII	[36]
39	Flavan-3-ol	Catechin		XXVIII	[37]
40	and proantho-	Epicatechin		—	[38]
41	cyanidin	Gallocatchin		XXIX	[38]
42		Epigallocatechin		—	[38]
43		Procyanidin		—	[37]
44		Prodelphinidin		—	[38]
45		Epigallocatechin-catechin		—	[38]
46	Flavanonol	Dihydromyricetin		—	[38]

CN, Flavonoid identifiers in the text.
[a]Popular names.

from ginkgo leaves after staining with the nitroso reagent, and showed the products stained with the reagent plus NaOH were clearly localized either in one or more vesicles of varying sizes in presumably young cells near the end of chains of cells, or in a large central vacuole of mature cells [40].

5 Conclusion

More than 40 flavonoids have been identified in the leaves of *Ginkgo biloba* L., listed in Table 4. Flavonol glycosides and biflavones are major constituents in the flavonoids although flavones, flavan-3-ols, and proanthocyanidins are also found as minor components. It is noted that the content of each flavonoid varies with season; and various biological functions of flavonoids such as UV-protection, radical scavenging, antimicrobial activities, and insecticidal properties have been identified. It is necessary to investigate further the geographical variation in flavonoid constituents, from chemosystematical and pharmacological points of view. There are few papers on the biosyntheses of flavonoids in *Ginkgo* leaves. *Ginkgo* leaves seem to be good material for studies of flavonol glycosylation and acylation, since unusual various acylated flavonol glycosides are contained in the leaves.

References

1. Baker W, Ollis WD (1961) Biflavonyls. In: Ollis WD (ed) Recent developments in the chemistry of natural phenolic compounds. Pergamon, New York pp 152–184
2. Furukawa S (1932) Studies on the constituents of "*Ginkgo biloba* L." leaves. Part. I. Sci Papers Ins Phys Chem Res (Tokyo) 19:27–38
3. Furukawa S (1933) Studies on the constituents of "*Ginkgo biloba* L." leaves. Part. IV. Sci Papers Ins Phys Chem Res (Tokyo) 21:278–285
4. Nakazawa K (1941) A study of the structure of flavone Ginkgetin in the leaves from *Ginkgo biloba* (in Japanese). J Pharm Soc Jpn 61:174–184
5. Nakazawa K (1941) An alkaline degradation of Ginkgetin (in Japanese). J Pharm Soc Jpn 61:228–229
6. Baker W, Simmonds WHC (1940) Derivatives of 5:6:4′- and 5:8:4′-trihydroxyflavones, and a note on the structure of ginkgetin. J Chem Soc 1370–1374
7. Baker W, Flemons GF (1948) The structure of ginkgetin. Part I. Synthesis of 5:8-dihydroxy-4′-methoxyflavone. J Chem Soc 2138–2143
8. Baker W, Flemons GF, Winter R (1949) The structure of ginkgetin. Part II. New syntheses of 5:8-dihydroxy-4′-methoxyflavone. J Chem Soc 1560–1562
9. Baker W, Finch ACM, Ollis WD, Robinson KW (1959) Biflavonyls, a new class of natural product. The structures of ginkgetin, isoginkgetin and sciadopitysin. Proc Chem Soc 91–92
10. Baker W, Finch ACM, Ollis WD, Robinson KW (1963) The structures of the naturally occurring biflavonyls. J Chem Soc 1477–1490
11. Nakazawa K (1959) Synthesis of ginkgetin tetramethyl ether. Chem Pharm Bull 7:748–749
12. Nakazawa K, Ito M (1963) Syntheses of ring-substituted flavonoids and allied compounds. X. Synthesis of ginkgetin. Chem Pharm Bull 11:283–288
13. Miura H, Kihara T, Kawano N (1969) Studies on biflavones in the leaves of *Podocarpus macrophylla* and *P.nagi*. Chem Pharm Bull 17:150–154
14. Joly M, Haag-Berrurier M, Anton R (1980) La 5′-méthoxybilobétine, une biflavone extraite du *Ginkgo biloba*. Phytochemistry 19:1999–2002

15. Briançon-Scheid F, Guth A, Anton R (1982) High-performance liquid chromatography of biflavones from *Ginkgo biloba* L. J Chromatogr 245:261–267
16. Briançon-Scheid F, Lobstein-Guth A, Anton R (1983) HPLC separation and quantitative determination of biflavones in leaves from *Ginkgo biloba*. Planta Medica 49:204–207
17. Pietta PG, Mauri PL, Rava A (1988) Reversed-phase high-performance liquid chromatographic method for the analysis of biflavones in *Ginkgo biloba* L. extracts. J Chromatogr 437:453–456
18. Pietta PG, Mauri PL, Manera E, Ceva PL, Rava A (1989) An improved HPLC determination of flavonoids in medicinal plant extracts. Chromatographia 27:509–512
19. Xiao SC, Wu YZ, Wen MZ, Ren WJ, Chen PQ (1990) Preparative process of flavones from *Ginkgo bioba* L. leaves. Chin J Pharm 21:340–341
20. Pan JX, Zhang HY, Yang XB (1995) Biflavones from the testa of *Ginkgo biloba* L. (in Chinese). J Plant Resources and Environment 4:17–21
21. Geiger H, Beckmann S (1965) Über das Vorkommen von Rutin und Kämpferol-3-rhamnoglucosid in *Ginkgo biloba* L. Z Naturforsch 20b:1139–1140
22. Geiger H (1979) 3′-O-Methylmyricetin-3-rhamnoglucosid, ein neues Flavonoid aus dem Herbstlaub von *Ginkgo biloba* L. Z Naturforsch 34c:878–879
23. Victoire C, Haag-Berrurier M, Lobstein-Guth A, Balz JP, Anton R (1988) Isolation of flavonol glycosides from *Ginkgo biloba* leaves. Planta Medica 54:245–247
24. Nasr C, Haag-Berrurier M, Lobstein-Guth A, Anton R (1986) Kaempferol coumaroyl glucorhamnoside from *Ginkgo biloba*. Phytochemistry 25:770–771
25. Nasr C, Lobstein-Guth A, Haag-Berrurier M, Anton R (1987) Quercetin coumaroyl glucorhamnoside from *Ginkgo biloba*. Phytochemistry 26:2869–2870
26. Nasr CG (1987) "Eude chimique de substances originales du *Ginkgo biloba* L. (Ginkgoacées); dérivés flavoniques (hétérosides-esters) et quinoléins carboxyliques; essais pharmacologiques préliminaires.", Thése de Doctorat, Université Louis Pasteur, Strasbourg
27. Vanhaelen M, Vanhaelen-Fastré R (1989) Flavonol triglycosides from *Ginkgo biloba*. Planta Medica 55:202
28. Hasler A, Gross GA, Meier B, Sticher O (1992) Complex flavonol glycosides from the leaves of *Ginkgo biloba*. Phytochemistry 31:1391–1394
29. Vanhaelen M, Vanhaelen-Fastré R (1988) Countercurrent chromatography for isolation of flavonol glycosides from *Ginkgo biloba* leaves. J Liq Chromatogr 11:2969–2975
30. Verotta L, Lolla E, Moggi A (1991) Improvement in the separation of two *Ginkgo biloba* coumaroyl flavonoids. Fitoterapia 62:339–341
31. Hasler A, Sticher O, Meier B (1990) High-performance liquid chromatographic determination of five widespread flavonoid aglycones. J Chromatogr 508:236–240
32. Lobstein A, Rietsch-Jako L, Haag-Berrurier M, Anton R (1991) Seasonal variations of the flavonoid content from *Ginkgo biloba* leaves. Planta Medica 57:430–433
33. Pietta PG, Mauri PL, Bruno A, Rava A, Manera E, Ceva P (1991) Identification of flavonoids from *Ginkgo biloba* L., *Anthemis nobilis* L. and *Equisetum arvense* L. by high-performance liquid chromatography with diode-array UV detector. J Chromatogr 553:223–231
34. Pietta PG, Mauri PL, Rava A, Sabbatini G (1991) Application of micellar electrokinetic capillary chromatography to the determination of flavonoid drugs. J Chromatogr 549:367–373

35. Pietta P, Facino RM, Carini M, Mauri P (1994) Thermospray liquid chromatography-mass spectrometry of flavonol glycosides from medicinal plants. J Chromatogr 661:121–126
36. Hasler A, Sticher O, Meier B (1992) Identification and determination of the flavonoids from *Ginkgo biloba* by high-performance liquid chromatography. J Chromatogr 605:41–48
37. Schrall R, Becker H (1977) Produktion von Catechinen und oligomeren Proanthocyanidinen in Callus- und Suspensionkulturen von *Crataegus monogyna*, *C. oxyacantha* und *Ginkgo biloba*. Planta Medica 32:297–307
38. Stafford HA, Kreitlow KS, Lester HH (1986) Comparison of proanthocyanidins and related compounds in leaves and leaf-derived cell cultures of *Ginkgo biloba* L., *Pseudotsuga menziesii* Franco, and *Ribes sanguineum* Pursh. Plant Physiol 82:1132–1138
39. Stafford HA, Lester HH (1985) The conversion of (+)-dihydromyricetin to its flavan-3,4-diol (leucodelphinidin) and to (+)-gallocatechin by reductases extracted from tissue cultures of *Ginkgo biloba* and *Pseudotsuga menziesii*. Plant Physiol 78:791–794
40. Stafford HA, Lester HH, Weider RM (1987) Histochemical assay of proanthocyanidin heterogeneity in cell culture. Plant Science 52:99–104

Allergic Substances of *Ginkgo biloba*

Naohisa Kochibe

1 Introduction

Contact dermatitis is an inflammatory reaction of the skin, caused by foreign irritants (antigens) which are harmless on first contact but able to elicit specific effector T cells in the individual. If sensitization has been established, skin lesions develop several hours after contact with the specific material, and the reaction reaches a maximum in a few days (delayed hypersensitivity).

The causative agents (allergens) for contact dermatitis are many and varied. They include metals, relatively simple compounds, and some plant metabolites. The seed coat of *Ginkgo biloba* contains allergic substances, which have been isolated and characterized as presented in this article.

2 Structure of Seed Coat

The testa of the mature seed of *G. biloba* is made up of two parts, the outer seed coat, episperm, and the inner, endopleura. The episperm has two distinctive layers composed of characteristic cells. The outer layer is surrounded by epidermis and consists primarily of parenchymatous cells that accumulate many kinds of chemical constituents during ripening. The inner layer is made up of sclerenchymatous cells forming a hard shell. The endopleura is composed of layers of parenchymatous cells that were evacuated at the time of abscission and eventually become a thin film-like layer.

3 Chemical Constituents of Episperm

Some of the authors of the literature cited in this chapter refer to the seed of *Ginkgo* as a fruit and to the outer layer of episperm as the fruit pulp. This confusion is presumably because the appearance of the episperm resembles that

---

Department of Biology, Faculty of Education, Gunma University, 4-2 Aramakimachi, Maebashi, Gunma 371, Japan

of the sarcocarp in some angiosperms. We refer to the fruit and fruit pulp as the seed and outer layer of episperm, respectively.

Since the endosperm is edible and rich in nutrients, it is possible to assume that, where the plant has grown or been cultivated from ancient times, people would have been in contact with irritant and allergic substances in the episperm. However, we have been unable to find literature about the nature of causative substances and of dermatitis in humans before 1928 and 1929, respectively.

Kawamura [1] isolated and characterized three novel substances from the episperm. He extracted these using ether in the presence of aqueous alkali and carried out sequential fractionation consisting of solubility change after neutralization, insolubilization of an acid by lead salt formation, and ethanol extraction from the mother liquid after removal of ether. The acidic substance was named ginkgolic acid, and its elemental analysis gave a formula $C_{21}H_{32}O_3$, which liberated CO_2 after heating to 240°C to give $C_{20}H_{32}O$ (or $C_{20}H_{31}OH$), named ginkgol. Ginkgolic acid was hydrogenated using platinum as a catalyst to hydroginkgolic acid, which also liberated CO_2 on heating and eventually gave hydroginkgol, $C_{20}H_{33}OH$. From these experimental results and other organochemical findings, ginkgolic acid was assumed to be a phenolic acid with a pentadecenyl side chain. From an ethanol extract of the mother liquid, a phenolic substance, bilobol, was isolated and hydrobilobol was obtained after hydrogenation of bilobol; the formulae were $C_{21}H_{32}(OH)_2$ and $C_{21}H_{34}(OH)_2$, respectively. The ether layer left after extraction with neutralized aqueous solution contained an aliphatic alcohol, ginnol, which was crystallized and analyzed as having the formula $C_{27}H_{55}OH$.

A series of structural investigations was undertaken on some of these substances by Furukawa [2–5]. He determined a structure of a pentadecenyl side chain in ginkgol and bilobol, substituting positions of $-COOH$ and $-OH$ residues on the benzene ring in ginkgolic acid, and the position of two $-OH$ residues in bilobol molecule. The proposed structures of these substances are shown in Fig. 1. Instrumental analyses of these materials confirmed the structures and determined the configuration of a double bond in the pentadecenyl chain to be *cis* form [6, 7]. In a systematic isolation using silica gel or aluminium oxide column chromatography and then HPLC on an ODS column, the yields from 100 g of raw episperm were 1.4 g of ginkgolic acid, 0.31 g of bilobol, and 0.045 g of ginkgol [7].

An acidic substance in an ether extract of *Ginkgo* episperm was isolated and analyzed to be a fatty acid named ginkgoic acid, $C_{24}H_{48}O_2$ [8]. The nature of this substance, its identity with lignoceric (tetracosanoic) acid that has the same formula, and its effect on humans have not been further studied.

The biosyntheses of long-chained phenols and phenolic acids in *Ginkgo* seed are partially understood. Gellerman et al. reported that the salicylic moiety of ginkgolic acid was synthesized via a polyketide pathway [9], and noted the presence of hydroxyl ginkgolic acid in immature seed as a postulated precursor of bilobol [10]. Production of these allergenic substances seems to be limited to the

a

b

c

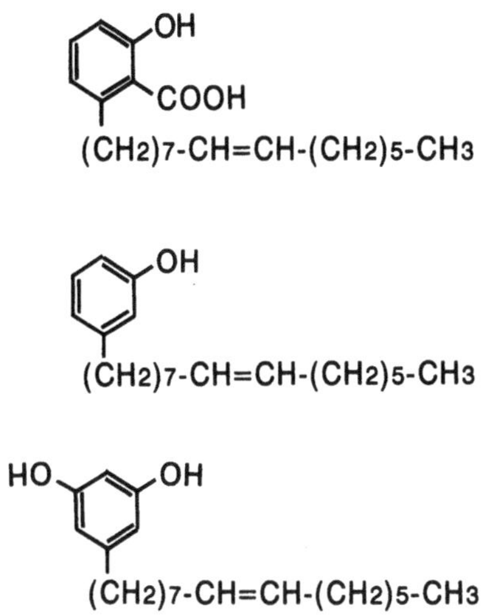

FIG. 1a–c. Major allergenic substances of *Ginkgo* episperm. **a**, ginkgolic acid; **b**, ginkgol; **c**, bilobol

seed coat. In the phenolic components extracted from the leaves, bilobol was completely absent when analyzed by gas chromatography and mass spectrometry [11].

4 Allergic Reaction Caused by Constituents of Episperm

Experimental dermatitis caused in guinea pigs by *Ginkgo* extract appeared in the abstract of an oral presentation by Tabayashi in 1927 [12]. Clinical description of contact dermatitis first appeared in 1929. Saito [13] reported four cases of contact dermatitis and experimented with skin reactions in rabbits and guinea pigs using a crude extract together with some purified materials. As a result of searching for causative substances among fractions from an ethanol extract and purified specimens supplied by Kawamura [1], Saito described the strength of reactivity as ginkgol > methylginkgol (a methylated derivative of ginkgol) > ginkgolic acid > hydroginkgol, hydroginkgolic acid, and wrote that bilobol might be the strongest substance, although hydrobilobol was less active. Some histological examinations of humans, rabbits, and apes also appeared in his report. Pharmacological effects in rabbits and frogs were studied by the same author after administering a mixture of ginkgolic acid and bilobol [14]. Afterwards, many case reports of dermatitis [15–19] and a case of contact reaction in mucous membranes after swallowing [20] appeared in the United States, Europe, and Japan.

Allergic contact dermatitis caused by the exudates of *Rhus* (poison ivy/oak, Japanese lacquer tree) and *Anacardium* (cashew-nut shell oil) is well known [21

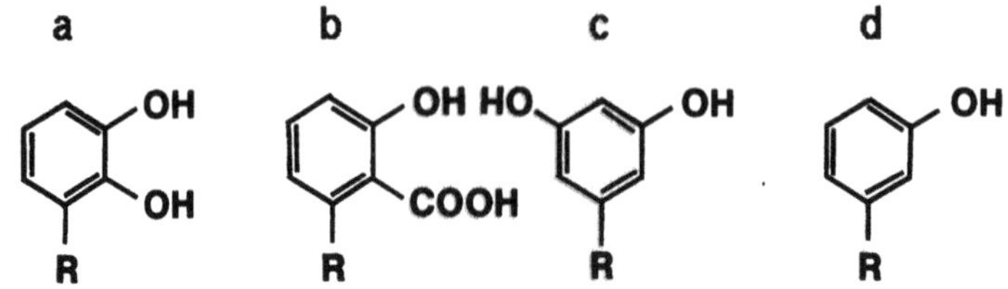

R : $(CH_2)_{14}CH_3$

$(CH_2)_7CH{=}CH(CH_2)_5CH_3$

$(CH_2)_7CH{=}CHCH_2CH{=}CH(CH_2)_2CH_3$

$(CH_2)_7CH{=}CHCH_2CH{=}CHCH{=}CHCH_3$

$(CH_2)_7CH{=}CHCH_2CH{=}CHCH_2CH{=}CH_2$

FIG. 2a–d. Major allergic substances of Anacardiaceae. **a**, urushiol(s); **b**, anacardic acid(s); **c**, cardanol(s); **d**, cardol(s)

and literature cited therein], and it was pointed out that the clinical symptoms are similar to those caused by *Ginkgo* [16]. The major allergic antigens have been isolated and characterized as urushiols from *Rhus* and as anacardic acids, cardanols, and cardols from *Anacardium* [22 and literature cited therein]. As shown in Fig. 2, chemical structures of these substances are identical or similar to those found in episperm of *Ginkgo* (see Fig. 1). Hence an immunological cross reaction between these plant groups was suspected. Sowers et al. reported that the reaction with *Ginkgo* was an allergic contact dermatitis by demonstrating a positive reaction with *Ginkgo* extract in patch tests on the skin of the patient, who had a history of poison ivy dermatitis [16]. Similar results were presented by Saito [19]. In spite of these clinical observations, the results of experimental dermatitis in guinea pigs did not completely support the cross reaction between these plants. In the animals sensitized either by a mixture of ginkgolic acid and its derivatives or by a single component in the mixture, positive reaction developed only after application of the same or related antigen, but not with bilobol or urushiol [19, 23].

Abundant reports concerning the pharmacological and physiological effects of *Ginkgo* leaf extract appeared during the 1990s, and those are discussed elsewhere in this volume. When extracts of fresh or dried leaves were applied to the skin of poison ivy sensitive patients, there was no reaction observed [24], supporting the analytical results of Verotta and Peterlongo [11]. Allergic reaction by airborne pollen of *Ginkgo* has been little reported, but Long et al. [25] tested fractions from a pollen extract as elicitors of *Ginkgo* pollinosis and defined some proteins as the antigens.

5 Effects on Cell Proliferation and Enzyme Activity

Allergic dermatitis develops when the activated T lymphocytes are exposed to the corresponding antigen. Although only a small proportion of the T lymphocytes infiltrated in the skin lesion are specific to the induced antigen, they could proliferate in the presence of a corresponding antigen and induce a variety of pathological skin reactions [26]. A T lymphocyte stimulation test is an in vitro measurement of ^{3}H-thymidine uptake into sensitized lymphocytes that proliferate by the addition of the antigen(s). This test was undertaken using the peripheral lymphocytes of *Ginkgo* dermatitis patients, and high levels of ^{3}H incorporation were observed especially when mixed antigen (ginkgolic acid, ginkgol, and hydroginkgol) was used as an inducer [19].

Negative results of tumorigenicity and antitumor activity of the allergic substances were reported. Bilobol induced strong inflammation in mice, but did not act as promoter of skin carcinogenesis [27]. On the other hand, no increase of life span was observed when ginkgolic acid, ginkgol, or bilobol was applied to mice that had been implanted with P-388 lymphocyte leukemia [28]. Inhibiting activities of long-chained phenols or phenolic acids were reported for lipoxygenase and prostaglandin endoperoxide synthase [29], and for glycerol-3-phosphate dehydrogenase [30].

6 Conclusion

One of the allergic reactions caused by plants is pollinosis triggered by the pollen of anemorphilous (wind-pollinating) plants. Another is contact dermatitis, as stated above, which develops on contact with certain plant metabolites, especially phenols or phenolic acids with a long unsaturated chain. These allergic substances are found in *G. biloba* and some species of Anacardiaceae; the latter are known to be poisonous plants. Analogous compounds are reported in the liverwort *Schistochila appendiculata* [31]. Presence of these substances might act as protective agents against herbivores, and it is of interest that a similar, if not identical, biosynthetic pathway has been found to function in phylogenetically distant organisms.

References

1. Kawamura J (1928) Ueber die chemischen Bestandteil der Frucht von *Ginkgo biloba* (I. Mitteilung). Jpn J Chem 3:89–108
2. Furukawa S (1934) Studies on the constituents of "*Ginkgo biloba* L." fruits. Part I. Sci Paper Inst Phys Chem Res 24:304–313
3. Furukawa S (1934) Studies on the constituents of "*Ginkgo biloba* L." fruits. Part II. Sci Paper Inst Phys Chem Res 24:314–319
4. Furukawa S (1934) Studies on the constituents of "*Ginkgo biloba* L." fruits. Part III. Sci Paper Inst Phys Chem Res 24:320–324

5. Furukawa S (1935) Studies on the constituents of "*Ginkgo biloba* L." fruits. Part IV. Sci Paper Inst Phys Chem Res 26:178–185
6. Morimoto H, Kawamatsu Y, Sugihara H (1968) Sterische Structur der Giftstoffe aus dem Fruchtfleisch von *Ginkgo biloba* L. Chem Pharm Bull 16:2282–2286
7. Itokawa H, Totsuka N, Nakahara K, Takeya K, Lepoittevin J-P, Asakawa Y (1987) Antitumor principles from *Ginkgo biloba* L. Chem Pharm Bull 35:3016–3020
8. Schwarzenbach (1857) Untersuchung des Fruchtfleisches von *Salisbulia adiantifolia* (*Ginkgo biloba*). Vierteljahresschrift prakt Pharm 6:424–430
9. Gellerman JL, Anderson WH, Schlenk H (1976) Synthesis of anacardic acids in seeds of *Ginkgo biloba*. Biochim Biophys Acta 431:16–21
10. Gellerman JL, Anderson WH, Schlenk H (1976) 6-(Pentadec-8-enyl)-2,4-dihydroxybenzoic acid from seeds of *Ginkgo biloba*. Phytochemistry 15:1959–1961
11. Verotta L, Peterlongo F (1993) Selective extraction of phenolic compositions from *Ginkgo biloba* extracts using supercritical carbon dioxide and off-line capillary gas chromatography/mass spectrometry. Phytochem Anal 4:178–182
12. Tabayashi T (1927) Experimental study of dermatitis owing to fruit of the *Ginkgo* (abstract in Japanese). Hifuka Hinyokika Zassi 27:962–964
13. Saito J (1929) Klinische und experimentelle Untersuchung der durch Ginkgofrucht verusachten Dermatitis und ihrer hautereizenden Bestandteile (in Japanese). Hifuka Hinyokika Zassi 29:105–129
14. Saito J (1930) Pharmakologische Untersuchung des Ginkgogifts, des Dermatitiserregers in *Ginkgo biloba* L. Tohoku J Exp Med 16:413–423
15. Bolus M, Raleigh NC (1939) Dermatitis venenata due to *Ginkgo* berries. Arch Derm Syph 39:530
16. Sowers WF, Weary PE, Collins OD, Cawley EP (1965) *Ginkgo* tree dermatitis. Arch Derm 91:452–456
17. Tomb RR, Foussereau J, Sell Y (1988) Mini-epidemic of contact dermatitis from *Ginkgo* tree fruit (*Ginkgo biloba* L.). Contact Derm 19:281–283
18. Jirasek L, Brozkova L (1981) Occupational contact hypersensitivity to *Ginkgo biloba* L (in Czech). Cesk-Dermatol 56:366–373
19. Saito F (1988) Contact dermatitis due to *Ginkgo* fruit pulp: Clinical survey, patch test, lymphocyte stimulation test and animal sensitization test (in Japanese with English summary). Hifu (Skin Res) 30:155–168
20. Becker LE, Skipworth GB (1975) *Ginkgo*-tree dermatitis, stomatitis, and proctitis. J Amer Med Asoc 231:1162–1163
21. Mitchell JC, Rook A (1979) Botanical dermatology. Greengrass, Vancouver
22. Benezra C, Ducombs G (1987) Molecular aspects of allergic contact dermatitis of plants. Recent progress in phytodermatochemistry. Dermatosen 35:4–11
23. Lepoittevin J-P, Benezra C, Asakawa Y (1989) Allergic contact dermatitis to *Ginkgo biloba* L.: relationship with urushiol. Arch Dermatol Res 281:227–230
24. Michell JC, Maibach HI, Guin J (1981) Leaves of *Ginkgo biloba* not allergic for *Toxicodendron*-sensitive subjects. Contact Derm 7:47–48
25. Long R, Yin R, Zhen Y (1992) Partial purification and analysis of allergenicity, immunogenicity of *Ginkgo biloba* L. pollen (in Chinese with English summary). J West China Univ Med Sci 23:429–432
26. Kalish RS (1991) Recent developments in the pathogenesis of allergic contact dermatitis. Arch Dermatol 127:1558–1563

27. Matsumoto K, Fujimoto M, Ito K, Tanaka H, Hirono I (1990) Comparison of the effects of bilobol and 12-0-tetradecanoylphorbol-13-acetate on skin, and test of tumor promoting potential of bilobol in CD-1 mice. J Toxicol Sci 15:39–46
28. Itokawa H, Totsuka N, Nakahara K, Maezuru M, Takeya K, Kondo M, Inamatsu M, Morita H (1989) A quantitative structure-activity relationship for antitumor activity of long-chain phenols from *Ginkgo biloba* L. Chem Pharm Bull 37:1619–1621
29. Grazzini R, Hesk D, Heininger E, Hildenbrandt G, Reddy CC, Cox-Foster D, Medford J, Craig R, Mumma RO (1991) Inhibition of lipoxygenase and prostaglandin endoperoxide synthase by anacardic acids. Biochem Biophys Res Commun 176:775–780
30. Irie J, Murata M, Homma S (1996) Glycerol-3-phosphate dehydrogenase inhibitors, anacardic acids from *Ginkgo biloba*. Biosci Biotech Biochem 60:240–243
31. Asakawa Y, Masuya T, Tori M, Campbell EO (1987) Long chain alkyl phenols from the liverwort *Schistochila appendiculata*. Phytochemistry 26:735–737

Food Poisoning by *Ginkgo biloba* Seeds

KEIJI WADA and MASANOBU HAGA

1 Introduction

Gin-nan is the Japanese word for the seed of *Ginkgo biloba* L. The albumen of the seed is used as a crude drug and food in China and Japan. In particular, it is used as an antitussive and expectorant in traditional medicine. However, when this substance has been taken to excess during food shortages, "*Gin-nan* food poisoning" has sometimes occurred in Japan [1–35] and China [36–37]. Figure 1 shows the numbers of patients with *gin-nan* food poisoning in Japan in our survey [1–38]. The symptoms of this poisoning are mainly convulsions and loss of consciousness. Infants and particularly children under 6 years of age made up about 74% of all patients (Fig. 2). The consequences are not serious for survivors, but mortality is about 27% in Japan.

At first, most investigators assumed that *gin-nan* food poisoning might be caused by cyanogenetic glycosides because of the kind of symptoms. However, it was shown that *Ginkgo* seeds did not contain cyanogenetic glycosides [39, 40], although several investigators reported on the cause of food poisoning [2, 11, 13, 15, 17, 18, 20, 24, 25].

2 Ginkgotoxin (4-O-methylpyridoxine, methoxypyridoxine, MPN)

Kobayashi reported that guinea pigs were highly sensitive to *Ginkgo* seed extract, and they showed typical symptoms of toxicity [25]. In our studies, we used guinea pigs to isolate the toxic principle (tentative name: ginkgotoxin) from *Ginkgo* seeds. After oral administration of the aqueous extract of *Ginkgo* seeds to guinea pigs, the signs of toxicity appeared within a few hours: paralysis of legs; opisthotonus, which is similar to strychninism; and tonic convulsions.

Faculty of Pharmaceutical Sciences, Health Sciences University of Hokkaido, Ishikari-Tobetsu, Hokkaido 061-02, Japan

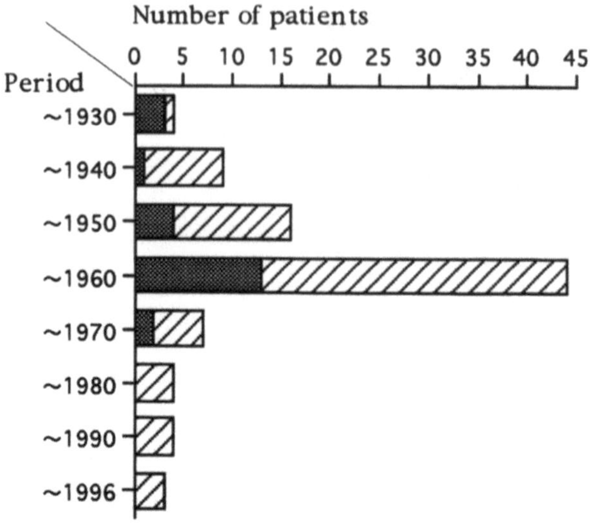

FIG. 1. Number of patients with *gin-nan* food poisoning in Japan. *Shaded areas*, number of deaths; *Striped areas*, number of survivors

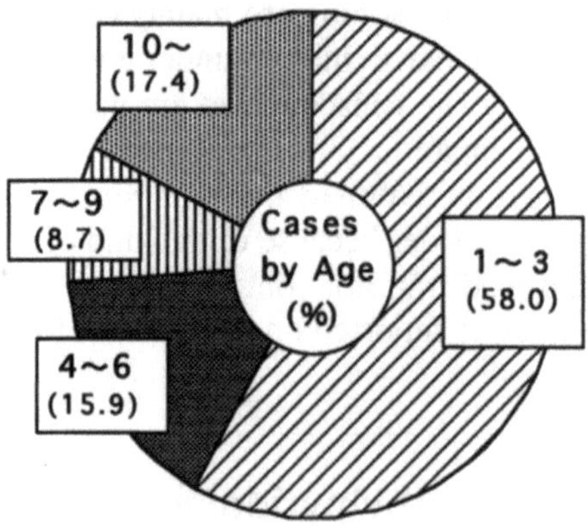

FIG. 2. *Gin-nan* food poisoning cases by age

The fractionation of the aqueous extract of *Ginkgo* seeds was done by partitioning successively between some organic solvents, such as petroleum ether, ethyl acetate, and butyl alcohol, and water. The resulting toxic fraction (butyl alcohol layer) was further separated into two fractions by alumina column chromatography [eluent: $CHCl_3$-methyl alcohol (MeOH), 19:1, v/v]. When the first fraction was given to the guinea pigs, it induced the characteristic convulsions at an oral dose of 11 mg/kg. The other fraction [149 mg/kg] did not show toxicities.

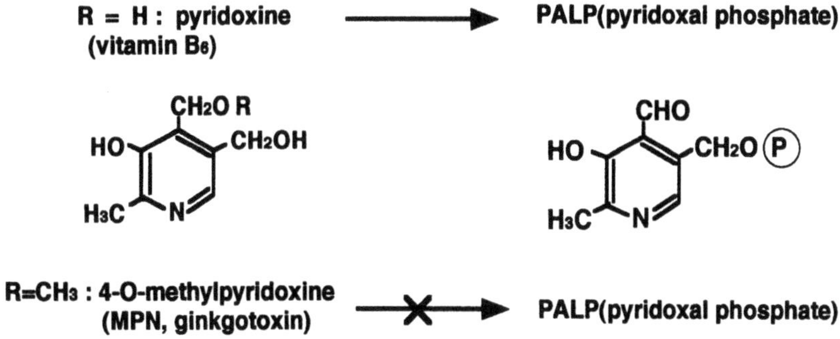

FIG. 3. Structures of 4-O-methylpyridoxine and pyridoxine derivatives. *MPN*, 4-O-methylpyridoxine

These isolation procedures afforded the "ginkgotoxin" in approximately 0.01% yield (from dry weight of *Ginkgo* seeds). Ginkgotoxin was found to have the molecular formula $C_9H_{13}NO_3$ [m/z 183.089 (M^+); $C_9H_{13}NO_3$ requires 183.089] by high-resolution mass spectroscopy. Ginkgotoxin was identified as 3-hydroxy-5-hydroxymethyl-4-methoxymethyl-2-methylpyridine (4-O-methylpyridoxine, MPN) from the evidence of spectral data containing ^1H and ^{13}C nuclear magnetic resonance (NMR) spectra, and some chemical reactions [30, 41, 42]. This is the first report of the isolation of MPN from natural sources (Fig. 3). Although MPN is often called methoxypyridoxine (or 4′-methoxypyridoxine), it seems reasonable to call it 4-O-methylpyridoxine [41, 42].

Synthetic MPN is known to be a potent convulsive agent having antivitamin B_6 activities in humans and in a variety of experimental animals such as mice, rats, cats, dogs, and monkeys [43–46]. It has already been proved that these convulsions caused by MPN can be prevented or stopped with pyridoxine (vitamin B_6) [47]. This suggests that pyridoxine will prevent the symptoms of *gin-nan* food poisoning.

MPN is also known to inhibit the formation of 4-aminobutyric acid (GABA) from glutamate in the brain. MPN may compete with vitamin B_6 which serves as a coenzyme of glutamate decarboxylase. As GABA is regarded as an inhibitory chemical transmitter, a deficiency of GABA in the brain may induce the seizures [48, 49]. Therefore, it is reasonable to speculate that MPN in the seeds of *G. biloba* might induce convulsions through the same mechanism. The lack of vitamin B_6 during times of food shortage may also be one of the causes of *gin-nan* food poisoning.

3 MPN-Induced Convulsions in Experimental Animals

A number of researchers have reported the relationship between MPN and pyridoxine deficiency or between MPN and convulsions [50–57]. As an another approach to clarify the mechanisms of MPN poisoning in guinea pigs,

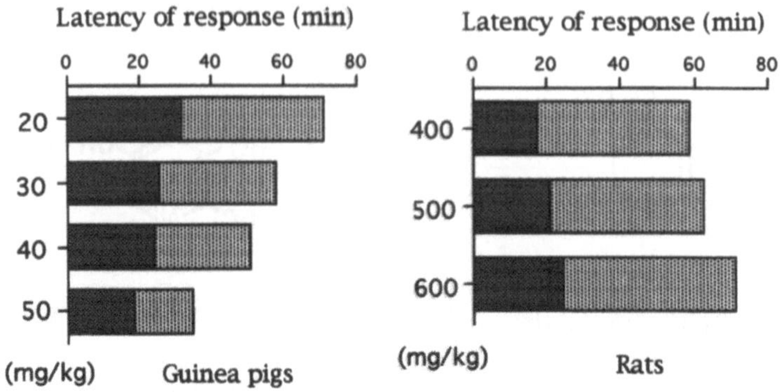

FIG. 4. MPN-induced convulsion in guinea pigs and rats. *Dark shade*, latency of clonic convulsion; dark and light shade, latency of death time. Death time was counted as the period from the administration of MPN until death

electroencephalogram (EEG) and cerebral blood flow were measured [58]. The administration of MPN (30–50 mg/kg, i.p.) to conscious guinea pigs caused a generalized epileptic seizure within 1 h, and then the EEG showed an epilepsy-like pattern. After exhibiting tonic and clonic convulsions, guinea pigs died of ventricular fibrillation. The latency to clonic convulsion, generalized seizures, and death were reduced dose-dependently (20–50 mg/kg). On the other hand, rats went into convulsions after the administration of MPN (400–600 mg/kg, i.p.) (Fig. 4). Thus higher doses of MPN were needed to induce convulsions in conscious rats than in guinea pigs. This species difference of MPN has also been reported by Julou et al. [45] and Kraft et al. [59].

Cerebral blood flow (by the laser Doppler method) increased during MPN-induced convulsions in the anesthetized guinea pig [58]. This confirmed that MPN-induced convulsions were not ischemic seizures. The cause of death from MPN poisoning was from atrioventricular block or ventricular fibrillation. These reactions were delayed from 40 to 120 min after intraperitoneal administration of MPN.

4 Seizure-related Factors

4.1 GABA, Glutamate, Glutamine, and Taurine Concentrations in Brain Tissues after the Administration of MPN in Guinea Pigs

MPN may compete with the pyridoxine that serves as a coenzyme of glutamate decarboxylase [48]. Thus, an attempt was made to clarify the mechanism of MPN action by determining cerebral tissue GABA and glutamate concentrations

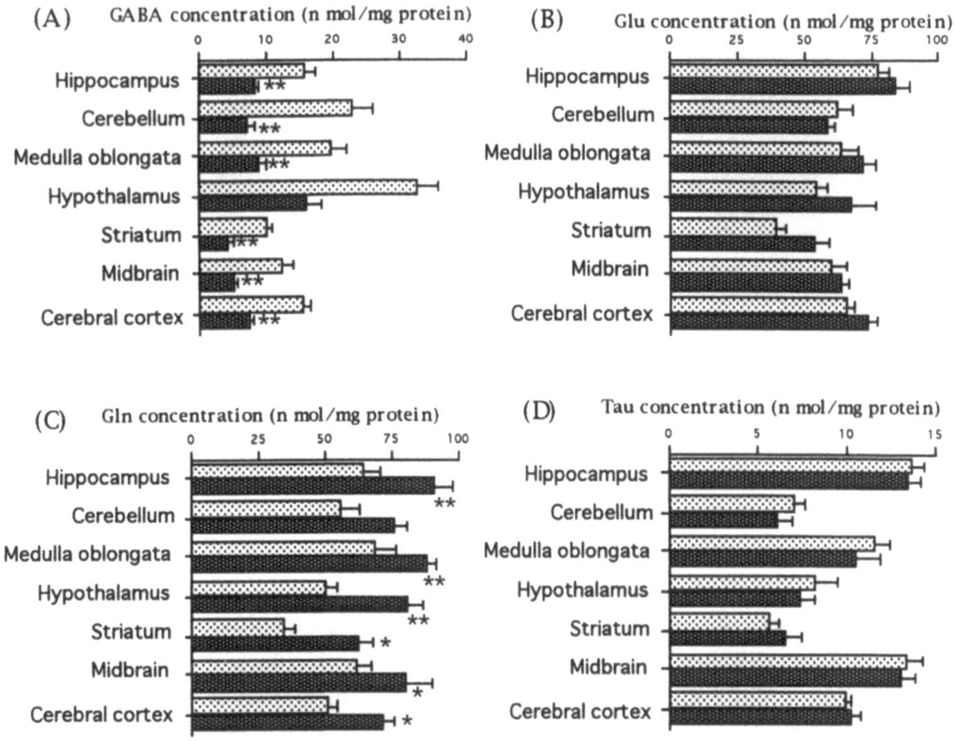

FIG. 5A–D. Amino acid concentrations in various brain regions after MPN administration in guinea pigs. Each value represents the mean ± SE (n = 6). *Light shade*, control; *dark shade*, treated with MPN (30 mg/kg, i.p., after 40 min); *GABA*, r-aminobutyric acid; *Glu*, glutamate; *Gln*, glutamine; *Tau*, taurine, $^{*}P < .05$; $^{**}P < .01$ *vs* control

using high-performance liquid chromatography with oxidative electrochemical detection (HPLC-ECD) [58]. The concentrations of amino acids and related compounds in various brain regions after MPN administration in guinea pigs are shown in Fig. 5. As GABA is regarded as an inhibitory chemical transmitter, a deficiency of GABA in the brain may induce the seizure. However, the practice of using the absolute level of GABA as an index of brain excitability must be questioned. Ozawa and Okada [48] and Nitsh and Okada [49] speculated that changes in concentration of GABA in specific areas of the brain might play a key role in epileptogenesis. In our results, cerebral GABA levels decreased significantly during convulsions compared with those of control guinea pigs. On the other hand, regional changes of GABA content after administration of MPN were uniform in all regions of the brain (Fig. 5A). We could not find a decrease in GABA content in any specific part of the brain. At the same time, cerebral glutamine concentrations increased significantly (Fig. 5C) and glutamate levels tended to increase (Fig. 5B). The amino

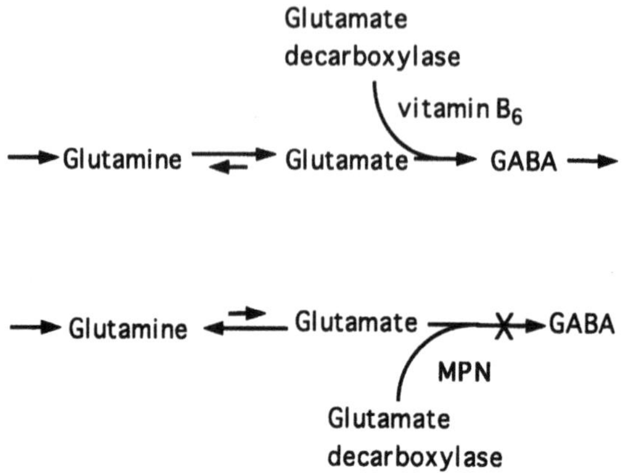

FIG. 6. Chemical pathways of glutamate metabolism. *GABA*, γ-aminobutyric acid

acid taurine has been proposed as an inhibitory transmitter in the central nervous system [60]. There were no differences in taurine content between control and MPN-treated guinea pigs (Fig. 5D). There is a pharmacological interest in specific agonists and antagonists for sub-types of central excitatory amino acid receptors operated by glutamate. According to current hypotheses, disorders involving neuronal death may often be associated with excitatory amino acid mechanisms.

In our results, glutamate tended to increase and the brain glutamine concentration increased significantly after administration of MPN. From these observations we postulated the stage in glutamic acid metabolism at which interruption results in the seizure. As shown in Fig. 6, glutamate has two types of metabolism. If the route from glutamate to GABA is blocked, another from glutamate to glutamine might occur. It is suggested that significant increase in glutamine content after MPN administration is a decompensation to protect the increase in MPN-induced glutamate content.

In conclusion, MPN in the seeds of the *Ginkgo biloba* might induce convulsions through a deficiency of GABA and an increase in glutamine. MPN is now used as a means to investigate the GABAergic mechanism in experimental animals [61].

4.2 Catecholamines, Serotonin, and 5-Hydroxyindole Acetic Acid Concentration in Brain Tissues after the Administration of MPN in Guinea Pigs

Figure 7 shows catecholamine concentrations in the guinea pig brain at 40 min after MPN administration [50]. In all the brain regions, norepinephrine concentrations were lower than those of control guinea pigs. At the midbrain and

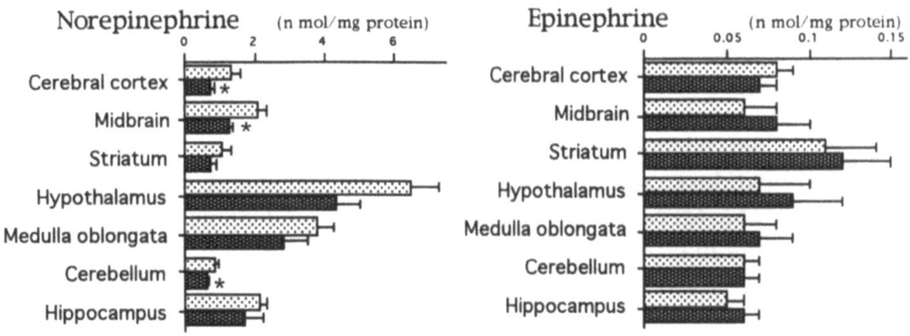

FIG. 7. Catecholamine concentrations in various brain regions after MPN administration in guinea pigs. Each value represents the mean ± SE ($n = 6$). *Light shade*, control; *dark shade*, treated with MPN (30 mg/kg, i.p., after 40 min); *$P < .05$ *vs* control

cerebral cortex, norepinephrine concentrations were significantly reduced after MPN treatment. No significant changes were observed in epinephrine and concentrations.

No marked changes in the levels of dopamine, serotonin or 5-hydroxyindole acetic acid (5-HIAA) were observed in various regions of the guinea pig brain after MPN treatment.

5 Determination of MPN in Serum of the Patient with *Gin-nan* Food Poisoning

Although there are many cases of *gin-nan* food poisoning [1–37], there are no reports to establish that the food poisoning in humans was caused by MPN. We were able to analyze the serum and urine of a 21-month-old male child with accidental *gin-nan* food poisoning [62]. The serum 4-O-methylpyridoxine (MPN) levels in the patient with *gin-nan* food poisoning were determined by HPLC. The blood of the patient was taken at 8.5 and 15.5 h after about 50 *Ginkgo* seeds were ingested. After deproteinization and centrifugation of the serum, the supernatant was pretreated with Sep-pak C-18 cartridge (Waters, Milford, MA) and was applied to HPLC. HPLC was performed with a reversed phase column using a fluorescence detector (wave length of excitation and fluorescence; 290 and 400 nm, respectively). The determination limit of MPN in the serum was 0.05 μg/ml. The serum MPN concentration was 0.09 μg/ml at 8.5 h after ingesting *Ginkgo* seeds, and was less than the determination limit of MPN (0.05 μg/ml) at 15.5 h.

When MPN (30/mg/kg, i.p., about LD_{50}) was administered to a rabbit, the lowest concentration of serum MPN that induced poisoning was 1.0 μg/ml [63]. In the rabbit, convulsions are repeatedly induced in 30–60 min, hence the rapid determination of serum MPN may predict the occurrence of symptoms. This method may be useful in determining *gin-nan* food poisoning in humans. This is the first report that the toxic principle (MPN) was found in the serum of a patient with *gin-nan* food poisoning.

We could not detect the free-form MPN in the urine under the same conditions. As 4-pyridoxic acid is excreted as a main metabolite of pyridoxine in humans [64, 65], MPN may be metabolized to 5-pyridoxic acid or any other conjugated form in urine [66, 67].

6 Detoxification of MPN with Bilobalide in *Ginkgo*-Leaf Extract

Ginkgo-leaf extract (GLE) has many components [68] and it has been prescribed for cases of senility-related chronic cerebral deterioration and related functional disturbances, and following cerebral vascular accidents and cranial trauma. Preliminary assays for MPN in GLE by HPLC were negative [69, 72].

Bilobalide is one of the characteristic components of *Ginkgo biloba*, and it induces the activity of 7-*O*-alkoxycoumarine dealkylases [38]. Oral administration of bilobalide (10–30 mg/kg per day, for 4 days) minimized the MPN-induced symptoms, because of the conversion of MPN to pyridoxine by demethylation of MPN.

It is very interesting that both the toxic factor (MPN) in the seeds and the detoxification factor (bilobalide) in the leaves exist in the same plant.

7 Treatment of MPN Poisoning

Gammon and Gummit have reported that 10–100 mg doses of GABA prevented the MPN-induced convulsions in mice [47]. However, the administration of GABA (10–50 mg/kg, i.p.) failed to prevent the seizure in guinea pigs. As it is supposed that GABA cannot pass through the blood brain barrier, it may be necessary to enhance the level of GABA endogenously [70].

As shown in Figure 8, diazepam may be capable of preventing the MPN-induced convulsions in two of six guinea pigs. The mean latency to convulsion was prolonged and duration of the convulsion was reduced by diazepam.

Diazepam is the one of the banzodiazepins, and their actions are a result of potentiation of the neural inhibition that is mediated by GABA. Therefore, diazepam did not prevent MPN-induced reduction of GABA concentration (Fig. 8). Diazepam countered the MPN-induced reduction of norepinephrine concentration, and it had no effect on indoleamine levels. The administration of pyridoxal phosphate is a more effective treatment for *gin-nan* food poisoning [71, and personal communication, J. Ishizawa, 1994].

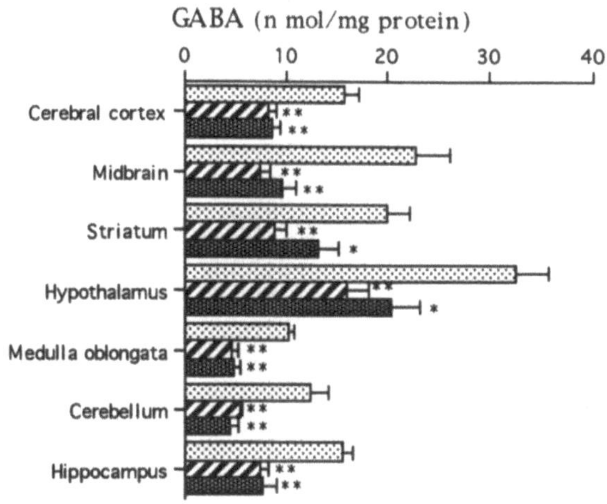

FIG. 8. GABA concentrations in various brain regions after diazepam and MPN administration in guinea pigs. Each value represents the mean ± SE (n = 6). *Light shade*, control; *dark shade*, treated with MPN (30 mg/kg, i.p., after 40 min); *stripes*, treated with MPN (30 mg/kg, i.p.) and diazepam (4.0 mg/kg, i.p.) after 40 min; $*P < .05$, $**P < .01$ *vs* control

8 Prevention of *Gin-nan* Food Poisoning

Gin-nan food poisoning is caused by MPN that has an antivitamin B_6 activity. As vitamin B_6 deficiency is induced not only by MPN but also by other medicines, i.e., isonicotinic acid hydrazide (isoniazid), eating *Ginkgo* seeds should be avoided by those with latent vitamin B_6 deficiency. Carbonyl reagents such as isoniazid react with pyridoxal and inhibit the role of pyridoxal as a coenzyme.

Because children are most susceptible to *gin-nan* food poisoning, it is recommended that consumption of *Ginkgo* seeds be limited to not more than five for children under 6 years of age.

9 Conclusion

Gin-nan is the seed of *Ginkgo biloba* L., and its albumen is used as a crude drug and food in China and Japan. However, when this substance has been taken to excess during food shortages, *gin-nan* food poisoning has sometimes occurred in Japan. Infants and particularly children under 6 years of age made up about 74% of all patients.

In our studies, a toxic substance, 4-O-methylpyridoxine (MPN), responsible for *gin-nan* food poisoning was isolated from the seeds of *G. biloba* MPN causes food poisoning not only by antagonizing vitamin B_6 (in the body) but also by inhibiting the formation of 4-aminobutanoic acid (GABA) from glutamate in the brain.

The MPN levels in the sera of patient with *gin-nan* food poisoning was determined by HPLC, which might be used for chemical diagnosis of *gin-nan* food poisoning.

As *gin-nan* food poisoning is caused by MPN that has an antivitamin B_6 activity, the administration of pyridoxal phosphate is an effective treatment.

References

1. Kudo K (1981) Does the *Ginkgo* seed contain large amounts of cyanogenetic glycosides (in Japanese)? Tokyo Iji Shinshi 149:19–21
2. Marui H (1928) Gin-nan food poisoning (in Japanese). Iji Koron 858:7–8
3. Kurashima S (1931) Three cases of Gin-nan food poisoning (in Japanese). Hokuetsu Igakukai Zasshi 46:525–532
4. Marui H (1933) Gin-nan food poisoning (in Japanese). Igaku Chishiki 1:98
5. Takahashi R (1934) Gin-nan food poisoning (in Japanese). Shindan To Chiryo 21:627–628
6. Hara M (1938) Gin-nan food poisoning (in Japanese). Shindan To Chiryo 25:1560–1561
7. Kobayashi S (1943) A case of Gin-nan food poisoning (in Japanese). Jika Shinryo 9:448–453
8. Nishimoto S (1944) A case of Gin-nan food poisoning (in Japanese). Jika Shinryo 10:79–80
9. Aida M (1944) A third case of Gin-nan food poisoning (in Japanese). Jika Shinryo 10:81–83
10. Hukushi K (1994) An autopsy case of Gin-nan food poisoning (in Japanese). Nihon Byorigakkai-kaishi 34:11
11. Shiomi T (1951) A case of Gin-nan food poisoning with a kind of baryencephalia (in Japanese). Shindan To Chiryo 39:392–395
12. Suzuki K (1951) A case of puerile Gin-nan food poisoning (in Japanese). Jika Shinryo 14:561–562
13. Fujiwara H (1952) Gin-nan food poisoning (in Japanese). Kyoto Igakukai Zasshi 3:260–262
14. Sakurada A (1952) A case of Gin-nan food poisoning (in Japanese). Jika Shinryo 15:428
15. Takano T, Kobayashi M, Wada I (1952) Gin-nan food poisoning in experimental animals (in Japanese). Monthly Report of Tokyo Metropolitan Research Laboratory of Public Health 41:46
16. Uruno K (1953) Gin-nan food poisoning (in Japanese). Jika Shinryo 16:195–197
17. Takano T, Kobayashi M, Wada I (1953) Study on Gin-nan food poisoning in experimental animals (in Japanese). Jui Chikusan Shinpo 109:353–357
18. Yano Y (1953) A case of Gin-nan food poisoning (in Japanese). Rinsho To Kenkyu 30:469–471
19. Ohara T (1955) A case of Gin-nan food poisoning (in Japanese). Shonika Shinryo 18:920

20. Kobayashi M (1956) Study on the toxic substance in *Ginkgo* seeds II (in Japanese). Nippon Eiseigaku Zasshi 11:41
21. Yoshida R, Asari Y (1957) A case of Gin-nan food poisoning (in Japanese). Shonika Shinryo 20:572
22. Okada K (1957) A case of Gin-nan food poisoning (in Japanese). Shonika Shinryo 20:572
23. Hosokai M, Ishikawa E (1957) A case of puerile Gin-nan food poisoning (in Japanese). Shonika Shinryo 20:1116–1117
24. Ito T, Kobayashi M (1958) Study on the toxic substance in *Ginkgo* seeds III (in Japanese). Nippon Eiseigaku Zasshi 13:54
25. Kobayashi M (1959) Study on the toxic substance in *Ginkgo* seeds (in Japanese). Nagano-ken Eisei Kenkyusho Chosa Kenkyu Hokoku 16:1–12
26. Ito H, Ebisu S (1959) A case of puerile Gin-nan food poisoning (in Japanese). Shonika Shinryo 22:101–104
27. Takeuchi K, Kokuho Y, Iijima S (1967) Two cases of Gin-nan food poisoning (in Japanese). Shonika Shinryo 30:987
28. Yokoi Y, Shimizu K, Inagaki H (1979) Two cases of Gin-nan food poisoning (in Japanese). Shonika Shinryo 42:1638
29. Tasumi M, Nakazawa A, Iwamoto H (1982) A case of Gin-nan food poisoning (in Japanese). Shonika Shinryo 45:2013
30. Wada K (1986) *Ginkgo biloba* and Gin-nan food poisoning (in Japanese). Kagaku To Yakugaku No Kyoshitsu 95:79–82
31. Nomoto F, Kubota F (1989) Gin-nan food poisoning (in Japanese). Seishin Igaku 31:535–538
32. Naito H (1990) Gin-nan (*Ginkgo* seeds) (in Japanese). Chugai Iyaku 43:50–51
33. Ishizawa J, Tsujikawa A, Ohashi T (1990) Gin-nan food poisoning (in Japanese). Gekkan Yakuji 32:2444–2445
34. Kokubo M, Terada A, Hayakawa S, Maeda K, Matumoto N, Wada K (1993) A case of Gin-nan food poisoning (in Japanese). Shonika Shinryo 56:488–491
35. The Epidemiological Data of Food Poisoning in Japan. (1955–1983) Koseisho Seikatsu-eisei-kyoku Shokuhinhoken-ka (ed) Nippon Shokuhin Eisei Kyokai, Tokyo (in Japanese)
36. Suzuki S (1975) Kokuyaku Honzoh Kohmoku, vol 8, *Ginkgo biloba* (in Japanese), Shunyodo Shoten Tokyo, pp 407–411
37. Nan S, Shu T (1958) Two cases of Gin-nan food poisoning (In Chinese), and references cited therein. Chuka Jika Zasshi 483–484
38. Wada K (1996) Studies on the constituents of edible and medicinal plants to affect the metabolizing system in mammals (in Japanese with English abstract). Natural Medicines 50:195–203
39. Suzu M (1959) An examination of plums and ginkgo-nuts for cyanophoric glycoside (in Japanese). Fukuoka Igaku Zasshi 50:5394–5398
40. Nishijima M, Kanmuri M, Takahashi S, Kamimura H, Nakazato M, Watari Y, Kimura Y (1975) Survey of cyanide in almond (in Japanese). Tokyo-To Eisei Kenkyusho Nempo 26:183–186
41. Wada K, Ishigaki S, Ueda K, Sakata M, Haga M (1985) An antivitamin B_6, 4′-methoxypyridoxine from the seed of *Ginkgo biloba* L. Chem Pharm Bull 33:3555–3557
42. Wada K, Ishigaki S, Ueda K, Take Y, Sasaki K, Sakata M, Haga M (1988) Studies on the constitution of edible and medicinal plants. I. Isolation and identification of 4-O-

methylpyridoxine, toxic principle from the seed of *Ginkgo biloba* L. Chem Pharm Bull 36:1779–1782
43. Ott WH (1947) Antipyridoxine activity of methoxypyridoxine in the chick. Proc Soc Exp Biol Med 66:215–216
44. Kopeloff LM, Chusid JG (1963) Methoxypyridoxine convulsions in epileptic and non-epileptic mice. Protective action of pyridoxine. Proc Soc Exp Biol Med 114:496–500
45. Julou L, Pasquet J, Ducrot R (1964) The antidotic effect of pyridoxine on one of the intermediary products of its manufacture which has a strong toxicity. Proc of European Society for the Study of Drug Toxicity 4:179–189
46. Mizuno N, Kawakami K, Morita E (1980) Competitive inhibition between 4′-substituted pyridoxine analogues and pyridoxal. For pyridoxal kinase from mouse brain. J Nutr Sci Vitaminol 26:535–543
47. Gammon GD, Gummit R (1957) Observations on the mechanism of seizures induced by a pyridoxine antagonist, methoxypyridoxine. Trans Amer Neurol Assoc 82:57–59
48. Ozawa S, Okada Y (1976) Decrease of GABA levels and the appearance of a depolarization shift in thin hippocampal slice in vitro. In: Roberts E, Chase TN, Tower DB (eds) GABA in nervous system function. Raven, New York, pp 449–454
49. Nitsh C, Okada Y (1976) Decrease of GABA levels in different parts of the rabbit brain after treatment with methoxypyridoxine. In: Roberts E, Chase TN, Tower DB (eds) GABA in nervous system function. Raven, New York, pp 455–460
50. Umbreit WW (1955) Vitamin B_6 Antagonists. Amer J Clin Nutrition 3:291–297
51. Kamrin RP, Kamrin AA (1961) The effects of pyridoxine antagonists and other convulsive agents on amino acid concentrations of the mouse brain. J Neurochemistry 6:219–225
52. Elliot KAC (1965) γ-Aminobutyric acid and other inhibitory substitutes. Brit Med Bull 21:70–75
53. Wiechert P, Herbst A (1966) Provocation of cerebral seizures by derangement of the natural balance between glutamic acid and γ-aminobutyric acid. J Neurochem 13:59–64
54. Steiner FA, Ruf K (1967) Interaction of L-glutamic acid gamma-amino butyric acid and pyridoxal-5′-phosphate at the neuronal level. Schweizer Archiv für Neurologie, Arch Neurol Neurochirurgie und Psychiatrie 100:310–320
55. Yoshida T, Tada K, Arakawa T (1971) Vitamin B_6-dependency of glutamic acid decarboxylase in the kidney from a patient with vitamin B_6 dependent convulsion (In Japanese). Tohoku J Exp Med 104:195–198
56. Cherayil GD, Cyrus AE Jr (1972) Effect of 4-methoxymethyl-pyridoxine on gangliosides, cholesterol and fatty acids in mouse brain. J Neurochem 19:1215–1219
57. Bukin Yu V, Sergeev AV, Kondrat'eva G Ya (1976) Effects of some 2-alkyl and 4′-O-methyl analogs of pyridoxol on the activity of pyridoxal kinase from mouse liver (in Russian). Biokhimiya 41:432–442
58. Minami M, Yanai A, Endo T, Hamaue M, Hamaue N, Monma Y, Wada K, Haga M, Morii K, Yoshioka M, Saito H (1990) Convulsion induced by 4-O-methylpyridoxine, from the seed of the *Ginkgo biloba* L., in guinea pigs and rats. Life Sci Adv 9:107–115
59. Kraft Von HG, Fiebig L, Hotovy R (1961) Pharmacology of vitamin B_6 and its derivatives (in German). Arzneimitel-Forsh 10:922–929
60. Oja SS, Kontro P (1978) In: Barbeau A, Huxtable RJ (eds) Taurine and neurological disorders, Raven, New York, pp 181–200

61. Hirai H, Okada Y (1993) Ipsilateral corticotectal pathway inhibits the formation of long-term potentiation (LTP) in the rat superior colliculus through GABAergic mechanism. Brain Res 629:23–30
62. Yagi M, Wada K, Sakata M, Kokubo M, Haga M (1993) Studies on the constituents of edible and medicinal plants. IV. Determination of 4-O-methylpyridoxine in serum of the patient with gin-nan food poisoning (in Japanese with English abstract). Yakugaku Zasshi 113:596–599
63. Take Y, Sasaki K, Wada K, Sakata M, Haga M (1987) Studies on the plant food poisoning III. The determination of toxic principle, 4′-methoxypyridoxine in the seed of *Ginkgo biloba* with HPLC. The 107th Annual meeting of Pharmaceutical Society of Japan, Kyoto
64. Huff JW, Perlzweig WA (1944) A product of oxidative metabolism of pyridoxine, 2-methyl-3-hydroxy-4-carboxy-5-hydroxy-methylpyridine (4-pyridoxic acid) I. Isolation from urine, structure, and synthesis. J Biol Chem 155:345–355
65. Schwartz R, Kjeldgaard NO (1951) The enzymic oxidation of pyridoxal by liver aldehyde oxidase. Biochem J 48:333–337
66. Shane B, Snell EE (1975) Metabolism of 5′-deoxypyridoxine in rats: 5′-deoxypyridoxine 4′-sulphate as a major urinary metabolite. Biochem Biophys Res Comm 66:1294–1300
67. Coburn SP, Mahuren JD (1976) In vivo metabolism of 4′-deoxypyridoxine in rat and man. J Biol Chem 251:1646–1652
68. Huh H, Staba EJ (1992) The botany and chemistry of *Ginkgo biloba* L. J Herbs, Spices Med Plants 1:91–124
69. Wada K, Sasaki K, Miura K, Yagi M, Kubota Y, Matsumoto T, Haga M (1993) Isolation of bilobalide and ginkgolide A from *Ginkgo biloba* L. shorten the sleeping time induced in mice by anesthetics. Biol Pharm Bull 16:210–212
70. Takagaki G (1981) Neurobiochemistry. In: Kyoritsu Zensho, vol 238 (in Japanese). Kyoritsu Shuppan, Tokyo, pp 72–75
71. Yoshimura I, Hayata M, Yamasaki F, Mori H (eds) (1996) Natural toxin. In: Acute toxicity information file (in Japanese). Hirokawa, Tokyo
72. Arenz A, Klein M, Fiehe K, Groβ J, Drewke C, Hemscheidt T, Leistner E (1996) Occurrence of neurotoxic 4′-0-methylpyridoxine in *Ginkgo biloba* leaves, *Ginkgo* medications and Japanese *Ginkgo* food. Planta Med 62:548–551

Results of Treatment with a Broad-Acting Nootropic Preparation on Chronic Diffuse Cerebral Disturbances

Ernst-Walter Fünfgeld

1 Introduction

This chapter presents personal experience obtained with patients and volunteers since 1987 [1–12]. The *Ginkgo* extract in use, orally or by infusions, was prepared after the standard methods delineated in EGb 761 by the companies Schwabe, Karlsruhe, and Intersan, Ettlingen, Germany. The same preparation is in use in France, manufactured by Ipsen (Paris). Another preparation manufactured in Switzerland was given to a volunteer under scientific/experimental conditions (Fünfgeld, unpublished data).

This chapter does not discuss the controversial aspects of the effectiveness of the preparation as a metabolic enhancing treatment. However, observations are presented on three independent levels: clinical observations by myself and the clinical staff, reports of the patients, and electrophysiological results, follow-up investigations, and computerized technique.

Objective figures from cerebral functions are not easy to obtain from the point of view of technique, what the patient can endure, and side effects, especially after several examination times, including the length of time and the cost. In comparison to the highly sophisticated techniques such as positron emission tomography (PET), single photon emission computed tomography (SPECT), and nuclear magnetic resonance (NMR), electrophysiological registrations are easy to obtain, cause no harm, and are less expensive. Progress introduced by the computerized storage of data and the further possibility for statistical evaluation is not yet common compared to the other technical methods. The computerized EEG method was even recommended by the famous PET group from Cologne (Heiss et al. in 1994 [13]) after their correlative findings of glucose uptake and delta activity.

It is possible to objectify the individual patterns and the different sequences of aging. The slowing of the frequencies reflected more the multifaceted functional changes of the central nerve tissue than the change of the frequency-related

Weintrautstrasse 10, D-35039, Marburg, Germany

electrical output (see Section 2, Technical Procedures and Documentation). Clinically oriented mapping programs give immediate information about the actual functional status of individuals. Even preclinically, these findings can facilitate individual therapeutical decisions.

Group investigations, e.g., involving 43 Parkinson patients suffering from so-called hypersensitive dopamine syndrome, revealed statistically significant electroclinical correlations [14]. These and other findings led to a very important conclusion: the applied computerized EEG method gives valid information and signifies progress in the sense of an easily applicable, economic, and safe objective documentation of cerebral function. The present observations with *Ginkgo* extract represent one part of a more extensive study comprising several other natural and chemical preparations or trace elements that were documented in a similar manner [14–18]. The numbers of reported experiences are large enough to validate the computerized EEG method as an objective evaluation for cerebral action.

2 Technical Procedures and Documentation

Electrodes for the computerized EEG recordings of frequencies were placed according to the 10/20 system. The records were taken simultaneously with the conventional apparatus (Model ED14 digital, Schwarzer GmbH, München) and the computerized technique. There are differences among computerized EEG techniques; most EEG systems measure the amount of electrical power of the conventional frequency bands. The system I use registers the percentages of different frequencies spread over cerebral regions, i.e., the cartographic distribution of the conventional frequency bands. The computerized color mapping programs show the localization and the percentages of the conventional four frequency bands using five different clinically orientated programs. In addition to this, the distributions of the four frequency bands are obtained separately, and each of the nine programs is available in color prints. Further, a 20-frequency band analysis gives more detailed information in steps of 1 Hz: two delta, four theta, six alpha, and eight beta bands. The apparatus and the software, the Brain Function Monitoring System, were developed by Dr. Itil in Tarrytown, NY, U.S.A. [19–22].

The frequency production and the distribution of these different frequencies over the head of each patient was stored in a big data pool. The different programs of the system show different patterns of frequency distributions. The visual comparison of these patterns over a period of time during the therapeutical intervention allows one to discern different reactions: more faster waves, more slower waves, or no change. A reduction of slow waves and an increase in faster waves, especially alpha waves, was seen together with an amelioration of the glucose uptake using the PET technique [13]. The acceleration reaction therefore signifies a positive reaction on the cerebral metabolism. Clinically, the possibility of this method of drug monitoring facilitates therapeutical decisions: continuing the drug with the same dosage, increasing the dosage, or changing to another drug. In the case of nonresponders, this early decision may save time and money.

From the data bank the registered frequencies were transformed for statistical analyses. A special network and the statistical program SPSS for Windows allowed (after 1993) evaluating individual and group responses following the medication (Wilcoxon, *t*-test, ANOVA, MANOVA) [23, 24].

3 Findings and Developments Documented by the Computerized EEG Method

The large database established by the HZI Research Center Tarrytown, the developer of the computerized EEG system, shows different patterns obtained from healthy young persons, healthy elderly persons, and patients suffering from cerebral diseases, e.g., Alzheimer's. Our own findings concerning healthy volunteers and patients suffering from cerebral abnormalities or progressive diseases are in accordance with the HZI data. In clinical practice, the patterns registered in the first and second mapping program are sufficient for the primary diagnostic screening and initial therapeutical decisions.

3.1 Comparison of Frequencies Between Parkinson Patients and Volunteers

A comparison of findings from healthy persons and Parkinson patients was done on the basis of statistical differences between the amounts of delta, theta, alpha,

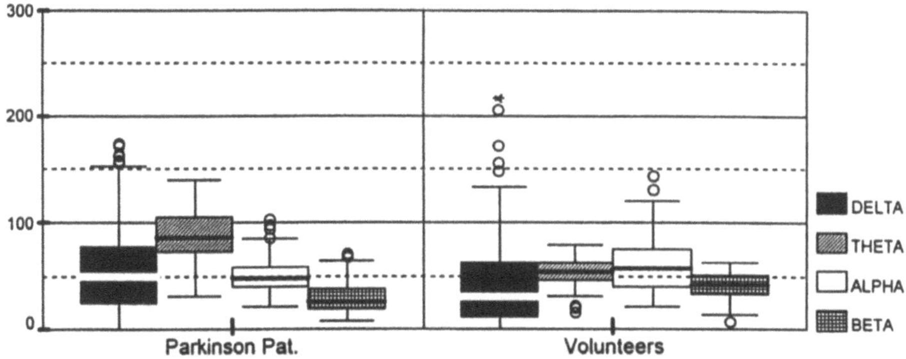

FIG. 1. Comparison of frequencies between volunteers ($n = 13$; age range, 60–89 years) and Parkinson patients ($n = 23$; age range, 44–81 years) with moderate senile dementia of Alzheimer's type (SDAT). Statistical analysis with the Mann–Whitney test for independent samples showed that the two groups are significantly different (delta, 2-tailed, $P = .0018$; theta, 2-tailed, $P << .0001$; alpha, 2-tailed, $P = .0005$; beta, 2-tailed, $P << .0001$)

and beta activity. Figure 1 shows box plots of the means and variance of a group of Parkinson patients ($n = 23$) and a group of healthy volunteers ($n = 13$).

3.2 Ginkgo biloba *Application Studies*

To bridge the gap between medical practitioners and scientists, case descriptions and group results with statistical confirmation are needed for the introduction of a "new" therapeutical principle, a preparation from a biological source. In accordance with national and international guidelines, the author personally informed the patients after the first clinical and computerized EEG examination about the aim, expected benefit, and possible risks of the *Ginkgo* medication. This was done, preferably, in the presence of a patient's adherent or that of another witness.

3.2.1 Case Studies

3.2.1.1 Treatment of a Volunteer with Ginkgo biloba

The healthy volunteer August Sch., 65 years old and a personal school friend of the author, was eager to know the results of the computerized EEG technique concerning his brain wave activity. In the first program (see Fig. 2), he had theta waves over the anterior half of his head (T). On the left occipital there was a slight beta activity (B). Alpha waves were present only over the right posterior part and left parietal (A). As the theta anteriorization was not a normal pattern in the first mapping program, I suggested that he take the EGb 761 preparation of *Ginkgo biloba*, 80 mg two times per day. He returned 5 months later, and his theta activity was nearly replaced by alpha. Subjectively he reported that he felt more active than he had 5 months earlier. Because of the curiosity of this person, preclinical EEG abnormalities (theta anteriorization) were found that were nearly abolished after 5 months of medication.

Figure 3 shows the data compression from the first two programs before and after *Ginkgo* therapy for the same person. With this technique, the means of frequencies are easier to compare but it is disadvantageous for the localized differences. In the second program, six portions of delta activity were abolished after 5 months of therapy. The lower part of Fig. 3 shows the statistical procedure (repeated measures ANOVA) and the levels of significance for each procedure.

3.2.1.2 Four-Year Documentation of the Brain Electrical Activity of Parkinson Patient Kurt Kn. Without and with Ginkgo *Therapy*

Kurt Kn. has suffered from parkinsonism since the age of 58. After he reached the age of 66, he stayed each year for several weeks in my hospital (1988–1992). Figure 4 shows development of occipital wave production as monitored by the 20-frequency band analyses (frequency plots; see legend on right side of figure). In July 1990, the slowing, especially of the delta activity, was more pronounced than in the preceding 2 years and the alpha activity was significantly reduced. This

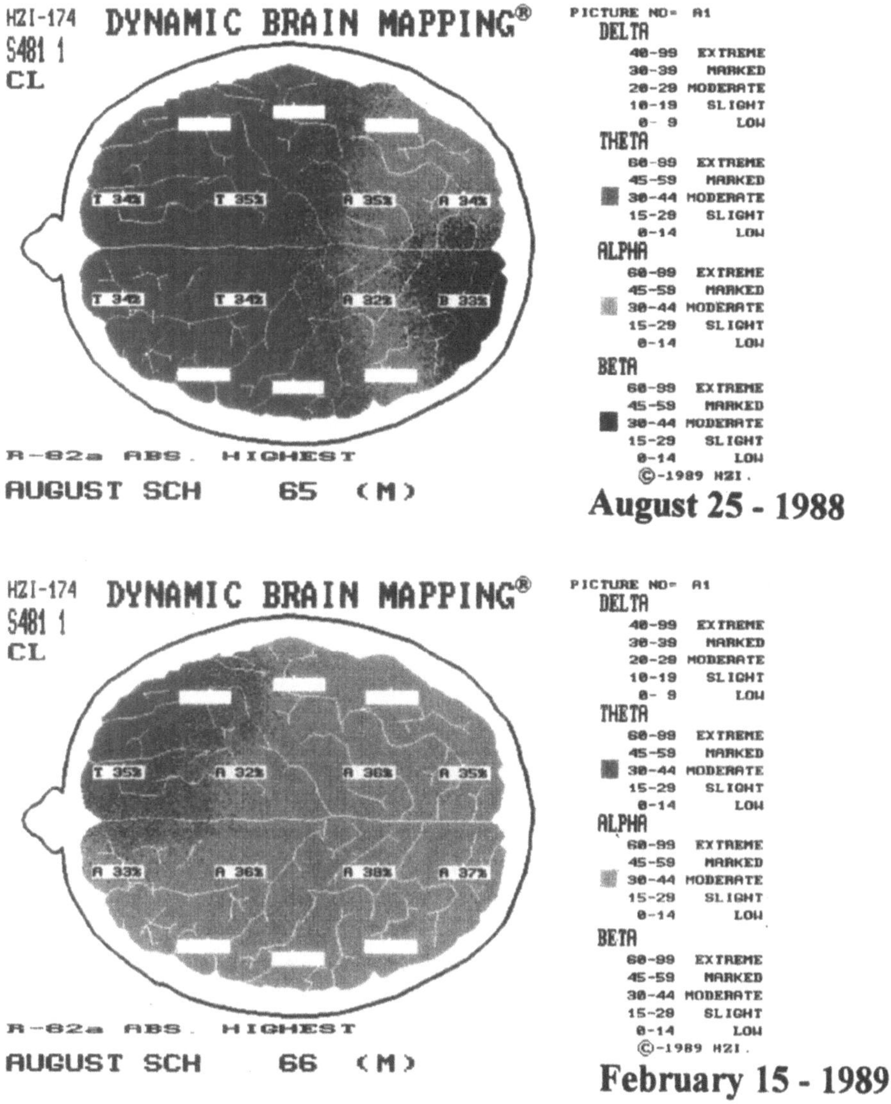

FIG. 2. Development of theta anteriorization before (*top*) and after 5 months of *Ginkgo* therapy (*bottom*) in a healthy volunteer. *I*, theta anteriorization (pathological preclinical sign); *A*, alpha; *B*, beta

progressive slowing, together with clinical deterioration, was the indication for *Ginkgo* treatment. Figure 5 shows the frequency plots before and after 10 infusions of 350 mg of *Ginkgo* extract. The lower part of the figure shows that over the anterior part of the head the delta activity is slightly more pronounced; over the posterior part, however, delta is reduced and theta activity is faster (the

	ABSOLUTE HIGHEST				SECOND HIGHEST			
%	DELTA	THETA	ALPHA	BETA	DELTA	THETA	ALPHA	BETA
1988	-	17	13	4	6	8	12	4
1989	-	4	31	-	-	11	4	16

Ginkgo biloba EGb 761	DELTA	THETA	ALPHA	BETA
Repeated Measures Analysis of Variance	⇓	↓	↑	⇑
Significance of averaged F	n. s.	p = 0,002	p = 0,023	n. s.

FIG. 3. Data compression (from first and second programs) and statistical analysis of computerized EEG data of August Sch., on August 25, 1988, compared to status on February 15, 1989. *Black arrows* indicate favorable results in alpha and theta brain wave activity

occipital maximum before therapy was 5.6–6.5 Hz; after therapy, it was 6.6–7.5 Hz). There was a slight increase in alpha over the occipital part (no statistical significance). Our advice to continue with the oral *Ginkgo* therapy was ignored by the home physician; 1 year later (June 1991) the slowing was even more pronounced (occipital region in Fig. 4). The statistically confirmed slowing (deterioration) is documented in Fig. 6.

As a consequence of this continuing deterioration, we again advised the patient to take the *Ginkgo* preparation continuously. The *Ginkgo*-related development between June 1991 and May 1992 (means) is documented in Fig. 7. After 11 months of *Ginkgo* treatment there is a statistically significant reduction of delta (a shift to theta) and a significant augmentation of alpha wave activity. This finding certifies the favorable *Ginkgo* effect even after 3 years without nootropic treatment, showing a progressive and statistically significant slowing, especially in delta activity.

These two case reports may serve as examples of individual development following *Ginkgo* treatment. Similar but not identical results were seen after treatment with other drugs or, in more advanced states, with a combined treatment of several drugs (a "therapy cascade"). It must be stated, however, that in the cases of nonresponders separate groups were formed for another therapeutical concept. There were no explanations or reasons for resistance to a certain therapy, except in very advanced status of the illness involving several (many) biological systems of the brain. However, in a *single* case of a very

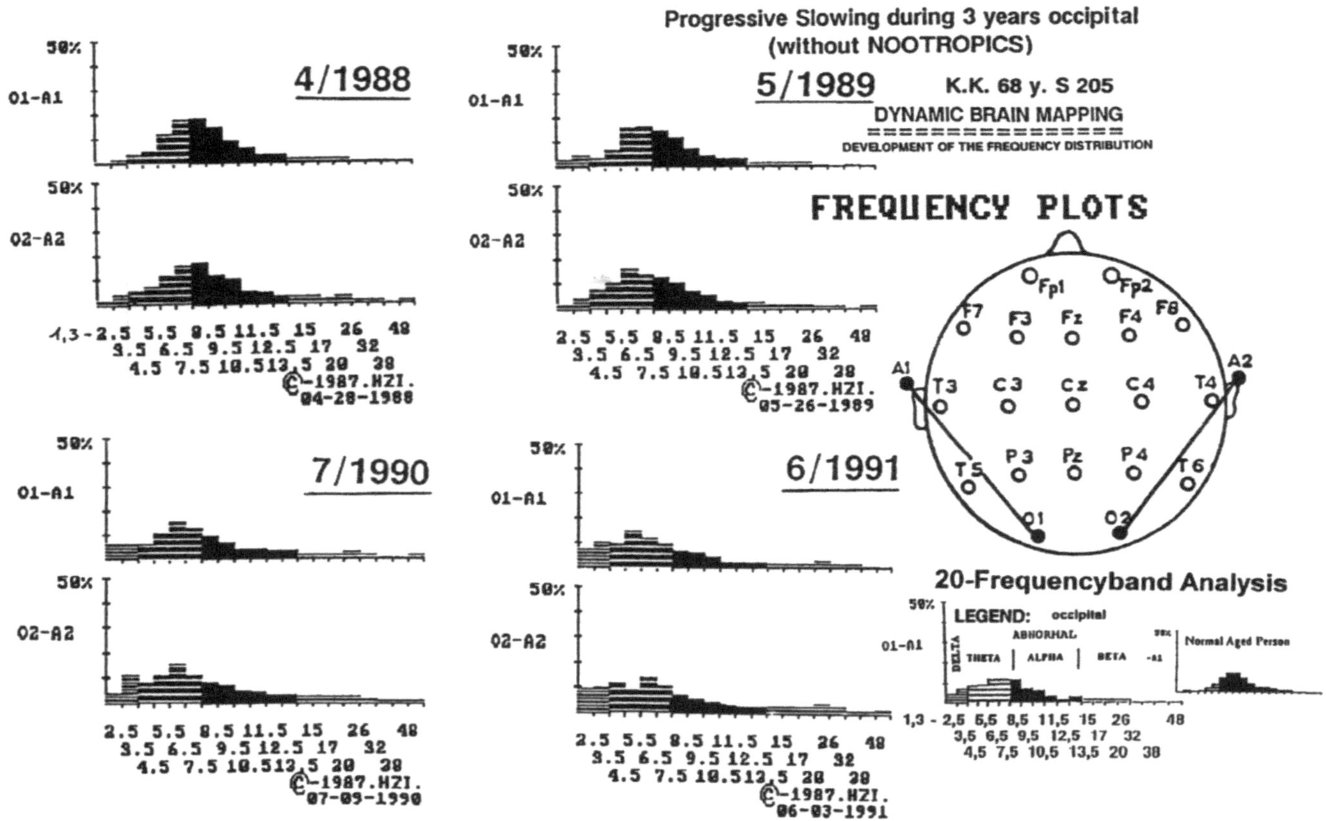

FIG. 4. Development of the 20-frequency band analysis over the occipital regions of Parkinson patient Kurt Kn., suffering from SDAT, within 3 years

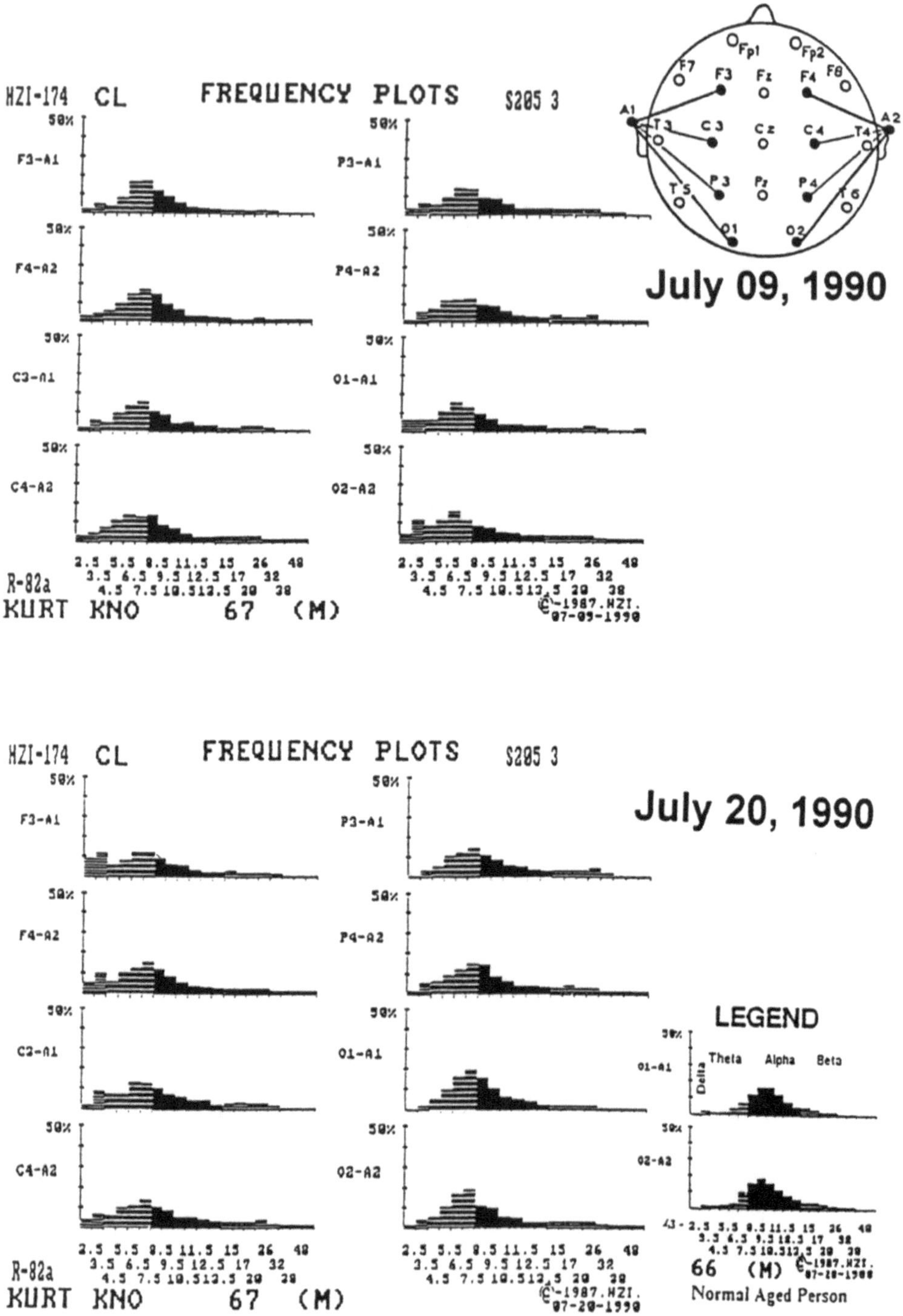

FIG. 5. The 20-frequency band distribution over eight regions of patient, Kurt Kn. before (*top*) and after (*bottom*) *Ginkgo* extract infusion therapy (preparation EGb 761, 10 × 350 mg)

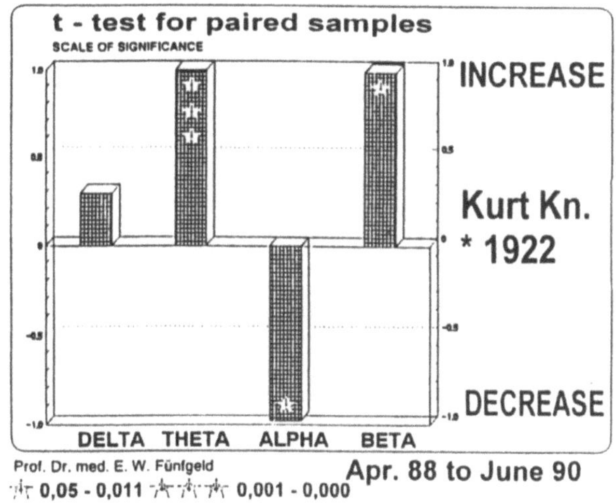

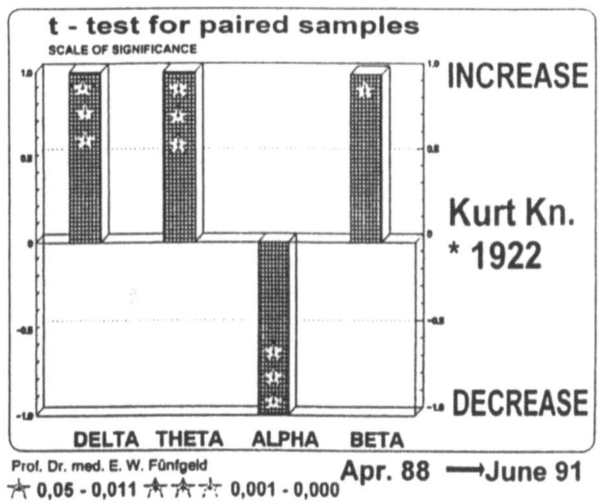

FIG. 6. Statistically confirmed deterioration of computerized EEG findings in patient Kurt Kn. from comparisons between April 1988 and June 1990 and between April 1988 and June 1991. The slowing in the delta band was highly progressive, and at the same time the alpha activity decreased

advanced status of presenile Alzheimer's disease (AD) (starting at the age of 48 years), a therapeutical trial was constantly requested by the patient's husband and sister, and after 6 months an unbelievable amelioration occurred (unpublished data).

At this point an explanation about the essentials of the therapeutic strategies seems suitable. During 8 years and except during the performance of two extensive drug studies, the main purpose was the broad individual-adapted drug trial explained and discussed with the patient and his or her adherents; it was only

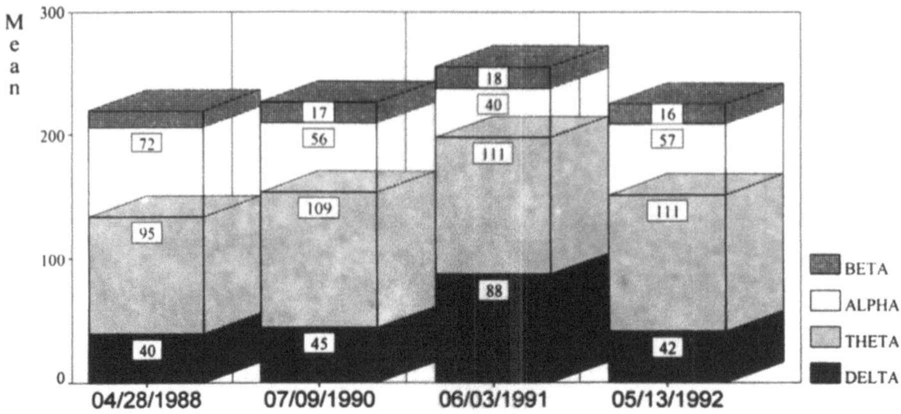

FIG. 7. Progressive slowing during 3 years of non-nootropic treatment is shown by change in distribution of frequencies. Between June 1991 and May 1992, significant reduction of delta and augmentation of alpha activity were seen after 80 mg *Ginkgo* three times daily

effected after the agreement of the patient (Declaration of Helsinki). The formation of the different therapeutical groups and the statistical evaluations were the second step for the scientific work.

3.2.2 Group Study of Parkinson Patients with Concomitant Disturbances of Cerebral Metabolism: Senile Dementia of Alzheimer's Type

During 1987–1992, pathological frequency patterns were seen in 90% of all cases ($n = 2500$). Among these, a group of patients for additional monotherapy with *Ginkgo* were selected.

3.2.2.1 Characterization of the Ginkgo*-Treated Group Between 1988 and 1991*

The group ($n = 84$) was separated into four age groups; 45 patients were older than 70 years (Table 1). The duration of their illness was between 3 and 25 years (mean, 12 years). The scale (after Hoehn and Yahr) of the initial severity of their illness was between 1.5 and 3, or mild to moderate disability [25].

TABLE 1. Age groups of patients who received *Ginkgo* treatment ($n = 84$)

Age group (years)	n
37–59	9
60–69	30
70–79	37
80–88	8

As for treatment strategy, in 1988 and 1989, 25 of 28 patients received a daily dosage of 120 mg, and only 3 patients received dosages up to 240 mg; the duration of the treatment ranged from 10 to 44 days. In 1990 and 1991, 46 of 56 patients received a mean of 130 mg and 10 patients received 10–15 infusions of 200 mg daily; the duration of treatment was from 10 to 37 days in 30 patients and from 6 months to 2 years in 26 patients. In these two groups the reactions (very good and significant computerized EEG acceleration) were similar, and therefore the results were combined.

3.2.2.2 Overview of the Different Significant Computerized EEG Findings Based on Individual Statistical Evaluations

The three main frequency bands showed different reactions, as evaluated using SPSS for Windows at a significance level of $P < .05$. (SPSS Inc. 444 North Michigan Avenue Chicago, IL 60611) The most favorable reaction, reduction of delta and theta plus increase of alpha activity, was observed in 26 patients. In 42 patients a delta reduction was seen combined with an increase of alpha, and 4 patients showed a delta reduction without a change of alpha (different development of theta). In 15 patients only the theta reduction and the alpha increase showed a statistically favorable reaction (see Section 3.2.1.1, case study of August Sch.) because of a low initial delta activity; 9 patients showed a significant rise in alpha without a significant change of delta and theta, and 4 patients showed a rise in theta and alpha combined with a reduction of delta (this may have resulted from a shift from the delta to the theta band). The 14 patients for whom we found no changes in either delta, theta, or alpha activity were classified as nonresponders (17%). Time- or dose-related accelerations were seen, especially with the infusions of *Ginkgo*, within 1–15 days.

The fast part of the frequencies, the beta sector, was not included because of the uncertain meaning of the changes in this sector. In addition, the detailed therapy-related interrelationships of the three frequency sectors were presented: some delta was transferred to theta and some theta activity to alpha. Table 2 shows the absolute figures of the changes from the three conventional frequency sectors at the end of the therapy.

Based on analysis of the frequencies, the action of the *Ginkgo* preparation is documented for the first time in a very differentiated manner. As mentioned earlier, the delta reduction was one of the most prominent and important changes in respect to the amelioration of the cerebral metabolism, in correlation with the

TABLE 2. Summary of the statistically significant changes for each of the three frequency bands

Changes in frequencies	Number of patients
Delta reduction	46
Theta reduction	42
Alpha increase	66

increase in glucose uptake found in PET by Heiss et al. in 1994 [13]. Of 84 patients, only 17 showed no reduction of the slow activities (17%). The responder rate was therefore 83% in the *Ginkgo*-treated Parkinson patients.

4 Discussion

Some reflections about cerebral illness during the aging process and the possibility of therapeutical intervention are necessary to understand the present scientific status. Reviewing the literature during the last 80 years, from Alois Alzheimer's publication to opinions about causes and therapeutical efforts, Coper [26] stated: "All the simple cause/effect models are inappropriate. New concepts are matter for discussion." The Scandinavian research group [27] concluded: "While AD is considered a small, relatively homogeneous group of presenile dementias, SDAT [senile dementia of Alzheimer's type] is considered a heterogeneous group including several subgroups." Our personal findings reached the same conclusions based on the differences concerning reactivity to the drug treatment [15].

Based on extensive PET findings, the functional deficits in "degenerative dementia of the Alzheimer type is characterized by a metabolic disturbance most prominent in the parieto-occipito-temporal association cortex and later in the frontal lobe. In Parkinson's disease, ... glucose metabolism is usually not altered.... Only on development of a dementia, a frequent concomitant of Parkinson's disease, are the metabolic changes typical of AD also in evidence" ([28], pp 473, 479). The same research group found a statistically significant correlation between reduction of glucose uptake and elevated delta activity. The glucose uptake was augmented after a 6-month medication with the nootropic substance phosphatidylserine and again was correlated with a reduction of delta waves [13]. A statistically significant acceleration related to this naturally derived preparation has been recently published [24].

The electrophysiological slowing, out of the normal range, and the special patterns in the elderly and in Parkinson patients with an incipient decline of higher cerebral functions (recent memory, drive, sustained alertness, reduced stress-compensating functions) therefore signify reduced metabolic functions, especially in glucose uptake, and certainly include and affect other metabolic and transmitter levels.

Comparison of the results between different diagnostic methods as EEG brain mapping, magnetic resonance imaging (MRI with 1.5 tesla), and SPECT taken from 11 SDAT patients and 11 patients with vascular dementia (VD) demonstrated the value of each of the methods [29]. Concerning power mapping, it is stated: "EEG brain mapping clearly demonstrates the slowing down of the basis rhythm in the SDAT patients, the main advantage in comparison with routine EEG lies in the statistical possibilities of this method." The value of a 2-year follow-up EEG study including 39 AD/SDAT patients and 33 age-matched normal (control) subjects was summarized as follows [30]:

"... sequential recording of EEG is a valuable tool in the confirmation of Alzheimer's disease in doubtful cases and in the differentiation of this disease from normal ageing."

An extensive research comprising 23 probable AD patients [EEG spectral analyses; cerebrospinal fluid (CSF) neurotransmitter markers] with 10 autopsies revealed [31] CSF acetylcholinesterase content correlated with delta power and alpha/delta ratio; a low level of cortical cholin acetyltransferase (ChAT); and reduced cell numbers of the nucleus basalis Meynert correlated with a higher delta power in four demented patients compared to six age-matched controls. "The lowest levels of ChAT as well as noradrenaline were observed in the two demented patients with largest delta power" ([31], p 442). These findings may support our impression that delta activities, even in a small amount as seen in the case of the volunteer (second program), were more associated with a (pathological?) aging process than the theta frequency band (except in the first mapping program as theta anteriorization).

A broad survey of the pharmacological activities of the *Ginkgo* extract and its clinical applications, provided by DeFeudis [32], served as a source book until 1990. It was concluded: "Two major 'concepts of action' have arisen from my analysis of the studies performed thus far. These may be termed 'cognition-enhancing' and 'vasoregulatory' actions. Taken together, these concepts appear useful in defining, although in a preliminary way, the polyvalent mechanism that underlies the beneficial actions of EGb 761 in managing cerebral insufficiency states, peripheral occlusive arteriopathies and associated neurosensory disturbances" (from Prologue).

These findings were not included in the third edition of a well-known pharmacological textbook [33]. Further confirmations of the favorable actions of the *Ginkgo* extract were found in cases of open-heart surgery in humans [34]: the reperfusion-induced free radicals were significantly diminished after a preceding 5-day treatment with 320mg/day *Ginkgo* orally.

Hofferberth [35] found in 24 preselected patients an elevated theta content in relation to alpha and, using a special blood test, an elevated cerebral vessel rish factor. As a third parameter the results of the reaction time test taken by the Vienna determination system were included. The study was effected under standard conditions (washout; 0, 4, and 12 weeks test of the verum drug against placebo). Favorable changes were observed after 4 and 12 weeks.

Because neuronal plasticity has been a matter of discussion during recent years, a review of the experimental data on this possible action of *Ginkgo* was carried out [36]. Among other observations, a dose-dependent effect on learning of aged rats (21 days) and an amelioration of the scores in three psychological tests in 156 elderly persons [37] emphasized the opinion: "Given the clinical and experimental findings, the hypothesis of EGb 761 having an effect on neuroplasticity is indeed probable and deserves to be the subject of further research to be undertaken using the tools of modern neurobiology" ([37], p 125). It might be possible that findings of an acceleration after a first *Ginkgo* infusion (see [24]) can be explained by an enlarged involvement of nerve cells (or nerve cell complexes) connected with the primary (leading) functional complex. Findings after chronic residuals of a stroke [38] may support this hypothesis.

In 1995, as mentioned previously [24], the acute and statistically significant acceleration 3 h after a first infusion with the *G. biloba* extract was published, and the favorable biological action on the cerebral functions was estimated as "an indispensable element for the clinical efficacy appearing subsequently." This raised the question whether double-blind studies may be taken as an indispensable part of the proof of a cerebral action of a drug. So-called drug monitoring by the highly sophisticated frequency analysis with Itil's adaptation (interface) and statistical programs may serve as a screening for the differentiation of the efficacy of cognitive enhancers in a short period of time and in a similar objective manner. The latest highly sophisticated, double-blind, placebo-controlled multicenter study [37] comprises 156 patients over a treatment period of 24 weeks. Pharmaco-EEG studies in 17 verum and 19 placebo cases were set into the power spectral analysis. It was stated "..., pre-to-post differences were calculated for the individual frequency ranges, but these showed no statistically verifiable treatment-related differences after the 24-week treatment period. In patients treated with EGb 761, however, the dominant frequency of the right occipito-temporal (O2–T6) lead increased under vigilance control (RT) conditions and under both five-minute periods of the resting record (RS1 and RS2). This did not occur under treatment with placebo. This effect reached statistical significance ($P = .0374$, descriptive). Under the resting condition RS1, with an average increase in frequency in the EGb 761 group from 8.87 Hz (SD = 0.77 Hz) to 9.15 Hz (SD = 0.79 Hz) and, at the same time, an average decrease in frequency in the placebo group from 9.5 Hz (SD = 1.34 Hz) to 9.32 Hz (SD = 1.37 Hz). Differential treatment effects were observed mainly in the theta frequency range, reaching a maximum over the central region" ([37], p 53). Kanowski's findings of the theta power [37] were arranged in a table and put in opposition, verum/placebo reaction; the Fisher–Yates test revealed no significance ($P = .1626$) (Table 3).

The differences concerning the pre and post findings to our study are apparently related to the mode of data storage and analysis, power mapping versus frequency mapping. The majority of our patients (46 of 84) showed a significant delta reduction, which is in concordance with the findings of Heiss et al. [13]; 42 of 84 patients showed a statistically significant reduction of delta combined with

TABLE 3. Principal results of the pharmacological EEG studies by Kanowski et al. [37]

	Theta increase	Theta reduction
Ginkgo EGb 761	7	10
Placebo	12	7

By Fisher–Yates test, results are not significant at $P = .1626$.

a significant rise in alpha. These findings are in concordance with our previous study showing the frequency changes 3h after a first *Ginkgo* infusion of 200–350mg ($n = 21$): 10 patients reacted with a significant delta reduction together with an augmentation of alpha [23]. When 2 nonresponder patients were included, the Wilcoxon matched-pairs signed ranks test (from SPSS for Windows) revealed delta-activity reduction ($P = .001$) and alpha augmentation ($P < .001$). In summary the frequency analysis is apparently much more effective in detecting treatment effects.

It is comprehensible that different preparation procedures of the *Ginkgo biloba* leaf extract may cause different biological reactions. A bioequivalence study of three different *Ginkgo* preparations revealed that only the preparation according to EGb 761 acts in the direction of a statistically significant alpha activation using the same frequency mapping method [39].

5 Conclusions

Ginkgo biloba extract prepared after the method of EGb 761 is a potent drug influencing chronic cerebral metabolic disturbances in elderly Parkinson patients. The preparation also acts on preclinical latent metabolic changes, which can be seen by the theta anteriorization in the first mapping program of the brain mapping system [19–22]. Drug monitoring by frequency analysis revealed much more objective information than methods using the power spectrum.

Acknowledgments. This publication was made possible by private grants to the "Verein zur Förderung der Erforschung und Behandlung von Hirnstoff wechselstörungen bei Patienten mit Bewegungsstörungen insbesondere der Parkinsonkrankheit e.V., Marburg."

This work is dedicated to Prof. Wolfgang Wesemann on the occasion of his 65th birthday and to the memory of the late Ingeborg Flechtner for her sustained encouragement.

References

1. Fünfgeld EW (1987) Neue therapeutische Möglichkeiten für Parkinson-Patienten. Praxis-Kurier 25:25–28
2. Fünfgeld EW (1987) Editorial: Multitherapie der Parkinson-Krankheit und der senilen Demenz vom Alzheimer Typ. Therapiewoche 37:4521–4522
3. Fünfgeld EW, Stalleicken D (1987) Dynamic-brain-mapping. TW Neurol Psychiatr 2:136–142
4. Fünfgeld EW (1988) Editorial: Dementias—a growing social and medical problem. In: Fünfgeld EW (ed) Rökan (*Ginkgo biloba*). Recent results in pharmacology and clinic. Springer, Berlin Heidelberg New York, pp 11–16
5. Fünfgeld EW, Stalleicken D (1988) The clinical effect of *Ginkgo biloba* extract in the case of cerebral insufficiency documented by dynamic-brain-mapping—a

computerized EEG evaluation. In: Fünfgeld EW (ed) Rökan (*Ginkgo biloba*). Recent results in pharmacology and clinic. Springer, Berlin Heidelberg New York, pp 302–313

6. Fünfgeld EW (1989) Nootropika-Wirkungen bei präseniler und seniler (Alzheimer) Demenz—Elektrophysiologische Reaktionen beim Dynamic Brain Mapping. In: Saletu B (ed) Biologische Psychiatrie. Thieme, Stuttgart, pp 203–206
7. Fünfgeld EW (1989) Dynamic Brain Mapping und 20-Frequenzband-Analyse: Therapie-Effizienz-Kontrolle und Vergleich mit konventionellen EEG-Befunden. TW Neurol Psychiatr 3:405–422
8. Fünfgeld EW(1989) Computergestützte EEG-Diagnostik-Möglichkeit einer kurzfristigen Erkennung hirnelektrischer Reaktionen bei Patienten mit zerebralen Abbauprozessen (SDAT). TW Neurol Psychiatr 3:520–529
9. Fünfgeld EW (1989) A natural and broad spectrum nootropic substance for treatment of SDAT—The *Ginkgo biloba* extract. In: Iqbal K, Wisniewski HM, Winblad B (eds) Alzheimer's disease and related disorders. Liss, New York, pp 1247–1260
10. Fünfgeld EW (1989) Dynamic brain mapping in dementia. In: Stefanis CN, Soldatos CR, Rabavilas AD (eds) Psychiatry: a world perspective, vol 2. Neuroscience in psychiatry: biological correlates of mental disorders. Elsevier, Amsterdam, pp 463–468
11. Fünfgeld EW (1990) Strategies for treatment in SDAT. In: Hasegawa K, Homma A (eds) Psychogeriatrics: biomedical and social advances. Excerpta Medica, Tokyo, pp 205–208
12. Fünfgeld EW, Stalleicken D (1992) Objektivierung der klinischen Effekte von *Ginkgo-biloba*-Extrakt bei cerebraler Insuffizienz mittels Dynamic-Brain-Mapping. In: Diehm C, Müller D (eds) Rökan. *Ginkgo biloba* EGb 761. Band 2, Klinik. Springer, Berlin Heidelberg New York, pp 117–128
13. Heiss WD, Kessler J, Mielke R, Szelies B, Herholz K (1994) Long-term effects of phosphatidylserine, pyritinol, and cognitive training in Alzheimer's disease—a neuropsychological, EEG, and PET investigation. Dementia 5:88–98
14. Fünfgeld EW (1995) Computerised brain electrical activity findings of Parkinson patients suffering from hyperkinetic side effects (hypersensitive dopamine syndrome) and a review of possible sources. J Neural Transm Suppl 46:347–361
15. Fünfgeld EW (1993) Monotherapy versus multitherapy—combined therapeutical approaches for treatment of senile dementia of Alzheimer's type (SDAT, AD). In: Beregi E, Gergely IA, Raiczi K (eds) Recent advances in aging sciences. Monduzzi, Bologna, pp 717–722
16. Fünfgeld EW (1992) Beobachtungen über den Einfluß von nativem Phosphor auf Verhalten und elektrophysiologische Parameter—insbesondere mit dem Dynamic Brain Mapping. In: Gaebel U, Laux G (eds) Biologische Psychiatrie, Synopsis 1990/1991. Springer, Berlin Heidelberg New York, pp 389–393
17. Fünfgeld EW (1993) The trace element zinc acting on different neuropsychiatric diseases. In: Anke M, Meissner D, Mills CF (eds) Trace elements in man and animals, tema 8. Proceedings of the eighth international symposium on trace elements in man and animals. Verlag Media Touristik, Gersdorf, pp 933–937
18. Fünfgeld EW (1994) Ein weiterer multitherapeutischer Zugang zur Therapie der senilen Demenz vom Alzheimer-Typ (SDAT): Die Therapie mit dem Spurenelement Zink. In: Kuhn W. Büttner T, Heinemenn W, Frey C, Schneider K, Zierden E, Przuntek H (eds) Altern, Gehirn und Persönlichkeit. Verlag Hans Huber, Bern, pp 345–349

19. Itil TM, Shapiro DM, Eralp E, Akman A, Itil KZ, Garbizu C (1985) A new brain function diagnostic unit, including the dynamic brain mapping of computer analyzed EEG, evoked potential and sleep (a new hardware/software system and its application in psychiatry and psychopharmacology). New Trends Exp Clin Psychiatry 1:107–177
20. Itil TM, Mucci A, Itil KZ (1990) The computer-analyzed electroencephalogram (CEEG) and brain mapping in the diagnosis of psychogeriatric patients. Int J Technol/Aging 3:29–37
21. Itil TM, Itil KZ (1990) Neurophysiological aspects of aging. In: Stefanis CN, Soldatos CR, Rabavilas AD (eds) Psychiatry: a world perspective, vol 2. Neuroscience in psychiatry: biological correlates of mental disorders. Elsevier, Amsterdam, pp 439–447
22. Itil TM, Slone HL, Itil KZ (1990) Brain function evaluation in normal and pathological aging and during its treatment. In: Bergener M, Finkel SI (eds) Clinical and scientific psychogeriatrics, vol 2. Springer, New York, pp 136–161
23. Fünfgeld EW (1995) Statistische Sicherung der zerebralen Wirkung von Nootropika mit der computerisierten 20-Frequenzband-Analyse des EEG bei präseniler bzw. seniler Alzheimerscher Erkrankung (AD/SDAT). 1. Mitteilung: Methoden, Beobachtungen unter der Medikation mit *Ginkgo-biloba*-Extrakt. Geriatr Forsch 5:41–54
24. Fünfgeld EW (1996) Statistische Sicherung der zerebralen Wirkung von Nootropika mit der computerisierten 20-Frequenzband-Analyse des EEG bei präseniler bzw. seniler Alzheimerscher Erkrankung (AD/SDAT). 2. Mitteilung: Beobachtungen unter der Medikation Phosphatidylserin. Geriatr Forsch 6:19–33
25. Hoehn MM, Yahr MD (1967) Parkinsonism: onset, progression, and mortality. Neurology 17:427–442
26. Coper H (1990) Drug treatment of dementia. In: Maurer K, Riederer P, Beckmann H (eds) Alzheimer's disease. Epidemiology, neuropathology, neurochemistry and clinics. Springer, Wien, pp 545–554
27. Gottfries CG, Blennow K, Regland B, Wallin A (1990) Alzheimer's disease—one, two or several? In: Maurer K, Riederer P, Beckmann H (eds) Alzheimer's disease. Epidemiology, neuropathology, neurochemistry and clinics. Springer, Wien, pp 343–356
28. Heiss W-D, Szelies B, Adams R, Kessler J, Pawlik G, Herholz K (1990) PET criteria for diagnosis of Alzheimer's disease and other dementias. In: Maurer K, Riederer P, Beckmann H (eds) Alzheimer's disease. Epidemiology, neuropathology, neurochemistry and clinics. Springer, Wien, pp 473–487
29. Lechner H, Niederkorn K, Logar CH, Schmidt R, Fazekas F, Horner S, Valetitsch H (1990) Results of EEG brain mapping and neuroimaging methods in senile dementia of Alzheimer's type (SDAT) and vascular dementia (VD). In: Maurer K, Riederer P, Beckmann H (eds) Alzheimer's disease. Epidemiology, neuropathology, neurochemistry and clinics. Springer, Wien, pp 429–435
30. Streifler M, Simanyi M, Fischer P, Danielczyk W (1990) EEG- and cognitive changes in Alzheimer's disease—a correlative follow-up study. In: Maurer K, Riederer P, Beckmann H (eds) Alzheimer's disease. Epidemiology, neuropathology, neurochemistry and clinics. Springer, Wien, pp 447–458
31. Riekkinen PJ, Riekkinen PJ Jr, Soininen H, Reinikainen KJ, Laulumaa V, Partanen VSJ, Halonen T, Paljärvi L (1990) Regulation of EEG delta activity by the cholinergic nucleus basalis. In: Maurer K, Riederer P, Beckmann H (eds) Alzheimer's disease.

Epidemiology, neuropathology, neurochemistry and clinics. Springer, Wien, pp 437–445
32. DeFeudis FV (1991) *Ginkgo biloba* extract (EGb 761): pharmacological activities and clinical applications. Elsevier, Amsterdam
33. Rang HP, Dale MM, Ritter JM (1995) Pharmacology, 3rd edn. Churchill Livingstone, Edinburgh
34. Culcasi M, Pietri S, Carriere I, d'Arbigny P, Drieu K (1993) Electron-spin-resonance study of the protective effects of *Ginkgo biloba* extract (EGb 761) on reperfusion-induced free-radical generation associated with plasma ascorbate consumption during open-heart surgery in man. In: Ferradini C, Droy-Lefaix MT, Christen Y (eds) *Ginkgo biloga* extract (EGb 761) as a free-radical scavenger. Advances in *Ginkgo biloba* extract research, vol 2. Elsevier, Paris, pp 153–162
35. Hofferberth B (1991) Simultanerfassung elektrophysiologischer, psychometrischer und rheologischer Parameter bei Patienten mit hirnorganischem Psychosyndrom und erhöhtem Gefäßrisiko—Eine Plazebo-kontrollierte Doppelblindstudie mit *Ginkgo-bioloba*-Extrakt EGb 761. In: Stodtmeister R, Pillunat LE (eds) Mikrozirkulation in Gehirn und Sinnesorganen. Enke Verlag, Stuttgart, pp 64–74
36. Cesaro P (1996) Neuronal plasticity: a mechanism in the clinical effects of the standardized *Ginkgo biloba* extract. In: Christen Y, Droy-Lefaix MT, Macias-Nunez JF (eds) Effects of *Ginkgo biloba* extract (EGb 761) on neuronal plasticity. Advances in *Ginkgo biloba* extract research, vol 5. Elsevier, Paris, pp 121–126
37. Kanowski S, Herrmann WM, Stephan K, Wierich W, Hörr R (1996) Proof of the *Ginkgo biloba* special extract EGb 761 in outpatients suffering from mild to moderate primary degenerative dementia of the Alzheimer type or multi-infarct dementia. Pharmacopsychiatry 29:47–56
38. Fünfgeld EW, Baggen M, Klose W (1988) Computergestützte neurophysiologische Initialbefunde und klinischer Verlauf. Psycho 14:383–384
39. Itil TM, Martorano D (1995) Natural substances in psychiatry (*Ginkgo biloba* in dementia). Psychopharmacol Bull 31:147–158

Recent Advances in *Ginkgo biloba* Extract (EGb 761)

Wiltrud Juretzek

1 Manufacture and Composition of EGb 761

A special standardized and purified extract from *Ginkgo biloba* leaves, described in the literature as EGb 761, is the only *Ginkgo* extract that has been subjected to comprehensive pharmacological, toxicological and clinical investigations. EGb 761 is made from dried *Ginkgo* leaves, extracted with acetone/water, drug/extract ratio 35–67:1, standardized to contain 22% to 27% flavone glycosides and 5% to 7% terpene lactones, of which approx. 2.8%–3.4% consist of ginkgolides A, B, and C, as well as approximately 2.6%–3.2% bilobalide; the content of ginkgolic acids is below 5 ppm.

The development of the special extract EGb 761 was targeted to enrich active and desirable substances and to remove or reduce inactive or unwanted elements (Table 1).

To guarantee the quality of EGb 761, a great deal of care is taken in selecting the necessary plant material. The conditions of cultivation and harvesting are subjected to accurate standardization and control. The leaves are supplied from *Ginkgo* plantations in South Korea, France (Bordeaux), or the United States (South Carolina). The plants receive no chemical treatment of any form; prior to processing, the dried leaves are tested for foreign matter, heavy metal, microbiological purity, pesticides, aflatoxins, and ethylene oxide.

The manufacturing process is strictly standardized. The dried leaves are subjected to a water/acetone extraction process under partial vacuum. After removal of the organic solvent, inactive and undesirable substances are eliminated or reduced by highly specific processes, and the active substances in the leaves, i.e., flavone glykosides, ginkgolides, and bilobalide, are enriched by more than 20 processing steps. The drug/extract ratio is 50:1, the pharmacological active substances are enriched by a factor of 50 to 100, and approximately 90% of the extractable substances are eliminated during production. Removal of the pharmaceutically or toxicologically problematic active substances is carried out in

Schwabe Pharmaceuticals, Willmar-Schwabe-Str. 4, D-76227 Karlsruhe, Germany

TABLE 1. Constituents of *Ginkgo biloba* leaves; their presence (+) or absence (−) in EGb 761

Flavonoids	
ginkgoflavone glycosides	+
ginkgoflavone aglycones	−
biflavonoids	−
proanthocyanidines	+
Catechins	−
Non-flavonoid glycosides	+
Terpene lactones	
ginkgolides A, B, C	+
bilobalide	+
Other terpenes	
polyprenols	−
Steroids	−
Organic acids	
ginkgolic acids	−
other organic acids	+

several production steps. Higher molecular tannins (catechin structures) are depleted in a precipitation phase. This also applies for protein compounds. These macromolecules are eliminated because, although they are not absorbed in the gastrointestinal tract, they are able to reduce the bioavailability of the active substances to a considerable extent. The lipophilic ginkgolic acids, which are toxicologically important, are eliminated. Finally, the bioflavonoids are eliminated, because they have a tendency to form precipitates that are difficult to dissolve. Among other factors, these purification steps contribute to the high tolerance of the extract.

2 Indications

The indications of EGb 761 are symptomatic treatment of mild to moderate cerebral insufficiency (demential syndromes in primary degenerative dementia, vascular dementia, and mixed forms of both) with the following symptoms: memory deficits, disturbances in concentration, depressive emotional condition, dizziness, tinnitus and headache [1]; symptoms of intellectual deficit in the elderly (disturbances in concentration and memory) [2–5]; improvement of pain-free walking distance in peripheral arterial occlusive disease such as intermittent claudication within the framework of physical therapeutic measures, in particular locomotory training [1]; vertigo or tinnitus of vascular and involutive origin [1].

Primary degenerative dementia (Alzheimer's disease) and vascular dementia are the most common causes of age-associated dementia [6]. The prevalence of dementia increases with age, particularly over age 75 years, with a prevalence of 20% or more over age 85 years [7]. In Japan, vascular dementia is the most common cause of dementia, responsible for up to 50% of all clinical cases [6]. Dementia is a clinical state defined as an acquired impairment of intellectual and memory functioning associated with disease of the brain not accompanied by disturbances on the consciousness level and that is of such severity that it interferes with social or occupational functioning. In the early stages of the disease, memory impairment may be an isolated difficulty [8]. Because Alzheimer's disease is based on heterogenous pathogenesis, a variety of different mechanisms may contribute to the characteristic brain damage [9]. Vascular dementia is due to cerebrovascular disease and cerebrovascular disease may worsen the dementia of the Alzheimer type [6].

Intermittent claudication, i.e., limited pain-free walking distance, is a symptom of chronic peripheral arterial occlusive disease and a common condition in the elderly, occurring in 3% to 20% of individuals over the age of 65 years [10]. Intermittent claudication is predominantly due to atherosclerosis of the legs [10].

No specific therapy exists for these complex disease states. The multifactorial etiopathologies necessitate a polyvalent therapy, involving neuroprotection, reduction of age-related neurotransmitter deficits, inhibition of oxidative damage by free radical mechanisms, effects on blood vessels including improvement of hemorheology, and microcirculation. EGb 761 has a complex range of pharmacological activities that are of particular interest with regard to its therapeutic use.

3 Experimental Pharmacology of EGb 761

3.1 Pharmacological Effects on the Brain

3.1.1 Neuro(cerebro)protection

In vivo. EGb 761 improved brain membrane fluidity in the aging mouse [11]. Mice and rats treated with the extract or the non-flavone fraction survived under hypoxic conditions for longer periods of time than untreated controls [12–15]; the concentration of glucose, creatinine phosphate, and ATP in the brain tissue was significantly higher, the content of lactate significantly lower than in untreated hypoxic animals [13, 14]. Bilobalide completely protected rat hippocampal cells from phospholipid breakdown under anoxia [16]. In various cerebral ischemia models, the administration of EGb 761 to rats and Mongolian gerbils reduced the mortality rate, the extent of neurological deficits, and the development of cerebral edema; the mitochondrial respiration and energy metabolism were improved [17–20]. The ginkgolides, potent PAF (platelet activating factor) antagonists, reduced postischemic neurological deficits and cerebral edema in Mongolian gerbils, the infarct area in the mouse brain, and neuronal damage in rat brain hippocampus and neocortex [21–24]. Bilobalide diminished the infarct volume in the mouse and rat brain [23]. An inhibitory effect of EGb 761 on the traumatic or cytotoxic cerebral edema induced by triethyltin; hexachlorophene, or bromethaline has been demonstrated in rats [25–31]. Bilobalide also had antiedematous effects on the cytotoxic triethyltin and hexachlorophene edema in the rat brain [28]. EGb 761 reduced brain damage following heat stress in the rat; the extract attenuated the hyperthermic response and reduced blood-brain permeability, brain edema, and neuronal cell changes [32]. The electroconvulsive shock-induced accumulation of free fatty acids and diacylglycerols in the hippocampus of rats was decreased by EGb 761 [33]. Long-term treatment with EGb 761 reduced age-dependent structural changes in the hippocampus of the mouse [34]. In the model of motor cortex ablation, EGb 761 and the terpene-free extract produced a dose-dependent acceleration of behavioral recovery and diminished cerebral ventricular dilation in rats [35].

In vitro. EGb 761 prevented the decrease in membrane fluidity of mouse cortex synaptosomes induced by free radicals [36]. The ginkgolides A, B, and bilobalide protected cultured neurons (hippocampal neurons from newborn rats or cultured neurons from embryonic chick telencephalon) against damage induced by glutamate (ginkgolides, bilobalide) or hypoxic damage (bilobalide) [22, 23]. Bilobalide completely prevented the anoxia-induced increase of choline efflux in rat hippocampal cells [16].

3.1.2 Influence on the Central Cholinergic System and on Age-Related Deficits in Neurotransmitters

In vivo. EGb 761 attenuated the amnesia induced by scopolamine [37]. In a way similar to that of cholinomimetics EGb 761 inhibited the spontaneous cage-climbing behavior of mice; scopolamine antagonized these effects of EGb 761 almost completely. Analgesic and hypoxic effects of cholinomimetics were potentiated by EGb 761, whereas cholinomimetic effects such as hypothermia, tremor, or salivation were not modified by the extract [38]. EGb 761 increased the high-affinity choline uptake into hippocampal synaptosomes after long-term administration to old rats in a statistically significant manner as compared to the reference group treated with tacrine [39]. In aged rats, chronic treatment with EGb 761 led to an increase in the density of muscarinic receptors in the hippocampus [40] and of α_2-adrenoceptors and 5-HT(hydroxytryptamine)$_{1A}$-receptors in the cortex [41, 42]. EGb 761 prevented the stress-induced desensitization of hippocampal 5-HT$_{1A}$-receptors in aged rats [43].

3.1.3 Improvement of Memory and Learning

In vivo. Continued administration of EGb 761 to young mice (9 and 13 weeks of age) facilitated learning ability and memory processes in an appetitive conditioning test [44]. In the passive avoidance test, chronic treatment with EGb 761 improved short-term memory in aged mice but not in young ones [11]. In the aged rat, chronic administration of EGb 761 ameliorated learning in a radial maze test [45].

3.2 Pharmacological Effects on the Vascular System

3.2.1 Effects on Blood Vessels

In vivo. EGb 761 suppressed the arteriolar vasospasm induced by the topical application of autologous serum on the brain surface of the rabbit [46] and reduced the increased permeability of the blood-brain barrier of rats, induced by hypertension, cerebral ischemia, or heat stress [47, 32].

In vitro. EGb 761 elicited a concentration-dependent relaxation of precontracted aortic strips by a mechanism involving endothelium-derived relaxing factor (EDRF) [48]. Incubation of rat aorta strips with EGb 761 strongly potentiated PGL_2 (prostacyclin) production [49]. The extract caused dose-related contractions in strips of isolated rabbit vena cava preparations that could be partly inhibited by the α-blocker phenoxybenzamine [50]. The extract and bilobalide protected endothelial cells from hypoxia-induced ATP decrease [51].

3.2.2 Cerebral and Peripheral Blood Flow

In vivo. Intravenous infusion of EGb 761 caused a significant increase in pial arteriolar diameter in cats [52]. In rats, EGb 761 increased the local blood flow in 39 analyzed anatomically defined brain structures by 50% to 100% [14, 53]. The active constituents of the extract responsible for increasing cerebral blood flow appear to be the non-flavonoid compounds, whereas administration of the flavone fraction produced only marginal effects on flow values [14]. Under ischemic conditions, the decrease in cerebral blood flow after embolization was less pronounced if the rats had been pretreated with EGb 761 [19]. Post-ischemic administration of EGb 761 to rats significantly increased cerebral blood flow and reduced hypoperfusion [15].

3.2.3 Blood Rheology

In vivo. Microthromboses induced in pial arteries by electrical stimulation or in small cortical veins by sodium lactate were reduced by EGb 761 pretreatment in rabbits [54].

In vitro. EGb 761 increased the osmotic resistance of rat erythrocytes and reduced the aggregation of human erythrocytes [55, 56]. The extract protected human erythrocytes from oxidative damage [57] and inhibited ADP-induced aggregation of rat platelets and PAF-induced aggregation of human platelets [49, 58]; the ginkgolides inhibited PAF-induced aggregation of rabbit platelets [59].

3.2.4 Free-Radical Scavenging Effects

In vivo. In rats EGb 761 inhibited the cisplatin-, bromethalin-, triethyltin- or Fe^{2+}/ascorbate-induced lipid peroxidation [30, 31, 60, 61].

In vitro. The free-radical scavenging properties of EGb 761 that protect membranes from peroxidative degradation and damage have been demonstrated in various in vitro models [62–68]; the lipid peroxidation induced by NADPH-Fe^{3+}/ADP systems [62, 68], doxorubicin [66], or ciclosporin A [63] was inhibited by EGb 761. The flavonoid constituents appear to be responsible for the free-radical scavenging activity [15, 62, 68], but the presence of the terpenes could contribute indirectly to this effect [67].

3.3 Pharmacological Effects on the Vestibular and Auditory System

In vivo. EGb 761 improved the ultrastructure qualities of vestibular sensory epithelia of mice when the tissue was fixed by vascular perfusion [69]. In unilateral vestibular deficits following labyrinthectomy in rats or vestibular neurectomy in cats, postoperative treatment with EGb 761 accelerated the compensation of postural, locomotor, and oculomotor disorders [70, 71]; EGb 761 accelerated recovery of the neuronal properties in deafferented neuronal cells [71].

4 Toxicological Effects of EGb 761

Therapy with EGb 761 may be regarded as very safe. This conclusion is supported by the results from toxicological studies on animals. It has not been possible to determine the acute toxicity (median lethal dose, LD_{50} value) of orally administered EGb 761 in the rat, and in the mouse the LD_{50} value is 7.73 g/kg. After intravenous administration in both rat and mouse, its LD_{50} value is about 1.1 g/kg, and LD_{50} values after intraperitoneal injection are 1.9 g/kg in mice and 2.1 g/kg in rats. Chronic toxicity studies in the rat (27 weeks) and dog (26 weeks), conducted with daily doses of 20 and 100 mg/kg initially, and then gradually increased to 800, then to 400 and 500 mg/kg in rats, and to 300, then 400 mg/kg in dogs, showed no evidence for organ damage and no impairment of hepatic or renal function. Also, daily oral administration of 100, 400, or 1600 mg/kg of EGb 761 to rats and 100, 300, or 900 mg/kg to rabbits did not elicit teratogenic effects or affect reproduction. EGb 761 exhibited no mutagenic activity in the usual screening tests. In a carcinogenicity study on rats, no increase in the incidence of pathological organ changes (macroscopic or microscopic) and no increase in mortality was recorded on continuous administration of EGb 761 in dosages of 4, 20, and 100 mg/kg daily over a period of two years [72].

5 Pharmacokinetics

An absorption rate of 60% was found after oral administration of [^{14}C] EGb 761 in the rat. Glandular, neuronal, and ocular tissues accumulated the labeled substances of the extract to the greatest extent. In healthy volunteers after oral application of EGb 761, the absolute bioavailability was 98%–100% for ginkgolide A, 79%–93% for ginkgolide B, and at least 70% for bilobalide [72].

6 Clinical Pharmacology and Efficacy

6.1 Clinical Pharmacology

In healthy volunteers, EGb 761 showed potent alpha-enhancing effects as well as a decrease in very slow activities in the quantitative pharmaco-EEG [73, 74], similar to other cognitive activators including tacrine [73]; event-related potentials were only slightly modified in amplitude whereas a clear decrease of P300 latency occurred [74]. Compared to other *Ginkgo* extracts EGb 761 was the only compound classified by the computer-analyzed electroencephalogram as a cognitive activator [75]. In healthy volunteers, EGb 761 increased tolerance to cerebral hypoxia [76], increased erythrocyte resistance to oxidative challenge [57], inhibited stress-induced leukocyte rigidification, [77] and reduced PAF-induced platelet aggregation [58, 78].

In patients with slight to moderate degree of dementia, both EGb 761 and tacrine produced EEG profiles similar to each other and other cognitive activators or anti-dementia drugs; the type of effects in the quantitative-pharmaco-EEG was characterized by an increase of alpha activity and a decrease of slow frequencies (Itil et al. 1996, unpublished). In patients suffering from cerebrovascular ischemic disturbances, EGb 761 improved the global and local cerebral blood flow without producing a steal effect [79, 80]; in patients with completed stroke, pathological hemorheological parameters, or peripheral arterial occlusive disease, the extract caused a reduction in enhanced blood viscoelasticity and fibrinogen levels, an increase in cutaneous microcirculation, and a reduction in erythrocyte membrane rigidity and erythrocyte aggregation [57, 81–85].

6.2 Clinical Trials

6.2.1 Cerebral Insufficiency

In placebo-controlled, double-blind clinical trials, the efficacy of EGb 761 was assessed in patients with mild to moderate dementia of the Alzheimer type or multi-infarct dementia or mixed type [86, 87], mild to moderate dementia of the Alzheimer type [88–90], mild to moderate dementia of vascular origin [91–93], cerebral insufficiency [94–96], cerebral insufficiency with special respect to the symptom depression [97, 98], and mild to moderate memory impairment in the elderly [2–5].

The patients received 120 to 240 mg EGb 761 orally per day for between 4 and 24 weeks. The therapeutic efficacy was evaluated by global assessment and ratings, e.g. Clinical Global Impressions (CGI) Item 2 [87, 89], Sandoz Clinical Assessment Geriatric Scale (SCAG) [86, 88, 90–92], and by objective cognitive

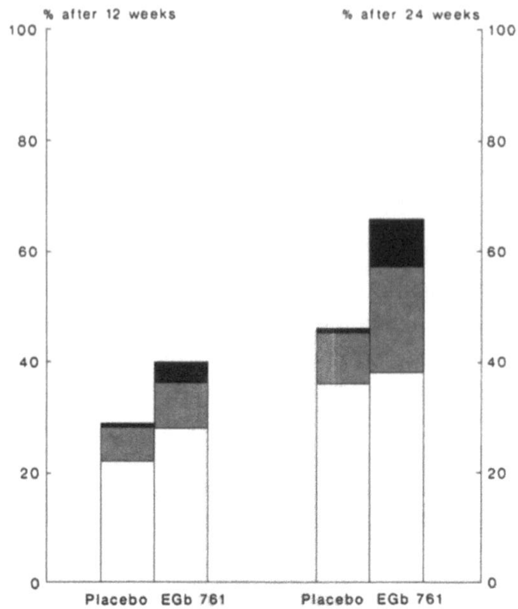

FIG. 1. Responders in 1 to 3 target variables. Response criteria were "much" and "very much improved" for Clinical Global Impressions (CGI), and improved by at least 4 and 2 units for the Syndrome Short Test (SKT) and Nuremberg Geriatric Observation Scale (NAB), respectively. Shown are the data of the valid cases for efficacy analysis (placebo: $n = 77$; EGb 761: $n = 79$). *White bar*, one variable; *shaded bar*, two variables; *black bar*, three variables. (From [87], with permission)

tests, e.g., trail-making tests (like the ZVT-G) [95], the digit symbol substitution test (ZST) and the digit span test (ZN) according to Wechsler [90, 94], the Syndrome Short Test (SKT) [87–89, 92] or the Vienna Test System [88, 95]; this was done by assessing the activities of daily living and quality of life, e.g., the Nuremberg Geriatric Observation Scale (NAB) [87] or the Crichton Scale [90].

The extract gave results superior to placebo; memory, intellectual performance, and depression were improved; vertigo, tinnitus, and headaches were attenuated. A direct comparison of 120 mg EGb 761 and 4.5 mg dihydroergotoxine (syn. co-dergocrine) showed similar improvements in both groups after 6 weeks [99].

In the most recent clinical study, 216 patients suffering from mild to moderate primary degenerative dementia of the Alzheimer type or vascular dementia were treated with a daily dosage of 240 mg EGb 761 for 24 weeks. In accordance with the recommended multidimensional evaluation approach, three primary variables were chosen: the Clinical Global Impressions (CGI) Item 2 for psychopathological assessment, the Syndrome Short Test (SKT) for the assessment of the patient's attention and memory, and the Nuremberg Geriatric Observation Scale (NAB) for behavioral assessment of activities of daily life. The frequency of therapy responders in the two treatment groups differed significantly in favor of EGb 761, with $P < .005$ in Fisher's Exact Test. Figure 1 shows the responders in 1 to 3 target variables after 12 and 24 weeks. The results of this clinical study led to the conclusion that EGb 761 is clinically efficacious in the treatment of outpatients with mild to moderate dementia of the Alzheimer type or vascular dementia [87].

6.2.2 Peripheral Arterial Occlusive Disease

The effectiveness of EGb 761 in the treatment of intermittent claudication (peripheral arterial occlusive disease, Fontaine stage II) has been demonstrated in placebo-controlled double-blind clinical trials by a statistically significant increase in the walking distance as compared to placebo [100–104, Peters 1996, unpublished]. The patients were treated with doses of 120 to 160 mg EGb 761 per day for up to 24 weeks. The treatment with EGb 761 in addition to physical training, or in maximum-level trained patients led to a further clear improvement of the walking distance [102, 103].

Sixty patients with a diagnosis of angiographically confirmed peripheral arterial occlusive disease, Fontaine stage IIb, and an upper pain-free walking distance limit of 150 m underwent excessive physical training until no further increase in walking distance could be achieved. The therapeutic groups were treated with either EGb 761 (160 mg/day) or placebo over a duration of 24 weeks following a two-week run-in phase. Whereas only a marginal increase in the walking distance was observed under placebo, a statistically significant effect of EGb 761 relative to that of placebo was confirmed after 8, 16, and 24 weeks of therapy ($P < .001$). In these maximum-level trained patients, the additional increase in pain-free

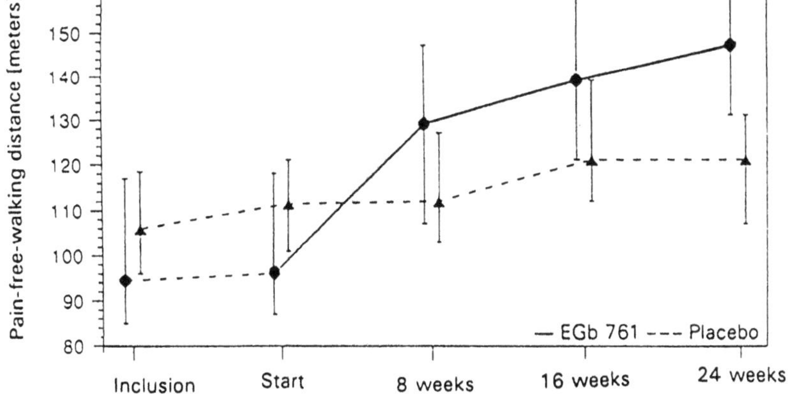

FIG. 2. Pain-free walking distance [m] in maximum-level trained patients with intermittent claudication (median values with 95% confidence interval). (from [103], with permission)

walking distance amounted to 41 m under active medication and to 8 m under placebo (Fig. 2). Thirty-one percent of the patients revealed an increase in walking distance of at least 60% compared with 0% in the placebo group (P = 0.002) [103].

The comparison of methods and results of clinical trials on the efficacy of EGb 761 and pentoxifylline in intermittent claudication showed no differences in the quality of documentation and in clinical efficacy; the additional percentage increase when compared with placebo was slightly higher under pentoxifylline (+57%) than under EGb 761 (+45%), but the scatter between different studies was substantially larger with pentoxifylline (−21% to +186%) than under EGb 761 (+17% to 78%) [105].

6.2.3 Vertigo and Tinnitus

In placebo-controlled, double-blind clinical trials, EGb 761 was examined in 67 patients with vertiginous syndrome of recent onset without definable etiopathology [106], in 35 patients with vestibular vertigo, refractory to standard medicaments [107], and in 33 elderly patients with vertigo and ataxia [108]. The patients received EGb 761 at a daily dosage of 120 to 160 mg for 4 to 12 weeks. After therapy, questioning on the intensity of vertiginous symptoms revealed a significantly lower total score for all symptoms in the EGb 761-treated patients ($P < .05$ *vs* placebo). This result was confirmed by visual analogue scale for self-rating of vertigo: a significantly higher improvement was reported in the EGb 761 group ($P < .05$ *vs* placebo) [106]. In patients with specific physical training to improve vestibular and proprioceptive reflexes, concomitant treatment with EGb 761 led to a futher decrease in sway amplitude (posturographic body sway measurements in Romberg's test) [107]. With EGb 761 treatment, craniocorpography showed a clear decrease in the lateral sway amplitude ($P < .005$ *vs* placebo) [108]. In two placebo-controlled, double-blind studies (3 months duration) involving 99 patients (Morgenstern 1996, unpublished) and 103 patients [109] with tinnitus, EGb 761 (120 mg respectively 160 mg daily) decreased intensity of tinnitus, and a significant improvement could be confirmed statistically ($P < .02$ *vs* placebo, Morgenstern 1996; $P < .05$ *vs* placebo [109]).

7 Contraindications

Hypersensitivity to *Ginkgo biloba* preparations is the only contraindication of EGb 761 [1].

8 Adverse Reactions

EGb 761 is well tolerated with a very low rate of side effects. Headache, mild gastrointestinal complaints, or allergic skin reactions may occur in very rare cases. No symptoms of overdosage are known [1].

Ginkgo leaves contain ginkgolic (anacardic) acids [110]. Ginkgolic acids are strong skin irritants, and ingestion results in gastrointestinal disturbances [111–114]. To minimize the allergenic risk of *Ginkgo* extracts, the ginkgolic acids must be eliminated; the content of these substances should be below 5 ppm [1].

9 Posology

The daily dose of EGb 761, as coated tablets or solution for oral administration, is 120–240 mg in two or three individual doses [1].

10 Conclusions

Pharmacological in vivo and in vitro investigations performed with the standardized *Ginkgo biloba* extract EGb 761 revealed the following pharmacological properties: EGb 761 has neuro(cerebro)protective effects by improving brain membrane fluidity, increasing hypoxic tolerance and inhibiting cerebral edema, thus protecting cerebral tissue from the effects of hypoxia and ischaemia. The extract inhibits a reduction of muscarinergic-, alpha$_2$- and 5-HT$_{1A}$-receptors due to aging, improves the uptake of choline into the hippocampus, and increases memory performance as well as learning capacity. EGb 761 improves the hemorheological properties of the blood and circulatory perfusion, primarily in the area of microcirculation, without producing a steal effect. EGb 761 has radical scavenging properties and acts antagonistically on the platelet-activating factor (PAF). Toxicological studies have revealed a high safety of EGb 761 without teratogenic, mutagenic, and carcinogenic effects. In double-blind clinical studies it has been possible to substantiate the clinical efficacy of EGb 761 in elderly patients with mild to moderate disturbances of cerebral function and its accompanying symptoms, both of vascular origin and on the basis of primary degenerative dementia of the Alzheimer type, as well as for a mixed form of both. In patients with peripheral arterial occlusive disease, Fontaine stage II (intermittent claudication), medication with EGb 761 was shown to be significantly superior to the placebo treatment with respect to the selected response variable, the increase in pain-free walking distance, even in fully trained patients. Additionally, it has been possible to demonstrate a significant effect of EGb 761 on tinnitus and vertigo. EGb 761 is well tolerated, with a very low rate of side effects.

References

1. Monographie (1994) Trockenextrakt (35–67:1) aus Ginkgo-biloba-Blättern extrahiert mit Aceton-Wasser. Bundesanzeiger 46:7361–7362

2. Israel L, Dell'Accio E, Martin G, Hugonot R (1987) Extrait de Ginkgo biloba et exercices d'entrainement de la mémoire. Evaluation comparative chez des personnes âgées ambulatoires. Psychol Méd 19:1431–1439
3. Wesnes K, Simmons D, Rook M, Simpson P (1987) A double-blind placebo-controlled trial of Tanakan in the treatment of idiopathic cognitive impairment in the elderly. Hum Psychopharmacol 2:159–169
4. Rai GS, Shovlin C, Wesnes KA (1991) A double-blind, placebo controlled study of Ginkgo biloba extract (Tanakan) in elderly outpatients with mild to moderate memory impairment. Curr Med Res Opin 6:350–355
5. Allain H, Raoul P, Lieury A, LeCoz F, Gandon M (1993) Effect of two doses of Ginkgo biloba extract (EGb 761) on the dual-coding test in elderly subjects. Clin Ther 15:549–558
6. Roman GC, Tatemichi TK, Erkinjutti T, Cumming JL, Masden JC, Garcia JH, Amaducci L, Orgogozo JM, Brun A, Hofman A, Moody DM, O'Brien MD, Yamaguchi T, Grafman J, Drayer BP, Bennett DA, Fisher M, Ogata J, Kokmen E, Bermejo F, Wolf PA, Gorelick PB, Bick KL, Pajeau AK, Bell MA, DeCarli C, Culebras A, Korczyn AD, Bogousslavsky J, Hartmann A, Scheinberg P (1993) Vascular dementia: Diagnostic criteria for research studies. Neurology 43:250–260
7. American Psychiatric Association (1994) Diagnostic and statistical manual of mental disorders, 4th edn. Washington DC, American Psychiatric Association, p 137
8. Friedland RP (1993) Alzheimer's disease: Clinical features and differential diagnosis. Neurology 43:545–551 (Suppl 4)
9. Hoyer S (1995) Moeglichkeiten und Grenzen der Therapie von Hirnleistungsstoerungen im Alter. Z Gerontol Geriatr 28:4547–4620
10. Bevan EG, Waller PC, Ramsay LE (1992) Pharmacological approaches to the treatment of intermittent claudication. Drugs Aging 2:125–136
11. Stoll S, Scheuer K, Pohl O, Mueller WE (1996) Ginkgo biloba extract EGb 761 independently improves changes of passive avoidance learning and brain membrane fluidity in the aging mouse. Pharmacopsychiatry 29:144–149
12. Chatterjee SS, Gabard B (1981) Protective effect of an extract of Ginkgo biloba and other hydroxyl radical scavengers against hypoxia. VII International Congress on Pharmacology, Tokyo, p 483
13. Karcher L, Zagermann P, Krieglstein J (1984) Effect of an extract of Ginkgo biloba on rat brain energy metabolism in hypoxia. Naunyn-Schmiedebergs Arch Pharmacol 327:31–35
14. Beck T, Abdel-Rahman MM, Bielenberg GW, Oberpichler H, Krieglstein J (1986) Comparative study on the effects of two extract fractions of Ginkgo biloba on local cerebral blood flow and on brain energy metabolism in the rat under hypoxia. In: Krieglstein J (ed) Pharmacology of Cerebral Ischemia. Elsevier, Amsterdam, pp 345–350
15. Krieglstein J, Oberpichler H (1989) Ginkgo biloba und Hirnleistungsstoerungen. Pharm Ztg 13:2279–2289
16. Klein J, Chatterjee SS, Loeffelholz K (1995) Bilobalide: A new centrally acting phospholipase A_2 inhibitor? Naunyn-Schmiedebergs Arch Pharmacol Suppl 35:R 86, Abstr 341
17. Larsen RG, Dupeyron JP, Boulu RG (1978) Modèles d'ischémie cérébrale expérimentale par microsphères chez le rat. Etude de l'effet de deux extraits de Ginkgo biloba et du naftidrofuryl. Thérapie 33:651–660

18. Rapin JR, Le Poncin-Lafitte M (1986) Consommation cerebrale du glucose. Effet de l'extrait de Ginkgo biloba. Presse Med 15:1494–1497
19. Le Poncin-Lafitte MC, Rapin J, Rapin JR (1980) Effects of Ginkgo biloba on changes induced by quantitative cerebral microembolization in rats. Arch Int Pharmacodyn 243:236–244
20. Spinnewyn B, Blavet N, Clostre F (1986) Effets de l'extrait de Ginkgo biloba sur un modèle d'ischemie cérébrale chez la gerbille. Presse Méd 15:1511–1515
21. Oberpichler H, Sauer D, Rossberg C, Mennel HD, Krieglstein J (1990) PAF-antagonist ginkgolide B reduces postischemic neuronal damage in rat brain hippocampus. J Cereb Blood Flow Metab 10:133–135
22. Prehn JHM, Krieglstein J (1993) Platelet-activating factor antagonists reduce excitotoxic damage in cultured neurons from embryonic chick telencephalon and protect the rat hippocampus and neocortex from ischemic injury in vivo. J Neurosc Res 34:179–188
23. Krieglstein J, Ausmeier F, EI-Abhar H, Lippert K, Welsch M, Rupallay K, Henrich-Noack P (1995) Neuroprotective effects of Ginkgo biloba constituents. Eur J Pharm Sci 3:39–48
24. Spinnewyn B, Blavet N, Clostre F, Bazan N, Braquet P (1987) Involvement of plateletactivating factor (PAF) in cerebral post-ischemic phase in mongolian gerbils. Prostaglandins 34:337–349
25. Chatterjee SS, Gabard B (1984) Effect of an extract of Ginkgo biloba on experimental neurotoxicities. Naunyn-Schmiedebergs Arch Pharmacol Supp 325, R 107, Abstr 425
26. Otani M, Chatterjee SS, Gabard B, Kreutzberg GW (1986) Effect of an extract of Ginkgo biloba on triethyltin-induced cerebral edema. Acta Neuropathol 69:54–65
27. Borzeix M (1985) Effects of Ginkgo biloba extract on two types of cerebral edema. In: Agnoli A, Rapin JR, Scapagnini V, Weitbrecht WV (eds) Effects of Ginkgo biloba extract on organic cerebral impairment. John Libbey, London, pp 51–56
28. Chatterjee SS, Gabard B, Jaggy H (1986) Pharmaceutical compositions containing bilobalid for the treatment of neuropathies. US Patent no 4,571,407 (Feb 18, 1986)
29. Sancesario G, Kreutzberg GW (1986) Stimulation of astrocytes affects cytotoxic brain edema. Acta Neuropathol 72:3–14
30. Dorman DC, Coté LM, Buck WB (1992) Effects of an extract of Ginkgo biloba on bromethalin induced cerebral lipid peroxidation and edema in rats. Am J Vet Res 53:138–142
31. Boulieu R, Munoz JF, Macovschi O, Pacheo H (1985) Implication de la peroxydation lipidique dans l'intoxication par le triéthyltin chez le rat. CR Soc Biol (Paris) 182:196–201
32. Sharma HS, Westman J, Nyberg F, Cervos-Navarro J, Dey PK (1994) Neuroprotective effects of an extract of Ginkgo biloba (EGb 761) in heat stress induced brain damage in the rat. In: Zeisberger E, Schoenbaum E, Lomax P (eds) Thermal Balance in Health and Disease. Birkhaeuser, Basel, pp 461–467
33. Rodriguez de Turco EB, Droy-Lefaix MT, Bazan NG (1993) Decreased electroconvulsive shock-induced diacylglycerols and free fatty acid accumulation in the rat brain by Ginkgo biloba extract (EGb 761): selective effect in hippocampus as compared with cerebral cortex. J Neurochem 61:1438–1444

34. Barkats M, Venault P, Christen Y, Cohen-Salmon C (1995) Effect of long-term treatment with EGb 761 on age-dependent structural changes in the hippocampi of three inbred mouse strains. Life Sci 56:213–222
35. Brailowsky S, Montiel T, Medina-Ceja L (1995) Acceleration of functional recovery from motor cortex ablation by two Ginkgo biloba extracts in rats. Restor Neurol Neurosci 8:163–167
36. Ramassamy C, Girbe F, Christen Y, Costentin J (1993) Ginkgo biloba extract EGb 761 or trolox C prevent the ascorbic acid/FE^{2+} induced decrease in synaptosomal membrane fluidity. Free Rad Res Commun 19:341–350
37. Chopin P, Briley M (1992) Effects of four non-cholinergic cognitive enhancers in comparison with tacrine and galanthamine on scopolamine-induced amnesia in rats. Psychopharmacology 106:26–30
38. Chatterjee SS, Noeldner M (1989) Behavioural observations demonstrating influence of the extract of Ginkgo biloba (EGb 761) on some specific central cholinergic systems. Naunyn-Schmiedebergs Arch Pharmacol Suppl 339:R 107, Abstr 425
39. Kristofikova Z, Benesova O, Tejkalova H (1992) Changes of high-affinity choline uptake in the hippocampus of old rats after long-term administration of two nootropic drugs (Tacrine and Ginkgo biloba extract). Dementia 3:304–307
40. Taylor JE (1986) Liaison des neuromédiateurs à leurs récepteurs dans le cerveau de rats. Presse Med 15:1491–1493
41. Huguet F, Tarrade Th (1992) α_2-Adrenoceptor changes during cerebral ageing. The effect of Ginkgo biloba-extract. J Pharm Pharmacol 44:24–27
42. Huguet F, Drieu K, Piriou A (1994) Decreased cerebral 5-HT_{1A} receptors during ageing: reversal by Ginkgo biloba extract (EGb 761). J Pharm Pharmacol 46:316–318
43. Bolanos-Jimenez F, Manhaes deCastro R, Sarhan H, Prudhomme N, Drieu K, Fillion G (1995) Stress-induced 5-HT_{1A} receptor desensitization: protective effects of Ginkgo biloba extract (EGb 761). Fundam Clin Pharmacol 9:169–174
44. Winter E (1991) Effects of an extract of Ginkgo biloba on learning and memory in mice. Pharmacol Biochem Behav 38:109–114
45. Blavet N (1992) Effects of Ginkgo biloba extract (EGb 761) on learning in the aged rat. In: Christen Y, Costentin J, Lacour M (eds) Effects of Ginkgo biloba extract (EGb 761) on the central nervous system. Elsevier, Paris, pp 119–127
46. Reuse-Blom S, Drieu K (1986) Effet de l'extrait de Ginkgo biloba sur le spasme artériolaire chez le lapin. Presse Med 15:1520–1523
47. Grosdemouge C, Le Poncin-Lafitte M, Rapin JR (1986) Effets protecteurs de l'extrait de Ginkgo biloba sur la rupture précoce de la barriere hémoencéphalique chez le rat. Presse Med 15:1502–1505
48. DeFeudis FV, Auguet M, Delaflotte S, Hellegouarch A, Baranes I, Chapelat M, Braquet M, Etienne A, Drieu K, Clostre F, Braquet P (1985) Some in vitro and in vivo actions of an extract of Ginkgo biloba (GBE 761). In: Agnoli A, Rapin JR, Scapagnini V, Weitbrecht WV (eds) Effects of Ginkgo biloba extract on organic cerebral impairment. John Libbey, London, pp. 17–29
49. Chatterjee SS, Gabard B (1982) Studies on the mechanism of action of an extract of Ginkgo biloba, a drug used for treatment of ischemic vascular diseases. Naunyn-Schmiedebergs Arch Pharmacol Suppl 321:R 52, Abstr 207
50. Hellegouarch A (1985) Comparison of the contractile effects of an extract of Ginkgo biloba and some neurotransmitters on rabbit isolated vena cava. Gen Pharmacol 16:129–132

51. Janssens D, Michels C, Delalive E, Eliaers F, Drieu K, Remacle J (1995) Protection of hypoxia-induced ATP decrease in endothelial cells by Ginkgo biloba extract and bilobalide. Biochem Pharmacol 50:991–999
52. Iliff LD, Auer LM (1982) The effect of intravenous infusion of Tebonin (Ginkgo biloba) on pial arteries in cats. J Neurosurg Sci 27:227–231
53. Krieglstein J, Beck T, Seibert A (1986) Influence of an extract of Ginkgo biloba on cerebral blood flow and metabolism. Life Sci 39:2327–2334
54. Borzeix MG, Labos M, Hartl C (1980) Recherches sur l'action antiagregante de l'extrait de Ginkgo biloba. Sem Hop Paris 56:393–398
55. Ernst E, Matrai A (1986) Haemorheologische in-vitro-Effekte von Ginkgo biloba. Herz Kreisl 18:358–360
56. Etienne A, Hecquet F, Clostre F, DeFeudis FV (1982) Comparaison des effets d'un extrait de Ginkgo biloba et de la chlorpromazine sur la fragilité osmotique, in vitro d'érythrocytes de rat. J Pharmacol Paris 13:291–298
57. Artmann GM, Schikarski C (1993) Ginkgo biloba extract (EGb 761) protects red blood cells from oxidative damage. Clin Hemorheol 13:529–539
58. Klein P (1988) Untersuchung ueber die Hemmwirkung von Ginkgo-biloba-Extrakt. Therapiewoche 38:2379–2383
59. Braquet P (1987) The Ginkgolides: potent platelet-activating factor antagonists isolated from Ginkgo biloba L.: Chemistry, Pharmacology and clinical Applications. Drugs Future 12:643–699
60. Inselmann G, Bloehmer A, Kottny W, Nellessen U, Haenel H, Heidemann HT (1995) Modification of cisplatin-induced renal p-aminohippurate uptake alteration and lipid peroxidation by thiols, Ginkgo biloba extract, deferoxamine and torbafylline. Nephron 70:425–429
61. Droy-Lefaix MT, Bonhomme B, Doly M (1991) Protective effect of Ginkgo biloba extract (EGb 761) on free radical-induced changes in the electroretinogram of isolated rat retina. Drugs Exp Clin Res 17:571–574
62. Pincemail J, Dupuis M, Nasr C (1989) Superoxide anion scavenging effect and superoxide dismutase activity of Ginkgo biloba extract. Experientia 45:708–712
63. Barth SA, Inselmann G, Engemann R, Heidemann HT (1991) Influences of Ginkgo biloba on Cyclosporin A induced lipid peroxidation in human liver microsomes in comparison to Vitamin E, Glutathione and N-Acetylcysteine. Biochem Pharmacol 41:1521–1526
64. Dumont E, Petit E, Tarrade T, Nouvelot A (1992) UV-C irradiation-induced peroxidative degradation of microsomal fatty acids and proteins. Protection by an extract of Ginkgo biloba (EGb 761). Free Radic Biol Med 13:197–203
65. Dumont E, D'Arbigny P, Nouvelot A (1995) Protection of polyunsaturated fatty acids against iron-dependent lipid peroxidation by a Ginkgo biloba extract (EGb 761). Meth Find Exp Clin Pharmacol 17:83–88
66. Chatterjee SS, Gabard B (1982) Protection of Doxorubicin toxicity by an extract of Ginkgo biloba. Naunyn-Schmiedebergs Arch Pharmacol Suppl 319:R 15, Abstr 57
67. Marcocci I, Packer L, Droy-Lefaix MT, Sekaki A, Gardes-Albert M (1994) Antioxidant Action of Ginkgo biloba extract EGb 761. Methods Enzymol 234:462–475
68. Pincemail J, Deby C (1986) Propriétés antiradicalaires de l'extrait de Ginkgo biloba. Presse Med 15:1475–1479
69. Raymond J (1986) Effets de l'extrait de Ginkgo biloba sur la préservation morphologique des épithéliums sensoriels vestibulaires chez la souris. Presse Med 15:1484–1487

70. Denise P, Bustany P (1989) The effect of extract of Ginkgo biloba (EGb 761) on central compensation of a total unilateral peripheral vestibular deficit in the rat. In: Lacour M, Toupet M, Denise P, Christen Y (eds) Vestibular compensation: facts, theories and clinical perspectives. Elsevier, Paris, pp 201–208
71. Lacour M, Ez-Zaher L, Raymond (1991) Plasticity mechanisms in vestibular compensation in the cat are improved by an extract of Ginkgo biloba (EGb 761). Pharmacol Biochem Behav 40:367–379
72. Spiess E, Juretzek W (1993) Ginkgo. In: Haensel R, Keller K, Rimpler H, Schneider G (eds) Hagers Handbuch der Pharmazeutischen Praxis. Springer, Berlin, Vol 5, pp 269–292
73. Itil TM, Eralp E, Tsambis E, Itil KZ, Stein U (1996) Central nervous system effects of Ginkgo biloba, a plant extract. Am J Ther 3:63–73
74. Luthringer R, D'Arbigny P, Macher JP (1995) Ginkgo biloba extract (EGb 761), EEG and event-related potentials mapping profile. In: Christen Y, Courtois Y, Droy-Lefaix MT (eds) Advances in Ginkgo biloba Extract Research. Elsevier, Paris, Vol 4, pp 107–118
75. Itil TM, Martorano D (1995) Natural Substances in Psychiatry (Ginkgo biloba in Dementia). Psychopharmacol Bull 31:147–158
76. Schaffler K, Reeh PW (1985) Doppelblindstudie zur hypoxieprotektiven Wirkung eines standardisierten Ginkgo-biloba-Praeparates nach Mehrfachverabreichung an gesunden Probanden. Arzneim Forsch 35:1283–1286
77. Ernst E, Marshall M (1992) Der Effekt von Ginkgo-biloba-Spezialextrakt EGb 761 auf die Leukozytenfilterabilitaet—Eine Pilotstudie. Perfusion 5:241–242
78. Guinot P, Caffrey E, Lambe R, Darragh A (1989) Tanakan inhibits platelet-activating-factor induced platelet aggregation in healthy male volunteers. Haemostasis 19:219–233
79. Heiss WD, Zeiler (1978) Medikamentoese Beeinflussung der Hirndurchblutung. Pharmakotherapie 1:137–144
80. Tea S, Celsis P, Clanet M, Marc-Vergnes JP, Boeters U (1987) Quantifizierte Parameter zum Nachweis von zerebraler Durchblutung- und Stoffwechselsteigerung unter Ginkgo-biloba-Therapie. Therapiewoche 37:2655–2657
81. Koeltringer P, Eber O, Klima W, Rothlauer W, Wakonig P, Langsteger W, Lind P (1989) Die Mikrozirkulation und Viskoelastizitaet des Vollblutes unter Ginkgo biloba Extrakt. Perfusion 1:28–30
82. Koeltringer P, Langsteger W, Klima G, Reisecher F, Eber O (1993) Haemorheologische Effekte des Ginkgo-biloba-Extraktes EGb 761. Fortschr Med 111:170–172
83. Anadere J, Chmiel H, Witte S (1985) Hemorheological findings in patients with completed stroke and the influence of a Ginkgo biloba extract. Clin Hemorheol 5:411–420
84. Rudofsky G (1987) Wirkung von Ginkgo-biloba-Extrakt bei arterieller Verschlußkrankheit. Fortschr Med 105:397–400
85. Witte S, Anadere I, Walitza E (1992) Verbesserung der Hämorheologie durch Ginkgo-biloba-Extrakt. Fortschr Med 110:247–250
86. Taillandier J, Ammar A, Rabourdin JP, Ribeyre JP, Pichon J, Niddam S, Pierart H (1986) Traitement des troubles du vieillissement cérébral par l'extrait de Ginkgo biloba. Presse Med 15:1583–1587
87. Kanowski S, Herrmann WM, Stephan K, Wierich W, Hoerr R (1996) Proof of efficacy of the Ginkgo biloba special extract EGb 761 in outpatients suffering from

mild to moderate primary degenerative dementia of the Alzheimer type or multiinfarct dementia. Pharmacopsychiatry 29:47–56
88. Hofferberth B (1994) The efficacy of EGb 761 in patients with senile dementia of the Alzheimer type. A double-blind, placebo-controlled study on different levels of investigation. Human Psychopharmacol 9:215–222
89. Ihl R, Froelich L (1995) Ginkgo-biloba-Spezialextract bei der Demenz vom Alzheimer Typ. Z Phytother Suppl 13–14
90. Weitbrecht WU, Jansen W (1986) Primaer degenerative Demenz: Therapie mit Ginkgo-biloba-Extrakt. Fortschr Med 104:199–202
91. Mancini M, Agozzino B, Bompani R (1993) Effetti clinico-terapeutici dell'estratto di Ginkgo biloba (EGb) in confronto a placebo, nel trattamento di pazienti affetti da demenza psicorganica senile su base arteriosclerotica. Gazz Med Ital Arch Sci Med 152:69–80
92. Halama P, Bartsch G, Meng G (1988) Hirnleistungsstoerungen vaskulaerer Genese. Fortschr Med 106:408–412
93. Eckmann F, Schlag HZ (1982) Kontrollierte Doppelblind-Studie zum Wirksamkeitsnachweis von Tebonin forte bei Patienten mit zerebrovaskulaerer Insuffizienz. Fortschr Med 100:1474–1478
94. Dieli G, La Mantia V, Saetta M, Costanzo E (1981) Studio clinico in doppio cieco del tanakan nell'insufficienza cerebrale cronica. IL Lavoro Neuropsichiatrico 68:3–15
95. Hofferberth B (1989) Einfluβ von Ginkgo biloba-Extrakt auf neurophysiologische und psychometrische Meβergebnisse bei Patienten mit hirnorganischem Psychosyndrom. Arzneim Forsch 39:918–922
96. Graessel E (1992) Einfluβ von Ginkgo-biloba-Extrakt auf die geistige Leistungsfaehigkeit. Fortschr Med 110:73–76
97. Stocksmeier U, Eberlein M (1992) Depressive Verstimmung bei Hirnleistungsstoerungen. TW Neurologie Psychiatrie 6:74–76
98. Schubert H, Halama P (1993) Primaer therapieresistente depressive Verstimmung aelterer Patienten mit Hirnleistungsstoerungen: Wirksamkeit der Kombination von Ginkgo-biloba-Extrakt EGb 761 mit Antidepressiva. Geriatr Forsch 3:45–53
99. Gerhardt G, Rogalla K, Jaeger I (1990) Medikamentoese Therapie von Hirnleistungsstoerungen. Randomisierte Vergleichsstudie mit Dihydroergotoxin und Ginkgo biloba-Extrakt. Fortschr Med 108:384–388
100. Bauer U (1984) Six month double-blind randomized clinical trial of Ginkgo biloba extract versus placebo in two parallel groups in patients suffering from peripheral arterial insufficiency. Arzneim Forsch 34:716–720
101. Boehmer D, Kalinsky S, Michaelis P, Szoegy A (1988) Behandlung der PAVK mit Ginkgo-biloba-Extrakt (GBE) oder Pentoxifyllin. Herz Kreisl 20:5–8
102. Bulling B, Bary S (1991) Behandlung der chronischen peripheren arteriellen Verschluβkrankheit mit physikalischem Training und Ginkgo-biloba-Extrakt 761. Med Welt 42:702–708
103. Blume J, Kieser M, Hoelscher U (1996) Placebokontrollierte Doppelblindstudie zur Wirksamkeit von Ginkgo-biloba-Spezialextrakt EGb 761 bei austrainierten Patienten mit Claudicatio intermittens. VASA 25:265–274
104. Schneider B (1992) Ginkgo-biloba-Extrakt bei peripheren arteriellen Verschluβkrankheiten. Arzneim Forsch 42:428–436
105. Letzel H, Schoop W (1992) Ginkgo-biloba-Extrakt EGb 761 und Pentoxifyllin bei Claudicatio intermittens. Sekundaeranalyse zur klinischen Wirksamkeit. VASA 21:403–410

106. Haguenauer JP, Cantenot F, Koskas H, Pierat H (1986) Traitement des troubles de l'equilibre par l'extrait de Ginkgo biloba. Presse Med 15:1569–1572
107. Hamann KF (1985) Physikalische Therapie des vestibulären Schwindels in Verbindung mit Ginkgo-biloba-Extrakt. Therapiewoche 35:4586–4590
108. Claussen CF, Kirtane MV (1985) Randomisierte Doppelblindstudie zur Wirkung von Extraktum Ginkgo biloba bei Schwindel und Gangunsicherheit des aelteren Menschen. In: Claussen CF (ed) Presbyvertigo, Presbyataxie, Presbytinnitus. Springer, Berlin, pp 103–115
109. Meyer B (1986) Etude multicentrique randomisée a double insu face au placebo du traitement des acouphenes par l'extrait de Ginkgo biloba. Presse Med 15:1562–1564
110. Gellerman JL, Schlenk H (1968) Methods of isolation and determination of anacardic acids. Anal Chem 40:739–743
111. Choukchou-Braham N, Asakawa Y, Lepoittevin JP (1994) Isolation, structure, determination and synthesis of new dihydroisocoumarins from Ginkgo biloba L. Tetrahedron Letters 35:3949–3952
112. Becker ML, Skipworth GB (1975) Ginkgo-Tree dermatitis, stomatitis and proctitis. JAMA 231:1162–1163
113. Lepoittevin JP, Benezra C, Asakawa Y (1991) Allergic contact dermatitis to Ginkgo biloba L.: relationship with urushiol. Arch Dermatol Res 281:227–230
114. Tomb RR, Fousserreau J, Sell Y (1988) Mini-epidemic of contact dermatitis from ginkgo tree fruit (Ginkgo biloba L.). Contact Dermatitis 19:281–283

Platelet-Activating Factor and Its Antagonists: Scientific Background and Clinical Applications of Ginkgolides

PIERRE G. BRAQUET

1 Introduction

Platelet-activating factor (PAF), a potent phospholipid inflammatory mediator, has been implicated in several pathophysiological disorders. The discovery of selective PAF receptor antagonists has greatly facilitated the accumulation of knowledge on the role of PAF in animal models of these disorders (see, for reviews, [1–3]). The use of selective, potent, orally active PAF receptor antagonists may, at least in part, allow us to successfully explore the problematic points of PAF research.

The most widely examined PAF antagonists are the ginkgolides [4–7], particularly ginkgolide B (BN 52021) (Fig. 1). Ginkgolides are not selective PAF receptor antagonists as they exhibit a serine proteinase inhibitor property as well [8, 9], which may contribute to their biological activity. The first total synthesis of ginkgolides was made in 1988 [10, 11]. To increase potency, several analogues of ginkgolides were proposed, recognizing that the *tert*-butyl (*t*-butyl) substituent is crucial for activity [12].

In this chapter, we pay particular attention to the inflammatory response, which is a fundamental endogenous defensive mechanism that has been implicated in a great variety of pathophysiological conditions. The term systemic inflammatory response syndrome, or SIRS, has been introduced recently to reflect the importance of exaggerated inflammatory reaction as a mechanism of septic shock and related conditions. This field has become the hottest point in PAF research, leading to the first appearance of PAF antagonists in clinical practice.

2 Platelet-Activating Factor Receptors

Because PAF receptors are ubiquitous in the organism and carry the powerful effect of phospholipid mediator in various organs, exploration of their structure is of great significance to understand the effect of specific receptor antagonists

Bio-Inova, 48–52, rue de la Gare, 78370 Plaisir, France

FIG. 1. Stuctures of ginkgolides extracted from the *Ginkgo biloba* tree

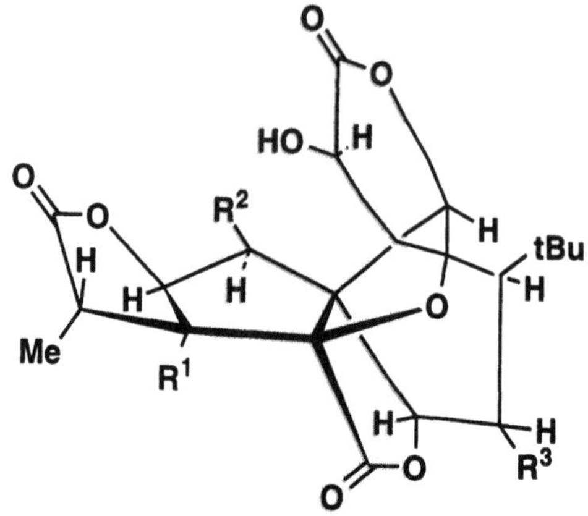

Ginkgolide	Nomenclature	R^1	R^2	R^3
A	BN 52020	OH	H	H
B	BN 52021	OH	OH	H
C	BN 52022	OH	OH	OH

that may competitively or noncompetitively displace PAF from these binding sites. [^{3}H]PAF binding and its antagonism with BN 52021 in washed rabbit platelets have revealed an interaction between PAF and PAF receptor antagonists that has a reversible and an irreversible component [13].

For an agonistic effect, strong hydrophobic interactions and an ether function are required in position 1 of the glycerol backbone; chain length limitations and steric hindrance demand a small group in position 2 [14]. The unusual structural properties of non-PAF-like antagonists required three-dimensional electrostatic potential calculations. This method, applied to seven PAF receptor antagonists, led to the conclusion that the structure of the PAF receptor is a bipolarized cylinder of 10–12 Å in diameter. More recent studies have suggested that the receptor may in fact be a multipolarized cylinder. Synthesized PAF antagonists with a simple framework were shown to exhibit a negative electrostatic potential generated by two trimethoxyphenyl groups at various distances between the negative clouds [15].

The human cDNA that encodes a functional receptor for PAF has been cloned by screening an HL-60 cell line cDNA library with degenerate oligonucleotide probes based on conserved sequences of rhodopsin-type receptors [16]. The 342-amino-acid receptor contains seven putative transmembrane domains but lacks sites for *N*-linked glycosylation at the N terminus. Stable transfected fibroblasts expressing the cloned cDNA responded to subnanomolar PAF stimulation with Ca^{2+} mobilization that could be inhibited by PAF antagonists. Using a 0.8-

kilobase-pair fragment of the guinea-pig lung PAF receptor cDNA, Nakamura et al. [17] have succeeded in cloning 342 amino acid residues with a calculated molecular mass of 39 203 Da. The coded receptor shared the amino acid sequence of seven putative spanning domains with 83% identity to the guinea-pig lung PAF receptor. Inositol 1,4,5-triphosphate (IP_3) production was demonstrated in both transfected CDS-7 cells and oocytes, and this effect was inhibited by guanosine 5′-*O*-(2-thio)biphosphate injection, providing evidence that PAF stimulates phosphoinositol turnover through a mechanism that involves G-proteins. PAF receptor mRNA was abundant in leukocytes and was expressed less in an undifferentiated human eosinophil cell line [18].

3 Biochemical Background of PAF Effect

PAF stimulates phosphoinositide turnover, resulting in IP_3 formation and consecutive release of $[Ca^{2+}]_i$ in a great variety of inflammatory cells, including polymorphonuclear neutrophils (PMN), macrophages, and endothelial cells. This process leads to activation of phospholipase A_2 with a consecutive release of arachidonic acid and eicosanoids.

4 PAF in the Inflammatory Response

Inflammation is an important natural defense mechanism, and PAF mimics the symptoms of the acute inflammatory response. The in vitro effects of PAF on various inflammatory cells, such as PMNs, eosinophil and basophil leukocytes, mast cells, and monocytes/macrophages have been extensively studied. Our knowledge in this field has been greatly increased, and, based on the interaction of PAF with various cytokines, has led to the development of a concept of autogenerated, regulatory feedback network of inflammatory mediator release [19, 20].

4.1 Amplification by PAF of Inflammatory Mediator Release

The concept that an autogenerated, autoregulated feedback network contributes to the regulation of inflammatory mediator release has been confirmed by recent results. For example, granulocyte-macrophage colony-stimulating factor, which regulates the growth of hemopoietic progenitor cells, has been found to prime mature human phagocytes to PAF that leads to respiratory burst and increased leukotriene (LT) B_4 synthesis [21]. When activated by inflammatory stimuli, macrophages produce tumor necrosis factor-α (TNF-α) and interleukine-1 (IL-1), which then induce PAF synthesis in endothelial cells, neutrophils, and macrophages so that an increased respiratory burst occurs [19, 22, 23]. Not only can growth factors and cytokines induce PAF release, but treatment with PAF also has been shown to increase cytokine production. In rats, long-term parenteral treatment with PAF dose-dependently stimulated IL-1 and IL-2 production in isolated spleen mononuclear cells [23, 24]. Mediator production declined after large doses of PAF, and BN 52021 inhibited either the stimulatory or the inhibitory effect of PAF.

When injected locally, PAF induces an inflammatory response characterized by increased blood flow, extravasation of plasma proteins, and transient intravascular cellular accumulation followed by perivascular mixed cellular infiltration. This similarity has led to the concept that PAF is involved in the mechanism of acute inflammation. Local applications of PAF antagonists such as BN 52021 have been shown to reduce the acute nonimmune inflammatory response induced by several irritants in the mouse ear [25].

Recent studies have shown that PAF contributes to the early phase of tissue damage in antigen-induced arthritis in healthy rabbits and TNF can amplify the inflammatory response induced by PAF [26], pointing to the importance of PAF/cytokine interaction. To develop a more uniform set of definitions, it has recently been proposed that sepsis and similar disorders may be called the systemic inflammatory response syndrome (SIRS) [27]. As discussed later, there is circumstantial evidence that PAF is also involved in the pathogenesis of SIRS.

4.2 PAF in Systemic Inflammatory Response Syndrome

Despite considerable advances in intensive clinical care in the past decade, septic shock has remained the most common cause of death in noncoronary intensive care units. The relative lack of effective pharmacological interventions highlights the complex pathophysiological events involved in sepsis. SIRS frequently leads to the development of acute lung injury, renal failure, and other end-organ dysfunction; therefore, the term multiple organ dysfunction syndrome has also been proposed. Although not completely understood, septic shock is believed to be the result of widespread endothelial damage caused by persistent inflammation from exaggerated release of inflammatory mediators [28]. Among various infections, gram-negative bacteria producing lipopolysaccharide (LPS) are known as a major cause of SIRS; therefore, attenuating the effect of LPS is a major issue in the treatment of septic shock. There is much evidence to suggest that, beyond various cytokines and eicosanoids, PAF also plays an important role in this general, exaggerated inflammatory response leading to microvascular failure [29].

4.3 Mediators of SIRS Induced by LPS: Effect of PAF Receptor Antagonists

LPS is a potent inflammatory stimulus for the host and triggers the release of endogenous mediators, including TNF, IL-1, PAF, eicosanoids, complement, contact activation system, and opioids [30]. The mediator role of cytokines, PAF, and eicosanoids with special regard of the sequence of their release in response to LPS, as well as the effect of some PAF antagonists on the release process, has been reviewed [31, 32] and is not discussed here.

Activated macrophages produce TNF-α, which is a principal candidate as mediator of septic shock [33]. Intramesenteric infusion of LPS in domestic pigs results in 50% early death [34, 35], and those animals that die within half an hour, termed nonsurvivors, respond with excess TNF-α production but do not release other mediators, suggesting a primary role for TNF-α as mediator of lethal septic shock. BN 52021 was ineffective against death in this critical early phase [34, 35]. In survivor pigs, sequential release of other mediators such as PAF and thromboxane B2 (TXB_2) occurs. The shock state in survivor pigs was indeed effectively alleviated by BN 52021.

The sequence of eicosanoid release also shows characteristic features. In survivor pigs, PAF release was associated with the generation of TXB_2, 6-keto-$PGF_{1\alpha}$ (prostaglandin) and LTB_4. The earliest increase in TXB_2 concentration was detected in pigs [34, 36], rabbits [37], dogs [38], and rats [39, 40], followed by an increase in plasma 6-keto-$PGF_{1\alpha}$; LTB_4 production was the last [35]. In accord, PAF receptor antagonists, such as BN 52021 [34, 35, 38, 39], can effectively block TXB_2 production and only moderately decrease the release of 6-keto-$PGF_{1\alpha}$ [35, 38, 39]. The mediator role of TXA_2 and other prostanoids is favored by the finding that indomethacin transiently restores blood pressure and increases survival in the critical early phase of LPS shock in pigs [41], and as well as ibuprofen alleviates many but not all the hemodynamic effects of LPS in dogs. Endogenous vasoconstrictors other than TXA_2, for example endothelin-1 (ET-1), have also been shown to be released by LPS in anaesthetized dogs and cultured endothelial cells [42]; the pathophysiological role of ET-1 in endotoxic shock, however, remains to be determined.

Beside attenuating LPS-induced inflammatory mediator release, PAF receptor antagonists are able to effectively reduce the symptoms of septic shock. Survival rate was increased and circulatory collapse was improved by BN 52021 in rats [39, 43, 44], in dogs [38], and in pigs [34]. PAF antagonists have been shown to attenuate other than LPS-induced changes in hemodynamic parameters. For example, LPS-induced pulmonary hypertension in sheep in vivo was reduced by BN 52021 [45] while this drug decreased pulmonary perfusion pressure in the rat in vitro. More recently, the priming effect of PAF in LPS-induced lung injury has also been demonstrated, and several PAF antagonists provided a marked protection [40]. It is not surprising that LPS and PAF produce similar pathophysiological alterations in homeostasis [46–48], and PAF antagonists attenuate these changes. All these findings strengthen the concept that PAF is involved in endotoxin-induced alterations.

5 Clinical Studies with PAF Antagonists

In contrast to the enormous mass of scientific articles published on experimental biology, biochemistry, pathophysiology, pharmacology, and toxicology of PAF antagonists, relatively few definitive data have been available from clinical trials.

Promising results were obtained in a clinical trial of BN 52021 in PAF- and allergen-induced brochoconstriction in asthmatic children [49]. Supporting the view that booster injection induces a profound decrease in the efficacy of PAF antagonists to reduce airway hyperresponsiveness in the guinea-pig [50, 51], a clinical study has clearly shown that another PAF antagonist, namely UK-74505, loses its potency to attenuate early and late asthmatic responses after repeated administration of allergen [52]. These negative experimental and clinical findings will certainly alter the notion that immediate hypersensitivity reactions, in particular bronchial asthma, may be the principal clinical indications for PAF antagonists, and may lead to a reevaluation of the strategy in the clinical development of these drugs.

It is now known that the PAF antagonist apafant exerts no beneficial effects in patients with chronic autoimmune thrombocytopenia [53], and even a modified bleeding time test has become positive in treated patients, suggesting that caution must be taken into account as to unexpected bleeding in thrombocytopenic individuals. BN 52021 has been demonstrated as an effective inhibitor of PAF-induced cutaneous weal and flare and of platelet responses in human healthy volunteers [54].

As already mentioned, septic shock has become one of the most promising clinical indications for PAF antagonists. Several compounds, including the natural antagonist BN 52021 and other synthetic agents, have been selected for clinical trials in patients with septic shock; the state of the art of these studies is summarized in Table 1. Although highly potent synthetic PAF antagonists have been developed, the highest potency may not be assured as the most important prerequisite for a successful clinical development, because some complementary effects may facilitate the benefit of a certain drug in the treatment of septic shock patients. Another possibility to elaborate a successful treatment for such a complex disorder as SIRS, based on a better elucidation

TABLE 1. Platelet-activating factor (PAF) receptor antagonists that have undergone clinical trials in septic shock

Compound	Pharmaceutical company	Present status of clinical research
BN 52021	IPSEN-Beaufour, France	Phase III
PCA 4280	Alter, Spain	Phase I
CV-3988	Takeda, Japan	Phase II
E-5880	Eisai, Japan	Phase II
CL-184005	American Cyanamid, USA	Phase I
TCV-309	Takeda, Japan	Phase II
UK-74505 (Modipafant)	Pfizer, USA	Phase II
BB-823	British Biotech, UK	Phase II

Phase I, tolerance study in healthy volunteers; Phase II, tolerance and efficacy in patients; Phase III, pivotal efficacy study in patients.

of the underlying pathophysiological mechanisms involved, is to use combinations of PAF antagonists with agents that may interfere with other mediators of septic shock.

6 Conclusions

In the past few years, research on PAF and PAF antagonists has led to important new theoretical and practical considerations. It is conceivable that PAF receptors exhibit different molecular characteristics in various organs, which may explain better why one drug may be more effective than another at a particular site. Experiments have provided evidence that PAF induces a unique cellular response even in highly differentiated cells, and PAF is able to increase phosphoinositide turnover leading to $[Ca^{2+}]_i$ release not only in inflammatory cells but also in isolated brain cells or neuronal cells in culture. These effects are coupled to G-proteins and appear to contribute to a great variety of cellular responses, including increased cellular excitability, eicosanoid release, and activation of the constitutive nitric oxide synthetase. Long ago, PAF was recognized as an important inflammatory mediator. Recent progress in research has elaborated the concept that PAF plays a mediator role not only in the local inflammatory response but also in SIRS. The fundamental mechanism through which PAF activates the inflammatory response involves important interactions with various cytokines, particularly TNF and interleukins. This autogenerated feedback network may lead to an exaggerated mediator release that finally results in SIRS and multiorgan failure in various shock conditions and ischemia. Inflammation has been recognized as an important defense mechanism in the organism, which may however be the source of an exaggerated response under certain pathophysiological conditions.

The extent of the inflammatory response depends on the balance between amplification and down-regulation of mediator release. These processes have been thought to involve mechanisms regulated by PAF; therefore, PAF antagonists can suppress amplification, providing protection against the consequences of exaggerated inflammatory mediator release, and exert relatively little influence on the down-regulation of mediator release. Systematic studies have provided circumstantial evidence that PAF plays a central role in endotoxin-induced inflammatory mediator release and therefore may be a major causative factor in the mechanism of septic shock. This recognition has changed the strategy in the clinical development of PAF antagonists, and explains why several PAF receptor antagonists have undergone clinical trials in patients with septic shock induced by gram-negative microorganisms. Septic shock is an extremely complex pathophysiological process that sometimes cannot be effectively treated with monoclonal antibodies raised against LPS or principal mediators such as TNF. PAF antagonists may be effective in sepsis because they selectively and potently block the effect of PAF at the receptor level, thus suppressing the deterioration induced by LPS and other factors through PAF release.

The failure of PAF antagonists to effectively influence bronchial asthma in clinical patients has elucidated important new aspects in experimental medicine. The finding that PAF antagonists lose their effectivity to block allergic bronchoconstriction points to the complexity of immediate-type hypersensitivity being fundamentally different from experimental models based on passive or active immunization. The alterations in PAF antagonist sensitivity after booster injection emphasize the importance of regulatory processes at the level of PAF receptors during the allergic process that may be brought about by inflammatory cells, most likely eosinophils. Ongoing research will clarify the appropriate applicability of PAF antagonists in acute and chronic allergic disorders. Applications of PAF antagonists in ischemic conditions of various organs, particularly in the heart and brain, have explored further details as to the relevance of these drugs as anti-ischemic agents. The results are rather divergent, and further research may clarify the requirements necessary for possible future clinical application of PAF antagonists in this field.

From the data summarized in this short review, it is obvious that most of the knowledge acquired on the role of PAF in pathophysiology was gained with the use of PAF antagonists. Among these antagonists, ginkgolide B (BN 52021) has been the most extensively studied and appears to have the greatest effects. Given that only a small number of components purified from *Ginkgo biloba* have been analyzed, the future possibilities of useful compounds from this tree are undoubtedly promising.

References

1. Braquet P, Touqui L, Shen TS, Vargaftig BB (1987) Perspectives in platelet-activating factor research. Pharmacol Rev 39:97–145
2. Koltai M, Hosford D, Guinot P, Esanu A, Braquet P (1991) Platelet-activating factor (PAF): a review of its effects, antagonists and possible future clinical applications. Drugs 42(part I):9–29
3. Koltai M, Hosford D, Guinot P, Esanu A, Braquet P (1991) Platelet-activating factor (PAF): a review of its effects, antagonists and possible future clinical applications. Drugs 42(part II):174–204
4. Braquet P, Hosford D (1991) Ethnopharmacology and the development of natural PAF antagonists as therapeutic agents. J Ethnopharmacol 32:135–139
5. Braquet P, Esanu A, Busine E, Hosford D, Broquet C, Koltai M (1991) Recent progress in ginkgolide research. Med Res Rev 11:295–355
6. Braquet P, Spinnewyn B, Braquet M, Bourgain RH, Taylor JE, Etienne A, Drieu K (1985) BN 52021 and related compounds: a new series of highly specific PAF-acether antagonists isolated from *Ginkgo biloba*. Blood Vessels 16:559–572
7. Huxtable JR (1992) The pharmacology of extinction. J Ethnopharmacol 37:1–11
8. Etienne A, Hecquet F, Guilmard C, Soulard C, Braquet P (1987) A potential link between platelet-activating factor and plasma protease activity. Int J Tissue React 9:19–26
9. Etienne A, Hecquet F, Soulard C, Spinnewyn B, Clostre F, Braquet P (1987) In vivo inhibition of plasma protein leakage and *Salmonella* enteritidis-induced mortality in the rat by specific PAF acether antagonist BN 52021. Agents Actions 17:368–370

10. Corey EJ, Gavai AV (1988) Enantioselective route to a key intermediate in the total synthesis of ginkgolide B. Tetrahedron Lett 29:3201–3204
11. Corey EJ, Kang MC, Desai MC, Ghosh AK, Houpis IN (1988) Total synthesis of (±)ginkgolide B. J Amer Chem Soc 110:649–651
12. Corey EJ, Rao KS (1991) Enantioselective total synthesis of ginkgolide derivatives lacking the *tert*-butyl group, an essential structural subunit for antagonism of platelet-activating factor. Tetrahedron Lett 32:4623–4626
13. Domingo MT, Piro F, Broquet C, Auclair E, Chabrier PE, Braquet P (1992) Reversible or irreversible modification of [^{3}H]PAF binding on rabbit platelet membranes differentiates various PAF receptor antagonists. Lipids 27:582–686
14. Godfroid JJ, Dive G, Lamotte-Brasseur J, Heymans (1991) PAF-receptor structure: a hypothesis. Lipids 26:1162–1166
15. Lamotte-Brasseur J, Heymans F, Dive G, Lamouri A, Batt JP, Reduilh C, Hosford D, Braquet P, Godfroid JJ (1991) PAF-receptor and "Cacheoreilles" effect: 2. Simple PAF-antagonists. Lipids 26:1167–1171
16. Ye RD, Prossnitz ER, Zou AH, Cochrane CG (1991) Characterization of a human cDNA that encodes a functional receptor for platelet activating factor. Biochem Biophys Res Commun 180:105–111
17. Nakamura M, Honda Z, Izumi T, Sakanaka C, Mutoh H, Minami M, Bito H, Seyama Y, Matsumoto T, Noma M, Shimizu T (1991) Molecular cloning and expression of platlet-activating factor receptor from human leukocytes. J Biol Chem 266:20400–20405
18. Honda Z, Nakamura M, Seyama Y, Shimizu T (1992) Properties of the guinea-pig lung platelet-activating factor receptor encoded by the cloned cDNA. J Lipid Mediators 5:105–107
19. Braquet P, Paubert-Braquet M, Bourgain R, Bussolino F, Hosford D (1989) PAF/cytokine autogenerated feedback networks in microvascular immune injury: consequences in shock, ischemia and graft rejection. J Lipid Mediators 1:75–112
20. Braquet P, Paubert-Braquet M, Koltai M, Bourgain R, Bussolino F, Hosford D (1989) Is there a case for PAF antagonists in the treatment of ischemic states? Trends Pharmacol Sci 10:23–30
21. McColl SR, Krump E, Naccache PH, Poubelle PE, Braquet P, Braquet M, Boergeat P (1991) Granulocyte-macrophage colony-stimulating factor increases the synthesis of leukotriene B_4 by human neutrophils in response to platelet-activating factor. Enhancement of both arachidonic acid availability and 5-lipoxygenase activation. J Immunol 146:1204–1211
22. Paubert-Braquet M, Hosford D, Koltz P, Guilbaud J, Braquet P (1990) Tumor necrosis factor primes PAF-induced superoxide production by human neutrophils: possible involvement of G proteins. J Lipid Mediators 2:S1–S14
23. Gatti S, Faggioni R, Echtenacher B, Ghezzi P (1993) Role of tumour necrosis factor and reactive oxygen intermediates in lipopolysaccharide-induced pulmonary oedema and lethality. Clin Exp Immunol 91:456–461
24. Pignol B, Hénane S, Sorlin B, Rola-Pleszczynski M, Mencia-Huerta JM, Braquet P (1990) Effect of long-term treatment with platelet-activating factor on IL-1 and IL-2 production by rat spleen cells. J Immunol 145:980–984
25. Merlos M, Gomez M, Giral M, Vericat ML, Garcia-Rafanell J, Forn J (1991) Effects of PAF-antagonists in mouse ear oedema induced by several inflammatory agents. Br J Pharmacol 104:990–994

26. Zarco P, Maestre C, Herrero-Beaumont G, Gonzales E, Garcia-Hoyo R, Navarro FJ, Braquet P, Egido J (1992) Involvement of platelet-activating factor and tumour necrosis factor in the pathogenesis of joint inflammation in rabbits. Clin Exp Immunol 88:318–323
27. Bone RC (1992) Towards an epidemiology and natural history of SIRS (systemic inflammatory response syndrome). JAMA 268:3452–3456
28. Bone RC (1991) The pathogenesis of sepsis. Ann Intern Med 115:457–469
29. Koltai M, Hosford D, Braquet P (1992) The role of PAF and cytokines in microvascular tissue injury. J Lab Clin Med 19:461–465
30. Zimmerman GA, Ringer TV (1992) Inflammatory host responses in sepsis. Prog Pediatr Crit Care 8:163–189
31. Koltai M, Hosford D, Braquet P (1993) PAF-induced amplification of mediator release in septic shock: prevention of downregulation by PAF antagonists. J Lipid Mediators 6:183–198
32. Koltai M, Hosford D, Braquet PG (1993) Platelet-activating factor in septic shock. New Horizons 1:87–95
33. Tracey KJ, Beutler B, Lowry SF, Merryweather J, Wolpe S, Milsark IW, Hariri RJ, Fahey TJ III, Zentella A, Albert DJ, Shires GT, Cerami A (1986) Shock and tissue injury induced by recombinant human cachectin. Science (Wash DC) 234:470–474
34. Mozes T, Heiligers JPC, Tak CJAM, Zijlstra FJ, Ben-Efraim S, Saxena PR, Bonta IL (1991) Platelet activating factor is one of the mediators involved in endotoxic shock in pigs. J Lipid Mediators 4:309–325
35. Mozes T, Zijlstra FJ, Heiligers JPC (1992) Interactions between platelet activating factor and eicosanoids during endotoxic shock in anaesthetized pigs. Mediat Inflamm 1:183–190
36. Klosterhalfen B, Horstmann-Jungemann K, Vogel P, Flohé S, Offner F, Kirkpatrick CJ, Heinrich PC (1992) The time course of various inflammatory mediators during recurrent endotoxemia. Biochem Pharmacol 43:2103–2109
37. Yue T-L, Rabinovici R, Farhat M, Feuerstein G (1990) Pharmacological profile of BN 50739, a new PAF antagonist in vitro and in vivo. Prostaglandins 39:469–480
38. Moore JM, Ernest MA, DiSimone AG, Abumrad NN, Fletcher JR (1991) A PAF receptor antagonist, BN 52021, attenuates thromboxane release and improves survival in lethal canine endotoxemia. Circ Shock 36:53–59
39. Fletcher JR, DiSimone AG, Earnest M (1990) Platelet activating factor antagonist improves survival and attenuates eicosanoid release in severe endotoxemia. Ann Surg 211:312–316
40. Rabinovici R, Yue TL, Farhat M, Smith EF III, Esser KM, Slivjak M, Feuerstein G (1990) Platelet activating factor (PAF) and tumor necrosis factor-α (TNF-α) interactions in endotoxemic shock: Studies with BN 50739, a novel PAF antagonist. J Pharmacol Exp Ther 255:256–263
41. Mozes T, Zijlstra FJ, Heiligers JPC, Tak CJAM, Ben-Efraim S, Bonta IL, Saxena PR (1991) Sequential release of tumour necrosis factor, platelet activating factor and eicosanoids during endotoxic shock in anaesthetized pigs: protective effect of indomethacin. Br J Pharmacol 104:691–699
42. Nakamura T, Kasai K, Sekiguchi Y, Banba N, Takahashi K, Emoto T, Hattori Y, Shimoda SI (1991) Elevation of plasma endothelin concentrations during endotoxic shock in dogs. Eur J Pharmacol 205:277–282
43. Pons L, Droy-Lefaix MT, Braquet P, Bueno P (1991) Role of free radicals and platelet-activating factor in the genesis of intestinal motor disturbances induced by *Escherichia coli* endotoxins in rats. Gastroenterology 100:946–953

44. Pons L, Droy-Lefaix MT, Braquet P, Bueno P (1991) Myoelectric intestinal disturbances in *Escherichia coli* endotoxic shock in rats. Involvement of platelet-activating factor. Lipids 26:1359–1361
45. Redl H, Vogl C, Schiesser A, Paul E, Thurnher M, Bahrami S, Schlag G (1990) Effect of PAF antagonist BN 52021 in an ovine endotoxin shock. J Lipid Mediators 2(suppl):S195–S201
46. Anderson BO, Poggetti RS, Shanley PF, Bensard DD, Pitman JM, Nelson DW, Whitman GJ, Harken AH (1991) Primed neutrophils injure rat lung through a platelet-activating factor-dependent mechanism. J Surg Res 50:510–514
47. Inoue Y, Kohno S, Miyazaki T, Yamaguchi K (1991) Effect of a platelet activating factor antagonist and antithrombin III on septicemia and endotoxemia in rats. Tohoku J Exp Med 163:175–185
48. Kucey DS, Kubicki EI, Rotstein DD (1991) Platelet-activating factor primes endotoxin-stimulated macrophage procoagulant activity. J Surg Res 50:436–441
49. Hsieh KM (1991) Effects of PAF antagonist, BN 52021, on the PAF-, methacholine-, and allergen-induced bronchoconstriction in asthmatic children. Chest 99:877–882
50. Desquand S, Lefort J, Dumarey C, Vargaftig BB (1990) The booster injection of antigen during active sensitization modifies the antianaphylactic activity of the PAF antagonist WEB 2086. Br J Pharamacol 100:217–222
51. Desquand S, Lefort J, Dumarey C, Vargaftig BB (1991) Interference of BN 52021, an antagonist of PAF, with different forms of active anaphylaxis in the guinea-pig: importance of the booster injection. Br J Pharmacol 102:687–695
52. Kuitert LM, Hui KP, Uthayarkumar S, Burke W, Newland AC, Uden S, Barnes NC (1993) Effect of the platelet-activating factor antagonist UK-74,505 on the early and late response to allergen. Am Rev Respir Dis 147:82–86
53. Giers G, Janzarik H, Kempe ER, Müller-Eckhardt C (1990) Failure of the platelet-activating factor antagonist WEB 2086 BS for treatment of chronic autoimmune thrombocytopenia. Blut 61:21–24
54. Guinot P, Braquet P, Duchier J, Cournot A (1986) Inhibition of PAF-acether induced weal and flare reaction in man by a specific PAF antagonist. Prostaglandins 32:160–163

Ginkgo and Humans

Resources and Prospects of *Ginkgo biloba* in China

HE SHAN-AN, YIN GU, and PANG ZI-JIE

1 History and Development

1.1 Is China the Last Distribution Area of Wild Ginkgo?

The natural distribution of *Ginkgo* has been restricted to China since the late Quaternary and early Tertiary when glaciation destroyed *Ginkgo* in reduced distribution areas in North America and Europe. Some parts of China untouched by the glacial epoch remained as refuges of certain archaic species such as *Ginkgo biloba*, *Metaseqouia glyptostroboides*, and *Liriodendron chinense*. However, it has long been a question whether wild *Ginkgo* trees survived in the wild or only as cultivated trees in gardens and temples.

According to our investigations and other reports, there are apparently many scattered distributed populations of *Ginkgo* in the wild: e.g., in the eastern part of China, in the West Tienmushan mountains, Zhejiang province, and southern mountainous area and Dabieshan mountains, Anhui province [1]; in central China, in Shengnongjia district and the Dahongshan mountains, Hubei province ([2], Zheng 1996, unpublished data); and in the western part of China, in Guizhou and Sichuan provinces (Xiang 1996, unpublished data).

Nevertheless, the existence of true wild *Ginkgo* is still a scientific problem. In fact, in the areas just mentioned, especially in those influenced by human activities, it is difficult to identify those naturally distributed individuals as either cultivated, escaped (seeds spread by animals), or true wild ones, except those located near temples, which are apparently cultivated trees. In the meantime, for the most probable wild population located in the West Tienmushan mountains, Wu et al. [3] suggested that because no significant difference has been found among 40 naturally distributed trees on the basis of research on genetic variation by means of isoenzyme analysis, the forest could be composed of progeny of plants cultivated at nearby temples by monks. In addition, Del Tredici et al. [4] have made a survey in that area and concluded that, because there have been

Institute of Botany, Jiangsu Province and Academia Sinica, P.O. Box 1435, Nanjing, 210014 Jiangsu, People's Republic of China

human activities in that area for more than 1500 years, it is difficult to be certain that the population is truly wild. For years scientists focused on these populations and were looking forward to identifying their properties, but it seems that to reach a final conclusion more convincing evidence is needed. We believe that if true wild *Ginkgo* trees remain, they are probably located in several locations in China.

1.2 History of Cultivation

In China there are many very old *Ginkgo* trees still living. Those as old as 1000 years or so are protected by the government. The oldest is reported to be more than 3000 years old (since the Shang dynasty), in Dinlinsi Temple, Fulaishan, Juxian county, Shandong province [5]. This tree is commonly recognized as a typical example of old *Ginkgo*. Other examples are (1) a tree more than 3500 years old in Panhu, Baoyacun village, Naxi district, Dongkouxian county, Hunan province [6], (2) a tree more than 2000 years old in Louguangtai, Zhouzhi county, Shaanxi province [5], and (3) one more than 2000 years old in Tancheng county, Shandong province. There are still about 100 trees more than 1000 years old and about 180 trees that are more than 500 years old. For example, in Hubei province there are 13 trees more than 1000 years old and 30 trees more than 500 years old (Zheng 1996, unpublished data). In Hebei province, individual trees more than 1000 years old were also reported [7]. Old trees are almost always located near temples or inside sites of historical interest and scenic beauty. Those considered as a part of the property of temples are conserved quite well. All these ancient trees should be regarded as the ancestors of cultivated *Ginkgo*. In this case, the true wild *Ginkgo* should be genetically different from these groups of ancient trees.

There is a great deal of literature about the history of *Ginkgo* cultivation. Early in the Three Kingdoms Period (A.D. 220–280) it was recorded that *Ginkgo* was widely cultivated in the south of the Yangtze River. It has been used as a source of edible nuts since at least the Han Dynasty [8]. In the Song dynasty (A.D. 960–1279), it was considered as an article of tribute to the emperor. In the poem by OuYang Xiu (A.D. 1007–1072) it was said that "Ginkgo grows south of the Yangtze River, the name coincides with its substance. Since the nuts have been used in tribute, then it is considered precious in the Capital." The record in *Shihuazonggui* (Xuan He period, Song Dynasty, A.D. 1119–1129) by Ruan Yue is that "In the Capital—now Kaifong city, Henan province—there was no *Ginkgo*. Since the coming of Mr. Li Wenhe, the emperor's son-in-law, from the south, he introduced it and planted it in his private house. Then it yielded and was propagated and developed there." All these records indicate that the cultivation of *Ginkgo* was initiated in south of the Yangtze River and then extended to the north, although there are even more ancient trees in the north at present.

1.3 Development of Cultivation

As mentioned, according to the written records the history of the cultivation of *Ginkgo* is more than 2000 years old. During the past a series of famous growing centers, including Taixing, Pizhou, Taixian, and Wuxian counties, Jiangsu province; Tancheng county, Shandong province; Zhuji, Changxing, and Linan counties, Zhejiang province; Xingan and Linchuan counties, Guangxi autonomous district; Anlu and Suizhou counties, Hubei province; and Nanzhang county, Hunan province, etc., have been developed [9]. Among these Taixing ranks first, and it has not only a history of 600 years but also its yield of nuts reaches one-third of the total of the country [9, 10]. During the past more than 40 years after the founding of People's Republic of China and because of the rapid increment in the consumption of nuts, the production of *Ginkgo* has greatly increased. The cultivation of *Ginkgo* has now extended to more than 20 provinces and cities. The total number of yielding trees is 0.7–0.8 million, and the total production of nuts is 5000–6000 tons per year on average.

As pharmaceutical usage developed, plantations for leaf harvest were initiated in the early 1980s in Europe but not until about 1990 in China. After 1990, the manufacture of extracts of leaves has rapidly developed, promoting the development of plantations for leaf harvest. Up to now there have been relatively more plantations in Jiangsu, Shandong, Anhui, Hubei, and Guangxi provinces, but the concentrated growing area is located in the adjacent areas of Jiangsu and Shandong provinces. For example, in Pizhou city, Jiangsu province, there are in total about 3000 hectares of plantations for leaf harvest. The potential of annual production could be very high, but actually in 1996 it was only about 1500 tons. It is estimated that the total production of dried leaves in China is about 3000 tons from regular dense plantations for leaf harvest and another 4000 tons from fruiting trees. For industrial usage, leaves collected from orchards after the harvest of nuts are not qualified resources because the content of active constituents is relatively low. It seems that this part of the resources should be totally replaced by leaves from young plants of dense plantations.

2 Distribution and Ecology

2.1 Geographic Distribution

Ginkgo is widely cultivated in China from latitude 25°N to 42°N and longitude 102°E to 122°E (Fig. 1). In the south it starts from the northeastern part of Guangdong province and extends northward to the southern part of Liaoning province. From eastern China the tree grows in Taiwan province and spreads down to the west in Guizhou and Yunnan provinces. It grows in more than 20 provinces. Among those Jiangsu and Guangxi provinces are in the lead. Vertical distribution is from below 50 m altitude in Pizhou city, Jiangsu province, to

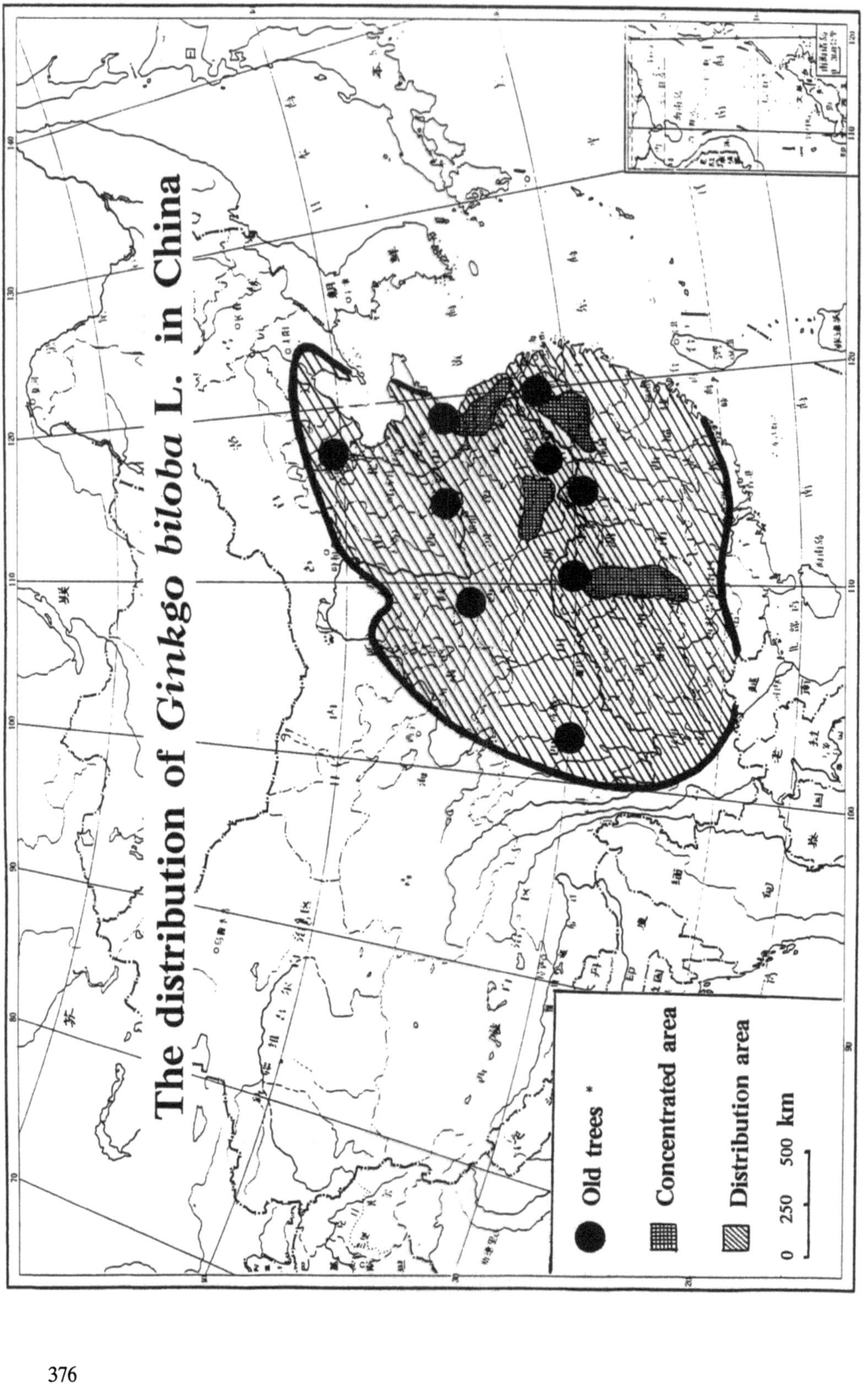
The distribution of *Ginkgo biloba* L. in China
Old trees *
Concentrated area
Distribution area
0 250 500 km

1000 m in Zhejiang province, and on the plateau of Yunnan and Guizhou at 2000 m.

2.2 Ecological Behavior

Ecologically, *Ginkgo* is quite adaptive. The concentrated growing areas have relatively different ecological conditions: annual mean temperature, 10°–18°C; annual mean precipitation, 600–1000 mm; absolute minimum temperature, 0° to −32.9°C; and absolute maximum temperature, 31.5°–42°C. Warm and moist weather suits *Ginkgo* but not very low temperatures and strong wind. Shengyang, Liaoning province is the northern limitation of its distribution. In that area, the survival rate of trees planted in the 1930s is only less than 10% after they were subjected to various extreme climatic conditions for more than 60 years. Down to the south in Guangzhou, the annual mean temperature is 21.9°C, annual mean precipitations is 1638 mm, maximum temperature is 38.7°C, and in winter time the mean temperatures in January and February are 14.3°C and 13.6°C, respectively. It seems that there is not enough chilling to meet the species requirement.

Ginkgo is a light-demanding tree growing well in exposed sites but not in shaded conditions. Long days and strong sunshine are favorable to its growth. Under short-day conditions, the seedling's growing season is shortened and it goes into dormancy early.

The tree is highly drought-resistant but sensitive to waterlogging. It is adaptive to various kinds of soil, but a deep soil with high fertility and moisture-holding capacity, and good drainage as well as a low underground water level, is most appropriate. The adaptability to soil acidity is relatively wide, from pH 4.5 to pH 8.0, but it is quite sensitive to soil salinity. It is satisfactory when salt content is less than 0.1%, but salt content is fatal when increased to more than 0.3%. Most large and healthy trees are located on sandy loam soil near houses, along dams, at roadsides, or at the foot of hill slopes. The climatic conditions of major growing areas in China are listed in Table 1.

2.3 Biological Characteristics

It is commonly known that the ginkgo tree does not grow very fast. According to analysis of the trunk growth, Liang [8] reported that the growth rate reaches its peak at the age of 30–40 years. In Jiangsu province, trees of 40 years old are about 12 m high with a breast height diameter (DBH) of about 30 cm; a 90-year-old tree in Tianmushan mountains, Zhejiang province, is about 20 m high and 35 cm in trunk diameter. The root system is very strong when the soil condition is favorable. For example, in Rugao county, Jiangsu province, a thousand-year-old tree

FIG. 1. The distribution of *Ginkgo biloba* L. in China. Old trees (*black circles*) are trees more than 1000 years old

TABLE 1. Climatic conditions of major growing areas

Locality	Latitude (°N)	Longitude (°E)	Annual temperature (°C)	Maximum temperature (°C)	Minimum temperature (°C)	Annual precipitation (mm)
Jiangsu						
Taixing	32°10′	120°00′	15.0	38.8	−11.7	1036.0
Wuxian	31°04′	120°26′	15.9	38.3	−8.7	1109.7
Anhui						
Ningguo	30°37′	118°58′	15.5	40.8	−14.5	1357.0
Xuancheng	30°56′	118°56′	16.0	40.7	−13.7	1293.0
Guangxi						
Xinan	25°38′	110°39′	18.0	38.5	−5.2	1767.7
Quanzhou	25°56′	111°04′	18.0	40.4	−6.3	1457.8

is 30m in height and 40m in crown diameter, but its root system is much larger and the radius of distribution of the root system is more than 37m. That means the range of the root system is about double its crown. Unless a *Ginkgo* tree is growing in a very dry region, there are usually a lot of sprouts surrounding the trunk.

Seedling trees commonly yield fruit late at the age of about 20 years; grafted trees often bear at the age of 8 or 10 years. Usually, the average yield of nuts is 3 tons per hectare. The full yield period starts at about 40 years old or older. The age for nut production can extend to 150–200 years. Because *Ginkgo* trees are usually cultivated in orchards that are not very uniform and regular, most records of yield of nut are based on individual trees. In general, the total amount of nuts was annually about 5000 tons before 1994; recently it has increased to 6000–7000 tons. In Wuxian county, Jiangsu province, a 130-year-old tree still bears 150–250kg per year. A 500-year-old tree in Taixing county, Jiangsu province, bears 25–50kg of nuts per year. The maximum yield record for a single tree is 600kg, corresponding to about 2kg nuts per square meter of the projected area of the tree crown.

There are two types of branches, long shoots and spurs. Each spur usually bears 2 or 3 and sometimes as many as 5–12 nuts. Physiologically, the auxin content of the long shoot and the spurs appear to be different.

Another interesting topic in *Ginkgo* research is the sexual identification of the juvenile stage, even from seed morphology. Female plants usually have less deeply divided leaves and often change color in autumn relatively earlier than do male trees. It is reported that the leaves of female plants present more bands on the zymogram of peroxidase isozyme [11], but this information still needs to be verified. Lee [12] reported that the difference in chromosome morphology of growth point cells can be a criterion for sexual identification. The chromosome number of *Ginkgo biloba* is $2\times = 24$. In somatic cells of the female plant, there are four chromosomes with satellites, but for males, only three have satellites. In matured macrospores during the diploid and haploid stages, there are four and two chromosomes with satellites, respectively. After meiosis of the matured microspores, half the pollen has only one chromosome with satellites and the other half has two chromosomes with satellites. Studying secondary root tips, Cheng et al. found that the mechanism of sex fixation is the WZ type [13]. To find a convenient and reliable technique for determining the sexual situation of seedlings or seeds is still an interesting challenge.

3 Genetic Resources and Cultivation

3.1 Varieties, Cultivars, and Conservation

Taxonomically, it is well known that in *Ginkgo biloba* L. eight varieties have been reported:

G. biloba var. *aurea* Beiss.
G. biloba var. *epiphylla* Mak.
G. biloba var. *fastigiata* Mast.
G. biloba var. *laciniata* Carr.
G. biloba var. *pendula* Carr.
G. biloba var. *rariegata* Carr.
G. biloba var. *stenocarpa* Hu.
G. biloba var. *variegata* Carr.

In 1935, three more varieties were identified by Tsen [14]:

G. biloba var. *apiculata* Tsen
G. biloba var. *typica* Tsen
G. biloba var. *huana* Tsen

Because these three varieties were not nominated following the International Nomenclature Code, Tsen's identification is considered invalid taxonomically but it is used practically in the horticultural cycle.

In *Flora of China*, Vol. 7 (1978), Zheng [15] discarded all identification of varieties published in the past but recorded 12 cultivars:

cv. Dongtinghuang
cv. Xiaofoshou
cv. Yaweiyinxing
cv. Fozhi
cv. Luanguofoshou
cv. Yuandifoshou
cv. Ganlanfoshou
cv. Wuxinyinxing
cv. Dameihai
cv. Tongzigao
cv. Mianhuaguo
cv. Damaling

In 1992, Song and Zhang [16] divided the species into ten ecotypes, but these are based only on geographic region of distribution without biological evidence, and also without representative cultivars or any common features of each ecotype, so that this division is not acceptable. *Ginkgo* is a multifunctioned species—its nuts, leaves, and timber are all economically important—but most of the cultivars described in China are mainly described after the shape, quality, and size of the nuts. All reported Chinese cultivars are divided into four groups:

1. Meihe-Yinxing, plum-stone-shaped ginkgo: Dameihe, Mianhuaguo, Nanhuiwuxin, Suanpanguo, Tongziguo, Ximeihe, Yuanzhu
2. Fushou-Yinxing, finger-citron-shaped ginkgo: Changbing-fushou, Dongtinghuang, Fuzi, Ganlan-fushou, Jiafushou, Jianchu, Jiangao-fushou, Jinguo-fushou, Luanguo-fushou, Xiao-fushou, Yuandi-fushou, Zaozi-fushou
3. Maling-Yinxing, horse-bell-shaped ginkgo: Damaling, Longyan, Huangpiguo, Qingpiguo, Ximaling, Zhongmaling
4. Others: Zhoupi, Dabaiguo, Jinzhuezi, Yuandou, Yanzhuzi, Bianfuzi Qixingguo, Nuomibaiguo, Dayuanling, Ganlanguo, Putaoguo Maotoubaiguo, Fuyanbaiguo, Liziguo, Dachaguo, Miguo zhongguo, Dayaoguo, Xiaoyaoguo, Gaoshenguo, Chuiziba, leafynut

The last cultivar, named "leafynut," has a special feature in that the nuts are borne on the leaves. A similar variation has been reported from Japan and also called Ohatsuki.

In addition, there are still many local cultivars, but these cultivars have no exact description or origin, etc., and thus they are not reviewed in this chapter. So far there is no cultivar in China named after ornamental features. In Western countries, however, *Ginkgo* trees are mostly used as street trees and there are many named cultivars [17].

In usage of the leaves for medical purposes, the quality of the leaves probably also differs among cultivars and provenances. This however has not carefully researched in China, although some results have been verified but not yet published.

The species has been recognized as a first grade of rare species to be recorded in the Chinese Red Data Book, Vol. I, but the collection and conservation of genetic resources is still a great task to be done in the future. At present only a few collections are considered to be repositories of genetic resources, including Nanjing Botanical Garden Mem. Sun Yat-Sen, Jiangsu Agricultural University, the Taixing collection at the county's forestry station, and the collection in pizhou county in Jiangsu, Guangxi Institute of Botany and Fuyang Institute of Chinese Academy of Forestry, Zhejiang. Unfortunately, all these collections lack a long-term project and qualified scientific maintenance.

3.2 Cultivation for Nuts and Leaves

3.2.1 Modern Intensive Cultivation

"Tree of grandfather and grandson" is the nickname of *Ginkgo*, which means that the tree bears fruits very late so that the grandfather plants the tree and the grandson harvests. Based on the traditional cultural practice the trees usually yield at 15–20 years old and enter a full yield period after 30–40 years of age.

Since the 1980s, under modern pomological cultivation the new technique of dense planting and early fruiting has been a hot topic, and a series of achievements have been made. Trees can start to yield at 3 years, and a record output of 24000 kg per hectare has been recorded in a matured orchard. The main measures of cultural practice are as follow.

1. Using cultivars of precocity: a selection named TC-9 can bear nuts 3 years after grafting. The average yield per plant was 1.37 kg at 5 years of age [18]. In Guangxi, a record 90% of plants in the orchard bore nuts at 5 years old for a yield of 1800 kg per hectare [19].
2. Early grafting: traditionally, grafting was made at 10 years after planting. Modern cultural practice has changed this to the use of 2- or 3-year-old seedlings as rootstocks.
3. Dense planting: according to the fertility of the soil, different degrees of the dense-planting method from 645 to 13320 plants per hectare are used for the early stages of orchards. After 5 years, the orchards are thinned gradually to 240–330 individuals per hectare. This measure increases the yield, especially in early years.
4. Tipping, girdling, and bark inversion: summer tipping, carried out by hand in July and August, is effective in promoting branching and stimulating the development of the plant to enter its reproductive stage. Making a girdle or bark inversion is also another essential measure to promote the tree to yield rapidly. Bark inversion is more effective than girdling; usually this is done by cutting a circular piece of bark with a width of one-tenth or one-fifteenth of the diameter of the trunk and turning it upside down. This technique can raise the percentage of fruiting trees from 6% to 92% at the age of 4 years.
5. Soil management: intensive cultural practice, with more watering and fertilizing, provides these critical factors to promote the yield at a young stage. Usually it is necessary to plough the soil to a depth of 60–80 cm or to prepare a planting space of 80 cm^3.
6. Fertilization: within the growing season, three- or fourfold fertilization is necessary. Although a complete fertilizer is necessary, growers mostly can supply only nitrogenous fertilizer. In general, fertilizations are carried out in March–April for leaves and root system, in May–June for young fruits, in the first half of July for development of fruits, and in October for the root system and next year's growth. Fertilization is important, but so far not enough research on the quantity and variety of fertilizers has been undertaken. The general concept accepted by growers is that organic manure should be provided for a tree twice a year in spring and winter, respectively, and that the quantity each time should be at least fourfold the amount of nuts born by the tree. Thus, chemical nitrogenous fertilizer must equal 5% of the amount of nuts for the tree during summer and autumn, respectively. In actual practice, very much smaller amounts of potassium and phosphorous fertilizers are used for *Ginkgo*.
7. Appropriate management: to ensure timely irrigation and drainage, it is important to avoid drought and waterlogging, especially during the periods of sprouting in spring, blooming, germination, and the overwinter periods.

8. Artificial pollination: it has been verified by experiments that low yield could be caused by insufficient pollen. Thus, artificial pollination with collected pollen or simply grafting some male tree branches on the female tree is recommended, and these methods appear to be effective cultural practices for high yield.

3.2.2 Plantation for Leaf Production

Dense plantation for leaf production has been practiced for about 15 years since the first project in France and then in the United States. Since the 1990s, China has begun to establish leaf plantations in Jiangsu and Shandong provinces but no attention has been paid to cultivars for leaf production. The density of a leaf plantation is from 50000 to 150000 plants per hectare. The density of plants is very variable because most plantations originate from nurseries. Because hand harvesting is used, in China output is 1 year earlier, from the second year after planting, than that of machine harvesting. In the third year it reaches a high peak of production and the yield could be 3 tons dried leaves per hectare. For economic reasons Chinese farmers prefer an intercropping system for leaf plantation, but this is not appropriate because it induces more pests and diseases than in a pure *Ginkgo* plantation. The total acreage of leaf plantation in China is now estimated to be about 5000 hectares.

5 Prospects

China is the origin of the live *Ginkgo* and there are still about 100 trees of more than 1000 years of age. The history of utilization and cultivation of this species covers more than 2000 years, and in China there is a rich biological diversity of the species and much experience in its utilization.

The species has multiple uses and functions. For example, it is an ornamental tree with many fewer pests and diseases and an air-pollution-hardy tree for the urban environment. It is also an excellent tonic and nut tree and provides valuable medicinal resources.

Although *Ginkgo* is a species of worldwide distribution, at present it is certain that so far the most suitable ecological environment for it has been found in China. Compared with most huge plantations in the world, the growing area in the northern part of Jiangsu province is at least one of the first-grade, if not the best, regions for the cultivation of this species.

Whether from a scientific, economic, environmental, or conservation point of view, the species combined with its habitats is worthy of further intensive research. Its achievements are surely significant for evolution, conservation, development, and economy in the future as well as for the sustainable survival of humans and our Earth.

References

1. Li HL (1956) A horticultural and botanical history of *Ginkgo*. Bull Morris Arbor 7:3–12
2. Jiang MX, Jin YX, Zhang QF (1990) A preliminary study on *Ginkgo biloba* in Dahongsan region, Hubei (in Chinese). J Wuhan Bot Res 8:191–193
3. Wu JY, Cheng PL, Tang SJ (1992) Isozyme analysis of the genetic variation of *Ginkgo biloba* L. population in Tian-Mu Mountain (in Chinese). J Plant Resour Environ 1:20–23
4. Del Tredici P, Ling H, Yang G (1992) The *Ginkgos* of Tian Mu Shan. Conserv Biol 6:202–209
5. Sun YW (1982) History of fruiticulture and fruit resources of China (in Chinese). Shanghai Science and Technology Press, Shanghai
6. Zhang ST, Zeng YY, Xie WB (1992) The distribution and production of ginkgo in Hunan province (in Chinese). In: Bulletin of the first national symposium on *Ginkgo*. Hubei Science and Technology Press, Wuhan
7. Yang Z, Yang HS (1994) Resources and utilization of *Ginkgo* in Hebei province. In: Bulletin of the Second national symposium on *Ginkgo* (in Chinese). Hubei Science and Technology Press, Wuhan
8. Liang LX (1993) The contemporary ginkgo encyclopedia of China (in Chinese). Beijing Agricultural University Press, Beijing
9. Song CS (1992) The situation of ginkgo production and its development (in Chinese). In: Bulletin of the first national symposium on *Ginkgo*. Hubei Science and Technology Press, Wuhan
10. Zong JZ (1992) Large ginkgo nuts of Taixin—the supreme of *Ginkgo biloba* (in Chinese). In: Bulletin of the first national symposium on *Ginkgo*. Hubei Science and Technology Press, Wuhan
11. Zhong HW, Yang ZH, Zhu GL, Cao ZX (1982) Peroxidase isozyme pattern as a biochemical test to distinguish the sex of individual plant in *Ginkgo biloba* L. (in Chinese). Sci Silvae Sin 18:1–5
12. Lee CL (1954) Sex chromosomes in *Ginkgo biloba*. Amer J Bot 41:545–549
13. Cheng RY, Song WQ, Lee XL (1993) A study on the sex chromosome of *Ginkgo biloba*. Cathaya 5:41–48
14. Tsen M (1935) *Ginkgo* of Zhuji, Chekiang (in Chinese). Horticulture 1:157–165
15. Zheng WJ (1978) Flora republicae popularis Sinicae, tomus 7 (in Chinese). Science Press, Beijing
16. Song CS, Zhang QH (1992) Exploration on the determination of geo-ecotypes and assessment of ginkgo cultivars (in Chinese). In: Bulletin of the first national symposium on *Ginkgo*. Hubei Science and Technology Press, Wuhan
17. Santamour FS Jr, He SA, McArdle AJ (1983) Checklist of cultivated ginkgo. J Arboric 9:88–92
18. Huangpu GY, Hou JH (1992) Short, dense, early and good harvest of ginkgo (in Chinese). In: Bulletin of the first national symposium on *Ginkgo*. Hubei Science and Technology Press, Wuhan
19. Guangxi Institute of Botany (1992) Early and good harvest of *Ginkgo* (in Chinese). In: Bulletin of the first national symposium on *Ginkgo*. Hubei Science and Technology Press, Wuhan

A Cultural History of *Ginkgo biloba* in Japan and the Generic Name *Ginkgo*

Shihomi Hori[1] and Terumitsu Hori[2]

1 Introduction

The Japanese people today are surrounded by many trees of the species *Ginkgo biloba*. They appreciate the taste of the nuts and use the shape of *Ginkgo* leaves in logos. How old is the oldest *G. biloba* in Japan? Is it native to Japan? If not, when and how did it come here? When did *G. biloba* become so popular in Japan? What about the name? It is known that the generic name *Ginkgo* originates from a Japanese name for this plant, but there have been many discussions on the name and spelling both inside and outside Japan: is it a misspelling or a misprint? Believing that we could expect to uncover the history of *G. biloba* in japan if we searched as many sources as possible, we gathered information from reports on food found in archaeological excavations, from legends, from dictionaries and books, from arts and crafts, and from other diverse sources.

2 Living and Nonliving Objects That Have Ginkgo in Their Name

In the Japanese language, some living and nonliving things are called *icho*,[1] the Japanese name for *G. biloba*, and there are also compound names containing that word (Table 1).

Hakone-shida (the Japanese name for *Adiantum monochlamys*) was nicknamed *Icho-shinobu*, or Ginkgo-hare's foot fern, because its leaflets resemble the leaves of *G. biloba* [4, 5]. This plant was collected by Kaempfer (1651–1716) [4, 5, 13], who is the author of *Amoenitatum Exoticarum* [14] and worked as a Dutch surgeon from 1690 to 1692 at Dejima,[2] which was the only port open to foreign

[1] 2-5-23 Higashi, Tsukuba, Ibaraki 305, Japan
[2] Institute of Biological Sciences, University of Tsukuba, 1-1-1 Tennodai, Tsukuba, Ibaraki 305, Japan
Correspondence to: T Hori

TABLE 1. Examples of living things with the word *icho* (ginkgo) in their names

Japanese name	English translation	Category	Latin names (and part)	References
Hime-icho	Small-Ginkgo	Green alga	*Udotea javensis* (whole body)	1
Icho-take	Ginkgo-mushroom	Fungus	*Paxillus panuoides* (whole body)	2
Icho-ukigoke	Ginkgo-floating moss	Bryophyte	*Ricciocarpus natans* (whole body)	3
Icho-shinobu	Ginkgo-hare's foot fern	Pteridophyte	*Adiantum monochlamys* (leaflet)	4
Icho-no-chichi	Nipples of *Ginkgo*	Gymnosperm	*Ginkgo biloba* (lignotuber)	4, 5
Icho-gusa	Ginkgo-grass	Angiosperm	*Lactuca repens* (whole body)	4
Icho-baika-mo	Ginkgo-white-flower alga	Angiosperm	*Rananculus nipponicus* (whole body)	3
Icho-imo	Ginkgo-potato	Angiosperm	*Batatatis rhizoma* (whole body)	6, 7
Icho-gani	Ginkgo-crab	Crab	*Cancer japonicus* (whole body)	8
Icho-gai	Ginkgo-shellfish	Shellfish	*Homalocantha anatomicus* (whole body)	9
Icho-shira-tori	Ginkgo-white bird	Shellfish	*Arcopagia (Merisca) diaphana* (whole body)	9
Icho-zame	Ginkgo-shark	Fish	*Sphryna zygaena* (whole body)	10
Icho-ba(ne)	Ginkgo-feather	Bird	*Aix galericulata* (tertiaries) (feathers)	6, 11
Icho-ha-kujira	Ginkgo-tooth whale	Mammal	*Mesoplodon ginkgodens* (teeth)	12

ships during the period when Japan closed its doors to the world. He traveled twice to Edo[3] (the old name of Tokyo) and found this fern at Hakone[4] near Mount Fuji [13]. It is said that Kaempfer mentioned that the fern was used to help a woman to maintain good health before and after giving birth [4, 5].

The name *icho no chichi*, nipples of *G. biloba*, appears with an explanation in *Honzo Komoku Keimo* [5], a textbook about plants and medicine written at the end of the Edo period[5] (1603–1868). Mandarin ducks (*Aix galericulata*) have been adored and featured in poems both in China and Japan because they are regarded as lovebirds. Chinese people noticed that the leaf of *G. biloba* resembles a duck's foot and named the tree after it: 鴨脚 or "duck foot" [15]. The Japanese, seeing that their beautiful orange tertiaries are *Ginkgo*-leaf-shaped, called them *icho-ba(ne)*[6] or ginkgo-feather [6, 11]. These are often mentioned in haiku,[7] 17-syllable poems [16, 17].

A dictionary of *zappai*,[8] playful poems, cites a *senryu*[9] (a satirical poem) written in 1773 in which the plectrum of a *shamisen*,[10] a traditional three-stringed

TABLE 2. Examples of nonliving objects with the word *icho* (ginkgo) in their names

Japanese name	Translation	Category	Reason for the name	References
Icho-man'no	Ginkgo-rake	Farming tool	*Ginkgo*-leaf-shaped blade	18
Icho	Ginkgo	Musical instrument	*Ginkgo*-leaf-shaped plectrum	19
Icho	Ginkgo	Family name	Unknown	20
Icho	Ginkgo	Tool	*Ginkgo*-leaf-shaped arrowhead	6
Icho-guchi	Ginkgo-mouth	Flower vase	*Ginkgo*-leaf-shaped sides	21
Icho-daisu	Ginkgo-cupboard	Furniture	As a material	22
Icho-gasshiki	Ginkgo-Noh-mask	Instrument	*Ginkgo*-leaf-shaped forelock	10
Icho-ashi	Ginkgo-feet	Furniture	*Ginkgo*-leaf-shaped table legs	23, 24
Gin'nan-uchi	Ginkgo-nuts-hitting	Game name	Playing with *Ginkgo* nuts	25
Icho-ba	Ginkgo-teeth	Footware	*Ginkgo*-leaf-shaped clog teeth	26, 27
Icho-gashira	Ginkgo-head	Hair style	*Ginkgo*-leaf-shaped topknot	28
Icho-ya	Ginkgo-shop	Name of a restaurant	Unknown	29
Icho-machi	Ginkgo-town	Placename	With *Ginkgo* trees	30
Icho-musume	Ginkgo-girl	Girl's nickname	A shopkeeper at a shop by a *Ginkgo*	31–34

musical instrument, is called *icho*[11] or ginkgo [19] (Table 2). It is recorded that some people whose family name was *Icho*[12] or Ginkgo lived in the Kyoto[13] area in the 1800s [20], and this name is also known today. *Icho-gasshiki*[14] is a mask with a *Ginkgo*-leaf-shaped forelock. It is a mask of a child's face that is worn by a boy actor in Noh,[15] Japan's oldest theater form, which began to flourish in the fourteenth century [10]. *Icho-ashi*[16] or ginkgo-feet are table legs. Before Western tables were introduced to Japan, individual small low tables were usually used at meals. The short legs of these tables are *Ginkgo*-leaf-shaped, broad and curved outward at the feet [23, 24].

There are towns called *Icho-machi*,[17] or Ginkgo-town, here and there in Japan. One of them, now part of Mito[18] City, Ibaraki Prefecture, has had the name since 1690 when an official command to name it was issued [30]. *Icho-musume*,[19] or ginkgo-girl, a beautiful shop girl at a toothpick shop at Asakusa in Edo, is pictured in ukiyo-e[20] prints [31] by Harunobu (1725?–1770) and described in *Han'nichi Kanwa* (1769–1823) [32], *Ameuri Dohei ga Den* (1769), and *Osen Ofuji Yuretsu no Ben* (1769) by Ota [33]. The *Nihon Denki Densetsu Dai-jiten* [34], a dictionary of legends and folk tales, gives good references for the ginkgo-girl, including a *kyogen*[21] or play, performed by Kabuki,[22] a popular dramatic art that has been a favorite since the seventeenth century, and a folk song, besides the stories.

3 Traditional Uses of *G. biloba*

3.1 Medicinal and Other Uses

In Japan, *G. biloba* has been used for centuries as a medicine. Japanese scholars once read Chinese books on medicine, such as the *Honzo Komoku* (1596) [35] and also the *Shokumotsu Honzo* (1638) [36]. Japanese books of medicine such as *Tashiki-hen* (1612) [37], *Yamato Honzo* (1709) [38], and *Wakan Sansai Zue* (1712) [39] were written in Chinese, based primarily on the *Honzo Komoku*. As seen in the *Hiden Kakyo* (1773) [40], some books added marks to help in reading the Chinese. Ono [5] wrote *Honzo Komoku Keimo* (1803–1806) in Japanese in the Edo period, citing many parts of *Honzo Komoku* and other Chinese books on medicinal plants and adding explanations of them. Therefore, it is considered that the Japanese way of using *G. biloba* as a medicine originates from the Chinese tradition.

The fruits and nuts of *G. biloba* were used for the following purposes: (1) getting rid of cough, sputum, fever, and the effects of poison [35, 36, 39, 41–43]; (2) facilitating recovery from hysteria and from the effects of alcohol [35, 36, 41, 42]; (3) as an invigorant [35, 36, 41–43]; (4) warming the lungs [35, 36, 39, 41, 43]; (5) reducing frequency of micturition, bed-wetting, cloudy urine, gonorrhea, and asthma [35, 36, 39–42]; (6) healing a swelling, boil, or other skin disease [35, 36, 41, 43]; and (7) stopping diarrhea and toothaches [35, 36, 41, 43]. Many books sternly warn against eating too many *Ginkgo* nuts because if people, especially young children, eat too many they often suffer from diarrhea [39, 40, 42] and sometimes die of it [35, 36, 38, 41, 44].

Juice made from the fresh leaves was used to warm chilblained parts [42]. Fallen leaves of *G. biloba* were used as insecticides [34, 35, 42] as well as for fertilizer [42]. It is reported that *Ginkgo* leaves inserted in books protect the pages from being stained [43, 44]. A record of annual events at the time of Emperor Gomizu-no-o[23] (1596–1680) [45] and a book of *haikai*[24] [46] mention that a *Ginkgo* leaf was used as a name card when a gift was presented to someone.

3.2 Wood

The wood of *G. biloba* is pale yellow, fine grained, shiny, and flexible. It does not shrink or crack or warp easily even if it repeatedly gets wet and dries out [44, 47]. Therefore, it is used for chopping boards, frames for making Japanese paper, tubs for brewing sake,[25] sawing benches, go[26] boards, chess pieces for *shogi*,[27] seal-chops, abacuses, tables, cupboards, home shrines, and so on [44, 47]. It is also used for lacquerware materials, such as for trays, bowls, and containers for tea powder [44]. These uses were developed in Japan. There are some shops in Fukui[28] and Fukuoka[29] prefectures that sell only *Ginkgo* timber and artifacts made of *Ginkgo* wood.

3.3 Food and Cooking

Ginkgo nuts have long been regarded as a delicious food by the Japanese people, and these have been cooked and served in various ways depending on the time. *Ginkgo* nuts appear as a fruit or as tea cakes in a textbook called *Shinsen Ruiju Orai* (1492–1521) [48] and in *Hisamasa Chakaiki* (1533–1596), which is the first volume of *Matsuya Kaiki* [49], a series of records of tea ceremonies. In the *Miyoshi-tei Onariki* [50], *Ginkgo* nuts are listed as sweets and as a dessert in the menu of a meal that was served in 1561 to Shogun[30] Ashikaga Yoshiteru[31] (1536–1565) and his followers.

In the Edo period, *Ginkgo* nuts came to be eaten by the common people. In a cookbook titled *Ryori Monogatari* (1643) [51], the nuts are categorized as vegetables and are also mentioned as ingredients for pickles. *Honcho Sokkan* (1697) [52], a book of edible and medicinal plants in Japan, and *Hiden Kakyo* [40] explain how to cook the nuts: take off both the shells and the thin inner skin, and grill them or boil them in soup. *Kaiseki-ryori-cho* (1785) [53] and *Kasen-shu* [54] regard *Ginkgo* nuts as a side dish when drinking sake. The most common ways of preparing *Ginkgo* nuts today are grilling them or boiling them in *chawan-mushi*,[32] a pot-steamed egg dish, or in *nabe-ryori*,[33] a sort of Japanese fondue. Furthermore, grilled nuts are still often eaten by the Japanese today when they drink sake.

Pieces of *Ginkgo*-leaf-shaped radish in soup were described in *Ukiyo-buro* (1813) [55]. Even today, slicing vegetables such as carrots and radishes into the shape of a *Ginkgo* leaf is one of the most common ways of preparing vegetables; it is called *icho-giri*,[34] ginkgo-cut.

4 Design

4.1 Symbolic Designs

4.1.1 Family Crests

Many plants and animals as well as other materials have been used in designs for family crests since the Middle Ages in Japan. Family crests are put on roof tiles, gates, and walls of structures such as castles, temples, shrines, and mansions, and are engraved on gravestones and swordguards. Family crests are also dyed or woven in flags and ceremonial kimonos[35] and so on. Nowadays we do not often put them on buildings, but crests are still put on gravestones and on black ceremonial kimonos, which are worn on special occasions such as weddings or funerals.

The leaves, rarely with the nuts, are used in more than 100 patterns of ginkgo crests [56, 57]. They are classified by the number of leaves, from 1 to 16, and also by the arrangement of the leaves (Fig. 1). Many of the *Ginkgo*-leaf designs are arranged to make a circle or are encircled, but some are in the shape of a square, a diamond, a pentagon, a hexagon, a heptagon, an octagon, etc. In some patterns, *Ginkgo* leaves are arranged clockwise. Some ginkgo heralds are desinged with other plants, fans, swords, etc. Some leaves are shaped into a butterfly or a crane; in fact, this is characteristic of ginkgo heraldry.

A family crest composed of three *Ginkgo* leaves is found among approximately 260 crests for families varying in social rank from shogun to commoner in *Kikigaki Shokamon* [58], a book of heraldry written at the time of Shogun Ashikaga Yoshimasa[36] (1436–1490), who caused Ginkakuji Temple[37] to be built in 1489. Later, Ihara [59] described a kimono decorated with ginkgo crests in *Nanshoku Okagami* (1687). It is known that the father of Shogun Tokugawa Ieyasu[38] used a crest of three *Ginkgo* leaves arranged with three swords, although later on the Tokugawas used a hibiscus crest [56, 57] (see Fig. 1).

4.1.2 Logos

The Nakamura family of the Kabuki theater had a logo that used a *Ginkgo* leaf shaped into a crane, and put it on a flag. This is described in *Nisemonogatari Tsuhosho* (1785) [60], a playful story, and is also pictured in an ukiyo-e print, *Nakamura-za Shibai-zu*, by Okumura [61] in 1731.

Today, many towns, cities, and prefectures, including Tokyo, use *Ginkgo*-leaf logos on printed materials and official vehicles, and universities and high schools use *Ginkgo*-leaf logos on badges, stationery, and so on.

4.2 Artistic Design

Kimonos decorated with *Ginkgo* leaves first appearred in the Momoyama period[39] (1585–1603) [62, 63]. From that time on, *Ginkgo* leaves have often been used in kimono patterns together with leaves of the maple, oak, pine, and so on. The leaf motif is called a *fukiyose*[40] or blown-together pattern because it sketches fallen leaves being blown by a strong autumn wind [63]. Designs of *Ginkgo* leaves are also found in books of *sho-es*[41] or paper patterns that show the exact length of the horizontal and vertical yarn and the arrangement of the colors required to weave cloth for kimonos [64]. *Fukiyose* patterns are seen today on ties, scarves, handkerchiefs, and so on as well as on autumnal kimonos and obis,[42] or sashes.

A picture of a flower vase called *icho-guchi*[43] or ginkgo-mouth is included in *Okazari no sho* (1523) [21], a record of the furniture, tools, and utensils that Shogun Ashikaga Yoshimasa kept in Ginkakuji Temple. Matsuya [49], who edited *Matsuya Kaiki*, a record of tea ceremonies held during three generations of the Matsuya family between 1533 and 1650, wrote that a plum blossom was arranged in an *icho-guchi* or ginkgo-mouth vase at the tea ceremony held on November 21, 1613.

A book titled *Ninsei* [65] reported *Nin'naji-gyoki*,[44] a record of Nin'naji Temple in Kyoto. The latter mentions that Nonomura Ninsei[45] demonstrated the use of a potter's wheel in the garden of the temple in 1655. He made two incense cases, and the lid of each had a *Ginkgo* leaf [65].

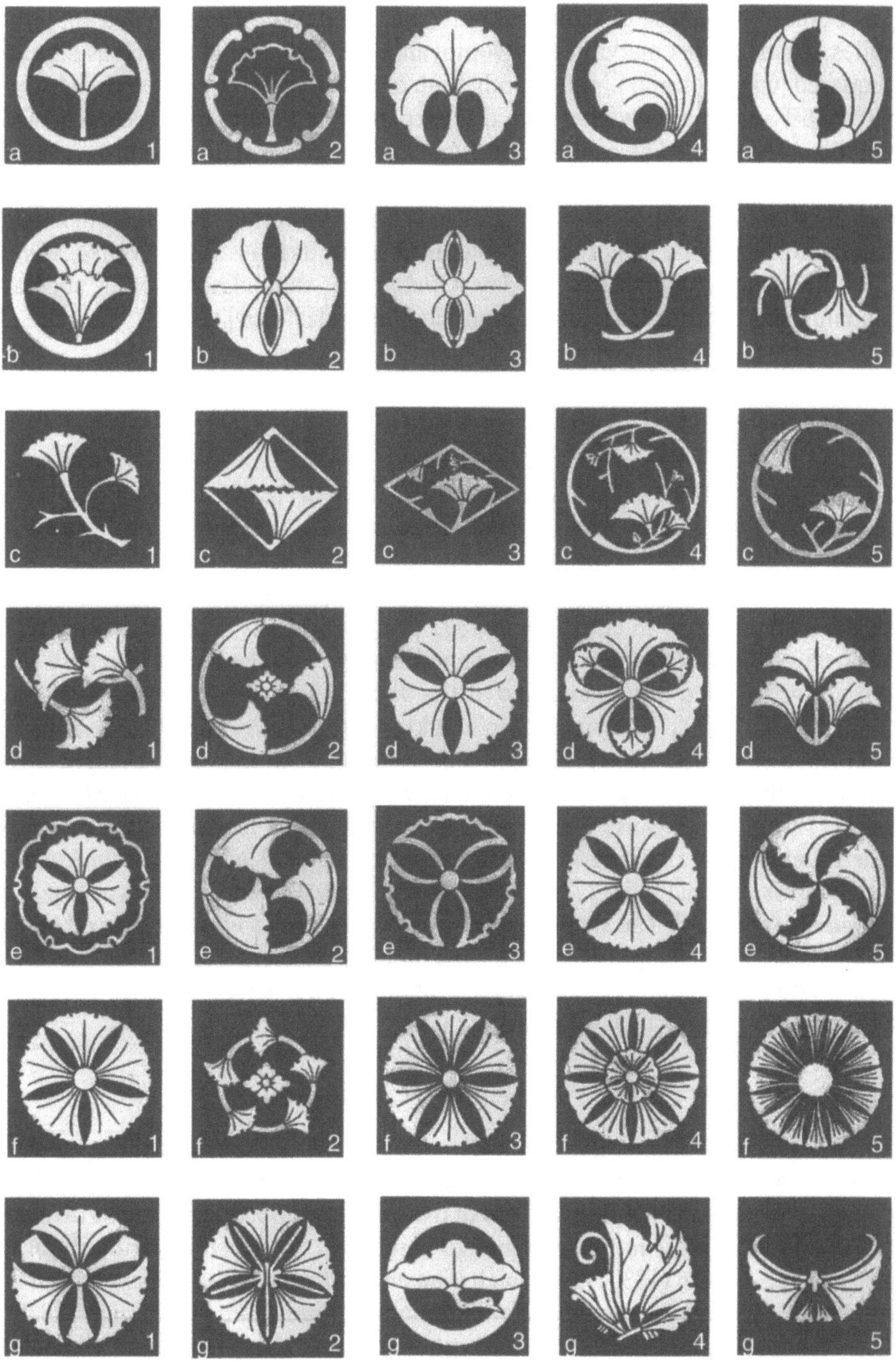

FIG. 1. Examples of ginkgo crests (**a.1–g.5**). *Center* (**d.3**): the oldest crest, composed of three *Ginkgo* leaves. *Left bottom* (**g.1**): crest made up of three *Ginkgo* leaves arranged with three swords (see Section 4.1.1 for detail). *Bottom*: crests shaped into a crane (**g.3**) and a butterfly (**g.4,5**) (from [56])

A book of Japanese designs and patterns, *Nihon no Monyo·Koyo* [66], includes artistic works made in the Momoyama and Edo periods, such as a gold-lacquered lunch box and kimono cloth, both of which are decorated with *fukiyose* patterns. Today many lacquerware dishes, plates, and containers are decorated with *Ginkgo* leaves. The catch of a *fusuma*[46] or Japanese sliding door decorated with a *Ginkgo* leaf is shown in a book of Japanese cloisonne [67]. A geisha[47] wearing a kimono decorated with a ginkgo crest and wearing *Ginkgo*-leaf-shaped ornamental hairpins in her hair is pictured in an ukiyo-e titled *Nanro Meigi* (1789–1800) by Utagawa [68]. Today, *Ginkgo*-leaf-shaped accessories, such as pendants and brooches, are often found in Higo-inlay[48] [69].

5 *Ginkgo*-Leaf-Shaped Hair Styles

5.1 *Seventeenth to Early Twentieth Centuries*

5.1.1 Coiffures for Men

During the Edo period, men wore topknots. This practice continued until they were legally allowed to cut them in 1871 [70], which was soon after the Meiji Restoration[49] of 1868, when Japan opened its doors to the world. Because the topknot was folded in two and tied at the back, it was called "*futatsu-ori*"[50] or "folded-in-two" [71]. It is pictured in an ukiyo-e titled *Sugata-e Hyakunin-isshu* by Hishikawa (1618?–1694) [72], and a description of women who thought men who wore their hair in that style were charming also appears in *Seken Musume-katagi* (1716) [73]. Because the tip of a topknot is *Ginkgo*-leaf-shaped, it was also called simply *icho*[51] or ginkgo. An example of this appears in *Soga Gonin Kyodai* [74], by Chikamatsu (1654–1724). *Icho-gashira*,[52] ginkgo-head, appears in *Shikido Okagami* (1688) [28]. The shape, size, and position of one's topknot showed one's social status. For example, samurais[53] and sumo[54] wrestlers wore *o-icho*[55] or big-ginkgo and merchants wore *ko-icho*[56] or small-ginkgo. These varieties were described in *Kazo Soga* by Chikamatsu [75].

5.1.2 Coiffures for Girls and Women

A hair style called *shimada*[57] has long been a formal coiffure; a bride today still wears her hair this way when she is dressed up in a bridal kimono. *Icho-mage*[58] or ginkgo-chignon is a simplified coiffure of the *shimada* style, and the bun is divided into two loops. This *icho-mage* or ginkgo-chignon is described in various books, such as *Kiyu Shoran* (1830) [29], which is an encyclopedic bibliography written in essay style. The *icho-mage* is also mentioned in *Shunshoku Eitaidango* (1838) [76], *Azuma Nagori* (1885) [25], and other sources. Kitagawa depicted social customs and manners between 1837 and 1853 in his essay *Morisada Manko* [27]. According to this essay, girls under 12 or 13 years of age and female instructors of Japanese dancing used to wear *icho-mage* or ginkgo-chignon in the beginning of the 1830s, and after about 1850, the custom spread to other women.

Icho-kuzushi,[59] a variation of ginkgo-chignon decorated with a greenish blue or purple crepe silk band at the foot of the chignon, became popular about 1850 among women under 30 in the geisha world [27]. *Icho-gaeshi*[60] or ginkgo-turning is another variation of *icho-mage* or ginkgo-chignon. This was called "butterfly" in the Kansai Distrct, the western part of Japan, so we will call it butterfly-ginkgo to distinguish it from the others. The butterfly-ginkgo has two *Ginkgo*-leaf-shaped chignons, one at each side of the head. *Azuma Nagori* [25] mentions that this hair style was popular first among girls and young women at the middle of the Edo period and then among artists and geishas. Between the end of the Edo and the beginning of the Showa period, women past middle age also wore their hair in this style. Girls and women wearing their hair in the butterfly-ginkgo style often appear in novels such as *Kokoro* [77] by Natsume (1867–1916) and *Gan* [78] by Mori (1862–1922).

5.2 The Sumo World

It is known that in early times in Japan a sumo match between Nomi-no-Sukune[61] and Taima-no-Kehaya[62] was held in the presence of Emperor Suinin.[63] As Nomi-no-Sukune won, he is called the God of Sumo. Ever since then sumo matches have been favored by emperors, shoguns, and powerful leaders [79, 80].

According to information provided by the Japan Sumo Association, the sumo wrestler's hair style has changed during its long history. Of the early hair styles only two types remain: *o-icho* and *chon-mage*.[64] The *o-icho* or big-ginkgo style is for sumo wrestlers from the rank of *juryo*,[65] or junior grade, to grand champion, *yokozuna*.[66] *Chon-mage* originates from *hoso-icho*[67] or thin-ginkgo, a style favored by shopkeepers in the Edo period, and today it is the hair style of sumo wrestlers below the *juryo* level. As time passed, the shape of the topknot of the *o-icho* style changed. In the Heian period[68] (794–1192) the topknot was straight up on the crown, and sometimes in the Edo period the forelock was shaven; these variations are pictured in *nishiki-es*[69] by Katsukawa Shunsho[70] and other ukiyo-e masters of the Katsukawa school [80]. From the Kansei time[71] (1789–1801) on, the whole hair came to be tied on the crown to make it like a big *Ginkgo* leaf, and the tip of the topknot was also shaped like a *Ginkgo* leaf. This style can still be seen today.

6 *Ginkgo* Described in Literature

Many plants appear in the *Manyo-shu* [81], the oldest anthology of *waka*,[72] 31-syllable poems, in Japan. It was compiled at the end of the Nara period[73] (710–794). There are a considerable number of *waka* about yellow leaves in this anthology. Are they *G. biloba*? Koshimizu remarks in *Manyo-shokubutsu to Kodai-jin no Kagaku-sei* [82] that among all the references to colored leaves in the *Manyo-shu* [81], only three specifically refer to red leaves. Otomo-no-Yakamochi,[74] one of the representative poets of that period, wrote poems containing the word chichi[75] (see the chapter by Del Tredici, this volume). Concerning this *chichi*, Kamo-no-Mabuchi [83] mentioned in *Kanji-ko* (1757) that it must be a *Ginkgo* tree because it forms chichi or nipples when it gets old. Shinmura [84] and Matsuda [85] studied about the plant called chichi by reading old dictionaries and books of plants and by studying dialects. As a result, they decided to follow *Kokon Yoranko* [86], which asserts that the plant is an angiosperm, *inu-biwa*[76] (*Ficus erecta*) [86]. Ashida [87] wrote that neither *G. biloba* nor a *Ginkgo*-like plant appears in the *Kokin Waka-shu* [88] or in the *Shin Kokin Waka-shu* [89], anthologies of *waka*s that were collected by Imperial command; in *The Tale of Genji*, one of the oldest novels in Japan, by Murasaki Shikibu [90]; or in the *Makura no Soshi*, the oldest collection of essays by Sei Shonagon [91], all of which were written in the Heian period. Therefore, he considers that *G. biloba* may not have existed in Japan at that time [87].

The poet Socho [92] wrote in his travel diary, *Socho Nikki* (1530–1531), that he gathered yellow *Ginkgo* leaves and gave them to someone along with a *waka*. This is the first description of *Ginkgo* that we have found in Japanese literature. In the Edo period the *Ginkgo* was described as being familiar in everyday life by Enomoto Kikaku[77] (1661–1707), by Yosa Buson[78] (1716–1783), by Shoha[79] (1727–1771) in haiku, 17-syllable poems, and by Ryokan[80] (1758–1831) in *waka*. A haiku almost always includes a *kigo*[81] or a season word, referring to an animal, plant, event, or custom of the season. Lists of *kigo*s or season words have been collected in books called *saijiki*.[82] Various *saijiki*s published in the Edo period have been collected in *Kinsei-koki Saijiki* [93]. Included are the *Ginkgo* tree, the *Ginkgo* flower, although it is not a real flower from the biological point of view, and the *Ginkgo* nut. Modern *saijiki*s list the flower and bud of *G. biloba* as season words for spring, the *Ginkgo* tree and its falling yellow leaves for autumn, while fallen leaves, withered *Ginkgo*, and ginkgo-tertiaries of mandarin ducks are listed for winter [16, 17]. Not only professional poets such as Yosano Akiko[83] (1878–1942) and Saito Mokichi[84] (1882–1953) but also ordinary people have composed verses about *Ginkgo*. All these show how keenly the Japanese people observe *Ginkgo* and how deeply they love the plant.

In prose, too, the *Ginkgo* has played an important role, such as that of a key by which people were soothed or given energy, etc., in various plots of novels or stories. Examples are found in novels such as *Hakai* by Shimazaki [94], *Sanshiro* by Natsume [95], and in short stories such as *Hana* by Akutagawa [96]. Shimazaki also wrote a juvenile story titled *Icho no Ki* [97]. Furthermore, Miyazawa depicted a mother tree with children nuts who are flying down in a fantasy story titled *Icho no Mi* [98].

7 Old *Ginkgo* Trees and Their Legendary Age

According to some legends, there are *Ginkgo* trees that have been alive for more than 2000 years. We chose *Ginkgo* trees that had been referred to in more than one legend as keys that might provide evidence to suggest the time *G. biloba* first

TABLE 3. Examples of old *Ginkgo* trees and their legendary age

Name of *Ginkgo*	Sex	Age (yr)	Century when *Ginkgo* appeared	Location (prefecture)
Jyonichiji-*Ginkgo*[a]	♀	1500	5th	Toyama
Nigatake-*Ginkgo*[b]	♀	1000–1200	8th to 10th	Miyagi
Horyo-*Ginkgo*[c]	♂	800–1200	8th to 12th	Aomori
Chosenji-big *Ginkgo*[d]	♂	1000	10th	Iwate
Senbon-*Ginkgos*[e]	♂	1000	10th	Chiba
Tsurugaoka-big-*Ginkgo*[f]	♂	800–850	12th	Kanagawa
Bodaij-*Ginkgo*[g]	♂	800	12th	Okayama
Zenpukuji-*Ginkgo*[h]	♀	700–800	13th	Tokyo

[a] Grows on the grounds of Jyonichiji Temple and is famous for "chichi" [99–102]. It is said that this *Ginkgo* tree was already big when the temple was founded in 682 [100].
[b] Grows on the grounds of Ubagami Shrine. It is also called Chichi-*Ginkgo* or Uba (wet nurse)-*Ginkgo*. When a wet nurse of the Emperor Shomu (701–756) was dying, she asked that a *Ginkgo* tree be planted by her grave. Later, the planted *Ginkgo* grew big and seemed to have long "chichis" [99–102].
[c] There are many stories about this *Ginkgo* [99, 100, 102]. It is said to have been planted in commemoration of the founding of a now-ruined temple, Zenshoji, in the Heian period (794–1192) [99, 102].
[d] Grows on the grounds of Chosenji Temple. It was the biggest *Ginkgo* until 1947 when a severe typhoon broke the trunk and branches [99–102].
[e] Grows on the grounds of Katsushika Hachiman Shrine. After the trunk was broken at 6m from the ground by lightning, many young shoots grew around it. Thus, it is called Senbon-*Ginkgo* or One-thousand-*Ginkgos*. In *Edo Meisho Zue* [103], an album of noted places in Edo, there is a description of this shrine reporting that this giant *Ginkgo* tree was already very big [99, 100, 102].
[f] Grows by the stone stairs that lead to the main building of the Tsurugaoka Hachiman Shrine. The third Shogun, Minamoto-no-Sanetomo, was assassinated there in 1219 by Kugyo, who was hiding behind the *Ginkgo* tree and watching for the moment to attack [99–102].
[g] This *Ginkgo* tree is said to have grown from a stick that Priest Honen (1133–1212) planted [99–102].
[h] Grows on the grounds of Zenpukuji Temple, which is reported to have been founded by Kukai in 824. It is also said that this *Ginkgo* tree grew from a stick that Priest Shinran put in the ground in 1232, wishing for the prosperity of the temple. The temple changed from Shingon to the Jodo-shinshu sect then. It is described in *Edo Meisho Zue* [103] as a big *Ginkgo* tree. When the main building of the temple was destroyed by a bomb toward the end of World War II, in 1945, the main trunk of the *Ginkgo* tree was also burnt. It has survived, however, with severe burn marks on the lower trunk [99–102].

existed in Japan or the time when it was introduced to Japan. For example, the aged *Ginkgo* tree at the Tsurugaoka Hachiman Shrine[85] (Table 3) has been referred to in various books. It is particularly well known because the tree is related to the assassination in 1219 of the third shogun, Minamoto-no-Sanetomo.[86] The Tsurugaoka Hachiman Shrine was built in Kamakura[87] in 1180 by the first shogun, Minamoto-no-Yoritomo.[88] He established the Kamakura-*bakufu*,[89] the first government controlled by the samurai or warriors, in 1192. The *Azuma Kagami* [104], which is an official record of incidents that occurred between 1180 and 1266, reported Sanetomo's assassination in detail, giving the date, the time, the weather, and the words Kugyo,[90] the murderer, uttered at the moment when he was attacking the shogun. The *Gukan-sho*, written about 1220 by Jien [105], a contemporary priest and poet, gives more detailed information about how Kugyo assassinated the shogun and even describes the costume the killer was wearing. According to the legend, Kugyo hid himself behind the ginkgo tree while waiting to ambush the shogun. These two books state that Kugyo attacked Sanetomo, who was coming down the shrine's stone steps, but make no mention of the *Ginkgo* tree standing beside it then. Uehara [100] points out in his book *Jumoku Zusetsu* that even if the *Ginkgo* tree had existed there at that time it would not have been big enough to conceal a man 19 years old, especially because the event occurred in the winter season when there would have been no leaves. He infers that the *Ginkgo* tree was planted later and suggests that by the time the *Kamkaura Monogatari* [106], in which the mention is first made of the *Ginkgo* tree that hid Kugyo, was published, about 1659, it might have grown big enough to make people believe the story [100]. From this research we believe it may be reasonably concluded that there is no reliable evidence to support the presence of *G. biloba* in Japan before the early part of the thirteenth century.

8 Japanese Names for *Ginkgo* and the Latin Generic Name

8.1 Japanese Names in Kanji *(Chinese Characters) and The Name in Chinese for Comparison*

In Japan, the word ginkgo first appears in two mid-fifteenth-century dictionaries, *Kagaku-shu* (1444) [15] and *Ainosho* (1446) [107] (Table 4) and a textbook, *Sekiso Orai*, written by Ichijo (1402–1481) [114]. Written in a catechetical style, *Ainosho* cites many old references to answer questions. Gyoyo [107], the author, asks "What is *Ginkgo*?" The answer is, "There is no name in *Wamyo*." We believe this refers to the *Wamyo-sho*,[91] a dictionary published about 934. In *Ainosho* and *Sekiso Orai* a Japanese name for *G. biloba* is written in *kanji*,[92] 銀杏, meaning "silver apricot", accompanied by the *katakana*,[93] phonetic syllabary, which shows the pronunciation as "*icho.*" *Kagaku-shu* [15] lists 銀杏 in the first place together with *icho* in *katakana*, but it also mentions that another name for *G. biloba* is 鴨脚 ("duck foot"), to which no *kana*[94] or pronunciation is added. Instead, it gives an explanation why duck foot became a name for the tree, citing a Chinese poem on duck foot.

In China, it is known that two famous Chinese poets, Ou-Yang Xiu[95] (1007–1072) and Mei Yao-Chen[96] (1002–1060) in the Song dynasty[97] (960–1279) exchanged a long series of poems about *G. biloba* [35, 115, 116]. In one of them Mei wrote that 鴨脚 grew in the wild in Jiangnan,[98] his home place, and Ou-Yang Xin mentioned that the nuts had been sent as tribute to the throne with a new name, 銀杏 (silver apricot), which was considered more appropriate for the plant. In the Yuan dynasty[99] (ca. 1279–1368) in China, many names for *G. biloba* appeared (see Table 4). Among them is 白果 or "white fruit" [35, 36], which is still used today as a colloquial name for *G. biloba* in China. Another Chinese name for *G. biloba*, 公孫樹 or "grandfather-grandchild tree," is mentioned in the *Jonanhoshi*

TABLE 4. The principal *kanji* (Chinese characters) for *G. biloba* that appeared in dictionaries and books in Japan, with approximate references in Chinese for comparison

Century	Japan	China
11th		鴨脚 (DF) → 銀杏 (SA) [35, 36]
12th		銀杏 (SA) > 鴨脚 (DF) [35, 36]
13th		銀杏 (SA) 白果 (WF), 鴨脚 (DF) [35, 36] 白眼 (WE), etc.
14th		
15th	銀杏 (SA) > 鴨脚 (DF) [15, 107]	
16th	銀杏 (SA) [48, 108]	銀杏 (SA), 白果 (WF), 鴨脚 (DF) [35, 36], etc.
17th	銀杏 (SA) > 鴨脚 (DF) [37, 109]	銀杏 (SA), 公孫樹 (GG) [110], 白果 (WF), etc.
18th	銀杏 (SA) > 白果 (WF), 鴨脚 (DF) 一葉 (OL) [39, 112]	白果 (WF) [111], 銀杏 (SA), 鴨脚 (DF), etc.
19th	銀杏 (SA) [27] > 公孫樹 (GG) [113]	白果 (WF), 銀杏 (SA), 鴨脚 (DF), etc.

DF, duck foot; SA, silver apricot; WF, white fruit; WE, white eye; GG, grandfather–grandchild tree; OL, one leaf. If "A" > "B", "A" is used more commonly than "B"; if "A" → "B", "A" was changed into "B".

(1620) [110], where the reason this plant got its name is explained: a *Ginkgo* tree planted by a man yields nuts in his grandchildren's generation. In the *Jibutsu Imei-roku* (1788) [111], 白果 appears as an entry.

In Japan, 白果 is mentioned in *Wakan Sansai Zue* (1713) [39] and 公孫樹 in *Somoku Seifu* (1827) [113]. *Wakun no Shiori* (1777–1887) [112] used a different name, *ichi-yo*[100] or "one leaf," for the entry. However, 銀杏 has been most commonly used as the name for *G. biloba* up to the present day (see Table 4).

8.2 The Pronunciation and Readings of 銀杏

In Japan today, the *kanji* (Chinese characters) for 銀杏 can be pronounced in two ways, "*icho*" and "*gin'nan*". In the early part of the twentieth century, Japanese linguists discussed how 銀杏 came to be pronounced that way. In the *Sekiso Manpitsu* [117], Kurokawa (1799–1866) wrote that the modern pronunciations originate from the Chinese ones of the Tang dynasty[101] (618–907). Otsuki [23] and Shinmura [118] were opposed to this idea. They considered that the modern Japanese pronunciations of "*icho*" and "*gin'nan*" are corruptions of the dialects of the southern regions of China between the Song and the Yuan dynasties. Li [119] reported that 鴨脚 is pronounced "*ai cho*" and 銀杏, "*nin an*," in the modern dialect of the southern Jiangsu[102] in China, although they are pronounced "*yajiao*" and "*yinxin*," respectively, in the standard Chinese. Therefore, he presumes that the Japanese pronunciations came from that area [119].

We investigated the pronunciation of 銀杏, which appeared in many Japanese dictionaries and books published in Japan between the fifteenth and eighteenth centuries. We omitted those published after the nineteenth century because they could not have influenced the generic name. When a *kanji* entry is accompanied with *kana* (either a *katakana* or a *hiragana*;[103] both are phonetic syllabaries), we

TABLE 5. The pronunciations of 銀杏 (silver apricot), which are shown by *katakana* or *hiragana* (Japanese syllabaries) in dictionaries and books published between the 15th and the 18th centuries

Century	Pronunciation			References
	icho[1]	ginkyo	gin'nan[2]	
15th	○	—	○	15, 107, 114
	—	—	○	120
16th	○	—	—	21
	—	—	○	49
	○	—	○	121, 122
17th	○	●	—	123
	—	●	○	124
	—	—	○	51
	○	—	—	28, 59
	○	—	○	37, 52, 109, 115
18th	○	—	○	39
	○	—	—	111

[1] Category "icho" includes the variations ichi-yau, ichi-eu, and itefu.
[2] Category "gin'nan" includes ki-an, kin'nan, and gin'an.
Circles represent usage in the references listed. Solid circles are used to emphasize the usage of "ginkyo."

know how to read it. Only the "*icho*" pronunciation (in *katakan*) for 銀杏 is given in some old references such as *Kagaku-shu* (1444) [15], *Ainosho* (1446) [107], and *Sekiso Orai* (1440?–1480?) [114]. The same *kanji* 銀杏, however, is accompanied with the pronunciation as "*ki'an*" (in *katakana*) in *Onkochishin-sho* (1484) [120]. Looking through old dictionaries and books that contain 銀杏 accompanied with either a *katakana* or a *hiragana* reading, we noticed that the *kana* letters show great variation: [ichi-yau], [ichi-eu], [itefu], [ki-an], [kin'nan], [gin'an], [gin'nan], [gin-kyau], etc. However, they can be classified into three groups of pronunciation: (1) "*icho*", (2) "*gin'nan*", and (3) "*ginkyo*". We found that two ways of reading, namely "*icho*" and "*gin'nan*", were the most common; a reading of "*ginkyo*" was seen only in books published in a period limited to the seventeenth century (Table 5).

8.3 Ginkgo *as the Latin Name of the Genus*

8.3.1 Discussions on the Spelling of the Genus *Ginkgo*

Linné adopted *Ginkgo* as the generic name for this plant in 1771 [125], following the name that Kaempfer gave in *Amoenitatum Exoticarum* [14] in 1712. During his stay in Japan (1690–1692), Kaempfer collected a number of materials and information on Japan, including plants: *G. biloba* is one of them. On p. 811 of *Amoenitatum Exoticarum*, Kaempfer writes: "杏銀 *Ginkgo*, vel *Gin an*, vulgo *Itsjo*." (At that time the characters and letters were written from right to left in Japan.)

There has been considerable discussion about this *Ginkgo*. We summarize here the main points: (1) the spelling of *Ginkgo* is strange [126–131]; (2) it must be a misprint for *Ginkyo* [126, 131] or *Ginkjo* [127]; (3) it must be a false transcription of 山杏 (Sankyo) or "mountain apricot" [129]; (4) it may be a transliteration of 銀果 (Ginka) or "silver fruit" [128]; and (5) the pronunciation of 銀杏 was slightly different at that time in Japan [119]. Because the Chinese characters 銀杏 are written in the text (p. 811) and on the plate (p. 813), it cannot be 銀果 as suggested by Barclay [128], who apparently failed to consult the original text.

Moule [129] studied Kaempfer's notes and manuscripts for the publication of *Amoenitatum Exoticarum*, which are kept in the British Museum. He found 10 or 11 mentions of the tree, which we now call *G. biloba*, among several volumes of Kaempfer's materials purchased and numbered by Sloane, a founder of the British Museum. Moule reported the result of his examination of the materials together with Sloane's numbering [129]. He found the term *Ginkgo* only in Sloane 2914, fol. 221 r^0 but nowhere else. Instead, the name "Sankyo" was found in Sloane 2915, fol. 126 r^0 and Sloane 2915, fol. 148 v^0. Therefore, Moule thought that *Ginkgo* might be a mistranscription of Sankyo [129].

Aside from the westerners' disputes, many Japanese [126, 127] have simply thought that *Ginkgo* is a misspelling by Kaempfer himself or a misprint of *Ginkyo*, because they know that 銀杏 can be phonetically pronounced as "*ginkyo*" in Japanese. Yo [127] mentioned that Kaempfer must have consulted *Kinmo Zui* [124], an illustrated encyclopedia published in 1666 in Japan, and transcribed the name of the tree from it. We do not know if Yo actually examined Kaempfer's manuscripts before he wrote this article, but we also consider that Kaempfer must have referred to *Kinmo Zui*, because Moule reported [129] that the Sloane 74 (index): "Gin-nan 32" is called "Catalogus Plantarum Iaponicarum in Kin mo chjju i". Moule mentions that the last three words, which we underlined, are so faint as to be doubtful. We consider that this is evidently *Kinmo Zui* and presume that Moule did not know of the existence of that work. Kaempfer took back many items from Japan including 54 books, among which was the *Kinmo Zui*. These are maintained as "Kaempfer's Japanese Library" in the British Library [132, 133]. Kitamura [134] compares the "Plantarum Japonicarum" in *Amoenitatum Exoticarum* with the *Kinmo Zui*, and concludes that the Chinese characters and the Japanese names for the plants in *Amoenitatum Exoticarum* were apparently taken from the *Kinmo Zui*.

Among intellectuals at the time when Kaempfer was in Japan, it seems to have been fashionable to use the Sino-Japanese reading of *kanji* (Chinese characters), as seen in 銀杏; neither "*icho*" nor "*gin'nan*" but "*ginkyo*". Kaempfer followed this fashion and used the Sino-Japanese readings of *kanji* for the plants in the first place except Chinese characters which can be seen in more than half of the plants listed in *Amoenitatum Exoticarum* [14].

8.3.2 The Resolution of the Mysterious Letter *g* in *Ginkgo*: Kaempfer Transcribed It Correctly

Why did Kaempfer give *Ginkgo* as the phonetic reading of 銀杏 in *Amoenitatum Exoticarum* [14]? According to the suggestion given by a German historian, a specialist of the Middle Ages in Europe, ginkgo can be pronounced "*ginkyo*". In Northern Germany, the sound of *g* can be *j*. Even today the same phenomenon can be found; for example, gut (= good) is often pronounced "*jut*" and gestern (=yesterday) is "*jestern*". Kaempfer was from Lemgo in North Germany, and *Amoenitatum Exoticarum* was published by a publisher in Lemgo. On the cover of the book it clearly states "Lemgoviae, Typis & Impensis Henrici Wilhelmi Meyeri, Aulae Lippiacae Typographi, 1712." There were no strict spelling rules in Kaempfer's time; standardization of writing systems was not established until the nineteenth century in Europe. In addition, Kaempfer explained in the preface of his book that he had taken great pains to transcribe the Japanese as accurately as possible, and described how he invented a transliteration system for this purpose, for example, the words spelled with Sĕo, Sĭo and Mĕo, Mĭo are more softly pronounced than those spelled Sjo and Mjo [14], all of which are *Sho* and *Myo* in the modern spelling. He carefully listened to how one word or one sound was pronounced, and noticed that the same word or sound can be pronounced in different ways, depending on how it connects to other words or where it is placed in compound words. Futher, Kaempfer gives *Gin an* and *Itsjo* for the colloquial names.

Because Kaempfer spelled with such meticulous care, the plant names transcribed in romanized letters in *Amoenitatum Exoticarum* show vividly how they were pronounced by the Japanese people at that time. If he had merely copied the plant names from books without hearing people saying the names, he would have used the same spelling to transcribe the same word or sound. Imai [133] analyzes Kaempfer's spelling of the place names and people's names in his posthumous work, *The History of Japan* edited by Scheuchzer [13] and its German edition by Dohm [135]. According to him, Kaempfer wrote **G**odo gawa for **Y**odo gawa[104] (River Yodo), Matzande**g**ro for Matsuda**i**ra (a family name), Tsina**g**os for Tsuna**y**oshi[105] (a first name), Isafa**g**e for Isaha**y**a[106] (a place name), **G**ao (a Chinese name) for **Y**ao, and so on [133]. In a book titled *Japan in Europa*, Kapitza [136] introduced articles and books on Japan in various fields that were written by Europeans. The authors introduced data from Marco Polo to Wilhelm von Humboldt, and Kaempfer is included. The years of publication of the books are considered to range from 1298 to 1826. In the appendix of this book, many Japanese words are listed together with the original spellings used in these books. Under *G* there are many words that are spelled with *g* instead of the modern spellings of i, e, u, or y: examples are **Gendo** for **E**do, **Gia**cata for **Ya**kata,[107] **Gia**mabusi for **Ya**mabushi,[108] and **G**rusi for **U**rushi.[109]

From this research, we believe it may reasonably be concluded that Kaempfer deliberately gave *Ginkgo* as the transcription for the pronunciation of "*ginkyo*", and that it was not a misspelling or a misprint. The reasons are that the same spelling, *Ginkgo*, is listed together with *Ginan* in the index [14], and it was not corrected in the Errata [129]. Kaempfer's transliteration most probably merely reflects the influence of his North German dialect on his spelling.

9 Conclusions

Thus far, the oldest pieces of evidence for the existence of *Ginkgo biloba* in Japan are the words written in dictionaries published in the 1400s [15, 107, 114]. Considering how information spread in premodern times, probably at least a hundred years would be needed before a word came to be listed in a dictionary. Considering the amount of time necessary to yield nuts, *G. biloba* must have existed in Japan in the 1300s. However, precise evidence is lacking at present. If futher discoveries are made through additional archaeological excavations and examination of the still-untouched treasures and materials of the old noble and imperial families, it might be possible to trace the history of *G. biloba* in Japan even further back in time.

Acknowledgments. We wish to express our deep appreciation to Dr. Harald Kleinschmidt at the University of Tsukuba for suggestions of the dialectal pronunciation of "g" in North Germany and for answering questions on German and Latin. We also wish to thank Mr. Akira Kanai, representing the group of Koten no Tsudoi, for giving suggestions and answering questions on the references of Japanese classical literature. Finally, we would like to thank Professor Nicholas Teele at Doshisha Women's College for reading the manuscript and giving various suggestions based on his deep knowledge of Japanese literature and culture.

References

1. Segawa S (1956) Coloured illustrations of the seaweeds of Japan (in Japanese). Hoiku-sha, Osaka
2. Imazeki R, Hongo T (1969) Coloured illustrations of fungi of Japan (in Japanese). Hoiku-sha, Osaka
3. Ito M (ed) (1978) Sekai no Shokubutsu·世界の植物: Plants of the world (in Japanese). Asahi Shinbun-sha, Tokyo
4. Makino T (1961) Makino's new illustrated flora of Japan (in Japanese). Hokuryu-kan, Tokyo
5. Ono R·小野蘭山 (1803–1806) Jutei Honzo Komoku Keimo·重訂本草綱目啓蒙: A text-book of plants and medicine based on Honzo-Komoku, 2nd edn vol 26 (in Japanese).
6. Nitsupon Daijiten Kankokai (1976) A Japanese–Japanese dictionary·日本國語大辭典, vol 2 (in Japanese). Shogaku-kan, Tokyo
7. Namba T (1980) The crude drugs in Japan, China and the neighbouring countries, vol 1 (in Japanese). Hoiku-sha, Osaka

8. Utsumi F (1974) Coloured illustrations of seashore animals of Japan (in Japanese). Hoiku-sha, Osaka
9. Kira T (1974) Coloured illustrations of the shells of Japan (in Japanese). Hoiku-sha, Osaka
10. Shimonaka K (1974) Dai-jiten·大辞典: An encyclopedic dictionary, 2nd edn, vol 1 (in Japanese). Heibon-sha, Tokyo
11. Oshima T, Omori S, Goto T, Saito S, Muratake S, Yoshida M (eds) (1971) Nihon wo shiru Jiten·日本を知る事典: A dictionary of Japanese culture (in Japanese). Shakaishiso-sha, Tokyo
12. Nishiwaki M (1965) Kujira-rui·Kikyaku-rui·鯨類・鰭脚類: Whales and seals (in Japanese). Tokyo University Press, Tokyo
13. Kaempfer E (1727) The History of Japan: Scheuchzer JG (ed) (1906) James MacLehose and Sons, Glasgow
14. Kaempfer E (1712) Amoenitatum Exoticarum. Henrici Wilhelmi Meyeri, Lemgoviae
15. Anonymous (1444) Kagaku-shu·下學集: An encyclopaedic dictionary of the Muromachi period, vol 2, chap 14: Plants (in Chinese and Japanese)
16. Mizuhara S, Kato S, Yamamoto K (eds) (1981) Kodansha-ban Kara-zusetsu Nihon Dai-saijiki· カラー図説日本大歳時記: Illustrated book of season words, vols winter, spring, autumn (in Japanese). Kodan-sha, Tokyo
17. Kadokawa-shoten (1973) Zusetsu Haiku Dai-saijiki·図説俳句大歳時記: Illustrated book of season words of haiku, vols winter, spring, autumn (in Japanese). Kadokawa-shoten, Tokyo
18. Okura N·大藏常永 (1823) Nogu Benri Ron·農具便利論: Theory of usefulness of gardening tools. In: Tsuzoku Keizai Bunko (1917) vol 11 (in Japanese). Nihon Keizai-sosho Kanko-kai, Tokyo
19. Suzuki K (1982) Zoku Zappai-go Jiten·続雑俳語辭典: A dictionary of haikai words (in Japanese). Meiji-shoin, Tokyo
20. Otaka T (1983) Jinbutsu Reference Jiten·人物レファレンス事典: Biographical reference dictionary (in Japanese). Nichigai Associates, Tokyo
21. Soami·相阿彌 (1523) Okazari no Sho·御飾書: A record of furniture, utensils and etc. kept by Shogun Ashikaga Yoshimasa. In: Sen S (1958) Sado Koten Zenshu: Tea Ceremony Classics, vol 2 (in Japanese). Tanko-shinsha, Kyoto, Tokyo
22. Tagawa M (1976) Cha no Yu Dogu Sunpo Zue·茶之湯道具寸法図会: Illustrated book on utensils for tea ceremony, 4th edn (in Japanese). Keiso-sha, Kyoto
23. Otsuki F (1932) Daigenkai·大言海: A Japanese–Japanese dictionary, vol 1 (in Japanese). Fuzan-bo, Tokyo
24. Shimoide G (1974) Kenchiku Dai-jiten·建築大辞典: A dictionary of architecture (in Japanese). Shokoku-sha, Tokyo
25. Okamoto K·岡本昆石 (1885) Kokon Hyappu·Azuma Nagori·古今百風・吾妻余波, vol 1: Ancient and modern various usages of Tokio, Japan (in Japanese). Manjiya-Yotaro, Tokyo
26. Ryutei T·柳亭種彦 (1813) Sasairo no Chokuhakoyomite·笹色猪口暦手: Illustrated story in the Edo period. In: Koten-sosho (1990) Ryutei Tanehiko-shu, vol 12 (in Japanese). Honpo-shoseki, Tokyo
27. Kitagawa M·喜多川守貞 (1837–1853) Morisada Manko·守貞漫稿: An illustrated encyclopaedic essay about Japan, with special reference to social manners and customs in the Edo period. In: Asakura H (ed) (1973), vols 1–3 (in Japanese). Tokyo-do Shuppan-sha, Tokyo

28. Fujimoto K·藤本箕山 (1688) Shikido Okagami·色道大鏡: An encyclopaedic guide book on how to behave in the courtesan quarters of the Edo period with maps. In: Noma M (exponent) (1974) vol 1 (in Japanese). Yagi-shoten, Tokyo
29. Kitamura N·喜多村信節 (1830) Kiyu Shoran·嬉遊笑覧: An encyclopaedic essay about Japan, with special reference to the Edo period. In: (1974) vols 1–2 (in Japanese). Meicho Kanko-kai, Tokyo
30. Shimonaka K (1982) Ibaraki-ken no Chimei·茨城県の地名: Placenames in Ibaraki Prefecture. In: Nihon Rekishi Chimei Taikei: Historical placenames in Japan, vol 8 (in Japanese). Heibon-sha, Tokyo
31. Zauho Press (1972) Masterpieces of ukiyo-e prints, ser 1, Harunobu (in Japanese). Shuei-sha, Tokyo
32. Ota N·大田南畝 (1769–1823) Han'nichi Kanwa·半日閑話: A collection of essays, vol 12. In: Hamada G (1988) Ota Nanpo Zenshu, vol 1 (in Japanese). Iwanami-shoten, Tokyo
33. Ota N·大田南畝 (1769) Ameuri Dohei ga Den·売飴土平伝 and Osen Ofuji Yuretsu no Ben·阿仙阿藤優劣弁 in Shosetsu Dohei-den·小説土平伝: A novel. In: Hamada G (1985) Ota Nanpo Zenshu, vol 1 (in Japanese). Iwanami-shoten, Tokyo
34. Inui K, Koike M, Shimura A, Takahashi M, Torikosi F (eds) (1986) Nihon Denki Densetsu Dai-jiten·日本伝奇伝説大事典: A dictionary of legends and folk tales in Japan (in Japanese). Kadokawa-shoten, Tokyo
35. Ri Jichin·李時珍 (1596) (1603 edn) Honzo Komoku·本草綱目: A dictionary of medicine of China, vol 29 (in Chinese)
36. Ri Ko·李杲 (13th–14th century) (revised by Ri Jichin·李時珍) (1638) Shokumotsu Honzo·食物本草: Edible plants and medicine (in Chinese)
37. Hayashi R·林羅山 (1612) (1631 edn) Shinkan Tashiki-hen·新刊多識編: A new encyclopaedia of the Edo period, vol 3 (in Chinese with Japanese)
38. Kaibara E·貝原益軒 (1709) Yamato Honzo·大和本草: A dictionary of medicine (in Chinese with Japanese). Ono Zensuke, Kyoto
39. Terashima R·寺島良安 (1712) Wakan Sansai Zue·和漢三才図會: Illustrated encyclopaedia of Japan and China (in Chinese with Japanese)
40. Hiraga G·平賀源内 (exponent) (1773) Jukoku·Hiden Kakyo·重刻・秘伝花鏡, 2nd edn, Secret dictionary of plants (in Chinese with Japanese)
41. Koso Shin-igakuin (1978) Chu-yaku Dai-jiten·中藥大辭典: A dictionary of Chinese medicine (in Chinese). Shanghai Kagaku Gijutsu Shuppan-sha, Hong Kong
42. Kimura K, Kimura T (1981) Medicinal plants of Japan in color, 2nd edn (in Japanese). Hoiku-sha, Osaka
43. Kawaguchi K (1982) Hana to Minzoku·花と民俗: Plants and folklore (in Japanese). Tokyo Bijutsu Sensho, Tokyo
44. Shibata K (ed) (1949) A cyclopedia of useful plants and plant products (in Japanese). Hokuryu-kan, Tokyo
45. Anonymous (16th–17th century) Gomizu-no-o-in Nenchu Gyoji-ki·後水尾院年中行事記: A record of annual events at the time of Emperor Gomizu-no-o. In: Mizuno T (1848–1854) Tankaku-sosho, vol 1 (in Japanese)
46. Riyu·李由, Kyoroku·許六 (eds) (1702) Uda no Hoshi·宇陀の法師: A book of haikai. In: Ono S (1913) Haikai-bunko, 3rd edn vol 5, Kyoroku-zenshu (in Japanese). Hakubun-kan, Tokyo
47. Noshomu-sho Sanrin-kyoku (ed) (1912) Uses of woodcraft·木材ノ工芸的利用. In: (1982) (in Japanese). Ringyo-kagaku Gijutsu Shinkojo, Tokyo

48. Anonymous (1492–1521) Shinsen Ruiju Orai·新撰類聚往來: Newly-collected text books. In: Ishikawa M (1970) Nihon Kyokasho Taikei; Orai-hen Ko-orai: Series of Japanese textbooks: Old textbooks (in Japanese). Kodan-sha, Tokyo
49. Matsuya H·松屋久重 (ed) (1533–1650) Matsuya Kaiki·松屋會記: Series of records of tea ceremonies, vol 1: Matsuya H (松屋久政) (1533–1596) Hisamasa Chakaiki·久政茶會記, vol 3 Matsuya H·松屋久重 (1604–1650) Hisashige Chakaiki·久重茶會記. In: Sen S (1964) Sado Koten Zenshu: Tea ceremony classics, vol 9 (in Japanese). Tanko-sha, Kyoto
50. Anonymous (1561) Miyoshi-tei Onariki·三好亭御成記: A record of the visitation of Shogun Ashikaga Yoshiteru to the Lord of Chikuzen, Miyoshi. In: Zoku Gunsho Ruiju, vol 23 part 1 (in Japanese). Zoku Gunsho Ruiju Kansei-kai, Tokyo
51. Anonymous (1643) Ryori Monogatari·料理物語: Stories of cooking. In: Ryori Taikan (1916) vol 1 (in Japanese). Ryori Chinsho Kanko-kai, Tokyo
52. Hitomi S·人見必大 (ed) (1697) Honcho Shokkan·本朝食鑑: A book of edible and medicinal plants in Japan, vol 1 (in Chinese with Japanese). Kanko-kai, Tokyo
53. Hageboshi 禿箒子 (1785) Kaiseki-ryori-cho 會席料理帳: A book of dinner menu. In: Ryori Taikan (1916), vol 16 (in Japanese). Ryori Chinsho Kanko-kai, Tokyo
54. Daigo-Sanjin·醍醐散人 (1801) Kasen-shu·華船集: A book of box meal. In: Ryori Taikan (1916) vol 16 (in Japanese). Ryori Chinsho Kanko-kai, Tokyo
55. Shikitei S·式亭三馬 (1809–1813) Ukiyo-buro·浮世風呂: Conversational stories of the public baths, depicting everyday life in the Edo period. In: Ishikawa T (1927) Nihon Meicho Zenshu: Series of Japanese literary masterpieces, Kokkeibon-shu: Comical stories, vol 14 (in Japanese). Nihon Meicho Zenshu Kanko-kai, Tokyo
56. Nosaka T (1979) Nihon Kamon Taikan·日本家紋大鑑: Illustrations of family crests (in Japanese). Shin Jinbutsu Orai-sha, Tokyo
57. Chikano S (1993) Nihon Kamon Sokan·日本家紋総鑑: Illustrated Japanese heraldry (in Japanese). Kadokawa-shoten, Tokyo
58. Anonymous (15th century) Kikigaki Shokamon·聞書諸家紋: Family crests of the Muromachi period. In: Hanawa H (1972) Gunsho Ruiju, 3rd edn vol 23, sec Buke-bu (in Chinese with Japanese). Zoku Gunsho Ruiju Kansei-kai, Tokyo
59. Ihara S·井原西鶴 (1687) Nanshoku Okagami·男色大鏡: Love stories of Samurai and Kabuki worlds, depicting their manners and customs. In: Ishikawa T (1929) Nihon Meicho Zenshu: Series of Japanese literary masterpieces, vol 1, Saikaku Meisaku-shu (in Japanese). Nihon Meicho Zenshu Kanko-kai, Tokyo
60. Shimizu E·志水燕十 (1785) Nisemonogatari Tsuhosho·仁勢物語通補抄, vol 1: A comical story. In: Hayakawa J (ed) (1911) Kinsei Bungei Sosho, vol 7 (in Japanese). Kokusho-kanko-kai, Tokyo
61. Okumura M·奥村政信 (1731) Nakamura-za Shibai-zu·中村座芝居図: An ukiyo-e of the Nakamura theatre. In: Kikuchi S (1968) Genshoku Nihon no Bijutsu: Series of Japanese art, vol 17, Block-printed ukiyo-e (in Japanese). Shogaku-kan, Tokyo
62. Anonymous (16th century) Naname-jima Icho-ha Yukiwa-chirashi-moyo Dofuku·斜縞銀杏葉雪輪散らし模様胴服: Kimono coat tie-dyed and embroidered with designs of Ginkgo-leaves, snow-crystals on a diagonal striped ground. Tokyo National Museum, Tokyo
63. Itakura T, Nomura K, Motoi T, Yoshikawa S, Yoshida M (1977) Genshoku Senshoku Dai-jiten·原色染織大辞典: Illustrated dictionary of dyeing and weaving (in Japanese). Tanko-sha, Kyoto

64. Yoshimoto K (ed) (1984) A popular edition of Wasarasa "shoe": collection of over 820 floral patterns in "sarasatic Japan style" (in Japanese with English). Gurafikku-sha, Tokyo
65. Nakagawa S (1974) Tojiki Taikei·Ninsei·陶磁器大系・仁清: Pottery series, a potter Ninsei, vol 23 (in Japanese). Heibon-sha, Tokyo
66. Masamune I, Nakata Y, Mitsuoka T, Motoi T, Kawara M (1984) Nihon no Monyo·Koyo·日本の文様・紅葉: Japanese designs and motifs·Colored leaves (in Japanese). Korin-sha, Kyoto
67. Suzuki N, Sakakibara S (1979) Nihon no Shippo·日本の七宝: Japanese cloisonne (in Japanese). Maria-shobo, Kyoto
68. Utagawa T·歌川豊広 (1789–1800) Nanro Meigi·南楼名妓: Ukiyo-e of geisha at Nanro. In: Kikuchi S (1968) Genshoku Nihon no Bijutsu: Series of Japanese art, vol 17, Block-printed ukiyo-e (in Japanese). Shogaku-kan, Tokyo
69. Kumamoto Nichinichi Shinbun-sha, Kumamoto-ken Daihyakkajiten Hensyu-iinkai (1982) Kumamoto-ken Dai-hyakka-jiten·熊本県大百科事典: An encyclopaedia of Kumamoto Prefecture (in Japanese). Kumamoto Nichinichi Shinbun-sha, Tokyo
70. Governmental Command (1871) Sanpatsu Datto-rei·散髪脱刀令: The government permission to cut topknots and not carry a sword. In: Horei-zensho (1889) (in Japanese). Hakubun-sha, Tokyo
71. Kanazawa Y (1961) Edo Keppatsu Shi·江戸結髪史: A history of hair styles in the Edo period (in Japanese). Seia-bo, Tokyo
72. Hishikawa M. 菱川師宣 (1696) Sugata-e Hyakunin-isshu 姿繪百人一首, vol 1: The hundred poems by one hundred poets with ukiyo-e prints. In: Kurokawa M (ed) (1914) Nihon Fuzoku Zue, vol 2 (in Japanese). Nihon Fuzoku Zue Kanko-kai, Tokyo
73. Eshima K·江嶋其磧 (1716) Seken Musume-katagi·世間娘氣質: The modern girls' way of thinking. In: (1928) Nihon Meicho Zenshu: Series of Japanese literary masterpieces, vol 9, Ukiyo-zoshi-shu (in Japanese). Nihon Meicho Zenshu Kanko-kai, Tokyo
74. Chikamatsu M·近松門左衛門 (ca. 1700) Soga Gonin Kyodai·曽我五人兄弟: A puppet drama: The five Soga brothers. In: (1986) Chikamatsu Zenshu, vol 3 (in Japanese). Iwanami-shoten, Tokyo
75. Chikamatsu M·近松門左衛門 (1707) Kazo Soga·加増曽我: A puppet drama: The Sogas increased. In: (1986) Chikamatsu Zenshu, vol 4 (in Japanese). Iwanami-shoten, Tokyo
76. Tamenaga S·為永春水 (1838) Shunshoku Eitaidango·春色·英對暖語: An illustrated story of licenced quarters in Edo. In: Ishikawa T (1928) Nihon Meicho Zenshu: Series of Japanese literary masterpieces, vol 15, Ninjobon-shu (in Japanese). Nihon Meicho Zenshu Kanko-kai, Tokyo
77. Natsume S·夏目漱石 (1914) Kokoro·心: A novel. In: (1956) Kokoro (in Japanese). Iwanami-shoten, Tokyo
78. Mori O·森鷗外 (1916) Gan·雁: A novel. In: Mori R (1936) Ogai Zenshu Chosaku-hen, vol 4 (in Japanese). Iwanami-shoten, Tokyo
79. Anonymous (1773) Kokon Sumo Daizen·古今相撲大全, vol 2: A book of sumo history (in Japanese). Buko-shorin, Edo (Tokyo)
80. Ishiguro G·ジョージ石黒 (1994) Sumo Nishiki-e Shushu-tan·相撲錦絵蒐集譚: A collection of sumo wood prints (in Japanese). Nishida-shoten, Tokyo
81. Anonymous (8th century) Manyo-shu·萬葉集: The oldest anthology of waka poems. In: Sasaki N (1930) Shin-kun Manyo-shu (in Japanese). Iwanami-shoten, Tokyo

82. Koshimizu T (1948) Manyo-shokubutsu to Kodai-jin no Kagaku-sei·万葉植物と古代人の科学性: Plants of the Manyo-shu and science in ancient times (in Japanese). Osaka Jiji Shinpo-sha, Osaka
83. Kamo-no-Mabuchi·賀茂真淵 (1757) Kanji-ko·冠辭考: A dictionary of pillow words. In: Hisamatsu S (ed) (1978) Kamo-no-Mabuchi Zenshu, vol 8 (in Japanese). Zoku Gunsho-ruiju Kansei-kai, Tokyo
84. Shinmura I·新村出 (1925) Chichi no mi to Icho 智智乃実と銀杏: An essay on *Ginkgo biloba*. In: Ogiwara K (ed) (1943) Toa Gogen-shi (in Japanese). Ogiwara Seibun-kan, Tokyo
85. Matsuda O (1970) Zotei·Manyo-shokubutsu Shin-ko·増訂・万葉植物新考: On plants of the Manyo-shu (in Japanese). Shakai Shiso-sha, Tokyo
86. Yashiro K·屋代弘賢 (1842) Kokon Yoranko·古今要覽稿, vol 395: A reference book on historical events in Japan. In: Ichijima K (ed) vol 5 (in Japanese). Kokusho Kanko-kai, Tokyo
87. Ashida T (1989) Kusaki Yo-bahashi·Ima ya Mukashi·草木夜ばなし・今や昔: Stories of plants: Past and today (in Japanese). Soshi-sha, Tokyo
88. Ki-no-Tsurayuki·紀貫之, Ki-no-Tomonori·紀友則, Oshikochi-no-Mitsune·凡河内躬恒, Mibu-no-Tadamine·壬生忠岑 (eds) (ca. 913) Kokin Waka-shu·古今和歌集: The first anthology of waka poems by Imperial command. In: Matsushita D, Watanabe F (eds) (1942) Kokka Taikan, 4th edn, vol Kashu: Poetry (in Japanese). Chubunkan-shoten, Tokyo
89. Minamoto-no-Michitomo·源通具, Fujiwara-no-Ariie·藤原有家, Fujiwara-no-Teika·藤原定家, Fujiwara-no-Ietaka·藤原家隆, Fujiwara-no-Masatsune·藤原雅経 (eds) (ca. 1205) Shin Kokin Waka-shu·新古今和歌集: The 8th anthology of waka poems by imperial command. In: Matsushita D, Watanabe F (eds) (1942) Kokka Taikan: Japanese literary classics, 4th edn vol Kashu: Poetry (in Japanese). Chubunkan-shoten, Tokyo
90. Murasaki Shikibu·紫式部 (ca. early 11th century) Genji Monogatari·源氏物語: The tale of Genji. In: Arikawa T (exponent) (1936) Genji Monogatari·Kogetsu-sho, vol 2 (in Japanese). Kobun-sha, Osaka, Tokyo
91. Sei Shonagon·清少納言 (ca. late 10th–early 11th century) Makura no Soshi·枕草子: Pillow book of Sei Shonagon. In: Ikeda K (exponent) (1940) (in Japanese). Iwanami-shoten, Tokyo
92. Socho·宗長 (1530–1531) Socho Nikki·宗長日記: A travel diary. In: Shimazu T (exponent) (1975) (in Japanese). Iwanami-shoten, Tokyo
93. Ogata C, Kobayashi S (1984) Kinsei-koki Saijiki: Honbun-shusei narabini Sogo-sakuin·近世後期歳時期：本文集成並びに総合索引: A reference book of saijikis (in Japanese). Bensei-sha, Tokyo
94. Shimazaki T·島崎藤村 (1906) Hakai·破戒: A novel. In: Toson Zenshu (1968) vol 15 (in Japanese). Chikuma-shobo, Tokyo
95. Natsume S·夏目漱石 (1914) Sanshiro·三四郎: A novel. In: Soseki Zenshu (1924) vol 4 (in Japanese). Iwanami-shoten, Tokyo
96. Akutagawa R·芥川龍之介 (1916) Hana·鼻: A short story. In: Akutagawa Ryunosuke Sakuhin-shu (1952) vol 1 (in Japanese). Iwanami-shoten, Tokyo
97. Shimazaki T·島崎藤村 (1917) Icho no Ki·銀杏の樹: A juvenile story: A *Ginkgo* tree. In: Toson Zenshu (1987) vol 8 (in Japanese). Chikuma-shobo, Tokyo
98. Miyazawa K·宮澤賢治 Icho no Mi·いてふの実: A juvenile story: *Ginkgo* nuts. In: Miyazawa Kenji Zenshu (1979) vol 7 (in Japanese). Chikuma-shobo, Tokyo

99. Honda M (1957) Shokubutsu Bunka-zai·Ten'nen Kinen-butsu·Shokubutsu 植物文化財：天然記念物・植物: Properties of plants: Monumental plants (in Japanese). Sansei-do, Tokyo
100. Uehara K (1965) Jumoku Zusetsu, vol 2, Icho-ka·樹木図説・銀杏科: Illustrated series of trees, Ginkgoaceae (in Japanese). Kashima-shoten, Tokyo
101. Makino K (1986) Kyoju no Minzoku-gaku·巨樹の民俗学: Ethnobiology of giant trees (in Japanese). Kobun-sha, Tokyo
102. Numata M (1984) Nihon no Ten'nen Kinen-butsu·日本の天然記念物: Natural monuments of Japan, vol 5, Plants III (in Japanese). Kodan-sha, Tokyo
103. Saito G·齊藤月岑 (1835–1837) Edo Meisho Zue·江戸名所図會: Albums of noted places in Edo. In: Yamatoya Bunko (1984) vol 7 (in Japanese). Shinten-sha, Tokyo
104. Anonymous (1180–1266) Azuma Kagami·吾妻鏡: An official record of the Kamakura Bakufu. In: Kokusho Kanko-kai (1973) vol 2 (in Chinese with Japanese). Meicho Kanko-kai, Tokyo
105. Jien·慈円 (ca. 1220) Gukan-sho·愚管抄: Series of Japanese history from Emperor Jinmu to Emperor Juntok. In: Okami M, Akamatsu T (exponents) (1971) Nihon Koten Bungaku Taikei, vol 86: Series of Japanese classical literature (in Japanese). Iwanami-shoten, Tokyo
106. Nakamura K·中村喜雲 (ca. 1659) Kamakura Monogatari·鎌倉物語: Stories of Kamakura. In: Akagi Bunko (1975) Kinsei Bungaku Shiryo Ruiju: Kohan Chishi-hen, vol 12 (in Japanese). Bensei-sha, Tokyo
107. Gyoyo·行譽 (1445–1446) Ainosho·埃嚢鈔: A reference book of words, languages, religious and historical events (in Japanese)
108. Anonymous (1596–1615) Setsuyo-shu: Manjuya-bon·節用集饅頭屋本: A Japanese–Japanese dictionary, Manjuya-edition (in Japanese)
109. Anonymous (1611) Setsuyo-shu: Keicho-ban·節用集・慶長板: A Japanese–Japanese dictionary, Keicho-edition (in Japanese)
110. Shu Bunka·周文華 (1620) Jonanhoshi·汝南圃史, vol 4: A gardening calendar (in Chinese)
111. Reisen·厲荃 (1788) Jibutsu Imei-roku·事物異名錄: A reference book of the Qing dynasty of China (in Chinese). In: Go S (ed) (1991) Gakuroku Shosha Shuppan, Chosa, Konan
112. Tanikawa K·谷川士清 (1777–1887) Wakun no Shiori·和訓栞: A dictionary of Japanese, including dialects (in Japanese)
113. Kiyohara S·清原重巨 (1827) Somoku Seifu·草木性譜: A book of plants (in Japanese)
114. Ichijo K·一條兼良 (1422–1481) Sekiso Orai·尺素往來: A textbook of how to write letters. In: Hanawa H (1971) Gunsho-ruiju, 3rd edn, vol 9 (in Chinese with Japanese). Zoku Gunsho-ruiju Kansei-kai, Tokyo
115. Mukai G·向井元升 (1684) Hochu-biyo Wamyo-honzo·庖廚備用倭名本草: A book of food and medicine (in Japanese)
116. Chugoku Gakujutsu-rui (ed) (1977) Kokon Tosho Shusei vol 54 Somokuten 2·古今図書集成・草木典: Series of reference books of ancient and modern China (in Japanese). Kanae Bunsho-kyoku, Taipei
117. Kurokawa H·黑川春村 (1860) Sekiso Manpitsu·硯鼠漫筆: A posthumous essay, mainly on words. In: Kurokawa M (ed) (1905) (in Japanese). Yoshikawa Kobun-kan, Tokyo
118. Shinmura I·新村出 (1927) Icho no Ki no Wakan-mei 鴨脚樹の和漢名: Chinese and Japanese names of *Ginkgo biloba*. In: Ogiwara K (ed) (1943) Toa Gogen-shi (in Japanese). Ogiwara Seibun-kan, Tokyo

119. Li H-L (1956) A horticultural and botanical history of *Ginkgo*. Morris Arbor Bull 7:3–12
120. Otomono-H·大伴広公 (1484) Onkochishin-sho·温古知新書: An old Japanese–Japanese dictionary (in Japanese)
121. Anonymous (1596–1614) Setsuyo-shu: Ekirin-bon·節用集・易林本: A Japanese–Japanese dictionary, Ekirin-edition (in Japanese)
122. Anonymous (1596–1615) Setsuyo-shu: Sosho-bon·節用集・草書本: A Japanese–Japanese dictionary, Sosho-edition (in Japanese)
123. Anonymous (1617–1619) Kagaku-shu, 1617–1619-edn·下學集・1617–19年版: An encyclopaedic dictionary of the early Edo period (in Chinese with Japanese)
124. Nakamura T (中村惕齊) (1666)·Kinmo Zui 訓蒙図彙: Illustrated encyclopaedia, vol 18. Yamagata-ya (in Japanese)
125. Linné C (1771) Mantissa Plantarum altera Generum editionis VI & Specierum editionis II. Holmiae, Impensis Direct, Laurentiisal VII
126. Anonymous (1905) On the spelling of *Ginkgo* (in Japanese). Bot Mag Tokyo 19:224
127. Yo J (1905) About *Ginkgo* (in Japanese). Bot Mag Tokyo 19:252
128. Barclay JG (1944) The name *Ginkgo*. J Roy Hortic Soc 69:68–69
129. Moule AC (1937) The name *Ginkgo biloba* and other names of the tree. Toung Pao 33:193–219
130. Seward SA (1938) The story of the maidenhair tree. Sci Prog 32:420–440
131. Moule AC (1944) The name *Ginkgo*. J Roy Hortic Soc 69:166
132. Brown Y-Y (1992) The *History of Japan* and Kaempfer's legacy in Britain. In: Kreiner J (ed) Kenperu no mita Tokugawa Japan (in Japanese). ケンペルのみたトクガワ・ジャパン, pp 224–240, Rokko-shuppan, Tokyo
133. Imai T (1989) Engeruberuto·Kenperu Nihon-shi エンゲルベルト・ケンペル日本誌: Engelbert Kaempfer's "The History of Japan" (in Japanese). Kasumigaseki-shuppan, Tokyo.
134. Kitamura S (1953) Botanical descriptions of Japanese plants by Engelbert Kaempfer (in Japanese). J Hist Sci Jpn 26:19–24
135. Kaempfer E (1777–1779) Geschichte und Beschreibung von Japan. In: Dohm CW (ed) (1964). Brockhaus, Stuttgart
136. Kapitza P (1990) Japan in Europa. Indicium Verlag, Munchën

Notes

[1] Icho, 銀杏, the Japanese name for *Ginkgo biloba.*
[2] Dejima, 出島, now in Nagasaki Prefecture.
[3] Edo, 江戸, the old name of Tokyo.
[4] Hakone, 箱根, now in Shizuoka Prefecture.
[5] The Edo period, 江戸時代, from 1603 to 1868.
[6] Icho-ba(ne), 銀杏羽, a *Ginkgo*-leaf-shaped tertiary (feather).
[7] Haiku, 俳句, a 17-syllable poem.
[8] Zappai, 雑俳, a playful poem.
[9] Senryu, 川柳, a satirical poem.
[10] Shamisen, 三味線, a traditional three-stringed musical instrument.
[11] Icho, 銀杏 (silver apricot), a *Ginkgo*-leaf-shaped plectrum for a shamisen.
[12] Icho, 鴨脚 (duck foot), a family name.
[13] Kyoto, 京都, an old capital.

[14] Icho-gasshiki, 銀杏喝食, a Noh mask with a *Ginkgo*-leaf-shaped forelock.
[15] Noh, 能, Japan's oldest theater form.
[16] Icho-ashi, 銀杏脚, a *Ginkgo*-leaf-shaped table leg.
[17] Icho-machi, 銀杏町, a town name.
[18] Mito, 水戸, now in Ibaraki Prefecture.
[19] Icho-musume, 銀杏娘, a girl called ginkgo.
[20] Ukiyo-e, 浮世絵, a woodblock print.
[21] Kyogen, 狂言, a play.
[22] Kabuki, 歌舞伎, a popular traditional dramatic art.
[23] Emperor Gomizu-no-o, 後水尾天皇 (1596–1680), the 108th emperor.
[24] Haikai, 俳諧, one of the traditional short poems; haiku (see note 7) originates from it.
[25] Sake, 酒, rice wine.
[26] Go, 碁, a national board game of Japan.
[27] Shogi, 将棋, a Japanese chess.
[28] Fukui, 福井, a prefecture name.
[29] Fukuoka, 福岡, a prefecture name.
[30] Shogun, 将軍, a leader of Japan's feudal government or the shogunate.
[31] Ashikaga Yoshiteru, 足利義輝, the 13th shogun (see note 30) of Muromachi bakufu (see note 89) (1536–1565).
[32] Chawan-mushi, 茶碗蒸し, a dish made by steaming and setting egg soup with chicken or fish and vegetables in a pot.
[33] Nabe-ryori, 鍋料理, a sort of Japanese fondue. A dish of hot soup with salt/soy-sauce/ miso in which vegetables, fish, and meat are boiled.
[34] Icho-giri, 銀杏切り, slicing a vegetable into a *Ginkgo* leaf shape.
[35] Kimono, 着物, traditional Japanese clothing.
[36] Ashikaga Yoshimasa, 足利義政, (1435–1490), the eighth shogun (see note 30) of the Muromachi bakufu (see note 89).
[37] Ginkakuji Temple, 銀閣寺, a temple built in 1489 by the command of Ashikaga Yoshimasa (see note 36).
[38] Tokugawa Ieyasu, 徳川家康, the first shogun (see note 30) of the Tokugawa Shogunate (1542–1616).
[39] The Momoyama period, 桃山時代, 1585–1603.
[40] Fukiyose, 吹き寄せ, a design pattern that sketches fallen leaves being blown together by a strong autumn wind.
[41] Sho-e, 正絵, a paper pattern that shows the exact length of the horizontal and vertical yarn and the arrangement of the colors required to weave cloth for kimonos.
[42] Obi, 帯, a sash for a kimono (see note 35).
[43] Icho-guchi, 銀杏口, a *Ginkgo*-leaf-shaped vase.
[44] Nin'naji-gyoki, 仁和寺御記, a record of Nin'naji Temple in Kyoto (see note 13).
[45] Nonomura Ninsei, 野々村仁清 (unknown), a potter of the early Edo period (see note 5).
[46] Fusuma, 襖, a traditional sliding door.
[47] Geisha, 芸者, a professional beauty and entertainer.
[48] Higo-inlay, 肥後象眼, an inlay made now in Kumamoto Prefecture.
[49] The Meiji Restoration, 明治維新, which occurred in 1868.
[50] Futatsu-ori, 二つ折, men's hair style called "folded-in-two" in which the topknot was folded in two and tied at the back.
[51] Icho, 銀杏, another name for futatsu-ori (see note 50), in which the topknot is *Ginkgo*-leaf shaped.
[52] Icho-gashira, 銀杏頭, another name for futatsu-ori (see note 50), the *Ginkgo*-leaf-shaped topknot.
[53] Samurai, 侍＝武士, a warrior.

[54] Sumo, 相撲, Japanese wrestling.
[55] O-icho, 大銀杏, a big *Ginkgo*-leaf-shaped topknot.
[56] Ko-icho, 小銀杏, a small *Ginkgo*-leaf-shaped topknot.
[57] Shimada (coiffure), 島田, one of the traditional formal coiffures that was worn by unmarried women; now worn by a bride when she wears a bridal kimono (see note 35).
[58] Icho-mage (ginkgo-chignon), 銀杏髷, a simplified shimada coiffure (see note 57).
[59] Icho-kuzushi, 銀杏崩し, a variation of ginkgo-chignon (see note 58) decorated with a band at the foot of the chignon.
[60] Icho-gaeshi, 銀杏返し, a coiffure with two *Ginkgo*-leaf-shaped chignons.
[61] Nomi-no-Sukune, 野見宿禰, a retainer of Emperor Suinin, who was called the "God of sumo" (see note 54).
[62] Taima-no-Kehaya, 當麻蹶速, a man at the time of Emperor Suinin (see note 63) who was regarded as the "father of sumo" (see note 54) but was defeated by Nomi-no-Sukune (see note 61).
[63] Emperor Suinin, 垂仁天皇 (unknown), the 11th emperor.
[64] Chon-mage, 丁髷, one of the topknots of the Edo period (see note 5) which remains as a hair style of sumo (see note 54) wrestlers.
[65] Juryo, 十両, a junior grade of sumo (see note 54).
[66] Yokozuna, 横綱, a grand champion of sumo (see note 54).
[67] Hoso-icho, 細銀杏, a style hair favored by shopkeepers in the Edo period (see note 5).
[68] Heian period, 平安時代, 794–1192.
[69] Nishiki-e, 錦絵, a colorful ukiyo-e (see note 20).
[70] Katsukawa Shunsho, 勝川春章 (1726–1792), a master of ukiyo-e (see note 20) and *nishiki-e* (see note 69).
[71] Kansei time, 寛政年間, 1789 1801 in the Edo period (see note 5).
[72] Waka, 和歌, a 31-syllable poem.
[73] Nara period, 奈良時代, 710–794.
[74] Otomo-no-Yakamochi, 大伴家持 (718?–785), one of the representative poets of Manyo-shu [81].
[75] Chichi, 乳 (nipple), a lignotuber.
[76] Inu-biwa, イヌビワ, a Japanese name for an angiosperm, *Ficus erecta*.
[77] Enomoto Kikaku, 榎本其角 (1661–1707), one of the representative haiku (see note 7) poets in the Edo period (see note 5).
[78] Yosa Buson, 與謝蕪村 (1716–1783), one of the representative haiku (see note 7) poets in the Edo period (see note 5).
[79] Shoha, 召波 (1727–1771), one of the representative haiku (see note 7) poets in the Edo period (see note 5).
[80] Ryokan, 良寛 (1758–1831), one of the representative *waka* (see note 72) poets in the Edo period (see note 5).
[81] Kigo, 季語, a season word that appears in a haiku (see note 7).
[82] Saijiki, 歳時記, a book of *kigo* (see note 81).
[83] Yosano Akiko, 與謝野晶子 (1878–1942), a modern poetess.
[84] Saito Mokichi, 齊藤茂吉 (1882–1953), a modern poet.
[85] Tsurugaoka Hachiman Shrine, 鶴岡八幡宮, a shrine built by Minamoto-no-Yoritomo (see note 88) in 1180 in Kamakura (see note 87).
[86] Minamoto-no-Sanetomo, 源実朝 (1192–1219), the third shogun (see note 30) of the Kamakura (see note 87) bakufu (see note 89).
[87] Kamakura, 鎌倉, now in Kanagawa Prefecture.
[88] Minamoto-no-Yoritomo, 源頼朝 (1147–1199), the first shogun (see note 30) of the Kamakura (see note 87) bakufu (see note 89).
[89] Bakufu, 幕府, Japan's feudal government controlled by samurai or warriors (see note 53); a shogunate.

[90] Kugyo, 公暁 (1200–1219), the third son of the second shogun (see note 30), Minamoto-no-Yori'ie, a steward of Tsurugaoka Hachiman Shrine (see note 85), who killed Minamoto-no-Sanetomo (see note 86).
[91] Wamyo-sho, 和名抄, a dictionary published ca. 934.
[92] Kanji, 漢字, a Chinese character.
[93] Katakana, カタカナ, one of the phonetic syllabries.
[94] Kana, 仮名, phonetic syllabaries: *katakana* (see note 93) and *hiragana* (see note 103).
[95] Ou-Yang Xiu, 欧陽修 (1007–1072), a famous poet and historian of the Song Dynasty (see note 97) in China.
[96] Mei Yao-Chen, 梅堯臣 (1002–1060), a famous poet of the Song Dynasty (see note 97) in China.
[97] The Song Dynasty, 宋代, 960–1279.
[98] Jiangnan, 江南, the southern regions of the Yangzijiang River in China.
[99] The Yuan Dynasty, 元代, 1280–1368.
[100] Ichi-yo, 一葉, one leaf.
[101] The Tang Dynasty, 唐代, 618–907.
[102] Jiangsu, 江蘇省, one of the southern regions of the Yangzijiang River in China.
[103] Hiragana, ひらがな, a cursive, phonetic syllabary.
[104] Yodogawa, 淀川, a river name.
[105] Tsunayoshi: Tokugawa Tsunayoshi 徳川綱吉 (1648–1709), the first name of the fifth shogun (see footnote 30) of the Tokugawa Shogunate.
[106] Isahaya, 諫早, a place name.
[107] Yakata, 屋形, a mansion or a houseboat.
[108] Yamabusi, 山伏, a mountain priest.
[109] Urusi, 漆, Japanese lacquer.

Scientific Contributions of Sakugoro Hirase

Toshiyuki Nagata

On September 9, 1996, the centenary ceremony to commemorate the discovery of the sperm of *Ginkgo biloba* was held at the Botanical Gardens, Faculty of Science, the University of Tokyo (Fig. 1). This date was chosen because Sakugoro Hirase [1] reported that he observed swimming *Ginkgo* sperm for the first time on the same date in 1896. Accompanying the ceremony, an international symposium and a public forum were held to commemorate the great discovery and discuss the scientific progress that arose from the finding. As I played a role in organizing the centenary events, I was asked by an editor of this special issue to give an account of the scientific contributions of Sakugoro Hirase (Fig. 1). As several articles on this subject have been published, I am concerned that this may partly duplicate previous reports and thus initially I hesitated to write this chapter. Nonetheless, I am prepared to undertake this task because some of his activities have simply been forgotten.

I would like to begin by illustrating the situation of Japanese scientists around the early 1890s. At that time, the process of fertilization in angiosperms had been largely understood, while that in the gymnosperms was just starting to be elucidated. Thus to follow the fate of fertilization in gymnosperms was a very hot issue at that time. Belajeff [2] presented a paper in which he reported that pollen tubes of *Taxus baccata* elongated and then penetrated into archegonial cells, resulting in the formation of embryos. He also showed that one smaller cell among three cells observed in the early stage of pollen development is responsible for fertilization as a generative cell. Subsequently, Strasburger [3] published a comprehensive monograph in which he demonstrated that the principle observed by Belajeff [2] can be generally applicable to almost all gymnosperms.

In this context, it is instructive to know how Japanese scientists at that time accessed the most recent results. Professor Jinzo Matsumura, who came back to Japan in 1889 after 2.5 years stay in Germany, is likely to have brought this news.

Department of Biological Sciences, Graduate School of Science, Hongo, Bunkyo-ku, Tokyo 113, Japan, and Botanical Gardens, Faculty of Science, University of Tokyo, 3-7-1 Hakusan, Bunkyo-ku, Tokyo 112, Japan

He was chairperson of Department of Botany at the time of the discovery and was simultaneously the first Director of the Botanical Gardens. Professor Manabu Miyoshi, who had been in the laboratory of Professor Wilhelm Pfeffer at Leibzig, Germany, and had witnessed the creation of a new scientific discipline of plant physiology there, came back to Japan and took up a teaching chair of Plant Physiology, which I now occupy a century later. Thus the news of the latest findings in Europe was immediately brought to Japan.

I found a few traces of this innovation when I opened a monograph written by Strasburger [3] found in a dusty corner of our library. I noted many penciled-in underlinings, most of which are confined to the descriptions on the fertilization on *Ginkgo*, *Cycas*, or *Zamia*. According to stamps placed in this monograph, it is presumed to have been first bought by Professor Matsumura at Heidelberg for his personal use and then donated to the library, probably at the time of his retirement from the university. In this case it is likely that Professor Matsumura did draw these lines himself. Although young scientists of Seiichiro Ikeno or Kenjiro Fujii, both of whom later became Professor of Botany at the Imperial University, could have done this, I am quite convinced that our fellow colleagues at that time were aware of the direction which they needed to take to go forward.

Another interesting point in the monograph by Strasburger [3] is that Professor Strasburger himself followed the fate of fertilization of *Ginkgo*. He described that from June to early September he followed the fate of development of *Ginkgo* pollen. *Ginkgo* seeds were sent every 2 weeks during this period from the Botanical Gardens of Vienna University by Professor Richard von Wettstein. He illustrated the early development of *Ginkgo* pollens; however, apparently he missed the later phase of pollen development.

FIG. 1. Sakugoro Hirase in 1912 (age 56)

Hirase followed this process every week, or even more frequently, with seeds collected from the big female *Ginkgo* tree that still grows vigorously in the center of the Botanical Gardens, because he and his fellow colleagues realized that this study was very important. Then unexpectedly he encountered the swimming sperm (spermatozoids) of *Ginkgo* on September 9, 1896, and what he observed was published first in Japanese in *Botanical Magazine Tokyo* [1]. Subsequently, the same finding was sent to *Botanisches Centralblatt* as a short communication [4], and a detailed report with many illustrations appeared in the *Journal of Science College of Imperial University* [5]. Except for the correction by Shimamura [6] that the first report that the swimming sperm have tails was an artifact, Hirase's observation was very precise and very detailed. He identified also which of the three cells in the pollen became the generative cells. This is the first discovery of sperm among spermatophytes, and *Ginkgo* has features suggesting an intermediate position between ferns and conifers. This also applies to *Cycas*, described here. As a consequence, Professor M. Engler established Ginkgoaceae, as separated from Pinaceae.

In this context, we should not forget that shortly after the discovery of *Ginkgo* sperms, Seiichiro Ikeno [7] reported that such sperm (or spermatozoids) were observed in the later stages of pollen development in *Cycas revoluta*. He added also that his discovery of the sperm of *Cycas* was delayed because he collected the materials in the Kagoshima Prefecture, in which *Cycas* grows wild, and he had to send them to Tokyo after fixing them for detailed observations. Simultaneously he also described that he was very much indebted to Hirase, because Hirase had published detailed conditions of the observation of *Ginkgo* sperms that facilitated the progress of his study. His report was communicated to *Botanisches Centralblatt* [8], and his detailed report also appeared in the *Journal of Science College of Imperial University* [9], in which he described the swimming sperm of *Cycas*. These two observations were communicated to *Annals of Botany* under the names of two authors, Ikeno and Hirase [10].

Regarding the discovery of *Ginkgo* sperm, it should also be noted that the discovery by Hirase did not arise by chance but was rather the result of enthusiasm, extraordinary efforts, and careful observations. Hirase [11] had published earlier detailed observations of the process of embryogenesis in *Ginkgo*, in which he described that after calculating the number of nuclei one could estimate how many times a cell divided. In fact, he estimated cells divided eight times, resulting in 256 cells. The illustrations that he drew himself were very accurate and sharp [12] in comparison with other works published many years later. Because of these major contributions, Hirase and Ikeno were awarded the Imperial Prize of the Japan Academy as the first representative of the biological sciences, in 1912, 16 years after their outstanding discoveries.

I have to add the fate of Hirase after this discovery. He resigned the research assistant position 1 year after his discovery and left the Imperial University for a job as a school teacher at the Hikone middle school of Shiga Prefecture. Although the detailed circumstances are not established, one likely explanation is that his situation may not have been stable and pleasant, although his scientific efforts were very much appreciated by his colleagues and by the Botanical Society of Japan. This is probably because he did not get a normal university education but was employed first as a technical assistant. In this connection, my personal supposition is that his resignation may have had been related to the resignation of Professor Ryokichi Yatabe and an administrator of Seitaro Hori of Botanical Gardens from the Imperial University, because it is apparent that both these persons, who played a role in Hirase's employment at the Imperial University, resigned successively in 1893. The exact situation of the resignation of Professor Yatabe at the age of 41 has not also been established. On the other hand, Seiichoro Ikeno later became Professor of Botany at the Agricultural College of Tokyo Imperial University; he wrote an obituary of Hirase when he died in 1925 at the age of 69 [13].

Last, I would like to add that we should not forget that the discovery of sperm in *Ginkgo* by Hirase and in *Cycas* by Ikeno took place shortly after the establishment of a modern education system in Japan, and this was the first scientific contribution of modern Japanese scientists to the world.

References

1. Hirase S (1896) Spermatozoids of *Ginkgo biloba* (in Japanese). Bot Mag Tokyo 10:367–368
2. Belajeff WC (1891) Zur Lehre von dem Pollenschlauche der Gymnospermen. Ber Dtsch Bot Ges 9:280–2296
3. Strasburger E (1892) Über des Verhalten des Pollens und die Befruchtungs vorgänge bei den Gymnospermen. Schwärmsporen, Gameten, pflanzliche Spermatozoiden und das Wesen der Befruchtung. Histol Beitr 4:1–158
4. Hirase S (1897) Untersuchungen über das Verhalten des Pollens von *Ginkgo biloba.* Bot Centralbl 66:33–35
5. Hirase S (1898) Études sur la fécondation et l'embryogénie du *Ginkgo biloba.* Second mémoire. J Sci Coll Imp Univ 12:103–149
6. Shimamura T (1937) On the spermatozoid of *Ginkgo biloba.* Cytologia (Tokyo) (Fujii Jubiloium Volume, pp 416–423)
7. Ikeno S (1896) Spermatozoids of *Cycas revoluta* (in Japanese). Bot Mag Tokyo 10:367–368
8. Ikeno S (1897) Vorläufige Mitteilung über die Spermatozoiden bei *Cycas revoluta.* Bot Centralbl 66:1–3
9. Ikeno S (1897) Untersuchungen über die Entwicklung der Geschlechtsorgane und den Vorgang der Befruchtung bei *Cycas revoluta.* J Sci Coll Imp Univ 12:151–214
10. Ikeno S, Hirase S (1897) Spermatozoids in gymnosperms. Ann Bot 11:33-35
11. Hirase S (1895) Études sur le *Ginkgo biloba.* Bot Mag Tokyo 9:238–239
12. Hirase S (1895) Études sur la fécondation et l'embryogénie du *Ginkgo biloba.* J Sci Coll Imp Univ 8:307–322
13. Ikeno S (1925) Sakugorô Hirase. Bot Mag Tokyo 36:96–99

Subject Index

MIX
Papier aus verantwortungsvollen Quellen
Paper from responsible sources
FSC® C105338

If you have any concerns about our products,
you can contact us on
ProductSafety@springernature.com

In case Publisher is established outside the EU,
the EU authorized representative is:
Springer Nature Customer Service Center GmbH
Europaplatz 3, 69115 Heidelberg, Germany

Printed by Libri Plureos GmbH
in Hamburg, Germany